Elements of Probability Theory

Elements of Probability Theory

Robert Fortet

Faculté des Sciences de Paris

GORDON AND BREACH SCIENCE PUBLISHERS

London New York Paris

Gordon and Breach Science Publishers Ltd.
42 William IV Street
London W.C.2

Editorial office for the United States of America

Gordon and Breach Science Publishers Inc.
1 Park Avenue
New York, N.Y. 10016

Editorial office for France

Gordon & Breach
7-9 rue Emile Dubois
Paris 75014

Library of Congress catalog card number 70-143219.
ISBN 0 677 02110 0.
 Printed in Great Britain.

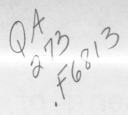

PREFACE

Probability theory is an important, well-nigh indispensable tool for many non-mathematical research workers, such as physicists, engineers, biologists, operations research workers. The aim of the first edition of this book, which appeared in 1950, was to provide these users with an exposition of the fundamentals of probability theory, at a level of mathematical sophistication which would not repel the non-specialist reader.

The first edition is now out of print. The present book is a second edition. Our aim in preparing the second edition was the same; nevertheless, the second edition has turned out quite different from the first.

First, probability theory, together with its fields of application, has expanded considerably, and will of necessity continue to develop. The second edition will therefore comprise *two* volumes.

Second, and more important, the utility of probability theory for physicists, engineers, biologists, operations researchers, etc. is no longer open to question. At the same time, the teaching of mathematics in general (and probability theory in particular) has developed considerably in regard to these scientific disciplines.

We therefore found it possible, and fruitful, to employ relatively advanced concepts of mathematics, in particular, the concepts of measure and Hilbert spaces. Introduction of these concepts, in themselves quite concrete and elementary, has great advantages, even though we have not discussed them in all detail.

The result is a gain in rigor; the reader may skip complete proofs, but he may not, and should not, disregard rigorous arguments. Thus, in Chapter VI, which is devoted to asymptotic laws of addition of independent random variables, with a fairly complete survey of the classical results, we have omitted all but the simplest proofs. However, we have taken pains throughout to explain *how* to formulate the problem of laws of large numbers, *how* to formulate limit laws, and so on. Similarly, we have not ignored the fact that the usual laws of large numbers are merely probability-theoretic versions of ergodic theorems.

Certain proofs have been given, such as that of the celebrated theorem of Bochner on continuous positive-definite functions. We deem it a psychological contradiction to make frequent use of a fundamental property, without ever having the curiosity to read a justification, especially when the latter involves no specialized knowledge.

On the other hand, only general concepts can give rise to general formulations: this is their *raison d'être*. One generally studies particular cases, and it is sometimes deduced that there is no need for general theories. My frequent collaboration with engineers, physicists, operations researchers, etc. has convinced me that this view, like many others, is ill founded. It is sufficient to glance at the technical reviews to see how often our engineers deal separately with slightly different cases, which are in effect amenable to a unified solution, if one only makes the effort to formulate it and set it down.

The present English translation is the translation of an original text, published in 1960 by the Centre National de la Recherche Scientifique at Paris. We took the opportunity to rectify a few errors, some quite serious, which disfigured the French text. We wish to thank the translation editor, Mr. D. Louvish, for his devoted efforts in this connection. We hope the reader will excuse the errors and inaccuracies which may still remain.

We are also indebted to the Centre National de la Recherche Scientifique, who authorized the English edition, and to Gordon and Breach Science Publishers, especially to Mme F. Chantrel-Riols, who have brought it to completion.

Professor R. Fortet
Paris

TRANSLATION EDITOR'S PREFACE

This is a translation of the author's revised version of *Eléments de la Théorie des Probabilités*, Vol. I, originally published by the *Centre National de la Recherche Scientifique* in their series *Applications des Théories Mathématiques*.

The terminology employed is that standard in English, with a few isolated exceptions which have been retained at the author's express request. At the same time, an attempt has been made to preserve the flavor of the original French text.

The translation has been checked by the author. We are indebted to him for his unflagging interest and willing cooperation, which were important factors in enabling us to complete the book.

TABLE OF CONTENTS

Preface v

Translation editor's preface vii

List of abbreviations, notations, symbols xvii

CHAPTER I - COMBINATORY ANALYSIS AND ITS APPLICATION TO CLASSICAL AND QUANTUM STATISTICS AND TO THE CHROMOSOME THEORY OF HEREDITY 1

I - Factorials and the Γ-function

1. Factorials and the Γ-function, Stirling's formula 2

II - Arrangements, permutations, combinations

2. Arrangements, permutations, combinations, properties of the numbers C_n^r, Newton's binomial formula 5

III - Arrangements, permutations and combinations with repetitions

3. Arrangements with repetitions, combinations with repetitions, permutations with repetitions, restricted repetitions 9

IV - Various problems and applications to the statistics of classical statistical mechanics and quantum mechanics

4. Problem 1, Problem 2; classical statistics of perfect gases; Bose-Einstein statistics; Fermi-Dirac statistics; Statistics of L. Brillouin 14

V - Chromosome theory of heredity

5. Chromosome theory of heredity 21

Bibliographic notes for Chapter I 26

Exercises 26

CHAPTER II - THE CONCEPT OF PROBABILITY. MEASURES OR MASS DISTRIBUTIONS. HILBERT SPACES. RANDOM ELEMENTS AND PROBABILITY LAWS 30

I - The concept of probability

1. Trials and random events, Example (2,1,1), systems of events 31

2. A first approach to the concept of probability; generalization, Example (2,2,1), dice; Example (2,2,2), Quantum Statistics 35

3. Frequency and probability; probability in the chromosome theory of heredity 44

4. Trials and classes of trials; first problem, Remark (2,4,1); second problem; events as subsets of the class of trials, Remark (2,4,2), limit of a sequence of sets or events 48

5. First axioms of probability theory, Axiom I, set functions, Remark (2,5,1), σ-algebra of subsets, Example (2,5,1), measure; Axiom II, Axiom III (Axiom of Total Probability), Remark (2,5,2); probability law of a class of trials; results 57

II - Measures or Mass Distributions

6. Measure and mass distribution; elementary properties of measures, Theorem (2,6,1), Theorem (2,6,2), sets of measure zero and properties true almost everywhere; point functions, Lemma (2,6,1); definite integral, Theorem (2,6,3), Theorem (2,6,4), m-equivalent point functions, m-equivalent mappings; integral over an arbitrary set, Remark (2,6,1), Theorem (2,6,5); indefinite integral, Theorem (2,6,6), Theorem (2,6,7), Remark (2,6,2) 64

7. Chebyshev's Theorem, Theorem (2,7,1), Theorem (2,7,2) (Chebyshev), Theorem (2,7,3) (Hölder), Schwarz's inequality, Theorem (2,7,4), Theorem (2,7,5), Theorem (2,7,6) (Minkowski), Theorem (2,7,7) 74

8. Types of convergence of point functions; convergence almost everywhere, convergence in measure; convergence in α-th mean, Theorem (2,8,1), Theorem (2,8,2), Theorem (2,8,3); Theorem (2,8,4) (Lebesgue), Theorem (2,8,5) (Fatou) 79

9. Totally discontinuous measures or mass distributions 85

III - Hilbert spaces

10. Hilbert spaces, Theorem (2.10.1), Theorem (2,10,2); Theorem (2,10,3), Theorem (2,10,4), Theorem (2,10,5); Theorem (2,10,6); Hilbert subspaces, Generalization (2,10,1), linear mappings of a Hilbert space into a Hilbert space, Remark (2,10,1) on isometries of Hilbert spaces; projections, canonical isomorphism; bases; Remark (2,10,2); example of a Hilbert space: the spaces L_2, Theorem (2,10,7) (Fischer-Riesz), Theorem (2,10,8), Remark (2,10,3) 87

IV - Random elements and probability laws

11. Random elements; probability law of a random element; induced measures, Remark (2,11,1), change of variable in an integral, Theorem (2,11,1); random elements which are functions of others; random variables, Example (2,11,1), Example (2,11,2), Example (2,11,3), Example (2,11,4); n-dimensional random vectors and variables, Example (2,11,5), Example (2,11,6); Remark (2,11,2); Lebesgue measure, absolutely continuous measures or mass distributions over $\mathfrak{X}_n$; Remark (2,11,3); uniform measures or mass distributions; Remark (2,11,4), Theorem (2,11,2) (Radon-Nikodym); convergence of a sequence of measures 109

12. Impossible and almost impossible events; certain and almost certain events, Example (2,12,1), Theorem (2,12,1); Remark (2,12,1); equivalent random elements; Remark (2,12,2) 125

Bibliographic notes for Chapter II 131

Exercises 132

CHAPTER III - DISTRIBUTION FUNCTIONS 135

1. Distribution functions; Remark (3,1,1); Remark (3,1,2); basic properties of distribution functions; distribution function associated with a measure, Theorem (3,1,1); measure or mass distribution defined by a distribution function, Theorem (3,1,2), Remark (3,1,3), Remark (3,1,4), Remark (3,1,5), Theorem (3,1,3); equivalent distribution functions, Theorem (3,1,4); totally discontinuous distribution functions; absolutely continuous distribution functions; integral of a point function with respect to a distribution function 136

2. One-dimensional distribution functions; Theorem (3,2,1); equivalent distribution functions, Remark (3,2,1); Remark (3,2,2); normalized totally discontinuous distribution functions, Dirac distribution functions; absolutely continuous normalized distribution functions; singular distribution functions; Theorem (3,2,2); Theorem (3,2,3) 146

3. Riemann-Stieltjes integrals; Theorem (3,3,1); integration by parts; extensions 152

4. Convergence of distribution functions; Compactness Theorem (3,4,1); Theorem (3,4,2), Remark (3,4,1); distance between two distribution functions; Remark (3,4,2) 156

5. Review of matrix theory; complex matrices; positive-definite matrices, Lemma (3,5,1) 161

6. The special role of exponential functions, Example (3,6,1), Remark (3,6,1), Remark (3,6,2); functions of classes (S_1) and (S_2) and their Fourier transforms, Lemma (3,6,1); Theorem (3,6,1); Remark (3,6,3), Example (3,6,2), Lemma (3,6,2), Theorem (3,6,2); convolution, Theorem (3,6,3); Fourier transform in $\mathcal{L}_2$; Theorem (3,6,4) (uniqueness), Theorem (3,6,5) (inversion) 168

7. Positive-definite functions, Lemma (3,7,1), Remark (3,7,1); Fourier transforms of bounded distribution functions, Lemma (3,7,2), Example (3,7,1), Theorem (3,7,1), Theorem (3,7,2), Remark (3,7,2), Theorem (3,7,3); Theorem (3,7,4) (Bochner) 186

8. Converse of Theorem (3,7,1), Theorem (3,8,1), Remark (3,8,1); convolution of two bounded distribution functions, Theorem (3,8,2), Remark (3,8,2), Remark (3,8,3) 197

9. Absolute and algebraic moments of a bounded distribution function, Theorem (3,9,1); relations between moments and Fourier transforms, Remark (3,9,1), Remark (3,9,2) 201

10. Distribution functions over $[0, +\infty)$ and their Laplace transforms, Theorem (3,10,1), Example (3,10,1); absolutely monotone functions, Lemma (3,10,1), Lemma (3,10,2), Lemma (3,10,3), Theorem (3,10,2), Theorem (3,10,3) (Karamata); convolution; generating functions 207

11. n-dimensional distribution functions, n-dimensional Riemann-Stieltjes integrals, Remark (3,11,1), Remark (3,11,2); separation of variables; F-equivalent functions 219

12. Convergence of a sequence of n-dimensional distribution functions, Fourier transform of an n-dimensional bounded distribution function, Theorem (3,12,1); convolution; moments of a bounded n-dimensional distribution function, interpretation of moments of order 1, central moments, moments of order 2 and moments of inertia, Theorem (3,12,2); application of the Hölder inequalities to moments; case of separated variables, Theorem (3,12,3); intrinsic study, notation, definition 228

Bibliographic notes for Chapter III 240

Exercises 241

CHAPTER IV - RANDOM VARIABLES, AXIOM OF CONDITIONAL PROBABILITY 245

I - Random variables

1. Distribution function of a random variable, Remark (4,1,1), discrete random variables, indicator, Example (4,1,1), continuous random variables, uniformly distributed random variables; Remark (4,1,2); certain number, almost certain number; types of random variables and distribution functions, symmetric random variables and distribution functions 246

2. Description of a random variable, mathematical expectation of a random variable, Remark (4,2,1), Example (4,2,1); computation of mathematical expectations, Example (4,2,2), Example (4,2,3); Remark (4,2,2); Theorem (4,2,1), Example (4,2,4); second moments and dispersion, Example (4,2,5); absolute and algebraic moments of a random variable, Remark (4,2,3); reduced random variables and laws; other central values and measures of dispersion 252

3. Complex random variables; characteristic function of a real random variable, Example (4,3,1), Example (4,3,2); second characteristic function; moment-generating function, generating function 268

4. Poisson laws; Remark (4,4,1); normal distribution, laws of normal type, Henry's line, degenerate normal distributions 275

II - Axiom of conditional probability

5. Conditional probability of one event given another; axiom of conditional probability (provisional formulation); Example (4,5,1); Example (4,5,2); lifetime of a radioactive atom and the exponential law; case of more than two events; independent events, Remark (4,5,1), Theorem (4,5,1), independent σ-algebras, case of any finite number of events, Remark (4,5,2); the general case; mutual independence of random elements, Remark (4,5,3); Example (4,5,3) 285

6. Application of the axiom of conditional probability; axiom of conditional probability (final formulation); Example (4,6,1), Remark (4,6,1), Remark (4,6,2), Example (4,6,2); Example (4,6,3): the axiom of conditional probability in Wave Mechanics; conditional mathematical expectation; Bayes' Theorem, Example (4,6,4), Theorem (4,6,1) (Bayes) 299

Bibliographic notes for Chapter IV 314

Exercises 314

CHAPTER V - n-DIMENSIONAL RANDOM VECTORS AND VARIABLES 319

1. Distribution function of an n-dimensional random variable; *a priori* or marginal probability laws of X and Y; characteristic functions and moments of a 2-dimensional random variable, correlation coefficient, Theorem (5,1,1), Example (5,1,1) (a theory of vision) 320

2. Conditional study of a random variable with respect to another, Bayes' formula, conditional mathematical expectation, Theorem (5,2,1); conditional variance, Theorem (5,2,2); Remark (5,2,1), case of independent X and Y, Theorem (5,2,3), Theorem (5,2,4); Theorem (5,2,5); Comment; estimation in the sense of minimum mean-square deviation, Problem 1, Problem 2, Theorem (5,2,6), the best conditional linear estimate 330

3. Intrinsic study of an n-dimensional random vector, Theorem (5,3,1), Theorem (5,3,2) 343

4. Complex random variables, independence of complex random variables; Hilbert space of second-order complex random variables; Remark (5,4,1); the covariance matrix, Theorem (5,4,1); Theorem (5,4,2); best linear estimate, Theorem (5,4,3) 346

5. n-dimensional normal random vectors and variables; Theorem (5,5,1), Theorem (5,5,2), Theorem (5,5,3), Theorem (5,5,4), rank of a normal random variable, distribution function and probability density of a normal n-dimensional random variable, Theorem (5,5,5); Remark (5,5,1), Remark (5,5,2), case of a degenerate normal random variable, Theorem (5,5,6); the case n = 2, Theorem (5,5,7); conservation of normality, Theorem (5,5,8); Example (5,5,1) 360

6. Conditional probability laws of normal n-dimensional random variables, Rule (5,6,1), the case n = 2; complex normal random variables, Remark (5,6,1) 381

Bibliographic notes for Chapter V 390

Exercises 390

CHAPTER VI - ADDITION OF INDEPENDENT RANDOM VARIABLES; STOCHASTIC CONVERGENCE, LAWS OF LARGE NUMBERS, ERGODIC THEOREMS; CONVERGENCE TO A NORMAL LAW, CONVERGENCE TO A POISSON LAW; GENERALIZATIONS 393

I - Addition of independent random variables

1. Addition of random variables; Theorem (4,2,1); Theorem (5,2,3), Theorem (6,1,1), Remark (6,1,1); the heads-or-tails scheme, Bernoulli law, Exercises (6,1,1) 394

2. Convolutions of distribution functions and probability densities of independent random variables, Theorem (6,2,1); Theorem (6,2,2), Corollaries (6,2,1), Remark (6,2,1); Exercise (6,2,1); Theorem (6,2,3), Theorem (6,2,4) (Raikov); Theorem (6,2,5), Theorem (6,2,6) (Cramer); closed types of laws, Theorem (6,2,7), Cauchy law and type 400

II - Stochastic convergence, laws of large numbers, ergodic theorems

3. Laws of large numbers and stochastic convergence 408

4. Convergence in quadratic mean; convergence in α-th mean 409

5. Convergence in probability; Theorem (6,5,1), Remark (6,5,1), Example (6,5,1), Example (6,5,2) 412

6. Almost sure convergence; Remark (6,6,1), Theorem (6,6,1); Theorem (6,6,2), Example (6,6,1), Example (6,6,2); Remark (6,6,2); application to the heads-or-tails scheme 415

7. Laws of large numbers; Theorem (6,7,1) (law of large numbers in quadratic mean); Theorem (6,7,2) (almost sure law of large numbers); application to the heads-or-tails scheme; Theorem (6,7,3) (Kolmogorov); Theorem (6,7,4) (Glivenko-Cantelli), Exercise (6,7,1) 420

8. Ergodic theory; Theorem (6,8,1), Remark (6,8,1), second-order stationary sequences, Theorem (6,8,2) 426

9. Birkhoff's ergodic theorem, Theorem (6,9,1) (Birkhoff); strictly stationary sequences, case of normal sequences; Theorem (6,9,2), Theorem (6,9,3), Lemma (6,9,1), Lemma (6,9,2); Theorem (6,9,4), Exercise (6,9,1) 432

III - Convergence to a law of normal type; Convergence to a Poisson law

10. Convergence to a normal law in the Bernoulli case, generalizations, Theorem (6,10,1), Theorem (6,10,2) 443

11. The general problem of convergence to a limit law, Definition (6,11,1); infinitely divisible laws and distribution functions, Theorem (6,11,1), Theorem (6,11,2), Theorem (6,11,3), Theorem (6,11,4), Corollary (6,11,1), Theorem (6,11,5), Theorem (6,11,6), Example (6,11,1), Example (6,11,2), Remark (6,11,1), Theorem (6,11,7); Theorem (6,11,8), Example (6,11,3), Exercise (6,11,1), Example (6,11,4), Remark (6,11,2) 451

12. Convergence to a limit law for cumulative sums of a large number of independent random variables, Definition (6,12,1), Theorem (6,12,1), Theorem (6,12,2), Theorem (6,12,3), Theorem (6,12,4), Exercise (6,12,1); special case: identically distributed X_j, Theorem (6,12,5), Theorem (6,12,6) 462

13. The Gauss law of measurement errors; normal approximation, Theorem (6,13,1) (A. Berry and C. Esseen); lattice random variables, Theorem (6,13,2), Theorem (6,13,3) 466

14. New approach, Theorem (6,14,1); convergence of probability densities to a normal density, Theorem (6,14,2), laws of the iterated logarithm, Theorem (6,14,3), Remark (6,14,1); Remark (6,14,2) 475

IV - Generalizations

15. Addition of independent n-dimensional random vectors; strictly stationary sequences of random elements; Theorem (6,15,1), Theorem (6,15,2), Theorem (6,15,3), Converse (6,15,1), Theorem (6,15,4); second-order random elements, Theorem (6,15,5), Remark (6,15,1), Theorem (6,15,6); convergence to an n-dimensional normal law; fluctuations of concentration in the atmosphere; the blue color of the sky, Remark (6,15,2) 480

Bibliographic notes for Chapter VI 497

Exercises 498

Numerical tables 506

Bibliography 509

Index 520

ABBREVIATIONS

r.e.	:	random element
m.e.	:	mathematical expectation
i.q.m.	:	in quadratic mean
c.f.	:	characteristic function
d.f.	:	distribution function
RS-integral	:	RIEMAN-STIELTJES integral
a.c.	:	almost certain, almost certainly
g.l.b.	:	greatest lower bound
a.e.	:	almost everywhere
l.u.b.	:	least upper bound
a.s.	:	almost sure, almost surely
r.v.	:	random variable

As usual in probability theory, we shall often use the adjective "stochastic" as a synonym for random, probabilistic, statistical, etc.

Sections are numbered by chapter; for instance: II.4 means Section 4 of Chapter II.

Formulas, theorems, lemmas, remarks, examples, etc., are numbered by section; for instance: formula (3,8,7) means formula 7 in Section 8 of Chapter III; Theorem (5,3,1) means Theorem 1 in Section 3 of Chapter V, etc.

The exercises at the end of each chapter are also numbered by chapter; for instance, Exercise 4.5 means Exercise 5 at the end of Chapter IV.

The sign ■ indicates the end of a proof, statement, remark, example, etc., and resumption of the text.

SYMBOLS

$>$	greater than
$<$	smaller than
$\geqq$ or $\geq$	greater than or equal to
$\leqq$ or $\leq$	smaller than or equal to
#	approximately equal to
(a,b)	interval with end-points a and b
$[a,b]$	closed interval (a,b)
$]a,b[$	open interval (a,b)
$]a,b]$	interval open at the left, closed at the right
$[a,b[$	interval open at the right, closed at the left
$A \subset B$ or $B \supset A$	A is a subset of B (or B contains A)
$\cup$, $\bigcup_{\mathfrak{F}}$	union, union of the sets of the family $\mathfrak{F}$
$\cap$, $\bigcap_{\mathfrak{F}}$	intersection, intersection of the sets of the family $\mathfrak{F}$
$\longrightarrow (x \longrightarrow \alpha)$	tends to (the parameter x tends to α)
$\lim_{n \to +\infty}$, $\lim_{n \to -\infty}$	limit as the integer n tends to positive infinity, negative infinity
$\lim_{x \to \alpha}$	limit as the parameter x tends to α
$\lim_{x \to \alpha + 0}$, $\lim_{x \to \alpha - 0}$	limit as the real number x tends to α from the right, from the left
$\emptyset$	empty set
$\check{A}$	complement of the set A, event complementary to A
$\{x \in \mathfrak{X} \mid P\}$	set of elements of the set $\mathfrak{X}$ possessing the property P
$[a]$	integral part of the real number a, i.e., the integer b defined by: $b \leqq a < b + 1$.
$\bar{a}$	conjugate of the complex number a
$\mathcal{R}(a)$	real part of the complex number a
$\mathcal{J}(a)$	imaginary part of the complex number a
$O(x)$	function of x such that the quotient $O(x)/x$ remains bounded as $x \to 0$ (or $x \to \infty$)
$o(x)$	function of x such that the ratio $o(x)/x$ tends to zero as $x \to 0$ (or $x \to \infty$)
$a \sim b$	a and b are equivalent

$n!$	n factorial
C_n^r or $\binom{n}{r}$	number of distinct combinations of n objects r at a time
exp. $\{A\}$	e^A
$^{\circ}A$	transpose of the matrix A
$\overline{A}$	complex conjugate of the matrix A
$\mathfrak{X}^*$	dual of the vector space $\mathfrak{X}$ (set of linear functionals on $\mathfrak{X}$)
$< z^* , z >$	number obtained by applying the linear functional z^* to the vector z
$\mathfrak{X}^{**}$	bidual of the vector space $\mathfrak{X}$ (dual of the dual $\mathfrak{X}^*$ of $\mathfrak{X}$)
$Pr(E)$	probability of the event E
$m(E)$	measure of the set E
$F * G$	convolution of F and G
${}_a m_\alpha$, ${}_{a,b} m_{\alpha,\beta}$	algebraic moment of order α (α,β) about a (a and b)
${}_a m^*_\alpha$, ${}_{a,b} m^*_{\alpha,\beta}$	absolute moment of order α (α,β) about a (a and b)
μ_α, $\mu_{\alpha,\beta}$	algebraic central moment of order α (α,β)
μ^*_α, $\mu^*_{\alpha,\beta}$	absolute central moment of order α (α,β)
$\vec{g}$	mean vector of mass distribution
$E(X)$	expectation of the random variable X
$\mathcal{V}(X)$	variance of the random variable X
$\sigma(X) = \sqrt{\mathcal{V}(X)}$	standard deviation of the random variable X
$\overline{\overline{X}}$	median of the random variable
r	correlation coefficient (of two random variables)
$Pr(B/A)$	conditional probability of the event B, given the event A
$E(X/A)$	conditional expectation of the random variable X with respect to the class of trials realizing A

CHAPTER I

COMBINATORY ANALYSIS
AND ITS APPLICATION TO
CLASSICAL AND QUANTUM STATISTICS
AND TO THE CHROMOSOME THEORY
OF HEREDITY

I. FACTORIALS AND THE Γ-FUNCTION

1. *Factorials and the Γ-function*

Let n be a positive integer; the product

$$1 \times 2 \times 3 \times \ldots \times p \times (p+1) \times \ldots \times n$$

of the first n integers is called "n factorial," usually denoted by

$$n\,!\ .$$

The notation n! is sometimes replaced by

$$\underline{n|}\ .$$

0! is actually undefined, but for various reasons it is convenient to define

$$0\,! = 1. \qquad (1,1,1)$$

Let x denote a positive real variable, and consider the function Γ(x) (the *gamma*-function, or Euler integral of the second kind), defined for $x > 0$ by the formula:

$$\Gamma(x) = \int_0^{+\infty} u^{x-1}\, e^{-u}\, du, \qquad (1,1,2)$$

where the integral is convergent for $x > 0$, but not for $x \leqq 0$.

It can be shown that

$$\Gamma(x+1) = x\,\Gamma(x), \qquad (1,1,3)$$

and hence one immediately obtains for integral $n > 0$

$$n\,! = \Gamma(n+1). \qquad (1,1,4)$$

This representation of n! as Γ(n+1) enables us to define n! even for non-integral values of n.

Moreover, the recurrence formula (1,1,3) enables us to define Γ(x) for negative x, with the exception of zero and the negative integers, since (1,1,3) is false for x = 0. It can also be proved that for any x

$$\Gamma(x) = \lim_{r \to +\infty} \frac{r^x \, r\,!}{x(x+1)\,\ldots\,(x+r)}, \tag{1,1,5}$$

implying that Γ(x) becomes infinite when x is zero or a negative integer.

The function Γ(x) can be defined even for complex arguments (other than the real integers $\leqq 0$) (see Table [6]), but we shall not need this extension here.

Note that by (1,1,2), Γ(1) = 1, which, among other things, justifies the convention (1,1,1). Also,

$$\Gamma(\tfrac{1}{2}) = \sqrt{\pi}. \tag{1,1,6}$$

The function B(x,y) of the two variables x and y defined by

$$B(x,\ y) = \int_0^1 u^{x-1}\,(1-u)^{y-1}\,du \tag{1,1,7}$$

is called the *Euler integral of the first kind* or B(beta)-function; it can be proved that

$$B(x,\ y) = \frac{\Gamma(x)\,\Gamma(y)}{\Gamma(x+y)}. \tag{1,1,8}$$

Stirling's formula. It can be proved that n! may be expressed in the form

$$n\,! = n^n\,e^{-n}\,\sqrt{2\,\pi\,n}\,(1+\varepsilon_n) \tag{1,1,9}$$

where

$$\varepsilon_n = \frac{1+\theta_n}{12n} = \frac{1}{12n} + \frac{1+\eta_n}{288n^2} \tag{1,1,10}$$

and $\varepsilon_n,\ \theta_n,\ \eta_n \longrightarrow 0$ as $n \longrightarrow +\infty$. It follows that for large n, n! may be approximated by any of the expressions

$$n! \# n^n e^{-n} \sqrt{2\pi n}, \tag{1,1,11}$$

$$n! \# n^n e^{-n} \sqrt{2\pi n} \left(1 + \frac{1}{12n}\right), \tag{1,1,12}$$

$$n! \# n^n e^{-n} \sqrt{2\pi n} \left(1 + \frac{1}{12n} + \frac{1}{288n^2}\right), \tag{1,1,13}$$

depending on the desired degree of accuracy; in practice (1,1,11) is almost always sufficient, and *in fact gives* n! *with a relative error of about* 0.01 *for* n *as small as* 8. Note that these approximation formulas, due to Stirling, are also valid for $\Gamma(n + 1)$, whether n is a positive real integer or not.

Log n! (natural logarithm) is needed as often as n!; for log (n!) we have the corresponding approximation formulas:

$$\log(n!) \# \left(n + \frac{1}{2}\right) \log n - n + \log \sqrt{2\pi}, \tag{1,1,14}$$

$$\log(n!) \# \left(n + \frac{1}{2}\right) \log n - n + \log \sqrt{2\pi} + \frac{1}{12n}, \tag{1,1,15}$$

$$\log(n!) \# \left(n + \frac{1}{2}\right) \log n - n + \log \sqrt{2\pi} + \frac{1}{12n} + \frac{1}{360n^2}. \tag{1,1,16}$$

Numerical values of n! (Table 1) and the function log $\Gamma(x)$ have been tabulated (for instance, Tables [1], [2], [3]). ■

In statistics one frequently uses the incomplete Γ-functions, defined by

$$\Gamma_\varepsilon(x) = \int_0^\varepsilon u^{x-1} e^{-u} du, \tag{1,1,17}$$

and the incomplete B-functions, defined by

$$B_\varepsilon(x, y) = \int_0^\varepsilon u^{x-1} (1 - u)^{y-1} du; \tag{1,1,18}$$

both of these functions have been tabulated (Tables [4] and [5]).

II. ARRANGEMENTS, PERMUTATIONS, COMBINATIONS

2. *Arrangements*

Consider a given collection of n *distinct* objects O; we can distinguish between them by using subscripts O_1, O_2,...O_n, different letters a, b, c..., or by representing them by different integers 1, 2, 3,..., n. Since the integers are naturally ordered (in order of increasing magnitude), as are the letters of the alphabet (in alphabetical order: a, b, c, d, etc.), we may also regard the n objects of the collection as being *ordered*; this order may be taken into account or disregarded, depending on the case at hand.

Any r of the n objects ($1 \leqq r \leqq n$) *constitute a set of cardinality* r. Two sets of different cardinalities are easily distinguished; two sets of the same cardinality can be distinguished by identifying their elements, e.g., the two sets ab and bd - both of cardinality 2; finally, if the n objects of the collection are regarded as *ordered*, we may distinguish between sets of the same cardinality consisting of the same objects according to the order in which the objects appear; from this point of view, the two sets adf and fad of cardinality 3 are distinct.

Any set of cardinality r ($1 \leqq r \leqq n$), taking the order into account, is called an *arrangement* of the n objects r at a time; i.e., the same r objects in two different orders form two *distinct arrangements*. Let us find the number A_n^r of distinct arrangements of n objects r at a time.

Taking any r of the n objects, consider an arbitrary arrangement α. Let us add to α any one of the n-r objects not contained in α, putting it in the last position, i.e. *after* the last object in α. We obtain n-r arrangements of the objects r + 1 at a time, which are obviously distinct. This procedure, applied to all arrangements taken r at a time, yields all arrangements taken r + 1 at a time, and *each is obtained exactly once*. Hence

$$A_n^{r+1} = (n - r)\, A_n^r .$$

Hence, by induction (since obviously $A_n^1 = n$),

$$A_n^r = n(n - 1)\,(n - 2) \cdots (n - r + 1) = \frac{n\,!}{(n - r)\,!} . \qquad (1,2,1)$$

Permutations

A *permutation* of n objects is an arrangement of these objects n at a time; by (1, 2, 1), the number P_n of distinct permutations of n objects is therefore

$$P_n = n\,! . \qquad (1,2,2)$$

Combinations

A *combination* of n objects r at a time ($1 \leqq r \leqq n$) is any set of cardinality r, *disregarding their order.* Two combinations of n taken r at a time are therefore not distinct if they contain the same objects; for instance, abc and bac are two distinct arrangements 3 at a time, but they represent the same combination.

The number C_n^r of distinct combinations of n objects r at a time is:

$$C_n^r = \frac{A_n^r}{r\,!} = \frac{n(n - 1) \ldots (n - r + 1)}{r\,!} = \frac{n\,!}{r\,!\,(n - r)\,!} . \qquad (1,2,3)$$

Indeed, let us choose any r of the n objects; these r objects yield only one combination, but by performing all r! permutations of these objects we obviously obtain all possible arrangements of them.

Another notation for C_n^r is

$$\binom{n}{r} .$$

Properties of the numbers C_n^r

An obvious result of (1,2,3) is

$$C_n^r = C_n^{n-r}. \tag{1,2,4}$$

Since $C_n^n = 1$, this formula implies the useful convention

$$C_n^0 = 1.$$

It may also be easily verified that

$$C_n^r = C_{n-1}^r + C_{n-1}^{r-1}.$$

This formula yields an easy inductive computation of C_n^r, provided n is not too large (this is the idea of Pascal's "arithmetical triangle").

Newton's Binomial Formula

One of the most important applications of the numbers C_n^r is Newton's expansion of the binomial, that is, the nth power (n a positive integer) of the sum of two terms A and B, $(A + B)^n$, or the product of n factors all equal to (A + B):

$$(A + B)^n = \underbrace{(A + B)\,(A + B)\,\ldots\,(A + B)}_{n \text{ factors}}\,. \tag{1,2,5}$$

The expansion of the n factors in the right-hand side of (1,2,5) yields the sum of a certain number of terms, each obtained in the following way. In α of the factors ($0 \leqq \alpha \leqq n$) the term A is chosen, while in the remaining β factors ($\beta = n - \alpha$) the term B is chosen. Multiplication of these terms gives the monomial $A^\alpha B^\beta$; it is clear that this monomial is obtained more than once in the expansion of (1,2,5), to be precise, $C_n^\alpha = C_n^\beta$ times, since there are C_n^α ways of choosing α (or β) factors from the n factors of (1,2,5).

Since multiplication is commutative, the order of the chosen factors is immaterial; the monomials appearing in the expansion of (1,2,5) are obviously

$$A^n B^o,\ A^{n-1} B^1,\ \ldots,\ A^1 B^{n-1},\ A^o B^n,$$

and thus

$$(A + B)^n = \sum_{r=0}^{n} C_n^r A^{n-r} B^r. \qquad (1,2,6)$$

Apart from its direct applications, this formula yields various useful properties of the C_n^r. Thus, putting A = B = 1 we see that

$$\sum_{r=0}^{n} C_n^r = 2^n . \qquad (1,2,7)$$

Similarly, putting A = -B, we obtain

$$0 = \sum_{r=0}^{n} (-1)^r C_n^r \text{ ou : } \sum_p C_n^{2p} = \sum_p C_n^{2p-1}. \qquad (1,2,8)$$

Formulas (1,2,7) and (1,2,8) give

$$\sum_p C_n^{2p} = \sum_p C_n^{2p-1} = 2^{n-1} . \qquad (1,2,9)$$

The summation in (1,2,8) and (1,2,9) extends over all p such that $0 \leqq 2p \leqq n$, and $0 \leqq 2p - 1 \leqq n$, respectively.

III. ARRANGEMENTS, PERMUTATIONS AND COMBINATIONS WITH REPETITIONS

3. *Arrangements with repetitions*

Consider again a collection of n distinct objects, but now suppose that each of these objects is available in infinitely many identical copies; thus a set of cardinality r may contain any number (at most r, of course) of copies of the same object*. If we agree to represent the different copies of an object g by the same letter g, one set of this kind is, for instance, aab, of cardinality 3, which contains two copies of the object a. Note that by assuming that all copies of the same object are *identical* we render them indistinguishable. Thus, if for instance we permute the two copies of a in aab, the set is *not altered*; of course, we can still distinguish between, say, aab and aba, by taking the order of the objects a, b into account.

Arrangements, permutations, and combinations formed under these conditions are said to be with *repetitions*. Of course, at any stage of an argument we can still distinguish mentally between indistinguishable copies, provided that these distinctions can again be disregarded at will; this procedure will occur frequently.

Note that owing to the multiplicity of the copies we can form sets of cardinality exceeding n.

Arrangements of n objects r at a time ($r \geqq 1$) with repetitions are defined in the same way as arrangements without repetitions; by a similar argument, the number A'^{r}_{n} of these arrangements is

$$A'^{r}_{n} = n^{r}. \qquad (1,3,1)$$

---

*[*Translation editor's note*: This is a departure from normal usage, in that the elements of a set are usually considered distinct. The original French has "groupe"; we have avoided the word "group" because of its usual connotation in mathematics.]

Combinations with repetitions

Similarly, a combination of n objects r at a time ($r \geqq 1$) with repetitions is any set of cardinality r with repetitions, where the order is disregarded; thus a combination of r of the objects a, b,..., l is *completely* characterized by the number α of occurrences of a ($\alpha \geqq 0$), the number β of occurrences of b ($\beta \geqq 0$), and so on; obviously $\alpha + \beta + \ldots = r$. Let us evaluate the number C'^{r}_{n} of distinct combinations of cardinality r that can can be formed from n objects a, b, c,...l with repetitions. To this end, let us set up a table of these combinations; it contains rC'^{r}_{n} letters in all; each of the letters a, b, c,..., l obviously occurs the same number of times; therefore, the letter a, for instance, occurs

$$\frac{r}{n} C'^{r}_{n}$$

times. Among these combinations, consider those in which a occurs *at least once*; there are C'^{r-1}_{n} of these, since if we omit a once in each we obtain a table of combinations with repetitions of n objects r - 1 at a time; in this table a occurs

$$\frac{r-1}{n} C'^{r-1}_{n}$$

times. Hence the relation:

$$\frac{r}{n} C'^{r}_{n} = C'^{r-1}_{n} + \frac{r-1}{n} C'^{r-1}_{n} = \frac{n+r-1}{n} C'^{r-1}_{n} .$$

Hence, since obviously $C'^{1}_{n} = n$, induction yields the formula

$$C'^{r}_{n} = \frac{(n+r-1)!}{r!\,(n-1)!} = C^{r}_{n+r-1} . \qquad (1,3,2)$$

Permutations with repetitions

The concept of a permutation of n objects with repetitions has no meaning unless the number of times each object occurs in the permutations is fixed. Therefore, let n distinct objects a, b, c..., l be given; take α copies of a, β copies of b,..., λ copies of l ($\alpha \geqq 0$, $\beta \geqq 0, \ldots, \lambda \geqq 0$), thus forming a set of $r = \alpha + \beta + \ldots + \lambda$ objects (*not necessarily distinct*): let us evaluate the number $P_n'(\alpha, \beta, \ldots, \lambda)$ of *distinct* permutations which can be formed from these r objects. Note first that r! permutations, obviously not all distinct, result when these objects are arranged in all possible ways. Thus, if we permute the letters a in one of these permutations *among themselves* in all possible ways, the same permutation is obtained $\alpha!$ times; the same applies to the letters b, c,..., l, so that

$$P_n'(\alpha, \beta, \ldots, \lambda) = \frac{r!}{\alpha!\, \beta! \ldots \lambda!} = \frac{(\alpha + \beta + \cdots + \lambda)!}{\alpha!\, \beta! \ldots \lambda!}. \quad (1,3,3)$$

This formula may also be established as follows. Suppose that r boxes are numbered in a given order and we wish to allocate r objects to these boxes in such a way that each box will contain exactly one object. Recalling that the r objects are not all distinct, we see that there are $P_n'(\alpha, \beta, \ldots, \lambda)$ distinct allocations of this kind. Proceeding step by step, let us first dispose of the objects a; this amounts to choosing α of the r boxes, which can be done in

$$C_r^\alpha = \frac{r!}{\alpha!\,(r - \alpha)!}$$

different ways. In the $r - \alpha$ remaining boxes we then place the β objects b; this can be done in

$$\frac{(r - \alpha)!}{\beta!\,(r - \alpha - \beta)!}$$

different ways. With each allocation of the objects a we can thus associate

$$\frac{(r-\alpha)!}{\beta!(r-\alpha-\beta)!}$$

allocations of the objects b. All in all, we therefore have

$$\frac{r!}{\alpha!(r-\alpha)!} \times \frac{(r-\alpha)!}{\beta!(r-\alpha-\beta)!}$$

allocations of the objects a and b. The same argument for the remaining objects c,..., l yields:

$$P_n'(\alpha, \beta, \ldots, \lambda) = \frac{r!}{\alpha!(r-\alpha)!} \times \frac{(r-\alpha)!}{\beta!(r-\alpha-\beta)!} \times \ldots \times \frac{\lambda!}{\lambda!\ 0!},$$

which reduces, after simplification, to (1,3,3).

The same argument shows that, given r distinct objects, there are $P_n'(\alpha, \beta, \ldots, \lambda)$ ways of dividing them into n sets of cardinalities $\alpha, \beta, \ldots, \lambda$ respectively, where $\alpha + \beta + \ldots + \lambda = r$ - this is just the operation performed above with the r boxes.

A generalization of Newton's binomial formula is easily deduced from (1,3,3). Consider n numbers A, B,..., L and expand the expression $(A + B + \ldots + L)^r$; this expansion is a sum of monomials of the type $A^\alpha \times B^\beta \times \ldots \times L^\lambda$ with $\alpha \geqq 0$, $\beta \geqq 0, \ldots, \lambda \geqq 0$ and $\alpha + \beta + \ldots + \lambda = r$.

To see this note that $(A + B + \ldots + L)^r$ is the product of r factors all equal to A + B + ... + L, and let the r factors play the role of the r boxes in the previous argument. We therefore have

$$(A + B + \ldots + L)^r = \sum_{\alpha, \beta, \ldots, \lambda} \frac{r!}{\alpha!\ \beta! \ldots \lambda!} A^\alpha B^\beta \ldots L^\lambda, \qquad (1,3,4)$$

where

$$\alpha + \beta + \ldots + \lambda = r.$$

Taking A = B = C ... = L = 1, we deduce:

$$n^r = r! \sum_{\alpha,\beta,\ldots,\lambda} \frac{1}{\alpha!\,\beta!\ldots\lambda!}, \qquad (1,3,5)$$

where $\alpha + \beta + \ldots + \lambda = r.$

Formula (1,3,5) is a generalization of (1,2,7).

Restricted repetitions

Under the same conditions as before, more difficult problems may be considered, assuming for instance that the number of copies of the same object that may occur in a set of cardinality r, instead of being any number $\leqq$ r, is a preassigned constant $\alpha(\alpha \leqq r)$ - possibly different for each object. We shall not present the relevant formulas, since they are seldom used. The interested reader will find them together with their proofs and various properties in P. Montel [1], which also contains other references on the subject.

IV. VARIOUS PROBLEMS AND APPLICATIONS TO THE STATISTICS OF CLASSICAL STATISTICAL MECHANICS AND QUANTUM MECHANICS

4. *Problem 1*

Given km (k and m integers $\geqq 1$) distinct objects and m distinct boxes, in how many ways can the km objects be allocated to the m boxes, so that there are exactly k objects in each box?

It is clearly sufficient to divide the km objects into m sets, each containing k; hence there are

$$P'(k, k, \ldots, k) = \frac{(km)!}{k!\, k! \ldots k!} = \frac{(km)!}{(k!)^m}$$

possible allocations.

Problem 2

Given N distinct objects, let $g_1, g_2, \ldots, g_s$ be s disjoint sets of boxes, each containing $m_1, m_2, \ldots, m_s$ boxes respectively. Assuming that $N = m_1 + 2m_2 + \ldots + sm_s$, in how many ways can the N objects be placed in these boxes so that for any i there are exactly i objects in each of the m_i boxes of the set g_i?

We need only divide the N objects into $m_1 + m_2 + \ldots + m_s$ sets of which m_1 are of cardinality 1, m_2 of cardinality 2, ..., m_s of cardinality s; hence there are

$$\frac{N!}{(1!)^{m_1}(2!)^{m_2} \ldots (s!)^{m_s}} \qquad (1,4,1)$$

possibilities.

Classical statistics of perfect gases

The statistical theories of Classical Mechanics or Quantum Mechanics may be reduced to the problem of placing N objects in a certain number of boxes. The differences between these theories depend on whether the N objects are distinct, or indistinguishable, or at least not all distinct; the same applies to the boxes. Any "allocation" forms what is usually called a *state*.

Thus, in the Classical Kinetic Theory of Perfect Gases, one regards the N objects (molecules) as indistinguishable. As for the boxes (domains of extension in the phase space), these belong to different types T_1, T_2,...; let g_i (i = 1, 2,...) denote the number of boxes of type T_i. Given numbers n_1, n_2,..., n_i,..., where $\sum n_i = N$, we wish to know:

a) the number $\mathfrak{N}$ of distinct ways of placing the N objects in the boxes such that for any i exactly n_i objects are in the set of g_i boxes of type T_i;

b) the total number of possible states $\mathfrak{C}$;

c) the quotient $\mathfrak{P} = \dfrac{\mathfrak{N}}{\mathfrak{C}}$ which, interpreted as a probability, plays an essential role in this context [cf. Example (2,2,2)].

To evaluate $\mathfrak{N}$, assume first that the N objects are distinguishable. Note that to obtain an allocation of the desired kind we can first divide the N objects into sets G_1, G_2,..., G_i,..., of cardinalities n_1, n_2,..., n_i, respectively; this can be done, as we have seen, in

$$\frac{N\,!}{n_1\,!\; n_2\,!\;\ldots\; n_i\,!\;\ldots}$$

different ways. The n_i objects of the set G_i are then distributed among the g_i boxes of type T_i; this can obviously be done in $g_i^{n_i}$ different ways, since no conditions are imposed on the distribution. When this is done for each set G_i we get $\mathfrak{N}'$ possible allocations, where

$$\mathfrak{N}' = N\,!\prod_i \frac{g_i^{n_i}}{n_i\,!}\,. \qquad (1,4,2)$$

But these $\mathfrak{N}'$ allocations are distinct only if the N objects are distinguishable. To clarify the argument, consider a simple example: N = 3, two types of boxes T_1 and T_2, $g_1 = 1$ and $g_2 = 1$, i.e., two boxes in all, which we denote by B_1 and B_2. The three objects are indistinguishable and we represent them by the same letter a, but since in the first analysis we wish to distinguish between them, we provide them with subscripts: a_1, a_2, a_3. The three allocations

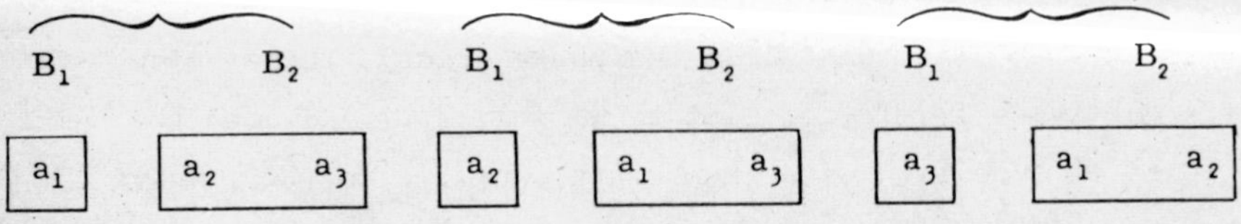

are counted as distinct in (1,4,2); but if we drop the distinction between the three objects, they become identical.

In the general case, consider an allocation in which m_1 boxes contain 1 object, m_2 boxes 2 objects, ..., m_s boxes s objects. All allocations obtained by permutation of the N objects among themselves are indistinguishable; their number is given by (1,4,1). Thus, if $\mathfrak{N}'(m_1, m_2, \ldots, m_s)$ denotes the number of the allocations counted as distinct in (1,4,2) where m_1 boxes contain 1 object, m_2 boxes 2 objects,..., m_s boxes s objects, $\mathfrak{N}$ is given by the formula

$$\mathfrak{N} = \sum_{m_1, m_2, \ldots, m_s} \frac{\mathfrak{N}'(m_1, m_2, \ldots, m_s)\,(1\,!)^{m_1}\,(2\,!)^{m_2}\ldots(s\,!)^{m_s}}{N\,!}, \qquad (1,4,3)$$

while obviously

$$\mathfrak{N}' = \sum_{m_1, m_2, \ldots, m_s} \mathfrak{N}'(m_1, m_2, \ldots, m_s).$$

However, evaluation of $\mathfrak{N}'(m_1, m_2, \ldots, m_s)$ is difficult. In fact, in the Kinetic Theory of Gases one assumes (1) that states in which at least one box contains more than one object are relatively rare; neglecting these states, (1,4,3) yields

$$\mathfrak{N} = \frac{1}{N\,!} \sum_{m_1} \mathfrak{N}'(m_1, 0, \ldots, 0) = \frac{\mathfrak{N}'}{N\,!} = \prod_i \frac{g_i^{n_i}}{n_i\,!}\,. \qquad (1,4,4)$$

Under the same conditions and to the same degree of accuracy, the total number of possible distinct states is

(1) Since the boxes are of arbitrarily small dimensions, which is equivalent to the fact that the g_i are arbitrarily large in relation to the n_i; cf. L. Brillouin [1], Vol. I, p. 120.

$$\mathcal{C} = \frac{G^N}{N!}, \qquad (1,4,5)$$

where G denotes the total number of boxes, i.e., $G = \sum_i g_i$. For effective utilization of formulas (1,4,4) and (1,4,5), the factorials are approximated by Stirling's formula; the numbers g_i and n_i may all be considered large.

The value of $\mathcal{P}$ results immediately from (1,4,4) and (1,4,5).

Bose-Einstein Statistics

In the *quantum* statistics of Bose-Einstein, the problem is the same as before and the same number $\mathfrak{N}$ is to be evaluated, but it is no longer assumed that except for a small number of states the boxes contain at most one object (2). Instead of using formula (1,4,3), which is now not very convenient, we employ another method.

Divide the N objects again into sets $G_1, G_2, \ldots, G_i, \ldots$, of respective cardinalities $n_1, n_2, \ldots, n_i \ldots, \left(N = \sum_i n_i\right)$. Since the objects are *indistinguishable*, there is *only one* such partition. Now distribute the n_i objects of G_i among g_i boxes of type T_i. Since each box can hold any number of objects and the latter are indistinguishable, this is equivalent to forming a combination of the g_i boxes n_i at a time with repetitions. The number of combinations is, following (1,3,2),

$$\frac{(n_i + g_i - 1)!}{n_i!\,(g_i - 1)!}.$$

Thus we obtain

$$\mathfrak{N} = \prod_i \frac{(n_i + g_i - 1)!}{n_i!\,(g_i - 1)!}. \qquad (1,4,6)$$

---

(2) Here the g_i have fixed finite values and cannot be considered arbitrary large relative to the n_i.

Under the same conditions, the total number of possible distinct *states* is

$$\mathcal{C} = \frac{(N + G - 1)!}{N!\,(G - 1)!} \qquad (1,4,7)$$

by the same argument; the value of $\mathcal{P}$ follows immediately from (1,4,6) and (1,4,7).

Fermi-Dirac Statistics

For the Fermi-Dirac statistics the preceding problem is solved once more, but with the assumption, following from Pauli's exclusion principle, that states in which a box contains more than one object are impossible; therefore, they are counted neither in the evaluation of $\mathcal{N}$ nor in that of $\mathcal{C}$. The preceding argument may be repeated without changes, except that instead of combinations with repetitions we now have combinations without repetitions. The result is:

$$\mathcal{N} = \prod_i \frac{g_i!}{n_i!\,(g_i - n_i)!}, \qquad (1,4,8)$$

$$\mathcal{C} = \frac{G!}{N!\,(G - N)!}. \qquad (1,4,9)$$

We cannot expect formulas (1,4,4) and (1,4,5) to be valid in this case. Before, states in which one or more boxes contain more than one object were not excluded but rather neglected, in the course of a limit process in which the g_i are considered infinitely large in relation to the n_i. Here the situation is quite different - these *states* are rejected a priori and the g_i have fixed rather than infinitely large values.

Statistics of L. Brillouin

In the statistics of L. Brillouin the N objects are *distinct*: we still have types of boxes $T_1, T_2, \ldots T_i, \ldots$; the number of boxes of type T_i is again denoted by g_i, but each box is subdivided into

α compartments. Every object placed in a box is assigned to one compartment of the box, and no compartment can hold *more than one object.* We must again evaluate the number $\mathfrak{N}$ of distinct states in which exactly n_i objects are placed in the set of boxes of type T_i $(i = 1, 2,\ldots)$ and the total number $\mathcal{C}$ of possible distinct states; then we form the quotient

$$\mathcal{P} = \frac{\mathfrak{N}}{\mathcal{C}}.$$

To obtain one of these $\mathfrak{N}$ allocations, we begin by dividing the N objects into sets $G_1, G_2,\ldots, G_i,\ldots$ of cardinalities $n_1, n_2,\ldots, n_i,\ldots$, respectively; this can be done, as we know, in

$$\frac{N!}{n_1!\; n_2!\ldots\; n_i!\ldots}$$

different ways; we then have to place the n_i objects of the set G_i in the g_i boxes of type T_i $(i = 1, 2,\ldots)$. αg_i compartments, each of which can be occupied only once, are available. In addition, the n_i objects are distinct - they may be distributed one by one. For the first, there are αg_i possibilities; for the second only $\alpha g_i - 1$ compartments remain, which gives $\alpha g_i - 1$ possibilities; and so on. Therefore, all in all there are

$$\alpha g_i(\alpha g_i - 1)\;\ldots\ldots\ldots\;[\alpha g_i - (n_i - 1)]$$

possibilities; putting $\alpha = 1/b$, we write

$$\frac{g_i}{b}\left(\frac{g_i}{b} - 1\right)\;\ldots.\;\left[\frac{g_i}{b} - (n_i - 1)\right] = \frac{\Gamma\left(\frac{g_i}{b} - 1\right)}{\Gamma\left(\frac{g_i}{b} - n_i + 1\right)}.$$

By formula (1,1,3), we find

$$\mathfrak{N} = N!\prod_i \frac{\Gamma\left(\frac{g_i}{b} + 1\right)}{n_i!\;\;\Gamma\left(\frac{g_i}{b} - n_i + 1\right)}, \qquad (1,4,10)$$

and, similarly,

$$\mathscr{C} = N! \frac{\Gamma\left(\frac{G}{b} + 1\right)}{\Gamma\left(\frac{G}{b} - N + 1\right)} . \qquad (1,4,11)$$

In Brillouin's formulas[(1)] these numbers are multiplied by b^N; this superficial difference is related to our introduction of "compartments," which, we believe, makes the argument more comprehensible. Note that, at any rate, the b^N disappear when the quotient $\frac{\mathscr{N}}{\mathscr{C}}$ is formed; this ratio, regarded as a probability, is the paramount factor in applications.

The number b is the reciprocal of a positive integer; however, Brillouin was led to assign it arbitrary, even negative values; without goint into details as to the physical interpretation of this device, we point out the remarkable fact that if b is assigned the values 0, 1, -1, we recover the Classical, Fermi-Dirac, and Bose-Einstein statistics, respectively.

---

(1) L. Brillouin [1], Vol. I, p. 170, formula (22); the fact that in this formula $\sum n_i = N$ enables us to simplify Brillouin's result.

V. CHROMOSOME THEORY OF HEREDITY

5. *Chromosome Theory of Heredity*

The aim of the Chromosome Theory of Heredity is to explain the hereditary transmission of individual traits; it was formulated following the empirical laws of Mendel. Since it involves problems of Combinatorial Analysis and Probability Theory we present a very brief exposition of its principles.

Every cell of a living being I contains chromosomes, which are rods or threads visible at certain times in the life of the cell; their number 2N, which is always even since the chromosomes are grouped in pairs, is the same for all the cells of the individual I and for all individuals of I's species. It varies from species to species ; humans have 46 chromosomes, rats 40.

When the reproductive cells, male or female, mature, they are called *gametes*. During the process of maturation (meiosis) each of the N pairs of chromosomes separates. Two gametes, not a single one,[1] are formed; one chromosome of each pair enters one of the gametes, while the other enters the other gamete. The resulting gametes do not contain 2N chromosomes, like the other cells, but only N (of course, not grouped in pairs). *The distribution of the two chromosomes of the same pair between the two gametes occurs at random, and for each pair independently of the other pairs.*

When they unite to form a single cell (the egg), the male and female gametes each contribute N chromosomes - a total of 2N; the pairs are formed again, so that the egg and the evolving cells possess the normal number of 2N chromosomes, grouped in N pairs.

The characteristics of an individual are determined by the genes. The actual nature of these genes, whose existence is assumed in the theory, is immaterial here; the essential fact for our pur-

---

(1) This is absolutely true only for males; in the case of females, not two gametes, but a gamete and another cell called a polar globule, which is not a true gamete, are formed; from the chromosomic point of view, the result is the same.

poses is that they are located on the chromosomes and are therefore transmitted with them from parents to children. The totality of genes that an individual possesses is called its *genotype*; the genotype defines the totality of the *hereditary* traits of the individual, i.e., traits that he has inherted from his parents and is capable of transmitting to his children.

Very often a gene may exist in two or more different states; these different possible states of a gene are called *alleles* or allelomorphic states; in the sequel we shall consider in principle only the simplest case, that of genes having only two alleles.

As mentioned above, the genes are located on the chromosomes. Now, the latter are grouped in pairs; let us stipulate that a single gene is carried by *one* pair of chromosomes, and that it appears at the same time on *both* chromosomes of this pair, but not necessarily in the same state. Thus, in the case of a gene with two possible alleles A and a, each cell contains the gene in duplicate: twice in state A, twice in state a, or once in state A and once in state a. It is convenient to symbolize these different possibilities by the formulas "AA" or "aa" or "Aa"; the individuals of the species thus fall into three categories: the Aa or *heterozygotes*, the *homozygotes* AA and the *homozygotes* aa.

The gametes, however, being special cells having only N chromosomes instead of 2N, possess each gene only once and not twice. They are therefore characterized with respect to a given gene by a formula containing one letter, A or a. By virtue of the process of meiosis and the separation of pairs of chromosomes, it is clear that the homozygotes AA form gametes all of type A, the homozygotes aa form gametes that are all a, while the heterozygotes Aa form both gametes A and gametes a, in exactly equal numbers - as follows from the meiosis process; this is the case for a male heterozygote. For a female heterozygote, A and a are distrubuted, as mentioned above, in a random fashion between gamete and polar globule. This means that of a large number of gametes produced by one or more females Aa about half will be A and the other half

a. Statistically, at least, male and female yield the same result. Conversely, the union of a gamete (male or female) A (or a) and of a gamete (female or male) A (or a) yields a homozygous egg AA (or aa), while the union of a gamete a and a gamete A yields a heterozygous egg Aa.

As regards simultaneous consideration of several genes, a new question arises. For instance, suppose that two genes may have alleles A, a and B, b respectively; the genotype of an individual will be defined, as far as these two genes are concerned, by a four-letter formula such as AAbb, AaBb, etc., and its gametes by two-letter formulas: Ab, AB, etc. It is clear that an individual AAbb, for instance, can form only gametes Ab, and individual AaBB can form gametes AB and aB in equal or almost equal numbers. For an individual AaBB, the possible formulas for the gametes are AB, aB, Ab, ab; the question is, are all these combinations effectively realizable, and in what proportions? If the two genes are not carried by the same pair of chromosomes, then, since the different pairs of chromosomes behave independently in meiosis, all four kinds of gametes will appear, and, at least on the average, in equal numbers; the two genes then behave independently. But one can imagine, without going into details, that the fact that the same pair of chromosomes carry the two genes may create a connection (*linkage*) between two gametes, making a certain type of gamete impossible, or at least rare.

An example will clarify the mechanism by which the traits of an individual result from its genotype. There are both colored mice (black, grey, and so on) and albinos; this is due to the existence of a coloration gene, capable of two alleles C and c, C implying coloration and c albinism. The homozygotes CC are colored, the homozygotes cc are albinos. What are the heterozygotes Cc, that possess both the color allele C and the albinism allele c? It turns out that C overcomes c; we say that C is *dominant*, or c is *recessive*, so that the heterozygotes Cc are colored, like the homozygotes CC. The properties of the heterozygotes are thus often regulated by the dominance or recessiveness of an allele in relation to another[1].

Each pair of alleles possessed by an individual produces its effects independently of other pairs of alleles (of other genes) present in the same individual: thus, the black or grey color of colored mice is due to a color gene capable of two alleles, state B resulting in black fur, state G in grey fur, and G dominates B; but a mouse with the pair cc is an albino, irrespective of the color pair (GG, BB or GB) it may have. Thus, as far as its fur is concerned, the genetic constitution of a mouse is characterized by a four-letter formula obtained by first writing its coloration pair, for instance Cc, then its color pair, for instance BB, resulting in formulas of the type CcBB (black mouse). Note that different genotypes may yield the same fur (for instance: CCGG and CcGB, grey mice).

Now let us see how the genes are transmitted by heredity. To this end, consider a gene capable of two alleles A and a. All the gametes of an individual are of type A, while those of an individual aa are all of type a; crossing of those two individuals can yield only heterozygotes Aa. On the other hand, consider a male Aa and a female Aa; the male Aa produces male gametes A and a in equal numbers, and the female Aa produces female gametes A and a in equal numbers, on the average. Crossing of this male and female may lead to one of the following four combinations:

male gam. A + fem. gam. A = egg AA,
male gam. A + fem. gam. a = egg Aa,
male gam. a + fem. gam. A = egg Aa,
male gam. a + fem. gam. a = egg aa.

---

(1) However, dominance or recessiveness is not always absolute; this obviously complicates the phenomenon. There are cases in which there is no dominance and the heterozygote presents characteristics intermediate between the two homozygotes.

It may happen that male gametes A, for instance, unite preferentially with a female gamete A rather than with a female gamete a. We shall not consider this rather complicated case and shall assume, as often happens, that *panmixion* occurs, i.e. absolutely random union of gametes. This means that for a large number of crossings male Aa + female Aa the four preceding combinations will occur in approximately equal proportions. Moreover, two of these lead to the same result - an egg Aa. Hence, of a large number of eggs obtained by such crossings, about half will be heterozygotes Aa, a quarter homozygotes AA and a quarter homozygotes aa.

As an application and an illustration of what happens when several genes are considered simultaneously, let us study the crossing of grey mice CCGG and albino mice ccBB, assuming that the coloration gene and the color gene are independent (without linkage). The first generation from these crossings can have only Cc as coloration pair and GB as color pair; hence we have grey mice CcGB. If the latter are crossed, we obtain, in more or less equal proportions, the four coloration pairs CC, Cc, cC, cc, of which two are not distinct. We also obtain in equal proportions the four color pairs GG, GB, BG, BB, of which two are not distinct. Each coloration pair may be associated without preference with each color pair - hence we have, *in equal proportions*, the following $4 \times 4 = 16$ following formulas: CCGG, CCGB, CCBG, CCBB, CcGG, CcGB, CcBB, cCGG, cCGB, cCBG, cCBB, ccGG, ccGB, ccBB, CcBG, ccBG, of which only nine are distinct. Of these 16 formulas, 9 correspond to grey mice, 3 to black, 4 to albinos; therefore a large number of crossings of grey mice CcGB result in grey, black, and albino mice in proportions of about 9/16, 3/16, 4/16.

It is now clear that the Chromosome Theory may lead to fairly complicated problems of Combinatorial Analysis.

BIBLIOGRAPHIC NOTES FOR CHAPTER I

Information on arrangements, combinations and permutations with or without repetitions may be found in most courses of higher mathematics, such as G. Cagnac, E. Ramis, and J. Commeau [1] (Vol. 1), A. Lentin and J. Rivaud [1]; on the Γ- and B-functions, in most Calculus treatises, such as E. Goursat [1] (Vols. 1 and 2), J. Favard [1]. A modern approach and extensive treatment of Combinatorial Analysis with a rich bibliography may be found in J. Riordan [1] and Marshall Hall [1].

An extensive treatment of the Statistics of Classical or Quantum Statistical Mechanics, is given in L. Brillouin [1]; see also F. Perrin [1], A. Blanc-Lapierre, P. Casal, and A. Tortrat [1], G. Bodiou [1].

For a treatment of the Chromosome Theory of Heredity the reader may consult M. Caullery [1]; for the relevant probability considerations, see G. Malecot [1], O. Kempthorne [1].

EXERCISES

1.1. If n is a positive integer and z a complex number, evaluate

$$f(z) = \sin z \sin\left(z + \frac{\pi}{n}\right) \sin\left(z + \frac{2\pi}{n}\right) \cdots \sin\left(z + \frac{k\pi}{n}\right) \cdots \sin\left(z + \frac{(n-1)\pi}{n}\right).$$

Hence deduce the value of the following function for z = 1:

$$g(z) = \Gamma\left(\frac{z}{n}\right) \Gamma\left(\frac{z+1}{n}\right) \cdots \Gamma\left(\frac{z+k}{n}\right) \cdots \Gamma\left(\frac{z+n-1}{n}\right).$$

Using (1,1,5), show that, for any z not an integer $\leqq 0$,

$$g(z) = (2\pi)^{\frac{n-1}{2}} n^{\frac{1}{2}-z} \Gamma(z).$$

Use this result to calculate the integral

$$\int_0^1 \log \Gamma(t)\, dt.$$

1.2. Suppose the n numbers $x_1, x_2, \ldots, x_n$ and the n numbers $y_1, y_2, \ldots, y_n$ satisfy the n relations

$$\sum_{k=0}^{n-j} (-1)^k C_{j+k}^k x_{j+k} = y_j \qquad (j = 1, 2, \ldots, n).$$

Find formulas expressing each x_j in terms of $y_1, y_2, \ldots, y_n$.

1.3. Let $\mathfrak{X}$ be a set of two distinct elements, denoted by 0 and 1, respectively; let $\mathfrak{Y}$ be the set of all ordered sequences $\{x_1, x_2, \ldots, x_n\}$ of n elements $x_1, x_2, \ldots, x_n$ (not necessarily distinct) of $\mathfrak{X}$. How many elements are in $\mathfrak{Y}$? A mapping f of $\mathfrak{Y}$ into $\mathfrak{X}$ is a function $f(x_1, x_2, \ldots, x_n)$, with values in $\mathfrak{X}$, of variables $x_1, x_2, \ldots, x_n$ in $\mathfrak{X}$. What is the number A_n of distinct mappings of $\mathfrak{Y}$ into $\mathfrak{X}$? A mapping $f(x_1, x_2, \ldots, x_n)$ of $\mathfrak{Y}$ into $\mathfrak{X}$ may be independent of some of the variables $x_1, x_2, \ldots, x_n$; for instance, $f(x_1, x_2, \ldots, x_n)$ is independent of x_n if for arbitrary fixed $x_1, x_2, \ldots, x_{n-1}$

$$f(x_1, x_2, \ldots, x_{n-1}, 0) = f(x_1, x_2, \ldots, x_{n-1}, 1);$$

what is the number B_n of mappings f of $\mathfrak{Y}$ into $\mathfrak{X}$ which depend *essentially* on all the variables $x_1, x_2, \ldots, x_n$? (Use the solution of Exercise 1.2).

1.4. Let

$$\Psi = \sum_{r=k}^{n} C_n^r p^r (1-p)^{n-r},$$

where $0 \leqq p \leqq 1$. By computing the derivative of Ψ with respect to p, show that

$$\Psi = \frac{B_p(k, n-k+1)}{B(k, n-k+1)},$$

where B_p denotes the incomplete B-function [cf. (1,1,18)].

1.5. Prove that for $0 \leqq k \leqq n$

$$\sum_{r=0}^{k+1} C_{n+2}^{r} - 4 \sum_{r=0}^{k} C_{n}^{r} = C_{n}^{k+1} - C_{n}^{k-1}.$$

1.6. Let S be sequence of n tosses $P_1, P_2, \ldots, P_n$ of a coin; let α_k be the number of tosses among the first $k - P_1, P_2, \ldots, P_k$ $(k = 1, 2, \ldots, n)$ which turn up "heads," β_k the corresponding number of "tails," $\alpha_k + \beta_k = k$. Let S' be another sequence of n tosses of the same kind with corresponding numbers α_n' and β_n'. S and S' are regarded as identical if $\alpha_n = \alpha_n'$ *for every* k $(k = 1, 2, \ldots, n)$, distinct otherwise.

a) What is the number of distinct sequences of tosses?

b) Let α and β be given integers with $\alpha \geqq 0$, $\beta \geqq 0$, $\alpha + \beta = n$; what is the number of distinct sequences of tosses such that $\alpha_n = \alpha$? (Answer: $C_n^{\alpha} = C_n^{\beta}$).

c) Assuming n odd, what is the number of distinct sequences of tosses such that $\alpha_n > \beta_n$? (Answer: 2^{n-1}). If n is even? (Answer: $2^{n-1} - 1/2\ C_n^{n/2}$).

d) Let α and β be given integers, $\alpha \geqq 0$, $\beta \geqq 0$, $\alpha + \beta = n$, $\alpha > \beta$; what is the number of distinct sequences of tosses such that $\alpha_n = \alpha$ and $\alpha_n > \beta_k$ for all k $(k = 1, 2, \ldots, n)$? [Use Exercise 1.5. Answer:

$$\frac{(\alpha + \beta - 1)!}{\alpha!\ \beta!} (\alpha - \beta)].$$

1.7. N subscribers are connected by a telephone exchange; a conversation may be represented by a pair (a, b) of two subscribers a, b. The pair (a, b) may be regarded as either unordered [(a, b) and (b, a) are not distinct] or ordered [(a, b) and (b, a) are distinct, e.g., when there are reasons to distinguish between caller and called]. No subscriber can carry on more than one conversation at a time.

The exchange is in a "state of level n" if exactly n $(0 \leqq n \leqq N/2)$ conversations are in process at the given instant. What is the number ν_n of distinct states of level n, for unordered pairs

$$\left(\text{Answer} : \nu_n = \frac{N!}{2^n\, n!\, (N-2n)!} \right),$$

and for ordered pairs

$$\left(\text{Answer} : \nu_n = \frac{N!}{n!\, (N-2n)!} \right).$$

What is the total number ν of distinct states of level n, $0 \leqq n \leqq T$, where $T \leqq N/2$ is given (distinguish between unordered and ordered pairs)? Assuming that N and T tend to $+\infty$ in such a way that the quotient T/N tends to a fixed limit $\rho > 0$, find a simple asymptotic expression for ν.

1.8. A player J plays an unlimited sequence of games $\mathcal{E}_1, \mathcal{E}_2, \ldots, \mathcal{E}_n, \ldots$; the only assumption is that, in each game $\mathcal{E}_n$, J can win or loose (algebraically) an amount equal to the stake M_n ($\gtrless 0$) which he puts up in the game $\mathcal{E}_n$. J determines his stake M_n by the following rule: he chooses M_1 arbitrarily; he then determines the M_n ($n > 1$) recursively: $M_{n+1} = M_n + 2$ if he wins the game $\mathcal{E}_n$, $M_{n+1} = M_n - 2$ if he loses ("D'Alembert ascent"). J stops playing at the game $\mathcal{E}_{2H}$, where H is the first integer > 0 such that in the games $\mathcal{E}_1, \mathcal{E}_2, \ldots, \mathcal{E}_{2H}$ he has won and lost exactly the same number H of times. Show that J's total winnings G are 2H (put $X_n = 1$ or -1 according to whether J wins or loses the game $\mathcal{E}_n$; express G as a function of M_1 and $X_1, \ldots, X_{2H}$, using the fact that $\sum_{n=1}^{2H} X_n = 0$).

CHAPTER II

THE CONCEPT OF PROBABILITY.

MEASURES OR MASS DISTRIBUTIONS.

HILBERT SPACES. RANDOM ELEMENTS

AND PROBABILITY LAWS

I. THE CONCEPT OF PROBABILITY

1. *Trials and random events*

Consider a random experiment, i.e., one in which chance is involved; in Probability Theory one usually calls such an experiment a *trial.*(1) For instance, throwing two dice in a definite way constitutes a trial. If the dice are thrown again under the same general conditions, this constitutes a *new* trial (sometimes called a *repetition* of the first, to indicate that the two trials are performed under identical conditions).

Games of chance are obviously the simplest examples of random experiments, since they are artificial; this is the reason we shall often refer to them, especially in the first chapters. This is also the historical reason that games of chance gave rise to Probability Theory. The reader should not suppose that applications of Probability Theory are confined to the study, admittedly very interesting, of games of chance.

We shall call any event(2) whose occurrence depends on chance, i.e., whose occurrence or non-occurrence is determined by the outcome of some trial, a *random*(3) event. Any event whose occurrence is logically necessary is called a *certain* event, while any event whose non-occurrence is logically necessary is called *impossible*. Strictly speaking, neither of these events is random, though it is clear (and this will be made precise later) that both may be considered as particular (limiting) cases of random events. For brevity's sake, we shall almost always say simply "event" rather than "random event"; we shall generally denote events by capital letters A, B,...

---

(1) German: Versuch, Probe; French: épreuve.

(2) German: Ereigis; French: événement.

(3) German: zufällig; French: aléatoire.

EXAMPLE (2,1,1). Consider a trial that consists of drawing a card at random from a deck of 32 cards;* the draw of a club is a random event. The event "the card drawn is one of the 32 cards in the deck" is certain, by the very definition of the trial. On the other hand, the event "the card drawn is the three of diamonds" is impossible, though it would be possible (and random) were the experiment performed with a full 52-card deck. ■

We now introduce a few useful definitions.

The events A, B,... of an arbitrary family of events are *pairwise mutually exclusive* if no two of them can occur simultaneously.

The notation

$$A = B$$

means that the events A and B are *identical*, i.e., whenever A occurs so does B, and vice versa. The notation

$$A \subset B \quad \text{or} \quad B \supset A$$

(read: A is *included* in B) means that the occurrence of A *necessarily implies* that of B, though B may occur without A. If $A \subset B$ and B cannot occur without A, i.e., we also have $B \subset A$, it is clear that $A = B$.

Given an arbitrary family $\mathfrak{F}$ of events A, the *union* of the events A of the family $\mathfrak{F}$, denoted by

$$\bigcup_{\mathfrak{F}} A,$$

is the event that occurs by definition if and only if at least one of the events of $\mathfrak{F}$ occurs. If the family $\mathfrak{F}$ is finite or countable so that the events A of $\mathfrak{F}$ may be numbered, $A_1, A_2, \ldots A_k, \ldots$, their union is denoted by

---

* [Translation editor's note: The author means a deck containing all cards from the seven up, as used, e.g., in the French game of bézique.]

$$\bigcup_k A_k ;$$

when $\mathfrak{F}$ contains only a finite number n of events A_1, A_2,..., A_k..., A_n, the following notation may also be used:

$$A_1 \cup A_2 \cup \ldots \cup A_k \cup \ldots \cup A_n.$$

The definition of union immediately suggests that of *difference* and subtraction. However, we shall not define the difference between two *arbitrary* events C and A. If A is *included in* C and B is defined as the event "C occurs while A *does not*," it is obvious that

$$A \cup B = C,$$

and A and B are mutually exclusive; we shall then call B the *difference* between C and A and write

$$B = C - A.$$

Then, obviously,

$$A = C - B.$$

If T is a certain event, $\emptyset$ an impossible event and A any event, we have $A \cup T = T$, $A \cup \emptyset = A$, $A \subset T$; this follows immediately from the relevant definitions. The difference T - A, which we denote by $\check{A}$, is called the event *complementary** to A, i.e., the event that occurs if and only if A does not. A and its complementary event are clearly mutually exclusive. The complementary event of a certain event is an impossible event, and, conversely, that of an impossible event is a certain event.

---

* [Translation editor's note: French original: *contraire*. The term "opposite event" is rarely used by English-speaking authors.]

System of events. One frequently has to deal with a set S, whose elements are *pairwise mutually exclusive* events C; such a set S will be called a *system of events*. At most one of the events $C \in S$ can occur. We shall call any subset A of S *an event defined by* S. Regarding the $C \in S$ which belong to this subset A of S, we can say that A, as an event, is the union of these events C.

S, as a subset of itself, is an event defined by the system S. If the event S is *certain*, the system S is said to be *exhaustive*: if S is exhaustive, one and only one of the $C \in S$ occurs. ∎

Given an arbitrary family $\mathfrak{F}$ of events A, the *intersection* of the events A of the family $\mathfrak{F}$, denoted by

$$\bigcap_{\mathfrak{F}} A,$$

is the event that occurs if and only if *all* the events A of $\mathfrak{F}$ occur. If $\mathfrak{F}$ is finite or countable the following notation, analagous to that for the union, is used:

$$\bigcap_k A_k, \quad A_1 \cap A_2 \cap \ldots \cap A_k \cap \ldots \cap A_n .$$

If T, $\emptyset$, A are certain, impossible, and arbitrary events respectively, $A \cap T = A$, and $A \cap \emptyset$ is impossible, as follows immediately from the definitions. For any events B and family $\mathfrak{F}$ of events A, if $\mathcal{Y}$ denotes the family of events $A \cap B$ where $A \in \mathfrak{F}$, then

$$\left(\bigcup_{\mathfrak{F}} A\right) \cap B = \bigcup_{\mathcal{Y}} (A \cap B).$$

If A and B are mutually exclusive, then $A \cap B$ is clearly impossible.

Note that the complementary event of $A \cup B$ *is* $\check{A} \cap \check{B}$ *while that of* $A \cap B$ is $\check{A} \cup \check{B}$.

The reader has probably noticed the analogy between the concepts of inclusion, union, intersection, etc., of events on the one hand, and those of inclusion, union, intersection, etc., of sets on the other. In fact, these concepts *coincide* for, as we shall see in II.4, every event can be identified with a suitable set. We assume the reader familiar with the elements of set theory; therefore, we shall not dwell any longer on the elementary definitions. At the end of II.4, however, we shall recall the less trivial definition of the limit of a sequence of sets or events.

2. *A first approach to the concept of probability*

Neither the occurrence nor the non-occurrence of a random event E is certain (except for the particular case of a certain or impossible event). In general, however, the mind does not remain neutral as to the possible occurrence of E, and formulates a *probabilistic judgment* expressed in terms of the probability[(1)] of the event E.

The concept of the probability of an event will be developed gradually; for the moment, we introduce the notation for the probability of an event E:

$$\Pr(E).$$

The human intellect evolved the concept of probability in various ways. The first approach, to which this section is devoted, allots an important role to the fact that in regard to their occurrence certain events are in some way equivalent or symmetric. This situation is rare in natural phenomena; on the other hand, it is the rule in the games of chance invented by man, which therefore play a role analogous to that of idealized laboratory experiments in physics.

Let us return to the example of drawing a card from a deck. Under normal conditions, when the deck has been shuffled and cut, different cards are, except for their values[(2)], indistinguishable by any means available when the trial is performed in the usual manner. There is therefore no apparent reason favoring the draw

---

(1) German: Wahrscheinlichkeit; French: probabilité.

(2) We use the term *value* for the combination of the rank (ace, king, queen, jack, ten, nine, eight, seven) and suit (spades, hearts, diamonds, clubs) of a card; ten of hearts and ten of clubs are two different values.

of one card or another; we describe this situation by saying that *the 32 cards and the corresponding 32 values are equiprobable, or have equal probabilities.*

We shall call the convention which assigns equal probabilities to different events, whenever there is no reason favoring the occurrence of one of them rather than that of others, the "principle of symmetry." Of course, sometimes one observer sees no reason that could favor one event at the expense of the others, while another more informed observer may discover one; but this is another question, related to the remarks of II.4. The principle of symmetry is not an absolute logical necessity - and is indeed by no means an axiom of Probability Theory. We shall simply use it whenever it seems natural, reasonable, and compatible with the objective circumstances - not to prove, but to justify certain axioms or applications of Probability from the intuitive viewpoint.

Consider again the draw of a card from a deck. The events whose occurrence or non-occurrence is determined by this trial may be very different. However, as a player usually does, let us confine ourselves to events whose occurrence or non-occurrence is defined solely by the value of the card drawn. Let $\mathfrak{F}$ denote the set of all these events. Two events A and B of $\mathfrak{F}$ are identical if the values that realize A are exactly the same as those that realize B. Thus $\mathfrak{F}$ contains a finite number of events.

With each event $E \in \mathfrak{F}$ we can associate the number r(E) of values that realize it. For instance, if E is the event "the card drawn is any king," then r(E) = 4, since this involves 4 values: king of spades, hearts, diamonds, clubs.

If T is the event "the card drawn is any one of the 32 cards," then T belongs to $\mathfrak{F}$, is obviously certain, and

$$r(T) = 32 ; \qquad (2,2,1)$$

T is the only certain event of $\mathfrak{F}$, and the only one for which r(T) = 32.

Let us introduce the following notation, where A and B are any two events of $\mathfrak{F}$:

1) ν denotes the number of values that realize both A and B, thus

$$\nu = r(A \cap B) ;$$

2) a denotes the number of values that realize A but not B, thus

$$a = r(A - A \cap B) ;$$

3) b denotes the number of values that realize B but not A, thus:

$$b = r(B - A \cap B).$$

Clearly

$$r(A) = \nu + a, \qquad r(B) = \nu + b,$$

$$r(A \cup B) = \nu + a + b = r(A) + r(B) - r(A \cap B). \quad (2,2,3)$$

We see immediately that:

1) A and B are mutually exclusive if and only if $\nu = r(A \cap B) = 0$; it is therefore convenient to regard the event realized by no value, i.e., the impossible event $\emptyset$, for which $r(\emptyset) = 0$, as an event of $\mathfrak{F}$.

With this convention $\mathfrak{F}$ contains exactly 2^{32} events, and the integer r(E) may take any value from 0 to 32.

2) $B \subset A$ if and only if $b = 0$; then

$$r(B) \leqq r(A) ;$$

more precisely,

$$r(A) = r(B) + a ; \qquad (2,2,4)$$

and $a = 0$ thus means that $B = A$.

3) If $B = \check{A}$ is the complementary event of A, $\nu = 0$ and $b = 32 - a$; in other words,

$$r(\check{A}) = 32 - r(A). \qquad (2,2,5)$$

4) Let $E_1, E_2, \ldots, E_k$ be pairwise mutually exclusive events of $\mathfrak{F}$ (of course, the E_k are finite in number); any value that realizes

one of them does not realize any of the others, but realizes the union $\bigcup_k E_k$, so that

$$r\left(\bigcup_k E_k\right) = \sum_k r(E_k). \tag{2,2,6}$$

Having made these remarks, note now that since the different values are equiprobable, if two events A and B (A and B $\in \mathfrak{F}$) are such that r(A) = r(B), i.e., if as many values realize A as realize B, the principle of symmetry implies that A and B have *equal probabilities*, i.e.,

$$\Pr(A) = \Pr(B).$$

For instance, let A be the event "the card drawn is black (spades or clubs)" and B the event "the card drawn is an honor (ace, king, queen or jack)." Then r(A) and r(B) are both 16, so that A and B have equal probabilities.

Now consider two events A and B such that $B \subset A$; from (2,2,4) it is clear that the occurrence of A is favored at least as much as that of B; we express this by saying that A is *at least as probable* as B, or that Pr(B) is not greater than Pr(A):

$$\Pr(B) \leqq \Pr(A).$$

More precisely, if $a > 0$, the occurrence of A is in fact more probable than that of B, and we express this by the inequality

$$\Pr(B) < \Pr(A).$$

For instance, the probability of drawing any king is greater than that of drawing the king of spades.

Now since one probability may be equal to, greater or smaller than another, probability is patently a *measurable quantity*. Our previous remarks make it quite evident that the probability Pr(E) of an event is *measured* precisely by the real number r(E), or, equivalently, by any increasing function of r(E). One such function is

$$\Pr(E) = \frac{r(E)}{32}, \tag{2,2,7}$$

The specific choice of (2,2,7) is not a logical necessity, but the following properties of this function attest to its convenience.

a) The probability Pr(E) of an event E is a real number $\geqq 0$ and $\leqq 1$.

b) For the certain event T "the card drawn is any of the 32 cards," formula (2,2,1) implies r(T) = 32, and therefore

$$Pr(T) = 1; \tag{2,2,8}$$

in other words, *the probability of the certain event is 1.*

c) Let $E_1, E_2, \ldots, E_k, \ldots$ be pairwise mutually exclusive events; from (2,2,6) it follows that

$$Pr\left(\bigcup_k E_k\right) = \sum_k Pr(E\ ). \tag{2,2,9}$$

In other words, the probability of a union of pairwise mutually exclusive events (of the family $\mathfrak{F}$, therefore finite in number) is the sum of their probabilities.

These properties imply the following:

α) If $B \subset A$, $A = B \cup (A - B)$, where B and (A - B) are mutually exclusive; therefore, by c):

$$Pr(A) = Pr(B) + Pr(A - B), \tag{2,2,10}$$

and, by a),

$$Pr(A) \geqq Pr(B), \tag{2,2,11}$$

in agreement with (2,2,4).

β) If A and B are arbitrary events, $A \cup B = A \cup (B - A \cap B)$, where A and $(B - A \cap B)$ are mutually exclusive; therefore

$$Pr(A \cup B) = Pr(A) + Pr(B - A \cap B).$$

Since $A \cap B \subset B$, formula (2,2,10) implies

$$Pr(B) = Pr(A \cap B) + Pr(B - A \cap B),$$

whence, finally,

$$Pr(A \cup B) = Pr(A) + Pr(B) - Pr(A \cap B), \qquad (2,2,12)$$

in agreement with (2,2,3); therefore

$$Pr(A \cup B) \leqq Pr(A) + Pr(B). \qquad (2,2,13)$$

γ) If A is an arbitrary event, $A \subset T$, therefore, by (2,2,10),

$$Pr(T) = Pr(A) + Pr(T - A), \qquad (2,2,14)$$

or:

$$Pr(\check{A}) = 1 - Pr(A)$$

in agreement with (2,2,5).

GENERALIZATION. The example studied hitherto may be generalized immediately as follows. Assume:

1) S is an exhaustive system of a finite number n of events $C_1, C_2, \ldots, C_n$; i.e., the C_j are pairwise mutually exclusive and $\bigcup_j C_j$ is certain.

2) The C_j are equiprobable (for reasons, say, of symmetry, etc.)

$$Pr(C_1) = Pr(C_2) = \cdots = Pr(C_n).$$

Consider the family $\mathfrak{F}$ of events defined by S, i.e., events that are unions of C_j. Then with each $E \in \mathfrak{F}$ we can associate the number r(E) of C_j whose union forms E; in analogy with (2,2,7), the following definition of Pr(E) suggests itself:

$$Pr(E) = \frac{r(E)}{n}. \qquad (2,2,15)$$

The above properties a), b), c) and their consequences α), β), γ) follow immediately. Formula (2,2,15) is the classical definition of probability: the probability of an event E is the ratio of the number r(E) of "favorable cases" (the number of C_j whose occurrence implies that of E) to the total number n of "possible cases" (total number of different C_j). Note that for $E = C_j$ formula (2,2,15) gives

$$Pr(C_j) = 1/n \quad (j = 1, 2, \ldots, n). \qquad (2,2,16)$$

For arbitrary E, evaluation of Pr(E) reduces to that of r(E), i.e., to a counting problem. This explains why problems of Probability frequently involve Combinatorial Theory.

EXAMPLE (2,2,1), *Dice*

Consider the trial performed by throwing two dice; let C_{ij} be the event "the number on the upturned face of the first die is i, the number on the upturned face of the second die is j." The possible values of i are 1, 2, 3, 4, 5, 6, and these are also the possible values of j; thus there are 6 × 6 = 36 possible pairs (i, j), therefore 36 C_{ij}'s. Moreover, the C_{ij} are pairwise mutually exclusive and their union is certain; they therefore form an exhaustive system S_1.

The event E_{10}: "i + J = 10" is realized by the pairs (4,6), (5,5), (6,4); therefore

$$E_{10} = C_{4,6} \cup C_{5,5} \cup C_{6,4} ,$$

i.e., E_{10} is an event defined by S_1. The same holds for the event E_5: "i + j = 5," since

$$E_5 = C_{1,4} \cup C_{2,3} \cup C_{3,2} \cup C_{4,1} .$$

In general, it is clear that the event E_s: "i + j has a fixed value s" is defined by S_1. The possible values of s are 2, 3, 4,..., 10, 11, 12; and their number is therefore 11. Moreover, they are pairwise mutually exclusive and their union (which is also the union of the events C_{ij}) is certain. They therefore form an exhaustive system S_2, different from S_1; while E_{10}, for instance, is defined by S_1 and also by S_2, the event $C_{2,3}$ is defined by S_1 but not by S_2. In general, any event defined by S_2 is also defined by S_1, but the converse may not be true.

If the dice are not loaded we may regard the C_{ij} as equiprobable by virtue of the principle of symmetry, so that, by (2,2,16),

$$\Pr(C_{i,j}) = \frac{1}{36}.$$

Formula (2,2,15) gives:

$$\Pr(E_{10}) = \frac{3}{36}, \quad \Pr(E_5) = \frac{4}{36}, \text{ etc.}$$

The elements C_{ij} of S_1 are equiprobable, but the elements E_s of S_2 are not.

EXAMPLE (2,2,2), *Quantum Statistics*

Classical or quantum statistics involves problems which do not differ essentially from drawing a card from a deck, throwing dice, etc., whose solution relies ultimately on formulas (2,2,15), (2,2,16).

Consider for instance a system Σ governed by Bose-Einstein statistics (we refer the reader to I.4 for the definitions, notation, and results recalled here). It is assumed that the state of the system Σ is random. The trial consists precisely in the choice of a state by Σ. Only distinguishable states are considered distinct. The set S of all possible distinguishable states (necessarily mutually exclusive) is an exhaustive system. By virtue of a hypothesis of Physics (not of Probability Theory) all distinguishable states are assumed equiprobable. By (2,2,16) and (1,4,7), their common probability is therefore

$$\frac{1}{\mathscr{C}} = \frac{N!\,(G-1)!}{(N+G-1)!}.$$

The event E: "the state assumed by Σ assigns n_i objects to the set of g_i boxes of type T_i $(i = 1, 2, \ldots)$" is realized by $\mathfrak{N}$ states; by (2,2,15), (1,4,6) and (1,4,7), its probability $\mathfrak{P}$ is therefore

$$\mathfrak{P} = \frac{\mathfrak{N}}{\mathscr{C}} = \frac{N!\,(G-1)!}{(N+G-1)!} \prod_i \frac{(n_i+g_i-1)!}{n_i!\,(g_i-1)}.$$

In applications, N, g_i and G are very large. It is easy to see that if at least one of the n_i is small, $\mathfrak{P}$ itself is small. The corresponding event E therefore has a small probability. Except for these events E, $\mathfrak{P}$ may be approximated with sufficient accuracy when the n_i are all large enough by replacing the factorials by their approximations according to Stirling's formula (1,1,11).

Moreover, the numbers 1 or 1/2 may be neglected in comparison with G, N + G, $n_i + g_i$, g_i or their logarithms. The result is:

$$\log \mathfrak{P} = N \log N + G \log G - (N + G) \log (N + G) \qquad (2,2,17)$$
$$+ \sum_i [- g_i \log g_i - n_i \log n_i + (n_i + g_i) \log (n_i + g_i)] + h ,$$

where h is a constant.

One of the fundamental problems here is to determine for fixed N, G, g_i those values n_i^0 of n_i for which $\mathfrak{P}$ is a maximum (to find the "most probable" state). Note that the n_i are not arbitrary; indeed we must have $\sum_i n_i = N$, and moreover the principle of conservation of energy becomes a condition of the form $\sum_i e_i n_i$ = constant, where the e_i are certain positive coefficients.

If we adopt the expression (2,2,17) for $\log \mathfrak{P}$, the classical method for finding the maximum of a function of several variables connected by two relations yields

$$n_i^0 = \frac{g_i}{e^{\alpha+\beta e_i} - 1} ,$$

where β has a definite physical meaning and α is determined by the condition

$$\sum_i \frac{g_i}{e^{\alpha+\beta e_i} - 1} = N$$

(cf. L. Brillouin [1], Vol. I, p. 134). It can be shown that $\mathfrak{P}$ decreases rapidly when the n_i deviate from the values n_i^0. The n_i therefore assume values very close to the n_i^0 with probability very close to 1, i.e., almost certainly. This is the physical significance of the numbers n_i^0.

These results justify the hypothesis - which is not directly verifiable - that the different distinguishable states are equiprobable.

3. *Frequency and Probability*

Another approach to the concept of probability is the statistical one. Suppose that a population $\mathcal{P}$ of n individuals I is under investigation. We are not necessarily implying a human, even an animate population - it may be a set of physical particles or arbitrary elements. To fix ideas, however, let us consider a human population, say the current population of France. Let E be an arbitrary property that an individual I of $\mathcal{P}$ may or may not possess; for instance, E might be membership in some professional group, such as "metal worker." A list of the n individuals of $\mathcal{P}$ can be drawn up (this is a difficult task if n is large, but we shall disregard this purely technical difficulty), which specifies for each individual whether or not he has the characteristic E.

Using this survey one can count the number r(E) of individuals in $\mathcal{P}$ possessing the property E, and evaluate the quotient

$$f(E) = \frac{r(E)}{n}, \qquad (2,3,1)$$

i.e., the actual *frequency* in $\mathcal{P}$ of the individuals possessing the property E.

Let us associate with every property E the event, denoted by the same letter E, that an individual I of $\mathcal{P}$ possesses the property E; the frequency f(E) given by (2,3,1) is thus the frequency of the occurrence of E.

Now suppose that a survey is prepared for some property E (say, "metal worker") in 1954, and again a few years later, say in 1959. By then the total population of France has increased to a value n' slightly greater than n. The number of individuals possessing the property E is now r'(E), which is different from r(E). Thus, in 1959 the frequency of the event E assumes a value $f'(E) = r'(E)/n'$ which may differ from f(E). Now it is remarkable that in spite of the different values of n and n' on the one hand, and r(E) and r'(E) on the other, f'(E) differs only slightly from f(E). This means that the socio-economic structure of the French population is changing quite slowly; in fact, it is practically constant for any relatively short period, say a few years, which involves no unusual disturbances such as wars, etc.

It is natural to express this objective observation by the assertion that the French population, regarded as an entity without reference to a specific time, has a definite structure with which we can associate, for any event E, a parameter or constant Pr(E) called the "probability of the event E," for which the actual frequency f(E) computed on the basis of a survey relating to a given year such as 1954 is a kind of empirical measure. Thus we have the *non-rigorous, approximate* equality

$$\Pr(E) = f(E). \tag{2,3,2}$$

Intuitively, several points are evident:

1) The validity of f(E) as a measure of Pr(E) is contingent on certain conditions, and is not admissible if the population is, say, too small, or disorganized.

2) In any event, (2,3,2) is only an approximation; the *quality* of this approximation depends on certain conditions. Suppose that the event E is the suicide of an individual. This is a rather rare event; a survey for a single year will indicate only a few suicides and the corresponding frequency f(E) may deviate considerably from Pr(E). It is reasonable to suppose that the quality of f(E) as a measure of Pr(E) will be better, the longer the period covered by the survey, provided the structure of the population does not change significantly during this longer period.

It goes without saying that the quality of the approximation may be considered sufficient or not according to the application in question.

3) Even if we are justified in considering f(E) as a measure of Pr(E), we are thereby interpreting the concept of measure in a way different from that customary in Deterministic Classical Physics; there the existence of the measured quantity is assumed absolute,

and the sole source of error is the deficiency of the observer and his tools, which may be reduced by perfecting the measurement techniques. On the other hand, an individual chooses, e.g., his profession freely and for reasons which are essentially personal; irrespective of the structure of the population and the value of Pr(E), it is therefore always possible that f(E) will deviate from Pr(E).

One of the main goals of Probability Theory and Mathematical Statistics is to throw light on these questions. We shall return to them after Chapter VI (the law of large numbers, etc.) and especially in Vol. II. For the time being we set them aside and confine ourselves to the consequences of (2,3,2). Formula (2,3,2) in fact implies that Pr(E) has the properties of the frequency f(E). In particular,

a) The probability Pr(E) of an event E is a real number $\geqq 0$ and $\leqq 1$.

b) Let T be the event that an individual of $\mathfrak{P}$ belongs to $\mathfrak{P}$. T is of course certain, and occurs for all n individuals of $\mathfrak{P}$, so that r(T) is necessarily equal to n and f(T) to n/n = 1. Thus the probability of a certain event is 1.

c) Let us associate with any event E the set $\tilde{E}$ of individuals of $\mathfrak{P}$ possessing the characteristic E, so that r(E) is just the number of individuals in $\tilde{E}$. Note, by the way, that since a *real* population $\mathfrak{P}$ is necessarily finite ($n < +\infty$) the number of distinct subsets $\tilde{E}$ of $\mathfrak{P}$ is finite; in other words, the number of distinct events E which we can actually associate with $\mathfrak{P}$ is necessarily finite. To mutually exclusive events E', E" correspond disjoint sets $\tilde{E}'$, $\tilde{E}''$; to $E' \cup E''$ corresponds the set $\tilde{E}' \cup \tilde{E}''$, so that $r(E' \cup E'') = r(E') + r(E'')$ if E' and E" are mutually exclusive. It follows immediately that if $E_1, E_2, \ldots, E_k, \ldots$ are pairwise mutually exclusive events, then

$$\Pr\left(\bigcup_k E_k\right) = \sum_k \Pr(E_k).$$

The reader may ask whether the concept of probability derived in this section from the statistical point of view coincides with that derived in II.2 from the principle of symmetry. We shall not deal with this problem, which is of greater interest for Philosophy than for Mathematics or the applications of Probability Theory. We remark only that the above properties a), b), c) are exactly those established and denoted also by a), b), c) in II.2, and recall that, as seen in II.2, properties a), b), c) necessarily imply properties α), β), γ) summarized in (2,2,10), (2,2,11), (2,2,12), (2,2,13), (2,2,14).

Probability in the Chromosome Theory of Heredity

In I.5, which dealt with the Chromosome Theory of Heredity, we repeatedly used the vague terms "at random," "on the average," etc., in relation to empirical assertions concerning frequencies. These become precise expressions if we replace frequencies by probabilities, defined in the above way. Thus, the statement "The distribution of two chromosomes of the same pair between the two gametes occur at random" (cf. I.5, pages 21-25) will become: "Any chromosome of a pair will pass into a given gamete with probability 1/2."

Consider, for instance, the crossing of two heterozygotes Aa; we saw that this crossing may occur in four ways:

male gam. A + fem. gam. A;
male gam. A + fem. gam. a;
male gam. a + fem. gam. A;
male gam. a + fem. gam. a.

We say that panmixion occurs if the empirical frequencies of these four types of crossing are approximately equal. This is now expressed by the statement that the four types are equiprobable. Therefore, each has probability 1/4, as follows immediately from properties a), b), c). Noting that two of these four types yield an egg Aa, then by c) we have:

probability 1/4 for an egg AA;
probability 1/4 for an egg aa;

probability 1/2 for an egg Aa.

4. *Trials and classes of trials*

We have already introduced the concept of a trial; we now introduce classes of trials. It is important to formulate each of the two concepts with precision and to distinguish clearly between them.

Recall that a trial is a specific experiment, therefore *unique*. It consists of the totality of *all* its characteristics and *all* the circumstances associated with its performance.

We shall call any set $\mathfrak{U}$ *of trials a class of trials.* Any trial belongs to infinitely many different classes of trials. We shall see later that the concept of a class of trials is fundamental in Probability Theory.

We can already make the following remarks.

Every language tends to distinguish between the following characteristics of a trial:

1) the general conditions, real or postulated, often controlled by the experimenter, under which the trial is performed;

2) the "outcomes" of the trial, i.e., those of its characteristics that are not necessary results of the above "general conditions." The latter are therefore peculiar to the specific trial in question, and are generally known only after the trial is performed. They often seem to be particular manifestations of chance.

Consider a definite trial u, performed by throwing an ordinary die D on a horizontal table on which an axis x'Ox is drawn. The rules of the game define the way the die is thrown and the mechanical properties, homogeneity, and symmetry of a fair die. These rules and properties constitute the main part of the "general conditions" under which the trial u is performed. They are an integral part of u, but they are not peculiar to it, being common to all regular throws of the die. Thus they do not define u, but they do define the class of trials $\mathfrak{U}$ of all regular throws of a die. To describe them means to state implicitly and imprecisely that $u \in \mathfrak{U}$, which should be stated explicitly.

Let ϕ be the angle between the horizontal edges of the die and x'Ox, and ν the point it determines when it has reached a definite equilibrium position on the table.

To fix ideas, suppose that for u and D, $\phi = 25°$ and $\nu = 3$. These two quantities are characteristics of u, but not necessarily of other throws of the die.

It is tempting to say that these are the "outcomes" of the trial u. However, as far as the trial u is considered in isolation, the quantities 25° and 3 are characteristics like any other characteristics: they are an integral part of u; the specification or definition of u involves, among other things, these quantities 25°, 3. To set them apart as "outcomes" would be arbitrary and meaningless.

On the other hand, consider the pair $\{u, \mathcal{U}\}$, taking account of the fact that $u \in \mathcal{U}$. If we wish, we can define an outcome of u to be any characteristic of u that is not a necessary consequence of the fact that $u \in \mathcal{U}$; but then the distinction is a property of the couple $\{u, \mathcal{U}\}$, not of u alone.

Having clarified these points, we shall resort to "everyday language" for the sake of brevity; the reader will be able to formulate the correct interpretation of these imprecise expressions without ambiguity.

Let us now see why the concept of a class of trials is necessary in Probability Theory. An important conclusion may be drawn from the preceding paragraphs; to disregard it may lead to serious difficulties, such as the so-called "paradoxes of Probability Theory"[1] which once engendered so much discussion and today seem to us purely false problems.

---

(1) cf., e.g., E. Borel [1], p. 80 ff. and p. 106 ff.

First recall the example of a card drawn from a deck. To fix ideas, let E be the event "ace of hearts." For a deck of 32 cards, we saw that Pr(E) = 1/32. Now let us restrict ourselves to the red suits of the deck, which contain only 16 cards including the ace of hearts, assuming them equiprobable; for this special class of trials it is reasonable to say that

$$\Pr(E) = \frac{1}{16},$$

a value differing from the previous one 1/32.

Now consider II.3 and a survey for the French population $\mathfrak{P}$ and the event E "an individual I is a metal worker"; for this event E, r(E) is about 500,000, while the cardinality n of $\mathfrak{P}$ is about 45,000,000 (in 1954); formulas (2,3,1) and (2,3,2) then give approximately

$$\Pr(E) = \frac{500{,}000}{45{,}000{,}000} = \frac{1}{80}.$$

Now let us assume that I belongs to the Paris district, whose population $\mathfrak{P}'$ has a cardinality n' of about 8,000,000; of these r'(E) are metal workers, where r'(E) is about 200,000; formulas (2,3,1) and (2,3,2) applied to $\mathfrak{P}'$ then give

$$\Pr(E) = \frac{200{,}000}{8{,}000{,}000} = \frac{2}{80},$$

a different value, twice that obtained above for $\mathfrak{P}$.

Thus, though an event E is well defined, as is the corresponding trial u, it does not have a well-defined probability. Its probability Pr(E) depends on the class of trials $\mathfrak{U}$ with which we agree to associate u: in the example of the deck of cards, $\mathfrak{U}$ may be the class or set of all cards of the deck, the set of all red cards, the set of all hearts, etc.; in the statistical example of II.3, $\mathfrak{U}$ may be the set of living Frenchmen (survey for the entire population $\mathfrak{P}$) or only the set of living Parisians (survey for the population $\mathfrak{P}'$), etc.

The probability of an event E depends on the class of trials with which it is associated; it is completely defined whenever this class is defined.

This obviously raises two problems.

First Problem. Among the various classes of trials with which an event E can be associated, which is the most convenient class $\mathfrak{U}$? The choice of $\mathfrak{U}$ is a function of the problem at hand and the desired objective. It is often obvious, and is therefore generally not specified explicitly. The disadvantage of this situation is that it obscures the fundamental importance of the concept of class. For instance, given a circle of radius R in the plane, let us choose a chord "at random"; what is the probability that its length will be less than R? Put in this way, the problem is meaningless, since there are many inequivalent ways of choosing a chord in a circle "at random," and they define different classes of trials, implying of course different probabilities for the event in question(1). In Chapter IV (Example (4,6,3)) we shall discuss an example of confusion between classes of trials in Wave Mechanics. However, if there is no danger of ambiguity, we ourselves shall conform to the custom of making no explicit and detailed mention of the class of trials involved.

The choice of the class of trials depends on the information that we possess, or (which is essentially the same) that we believe we possess.

Consider a definite individual I and the event E that I will still be alive in 10 years time; assume we know that I is a (male) Frenchman, x years of age. Taking only this information into account, we can define the probability P of E by referring to a survey which indicates all living (male) Frenchmen aged x and more, i.e., by defining our category of trials $\mathfrak{U}$ to be the set of all lives of living (male) Frenchmen aged x or more. However, an insurance agent selling I a life-insurance policy will not be satisfied with the probability P. In addition to the previous information, he will also take into consideration I's actual state of

(1) Cf. J. Bertrand [1], § 5, p. 4.

health s, which he will determine through a medical examination. He will then refer to the class of trials formed by the lives of all living (male) Frenchmen aged x or more whose state of health at age x was the same as that of I. These data, presumably at his disposal will yield the corresponding probability P' may differ significantly from P.

When a physicist wishes to repeat a previously performed experiment, he takes care to assign the various parameters (geometrical, mechanical, electric, etc.) involved in the experiment the same values as before. However, he will be able to achieve and verify equality of the parameters only within the limits of precision of his measurement instruments. He will deem it natural to adopt as a class of trials the set $\mathfrak{U}$ of all the experiments he can perform in this way with preassigned values of the parameters, up to the available precision. If convenient, he could adopt a wider class than $\mathfrak{U}$, for instance, that of all experiments of the same type involving arbitrary values of the parameters. However, were he to require a class more restricted than $\mathfrak{U}$, he could achieve this only "mentally," with no immediate experimental significance.

If, for instance, in tossing a coin we are presented with an apparently symmetric and homogeneous coin, we may entertain doubts as to its perfect symmetry and homogeneity, but we do not know whether the assymmetry is so marked as to have a significant effect, and, above all, we do not know which face of the coin is favored thereby.

Under such conditions, the procedure may depend on the circumstances:

Case 1: Suppose that we *must* bet at once (in real life we are generally obliged to bet willy-nilly, i.e., to adopt a definite attitude; to do nothing is an attitude like any other, generally implying some risk and therefore constituting a bet). Having no reason to prefer heads to tails or vice versa, it is reasonable to assign the same probability 1/2 to both events and place our bets accordingly. In so doing our class of trials $\mathfrak{U}$ is the set of all tosses of apparently symmetric and homogeneous coins - possibly loaded, but in such a way that the deception is not apparent and it cannot be decided at sight if the coin is loaded in favor of heads or tails. We are not disregarding the fact that this class $\mathfrak{U}$ is an expression, above all, of ignorance. If the game is played and 100 tosses result in 80 heads and only 20 tails, it becomes evident that for some reason heads are favored. We then naturally alter the basic assumption underlying our bets, and henceforth assign heads a probability greater than tails. In so doing, however, we adopt a new class of trials $\mathfrak{U}'$, the category of tosses of a coin loaded in favor of heads.

Case 2: We are not obliged to bet before having observed a great number of experiments. As the class of trials we may then adopt the set of actually performed tosses of the coin; it is this class of trials with which one usually has to do. This well-defined class assigns heads a definite probability p. For the moment this parameter is an unknown quantity; however, denoting it by the letter p, we can introduce it in computations. Once experiments have been performed, we shall be able to assign it a definite numerical value, chosen subject to the requirement that the results of the experiments be in good agreement with those of the computation. We shall see how to do this in Vol. II; the procedure is known as the *estimation* of p. Once p is estimated correctly, we have all the information necessary for a rational betting policy.

As implied by this example, experiments are often planned with the following objective in mind.

First, we may wish to verify some hypothesis (Case 1); a probability is often hypothetical, or at least conditioned by certain hypotheses concerning the trial. These hypotheses may prove false; i.e., the value chosen for the probability may prove inadequate with regard to the observed results. It is precisely in order to test the validity of this value and eventually to replace it by a more accurate one that the experiment is performed. But then changing the probability means changing the class of trials.

Alternatively, we may wish to estimate certain parameters, i.e., definite but unknown quantities (Case 2), such as probabilities.

REMARK (2,4,1)

Under any circumstances, care must be taken to ensure that any event under consideration be well defined; it would be pointless to assign a probability to a meaningless event.

Second Problem

If an event E has probability P in conjunction with a class of trials $\mathfrak{U}$, and probability P' in conjunction with another class $\mathfrak{U}'$, then, as we have just seen, in general $P \neq P'$. Now this does not necessarily mean that P' is absolutely independent of P. Is there any relation between P and P', and if so, what is it? These questions, which constitute the problem of changing the class of trials will be studied in IV.5 and IV.6.

Events as subsets of the class of trials

Let us return to the class $\mathfrak{U}$ of draws of a card from a 32-card deck. Any such draw is a trial of the class, i.e., an element u of the set $\mathfrak{U}$. Among the trials u of $\mathfrak{U}$, let us single out those in which the card drawn is a king; they form a subset E of $\mathfrak{U}$:

$$E \subset \mathfrak{U}.$$

It is clear that the set E of trials, which is a subset of $\mathfrak{U}$, and the event E: "the card drawn is a king," may be identified (and represented by the same symbol E).

Generally speaking, if E is any subset of $\mathfrak{U}$, it may be identified with the event E that the trial performed is an element of E; conversely, if E is an arbitrary event associated with $\mathfrak{U}$, the set of trials $u \in \mathfrak{U}$ in which the event E occurs form a subset $E \subset \mathfrak{U}$ that may be identified with the event E.

Thus, *once the class of trials* $\mathfrak{U}$ *has been selected in a definite way, there is a one-to-one correspondence which associ-*

ates with any event E *the subset of trials* u *of the class* $\mathfrak{U}$ *in which the event* E *occurs, and we systematically identify the event E with the corresponding subset* E. *The set of events is thus identical with the family of subsets of the class* $\mathfrak{U}$. All the concepts and general notations of set theory may therefore be applied to events, as already remarked at the end of II.1. The general terminology of set theory may also be employed for events, though by force of usage a few special terms concerning events are retained. For instance, we shall say that two events A and B are mutually exclusive rather than disjoint (as sets). Again, let A be an event; as a subset of $\mathfrak{U}$, we may speak of its *complement* $\check{A}$ with respect to $\mathfrak{U}$, i.e., the set of trials $u \in \mathfrak{U}$ that do not belong to A, or those for which A does not occur; $\check{A}$ is just the event complementary to A, and when speaking of events we shall use the term "complementary event" rather than "complement."

One of the subsets of $\mathfrak{U}$ is $\mathfrak{U}$ itself: any trial u belongs to $\mathfrak{U}$ by logical necessity; therefore $\mathfrak{U}$ *regarded as an event is just the certain event of the class* $\mathfrak{U}$, and $\emptyset$ is the complement or event complementary to $\mathfrak{U}$.

REMARK (2,4,2)

The last few lines require the following comment.

We have seen that the probability of an event E depends on the class of trials with which it is associated. In particular, the fact that E is certain and its probability is 1 is not an intrinsic property of E, but must be related to the appropriate class of trials. Recall the example of a card drawn from a deck, and let E be the event "the card drawn is red." As a first category of trials consider the class $\mathfrak{U}$ of all draws of a card from the deck. With respect to $\mathfrak{U}$, E is not certain, for the card drawn may be black; indeed, Pr(E) = 16/32 = 1/2. Now consider a second class, the class $\mathfrak{V}$ consisting only of draws of a red card; with respect to $\mathfrak{V}$, E is obviously certain. In the same way, two events A and B may not be mutually exclusive when associated with some class of trials $\mathfrak{U}$, but mutually exclusive when associated with some other class of trials $\mathfrak{V}$.

In general, let $\mathfrak{U}$ be a class of trials; if u denotes one of its elements, i.e., a trial, it is convenient to use the same notation for the subset of $\mathfrak{U}$ consisting of the single element u. With this convention, u denotes an event, namely, the event that the trial actually performed is the trial u. $\mathfrak{U}$ thus becomes the family of these events u, which are pairwise mutually exclusive by virtue of the very concept of trial. Thus $\mathfrak{U}$ is the *system* of the events u. Moreover, with respect to $\mathfrak{U}$ as a class of trials this system is *exhaustive,* since the union of the u, which is just $\mathfrak{U}$ considered as a subset of itself, is certain. On the other hand, $\mathfrak{U}$ may not be an exhaustive system relative to a class of trials other than $\mathfrak{U}$.

Limit of a sequence of sets or events

Before proceeding to the following paragraphs, we introduce some useful definitions.

Let $E_1, E_2, \ldots, E_k, \ldots$ be a countable sequence of sets E_k of elements u; let $\overline{L}$ be the set of elements u having the following property: for arbitrarily large K there exists *at least one* $k > K$ such that $u \in E_k$; it is clear that

$$\overline{L} = \bigcap_h \Big(\bigcup_{k>h} E_k \Big) \quad (h = 1, 2, 3, \ldots).$$

On the other hand, let $\underline{L}$ be the set of elements u having the property: there exists an integer K (which may depend on u) such that u belongs to *all* E_k for $k > K$; it is clear that

$$\underline{L} = \bigcup_h \Big(\bigcap_{k>h} E_k \Big) \quad (h = 1, 2, 3\ldots).$$

Obviously,

$$\underline{L} \subset \overline{L}.$$

We shall say that $\bar{L}$ is the *upper* and $\underline{L}$ the *lower limit* of the sequence E_k as $k \longrightarrow +\infty$. We shall say that the sequence E_k converges to a limit set L as $k \longrightarrow +\infty$,

$$\lim_{k \to +\infty} E_k = L,$$

if and only if

$$L = \underline{L} = \bar{L}.$$

The sequence of the E_k is said to be monotone nondecreasing if $E_k \subset E_{k+1}$ for all k; then

$$\bigcap_{k>h} E_k = E_h \; ; \; \underline{L} = \bigcup_h E_h .$$

$\bigcup_{k>h} E_k$ is a set L independent of h; consequently $\bar{L} = L$, and $L = \underline{L}$. Thus a monotone nondecreasing sequence $\{E_k\}$ has a limit L:

$$L = \bigcup_k E_k .$$

The sequence of the E_k is said to be monotone nonincreasing if $E_k \supset E_{k+1}$ for all k; then

$$\bigcup_{k>h} E_k = E_h \; , \; \bar{L} = \bigcap_h E_h .$$

$\bigcap_{k>h} E_k$ is a set L independent of h; consequently $\underline{L} = L$, and $L = \bar{L}$. Thus a monotone nonincreasing sequence $\{E_k\}$ has a limit L:

$$L = \bigcap_k E_k .$$

Since events are merely sets of trials, this terminology is directly applicable to sequences of events.

5. *First Axioms of Probability Theory*

Like any other branch of Mathematics, Probability Theory is a deductive theory based on a system of axioms, which we now proceed to state. The main purpose of the preceding sections was to indicate why this specific axiom system has been adopted, why it is consistent, and why it yields models suitable for the description of concrete phenomena. The examples do not and cannot serve as "proofs" of the axioms.

AXIOM I. Every problem of Probability Theory concerns some definite set $\mathfrak{U}$, called a *class of trials**, whose elements u, which are arbitrary in nature, are called trials (of the class).

Any subset E of $\mathfrak{U}$ is an *event*; we shall apply the general concepts, notation, and terminology of Set Theory to events, with the exception, as mentioned in II.4, of a few terms (mutually exclusive, complementary, etc.) customarily used specifically for events.

In particular, $\mathfrak{U}$ as a subset of itself is the *certain* event; to say that an event E is certain means that $E = \mathfrak{U}$. The empty subset ϕ of $\mathfrak{U}$ is the *impossible* event; to say that an event E is impossible means that it is the empty subset of $\mathfrak{U}$.

The main content of the following axioms is:

1) The probability Pr(E) of an event E, i.e., of a subset E of $\mathfrak{U}$, is a nonnegative real number.

2) In accordance with the property deduced in (2,2,9), if $E_1, E_2, \ldots, E_k, \ldots$ are finitely or countably many pairwise mutually exclusive events then

$$\Pr\left(\bigcup_k E_k\right) = \sum_k \Pr(E_k). \qquad (2,5,1)$$

The adoption of (2,5,1) for the case of finitely many E_k is natural in view of the considerations of II.2 and II.3. Of course, these intuitive, empirical considerations provide no clear indication concerning an infinite (countable) number of E_k; however, it has been recognized that the mathematical structure of Probability Theory requires the adoption of (2,5,1) even for the countable case. This strengthened form of (2,5,1) transcends the simple suggestions of experience or intuition; it can never contradict them.

The need for the following definitions should be apparent.

---

* [*Translation editor's note.* The more accepted term in English is *sample space.*]

Set functions

Consider a nonempty set $\mathfrak{U}$ of elements u, and the family $\mathfrak{G}$ of its subsets, including $\mathfrak{U}$ itself and the empty set $\emptyset$. Let $\mathfrak{F}$ be some family of subsets E of $\mathfrak{U}$ such that $\mathfrak{U} \in \mathfrak{F}$, $\emptyset \in \mathfrak{F}$; clearly $\mathfrak{F} \subset \mathfrak{G}$. Suppose that to every $E \in \mathfrak{F}$ corresponds a number m(E); we shall say that m(E) is a *function* of the set E, defined on $\mathfrak{F}$.

Let m(E) have the following property: For any finite or countable sequence $E_1, E_2, \ldots, E_k, \ldots$ of pairwise disjoint sets in $\mathfrak{F}$ such that $\bigcup_k E_k \in \mathfrak{F}$ we have

$$m\left(\bigcup_k E_k\right) = \sum_k m(E_k); \qquad (2,5,2)$$

then m(E) is said to be *completely additive* over $\mathfrak{F}$.

If in addition m(E) is a nonnegative real number and $m(\emptyset) = 0$, we shall say that m(E) is a *measure* over $\mathfrak{U}$, defined for all subsets of $\mathfrak{U}$ that belong to $\mathfrak{F}$.

The preceding concepts in themselves are quite simple, even trivial. Elementary examples of measures can be given. For instance, let $\mathfrak{U}$ be the set of points u on a straight line; with certain subsets E of this line, say with intervals, we associate their length l(E). It is clear that this length is a measure, and, in particular, that it possesses the property of complete additivity (2,5,2). Another quite analogous example is the 3-dimensional space $\mathfrak{U}$ of points of Elementary Geometry, if we assign a volume v(E) to certain subsets E of the space, such as parallelepipeds, etc.

REMARK (2,5,1)

The example of volume is noteworthy for the following reason. The volume v(E) is a measure, but with a special property: For any pair of congruent sets {E, E'} (in the sense of Elementary Geometry, i.e., sets obtainable from each other by a rigid motion), we have: v(E) = v(E'). Now it can be proved that no measure with this property can be defined for *all* subsets of the space of Elementary Geometry; the family $\mathfrak{F}$ of subsets E for which the volume v(E) is defined is smaller than the family of all subsets.

On the other hand, it is intuitively clear that the family $\mathfrak{F}$ of subsets E of a set $\mathfrak{U}$ on which a measure m(E) is defined is not arbitrary. For instance, because of complete additivity (2,5,2) it would be desirable, even necessary, to require that $\mathfrak{F}$ have the following property: For any finite or countable sequence of sets E_k such that $E_k \in \mathfrak{F}$ for every k, we have $\bigcup_k E_k \in \mathfrak{F}$. This leads to the following definition.

σ-algebra of subsets

Let $\mathfrak{U}$ be an arbitrary nonempty set of elements u; a Boolean σ-algebra (also: Borel field) of subsets of $\mathfrak{U}$ is any family $\mathfrak{F}$ of subsets of $\mathfrak{U}$ such that

1) $E \in \mathfrak{F}$ implies $\check{E} = \mathfrak{U} - E \in \mathfrak{F}$;

2) if $E_1, E_2, \ldots, E_k, \ldots$ is a finite or countable sequence of sets E_k in $\mathfrak{F}$, then

$$\bigcup_k E_k \in \mathfrak{F}, \quad \bigcap_k E_k \in \mathfrak{F} \quad ;$$

3) $\emptyset \in \mathfrak{F}$.

Note that 1) and 3) imply $\mathfrak{U} \in \mathfrak{F}$.

For the sake of brevity we shall say simply σ-algebra instead of Boolean σ-algebra.

Sub-σ-algebras: Let $\mathfrak{F}$ be a σ-algebra of subsets of $\mathfrak{U}$ and $\mathfrak{F}'$ a σ-algebra of subsets of $\mathfrak{U}$, such that $\mathfrak{F}' \subset \mathfrak{F}$; we shall say that $\mathfrak{F}'$ is a sub-σ-algebra of $\mathfrak{F}$.

Let m be a measure defined over $(\mathfrak{U}, \mathfrak{F})$, and m' the set function over $(\mathfrak{U}, \mathfrak{F}')$ defined by:

$$m'(E) = m(E) \text{ for every } E \in \mathfrak{F}'.$$

In other words, m' is the restriction of m to $\mathfrak{F}'$; it is clear that m' is a measure over $(\mathfrak{U}, \mathfrak{F}')$

EXAMPLE (2,5,1). Let $\mathfrak{U}$ be a class of trials u; if S is a system of events $C \subset \mathfrak{U}$ (cf. II.1), the family $\mathfrak{F}$ of the events defined by S is clearly a σ-algebra of subsets of the union of the C's; if this union is $\mathfrak{U}$, i.e. if S is exhaustive, $\mathfrak{F}$ is a σ-algebra of subsets of $\mathfrak{U}$.

Now, to say that S is exhaustive is equivalent to saying that the C's form a *partition* of $\mathfrak{U}$; hence every partition of $\mathfrak{U}$ defines a σ-algebra of subsets of $\mathfrak{U}$.

Now consider the converse. Let $\mathfrak{F}$ be an arbitrary given σ-algebra of subsets of $\mathfrak{U}$; the reader can easily find examples where *there does not exist* a partition of $\mathfrak{U}$, i.e. an exhaustive system S of subsets C of $\mathfrak{U}$, such that $\mathfrak{F}$ is the family of events defined by S.

Consider the case of an exhaustive system S of events $C \subset \mathfrak{U}$; let $\mathfrak{G}$ be the σ-algebra of all subsets of $\mathfrak{U}$. Clearly, the family of the events defined by S is a sub-σ-algebra of $\mathfrak{G}$.

Returning to Example (2,2,1), we see that the family of events defined by S_2 is a sub-σ-algebra of the family of events defined by S_1.

We can now give a rigorous definition of measure.

Measure. Let $\mathfrak{U}$ be an arbitrary nonempty set of elements u, and $\mathfrak{F}$ a σ-algebra of subsets E of $\mathfrak{U}$; a measure m(E) over $(\mathfrak{U}, \mathfrak{F})$ is a function of the set E, defined and completely additive on $\mathfrak{F}$, such that

1) m(E) is a nonnegative real number:

2) $m(\emptyset) = 0$.

The measure m is said to be bounded if $m(\mathfrak{U}) < +\infty$; if $m(\mathfrak{U}) = +\infty$ but there exists a countable sequence $\{E_k\}$ of sets $E_k \in \mathfrak{F}$ (k = 1, 2, 3,...) such that

a) $E_k \subset E_{k+1}$: $m(E_k) < +\infty$ (for all k),

b) $\lim_{k \to +\infty} E_k = \mathfrak{U}$,

then m is said to be σ-bounded.

We shall return to the study of measure in II.6. For the time being the definition makes it possible to state the following axioms.

AXIOM II. Certain events, called *probabilized events*, form a Boolean σ-algebra $\mathfrak{B}$ of subsets of the class of trials $\mathfrak{U}$. The probabilized events, and these alone, have a probability. The probability of a probabilized event E, denoted by Pr(E), is a well-defined real number, $\geqq 0$ and $\leqq 1$; in particular,

$$\Pr(\mathfrak{U}) = 1, \qquad (2,5,3)$$

i.e., the probability of the certain event is 1.

AXIOM III (*Axiom of Total Probability*). As a function of $E \in \mathcal{B}$, $\Pr(E)$ is a completely additive measure; that is, if $E_1, E_2, \ldots, E_k, \ldots$ are any finitely or countably many pairwise mutually exclusive events, then the probability of their union is the sum of their probabilities:

$$\Pr\left(\bigcup_k E_k\right) = \sum_k \Pr(E_k) \ . \qquad (2,5,4)$$

REMARK (2,5,2)

1) It is clear from Remark (2,5,1) that the σ-algebra $\mathcal{B}$ of probabilized events is generally *smaller* than the family of all subsets of the class of trials $\mathfrak{U}$; in other words, there exist non-probabilized events, which have no probability. This may seem surprising at first sight, but it is no more surprising than the fact that some figures of Elementary Geometry have no volume.

Whenever we wish to deal with the probability of an event E, we must first rigorously ascertain (if this is not explicitly assumed) that $E \in \mathcal{B}$. However, in practice one never has to deal with figures without volume, and in the same way, at least as far as the usual applications are concerned, one never encounters non-probabilized events.

Consequently, since this book does not claim to be a rigorous and complete mathematical exposition, but is oriented towards applications, we shall usually proceed as follows. We shall refer explicitly to σ-algebras, in particular, to σ-algebras of probabilized events, only when this is necessary to avoid confusion; we shall usually assume without proof that any event E under discussion is probabilized.

2) By (2,5,3), probability is a *bounded* measure. In the case of a countable number of E_k, formula (2,5,4) implies that the series

on the right-hand side of (2,5,4) is convergent and that its sum is equal to the left-hand side of (2,5,4).

Probability law of a class of trials

The *probability law* of the class of trials $\mathcal{U}$ is defined as the set of numbers Pr(E) for all $E \in \mathcal{B}$.

RESULTS

In II.2 we derived (2,2,10), (2,2,11), (2,2,12), (2,2,13), (2,2,14) from (2,2,8) and (2,2,9). In the same way, Axioms I, II, III imply the following results:

α) Given two events A and B such that $B \subset A$,

$$\Pr(A) = \Pr(B) + \Pr(B - A), \tag{2,5,5}$$

and therefore

$$\Pr(A) \geqq \Pr(B). \tag{2,5,6}$$

β) If A and B are any two events,

$$\Pr(A \cup B) = \Pr(A) + \Pr(B) - \Pr(A \cap B), \tag{2,5,7}$$

and consequently

$$\Pr(A \cup B) \leqq \Pr(A) + \Pr(B). \tag{2,5,8}$$

γ) For any event A, the probability of its complementary event $\check{A} = \mathcal{U} - A$ is

$$\Pr(\check{A}) = 1 - \Pr(A)\ ; \tag{2,5,9}$$

in particular, the probability of the *impossible* event, i.e., of the event complementary to the certain event, is *zero*.

6. *Measure and mass distribution*

Let $\mathcal{U}$ be an arbitrary set of elements u and $\mathfrak{F}$ a σ-algebra of subsets of $\mathcal{U}$. Let m(E) be a measure over $(\mathcal{U}, \mathfrak{F})$; not forgetting that m(E) is defined only for $E \in \mathfrak{F}$, we shall follow Remark (2,5,1) and omit explicit mention (or, according to the case, proof) of the assumption that the $E \subset \mathcal{U}$ in question belong to $\mathfrak{F}$. We mention in passing that a set $E \subset \mathcal{U}$ belonging to $\mathfrak{F}$ is often called a *measurable* set.

Suppose that a total *mass* M, finite or infinite, is *distributed* over the set or space $\mathcal{U}$. When $\mathcal{U}$ is the space of Elementary Geometry we thus obtain a "material system" - a commonplace concept in Mechanics. In Mechanics the mass is usually distributed only over a subset $\mathcal{U}_0$ of $\mathcal{U}$ ($\mathcal{U}_0$ is known as the "support" of the material system). However, we may always extend the distribution to the whole of $\mathcal{U}$, by assigning zero mass to the complement $\check{\mathcal{U}}_0$ of $\mathcal{U}_0$. Moreover, in Mechanics the total mass M is generally finite.

Consider, then, a distribution of a total mass M over $\mathcal{U}$. For any subset $E \subset \mathcal{U}$, let m(E) denote the corresponding mass; obviously,

$$m(E) \geqslant 0, \qquad m(E) \leqslant m(\mathcal{U}) = M ,$$

and if $E_1, E_2, \ldots, E_k, \ldots$ are disjoint subsets, finite or countable in number, then

$$m\left(\bigcup_k E_k\right) = \sum_k m(E_k). \tag{2,6,1}$$

As a function of the set E, m(E) is therefore a measure over $\mathcal{U}$; conversely, any measure over $\mathcal{U}$ may be interpreted as a mass distribution over $\mathcal{U}$. *The concepts of measure and mass distribution thus coincide.*

If M is finite, the measure m(E) is bounded; if moreover M = 1, as may always be ensured by a convenient choice of the unit mass, then m(E) becomes a probability law if $\mathcal{U}$ is regarded as a class of trials. *Thus the concepts of probability law and distribution of a bounded mass coincide.*

Since mass distributions are a familiar concept, it is often suggestive in Probability Theory to identify measures with mass distributions, and probability laws with distributions of a bounded mass.

Elementary properties of measures

Several general properties of measure will frequently be used in the sequel. Let m(E) be a measure over an arbitrary set $\mathfrak{U}$.

Let A and B be two subsets such that

$$A \supset B \; ;$$

then

$$A = B \cup (\check{B} \cap A),$$

where B and $(\check{B} \cap A)$ are disjoint. Thus, since the measure is completely additive,

$$m(A) = m(B) + m(\check{B} \cap A);$$

therefore

$$m(A) \geqq m(B). \qquad (2,6,2)$$

This was already proved in (2,5,6) for the particular case of a bounded measure.

Consider a countable sequence of arbitrary sets $A_1, A_2, \ldots, A_k, \ldots$; put:

$$B_1 = A_1,$$

$$B_k = \check{A}_1 \cap \check{A}_2 \cap \ldots \cap \check{A}_{k-1} \cap A_k, \qquad (k = 2, 3, \ldots),$$

$$C_k = \bigcup_{h=1}^{k} B_h \qquad (k = 1, 2, \ldots) \; ;$$

it is easy to see that the B_k are pairwise disjoint, $B_k \subset A_k$, and

$$\bigcup_{k=1}^{h} A_k = \bigcup_{k=1}^{h} B_k \; ; \; \bigcup_{k=1}^{+\infty} A_k = \bigcup_{k=1}^{+\infty} B_k .$$

Thus, since m(E) is completely additive, formula (2,6,2) gives

$$m\left(\bigcup_{k=1}^{\infty} A_k\right) = \sum_k m(B_k) \leqq \sum_{k=1}^{+\infty} m(A_k). \qquad (2,6,3)$$

On the other hand

$$m\left(\bigcup_{k=1}^{+\infty} A_k\right) = \sum_{k=1}^{\infty} m(B_k) = \lim_{h\to+\infty} \sum_{k=1}^{h} m(B_k) = \lim_{h\to+\infty} m\left(\bigcup_{k=1}^{h} B_k\right)$$

$$= \lim_{h\to+\infty} m\left(\bigcup_{k=1}^{h} A_k\right). \qquad (2,6,4)$$

In the particular case where $A_k \subset A_{k+1}$ for any k, so that $\bigcup_{k=1}^{h} A_k = A$ formula (2,6,4) gives:

THEOREM (2,6,1). If m(E) is a measure and the sets A_k (k = 1, 2,... are such that $A_k \subset A_{k+1}$ for all k, then

$$\bigcup_{k=1}^{+\infty} A_k = \lim_{h\to+\infty} A_h$$

and

$$m\left(\bigcup_{k=1}^{+\infty} A_k\right) = m\left(\lim_{h\to+\infty} A_h\right) = \lim_{h\to+\infty} m(A_h). \blacksquare \qquad (2,6,5)$$

Now consider a countable sequence $A_1, A_2, \ldots, A_k, \ldots$ such that $A_k \supset A_{k+1}$ for all k; put

$$B_k = A_k \cap \check{A}_{k+1} \qquad (k = 1, 2, \ldots),$$

$$C_h = \bigcap_{k=h}^{+\infty} A_k \qquad (h = 1, 2, \ldots).$$

The following assertions are easily proved:

1) $\bigcap_{l=1}^{+\infty} A_l$ and the B_k are pairwise disjoint.

2) If C_h is empty for some value of h, then C_h is empty for any h.

3) For any h,

$$A_h = \left(\bigcup_{k=h}^{+\infty} B_k\right) \cup \left(\bigcap_{l=1}^{+\infty} A_l\right).$$

Thus, if $m(A_k) < +\infty$ for at least one value of k, complete additivity gives

$$\lim_{h\to+\infty} m\left(\bigcup_{k=h}^{+\infty} B_k\right) = 0,$$

and therefore:

THEOREM (2,6,2). If m(E) is a measure, the sets A_k are such that $A_k \supset A_{k+1}$ for all k, and $m(A_k) < +\infty$ for at least one k, then

$$\bigcap_{l=1}^{+\infty} A_l = \lim_{h\to+\infty} A_h ,$$

and

$$\lim_{h\to+\infty} m(A_h) = m\left(\bigcap_{l=1}^{+\infty} A_l\right) = m\left(\lim_{h\to+\infty} A_h\right). \qquad (2,6,6)$$

Thus if $\bigcap_{l=1}^{+\infty} A_l = \emptyset$ it follows that

$$\lim_{h\to+\infty} m(A_h) = 0. \blacksquare \qquad (2,6,7)$$

Naturally, *since probability is a (bounded) measure the properties expressed in* (2,6,3) *and Theorems* (2,6,1), (2,6,2) *are valid* for *probabilities.*

Sets of measure zero and properties true almost everywhere

If m is a measure over $\mathfrak{U}$, a subset $E \subset \mathfrak{U}$ is said to be of *measure zero* if m(E) = 0. A property is said to hold almost *everywhere*[1] (with respect to the measure m) if the set E of elements u of $\mathfrak{U}$ not having the property is of measure zero. The importance of these concepts will become apparent in the sequel.

Point functions. Suppose that with any element u of a space $\mathfrak{U}$ we associate one (and only one) number λ (u), finite or not; then we say that a *point function* λ is defined over $\mathfrak{U}$. The term "point function" will be convenient in order to avoid any confusion with set functions.

The point function λ is said to be real (complex) if the numbers $\lambda(u)$ are real (complex). λ is a mapping of $\mathfrak{U}$ into the set R of the real numbers if λ is a real point function, into the set C of complex numbers if λ is a complex point function.

---

(1) Abbreviation, a.e.; German: fast überall; French: presque partout.

For the moment we shall limit ourselves to real point functions.

Consider a countable sequence $\{\lambda_k\}$ ($k = 1, 2, 3,\ldots$) of point functions defined over $\mathfrak{U}$; we shall say

a) that it is a nondecreasing sequence, if, for any $u \in \mathfrak{U}$

$$\lambda_k (u) \leqq \lambda_{k+1} (u) \text{ for every } k$$

(a nonincreasing sequence is defined similarly, *mutatis mutandis*);

b) *that it converges "in the ordinary sense" to a limit point function* λ *if for any* $u \in \mathfrak{U}$

$$\lim_{k \to +\infty} \lambda_k(u) = \lambda(u). \blacksquare$$

Given a subset $E \subset \mathfrak{U}$, the point function $\phi(E;u)$ defined as 1 if $u \in E$ and 0 if $u \notin E$ is called the *indicator*[1] of E.

A point function $\lambda(u)$ is a *simple function* if it assumes only a finite number s of distinct values $\lambda_1, \lambda_2, \ldots, \lambda_s$, finite or not. An indicator is a simple function.

LEMMA (2,6,1). Any point function λ is the limit (in the ordinary sense) of at least one nondecreasing sequence of simple functions.

In fact, consider the numbers $r/2^k$ (k a positive integer; r any integer); for arbitrary fixed k and any u such that $|\lambda(u)| < +\infty$, let r(u) denote the greatest integer such that $r(u)/2^k \leqq \lambda(u)$. We define:

$$\lambda_k(u) = \begin{cases} \dfrac{r(u)}{2^k} \text{ if } |\lambda(u)| < +\infty \quad \text{and } -2^k \leqq \lambda(u) < 2^k \;; \\ -\infty \text{ if } \lambda(u) < -2^k \;; \\ 2^k \text{ if } \lambda(u) \geqq 2^k . \end{cases}$$

---

[1] The term *characteristic function* is also used; we shall not employ this term, since in Probability Theory it usually denotes a quite different concept (cf. Chapter IV, § 3 p.).

It is immediate that λ_k is a simple function, the sequence $\{\lambda_k\}$ is nondecreasing, and it converges in the ordinary sense to λ. Note in addition that if $\lambda(u) \geqq 0$ for all u the λ_k may be defined in such a way that $\lambda_k(u)$ is bounded and nonnegative for any k. ■

We shall now see that a measure makes it possible to define an integral for certain point functions. As we know, a measure is a mass distribution i.e., a "material system." The integral we define is analogous to the concept of center of gravity [cf. (2.6.8)]; essentially, therefore, it is a very simple concept.

Definite integral

Let m be a measure over $\mathcal{U}$ and λ a point function defined over $\mathcal{U}$.

Case I. Let λ be a nonnegative simple function, i.e., it assumes only a finite number s of distinct values $\lambda_1, \lambda_2, \ldots, \lambda_s$, finite and nonnegative. Let E_j be the set of elements $u \in \mathcal{U}$ such that $\lambda(u) = \lambda_j$; then the definite integral of λ over $\mathcal{U}$ with respect to the measure m (when no ambiguity is possible we shall say simply: the integral of λ) is defined to be the nonnegative number (possibly infinite, but always well defined) $\sum_j \lambda_j \, m(E_j)$. This integral will often be denoted by $I(\lambda)$ or $\int_{\mathcal{U}} \lambda(u) \, m(du)$:

$$I(\lambda) = \int_{\mathcal{U}} \lambda(u) \, m(du) = \sum_j \lambda_j \, m(E_j). \qquad (2,6,8)$$

Case II. Suppose that λ, while not necessarily a simple function, is nonnegative. By Lemma (2,6,1) we know that there exists a nondecreasing sequence $\{\lambda_k\}$ of simple, bounded, nonnegative functions which converges to λ. The integrals $I(\lambda_k)$ clearly form a nondecreasing sequence of nonnegative numbers; they therefore have a limit $I(\lambda)$ (possibly infinite) as $k \longrightarrow +\infty$. It can be proved that this limit does not depend on the specific sequence $\{\lambda_k\}$ used, provided it is a nondecreasing sequence of simple, bounded, nonnegative functions converging to λ. This limit $I(\lambda)$ is defined to be the integral of λ (over $\mathcal{U}$ with respect to the measure m), denoted by

$$I(\lambda) = \int_{\mathfrak{U}} \lambda(u)\ m(du) = \lim_{k \to +\infty} \int_{\mathfrak{U}} \lambda_k(u)\ m(du). \qquad (2,6,9)$$

If $I(\lambda) < +\infty$, λ is said to be *integrable* over $\mathfrak{U}$ (with respect to m).

Case III. Now let λ be any real point function defined over $\mathfrak{U}$, and λ^+ and λ^- the following two functions:

$$\lambda^+(u) = \frac{|\lambda(u)| + \lambda(u)}{2} = \begin{cases} \lambda(u) \text{ if } \lambda(u) \geqq 0\ ; \\ 0 \text{ if } \lambda(u) < 0\ ; \end{cases}$$

$$\lambda^-(u) = \frac{|\lambda(u)| - \lambda(u)}{2} = \begin{cases} 0 \text{ if } \lambda(u) \geqq 0\ ; \\ -\lambda(u) \text{ if } \lambda(u) < 0. \end{cases}$$

Then:

$$\lambda(u) = \lambda^+(u) - \lambda^-(u),\ \lambda^+(u) \geqq 0,\ \lambda^-(u) \geqq 0.$$

λ is said to be *integrable* (over $\mathfrak{U}$, with respect to m), if λ^+ and λ^- are integrable (Case II); the integral $I(\lambda)$ of λ, denoted by

$$I(\lambda) = \int_{\mathfrak{U}} \lambda(u)\ m(du),$$

is defined as the number

$$I(\lambda) = I(\lambda^+) - I(\lambda^-) = \int_{\mathfrak{U}} \lambda^+(u)\ m(du) - \int_{\mathfrak{U}} \lambda^-(u)\ m(du). \qquad (2,6,10)$$

Case IV. Now consider the case of a complex point function $\lambda(u)$; it may be split into its real part $a(u)$ and imaginary part $b(u)$:

$$\lambda(u) = a(u) + ib(u).$$

Then λ is said to be *integrable* (over $\mathfrak{U}$, with respect to m), if the real point functions a and b are integrable (Case III above); by definition, the integral $I(\lambda)$ of λ, denoted by

$$I(\lambda) = \int_{\mathfrak{U}} \lambda(u)\ m(du),$$

is the complex number

$$I(\lambda) = I(a) + iI(b) = \int_{\mathfrak{U}} a(u)\ m(du) + i\int_{\mathfrak{U}} b(u)\ m(du). \quad (2,6,11)$$

The following discussion is valid for both complex and real point functions; from now on we shall not distinguish between the two cases. Clearly, we have:

THEOREM (2,6,3). A function λ is integrable (over $\mathfrak{U}$) if and only if $|\lambda|$ is integrable, and

$$\left|\int_{\mathfrak{U}} \lambda(u)\ m(du)\right| \leqq \int_{\mathfrak{U}} |\lambda(u)|\ m(du).$$

THEOREM (2,6,4).

a) If λ is integrable (over $\mathfrak{U}$) and a is a constant, then $a\lambda$ is integrable, and

$$\int_{\mathfrak{U}} [a\lambda(u)]\ m(du) = a \cdot \int_{\mathfrak{U}} \lambda(u)\ m(du).$$

b) If the functions λ and μ are integrable, their sum, difference, and product are integrable, and

$$\int_{\mathfrak{U}} [\lambda(u) + \mu(u)]\ m(du) = \int_{\mathfrak{U}} \lambda(u)\ m(du) + \int_{\mathfrak{U}} \mu(u)\ m(du).$$

c) If μ is an integrable point function, real and nonnegative, and λ is a point function such that

$$|\lambda(u)| \leqq \mu(u) \qquad \text{a.e.},$$

then λ is integrable; if λ and μ are real and integrable and

$$\lambda(u) \leqq \mu(u) \qquad \text{a.e.},$$

then

$$\int_{\mathfrak{U}} \lambda(u)\ m(du) \leqq \int_{\mathfrak{U}} \mu(u)\ m(du).$$

m-equivalent point functions

We shall say that two point functions λ and μ are equivalent with respect to the measure m, or *m-equivalent*, if

$$\lambda(u) = \mu(u)$$

almost everywhere with respect to m; we shall say that the point function λ is *m-equivalent to* 0, if

$$\lambda(u) = 0$$

almost everywhere with respect to m. Clearly, λ and μ are m-equivalent if and only if their difference $\lambda - \mu$ is m-equivalent to 0.

The point of these definitions is apparent from the following remarks.

1) If λ and μ are m-equivalent and μ is integrable, then λ is integrable, and

$$\int_{\mathfrak{U}} \lambda(u)\; m(du) = \int_{\mathfrak{U}} \mu(u)\; m(du).$$

2) if λ is real and nonnegative then λ is m-equivalent to 0 if and only if

$$\int_{\mathfrak{U}} \lambda(u)\; m(du) = 0.$$

m-equivalent mappings. More generally, let $\mathfrak{X}$ and $\mathfrak{Y}$ be two arbitrary spaces, and λ and μ two mappings,

$$x = \lambda(u) \in \mathfrak{X} \quad \text{and} \quad y = \mu(u) \in \mathfrak{Y}$$

of $\mathfrak{U}$ into $\mathfrak{X}$ and $\mathfrak{Y}$, respectively. λ and μ are said to be equivalent with respect to the measure m, or m-equivalent, if the set E_0 of $u \in \mathfrak{U}$ such that

$$\lambda(u) \neq \mu(u)$$

is of measure zero: $m(E_0) = 0$; i.e., there exists a space $\mathfrak{Z} \subset \mathfrak{X} \cap \mathfrak{Y}$ and a set $E_0 \subset \mathfrak{U}$ such that

$$m(E_0) = 0\;;\; \lambda(u) = \mu(u) \in \mathfrak{Z} \text{ if } u \notin E_0,\; u \in \mathfrak{U}.$$

In other words, $\lambda(u)$ and $\mu(u)$ are equal almost everywhere.

We shall see in II.12 that if m is a probability law this concept becomes that of *equivalent random elements*.

Integral over an arbitrary set

Again, let λ be a point function defined over $\mathfrak{U}$, and A an arbitrary subset of $\mathfrak{U}$, with indicator $\varphi(A;u)$. Then λ is said to be integrable over A (with respect to m) if the product $\lambda(u)\cdot\varphi(A;u)$ is integrable over $\mathfrak{U}$; if λ is integrable over A, the (definite) integral of λ over A with respect to the measure m, denoted by

$$\int_A \lambda(u)\ m(du),$$

is defined as the integral

$$\int_A \lambda(u)\ m(du) = \int_{\mathfrak{U}} \lambda(u)\ \varphi(A\ ;\ u)\ m(du).$$

$\int_A \lambda(u)\ m(du)$ may also be defined directly, as done previously for $\int_{\mathfrak{U}} \lambda(u)\ m(du)$, by simply letting A play the role of $\mathfrak{U}$; the two definitions are equivalent.

REMARK (2,6,1)

Nevertheless, it must be remarked that the second definition may also hold for a function λ which is not defined outside A. In general, it is quite obvious that in the previous discussion and the sequel any [measurable] subset A of $\mathfrak{U}$ may play the role of $\mathfrak{U}$. There is no need to study this simple modification explicitly, and from now on we shall always confine ourselves to the case where λ is defined over $\mathfrak{U}$.

The following theorem is easy to prove.

THEOREM (2,6,5). If m_1 and m_2 are two measures over $\mathfrak{U}$, then their sum $m = m_1 + m_2$, i.e., the set function $m(E)$ defined by $M(E) = m_1(E) + m_2(E)$, is also a measure over $\mathfrak{U}$. If the point function $\lambda(u)$ is integrable with respect to both m_1 and m_2, then it is integrable with respect to m; conversely, if it is integrable with respect to the sum m, it is integrable with respect to both m_1 and m_2, and

$$\int_{\mathfrak{U}} \lambda(u)\ m(du) = \int_{\mathfrak{U}} \lambda(u)\ m_1(du) + \int_{\mathfrak{U}} \lambda(u)\ m_2(du).$$

Indefinite integral

If λ is integrable with respect to a measure m over $\mathfrak{U}$, it is obviously integrable with respect to m over any measurable subset A of $\mathfrak{U}$; consequently, the formula

$$\Lambda(A) = \int_A \lambda(u)\ m(du)$$

defines a function $\Lambda(A)$ of the set A over $\mathfrak{U}$, which is called the *indefinite integral* of λ. We have the following theorems.

THEOREM (2,6,6). If λ is integrable over $\mathfrak{U}$, then for any $\varepsilon > 0$ there exists a set $A \subset \mathfrak{U}$ such that $m(A) < +\infty$ and

$$\left| \int_{\mathfrak{U}} \lambda(u)\ m(du) - \int_A \lambda(u)\ m(du) \right| = \left| \int_{\mathfrak{U}-A} \lambda(u)\ m(du) \right| < \varepsilon .$$

THEOREM (2,6,7). Let λ be integrable over $\mathfrak{U}$; then

a) if $m(A) = 0$, $\Lambda(A) = 0$;

b) if $\Lambda(A) = 0$ for all A, $\lambda(u) = 0$ a.e.;

c) $\Lambda(A)$ is a bounded and completely additive set function.

REMARK (2,6,2)

One can define the integral of a point function $\lambda(u)$ with respect to a completely additive set function of arbitrary sign (i.e., not necessarily nonnegative, therefore not necessarily a measure); as this extension is of little use in Probability, we shall not deal with it here.

7. *Chebyshev's Theorem*

Let $\mathfrak{U}$ be an arbitrary set or space of elements u, m a measure - bounded or not - defined over $\mathfrak{U}$, and α any nonnegative real number. For any point function $\lambda = \lambda(u)$ defined over $\mathfrak{U}$, we define

$$M_\alpha(\lambda) = \int_{\mathfrak{U}} |\lambda(u)|^\alpha m(du), \quad N_\alpha(\lambda) = [M_\alpha(\lambda)]^{\frac{1}{\alpha}}. \tag{2,7,1}$$

Note that since $|\lambda(u)|^\alpha$ is always real and nonnegative, the integral (2,7,1) that defines $M_\alpha(\lambda)$ always has a well-defined value: it is either finite and nonnegative, i.e., it "exists," or equal to $+\infty$. It follows that the same holds for $N_\alpha(\lambda)$. With this in mind, we have

THEOREM (2,7,1). If the measure m is bounded, and there exists a positive number α_0 such that

$$M_{\alpha_0}(\lambda) = \int_{\mathfrak{U}} |\lambda(u)|^{\alpha_0} m(du) < +\infty,$$

then

$$M_\alpha(\lambda) = \int_{\mathfrak{U}} |\lambda(u)|^\alpha m(du) < +\infty$$

for all α such that $0 \leqq \alpha < \alpha_0$; obviously, if $|\lambda(u)| \geqq 1$ almost everywhere, then

$$M_\alpha(\lambda) \leqq M_{\alpha_0}(\lambda).$$

THEOREM (2,7,2) (Chebyshev). If m is a measure, bounded or not, and for some real number $\alpha \geqq 0$ we have

$$M_\alpha(\lambda) < +\infty,$$

then, letting A denote the set of all u such that $|\lambda(u)| \geqq a$ (where a is any positive real number), we have

$$m(A) \leqq M_\alpha(\lambda)/a^\alpha.$$

Indeed,

$$a^\alpha . m(A) = \int_A a^\alpha . m(du) \leqq \int_A |\lambda(u)|^\alpha m(du) \leqq \int_{\mathfrak{U}} |\lambda(u)|^\alpha m(du) = M_\alpha(\lambda).$$

THEOREM (2,7,3) (Hölder). Let m be a measure over $\mathfrak{U}$, bounded or not, and α and β two real numbers > 1 such that $1/\alpha + 1/\beta = 1$. Let $\lambda(u)$ and $\mu(u)$ be two point functions. If $N_\alpha(\lambda) < +\infty$ and $N_\beta(\mu) < +\infty$, then (Hölder's inequality)

$$\left| \int_{\mathcal{U}} \lambda(u)\ \bar{\mu}(u)\ m(du) \right| \leqq N_\alpha(\lambda) \cdot N_\beta(\mu). \tag{2,7,2}$$

Inequality (2,7,2) implies, naturally, that the integral in the left hand side exists. We shall not prove this theorem (a short proof may be found in L. Loomis [1], p. 37; cf. also Exercise 3.10), limiting ourselves here to a particular case.

Schwarz's inequality

If $\alpha = \beta = 2$, inequality (2,7,2) is known as Schwarz's inequality:

$$\left| \int_{\mathcal{U}} \lambda(u)\ \bar{\mu}(u)\ m(du) \right|^2 \leqq M_2(\lambda) \cdot M_2(\mu). \tag{2,7,3}$$

An elementary proof of (2,7,3) is as follows. Put

$$H = \int_{\mathcal{U}} \lambda(u)\ \bar{\mu}(u)\ m(du)\ ,\ H' = \int_{\mathcal{U}} |\lambda(u)\ \mu(u)|\ m(du).$$

Assume first that $M_2(\mu) > 0$; let ρ be an arbitrary complex constant, and consider the two integrals

$$I(\rho) = M_2(\lambda - \rho \cdot \mu) = M_2(\lambda) - \bar{H}\rho - H\bar{\rho} + M_2(\mu)\ |\rho|^2,$$

$$J(\rho) = M_2(|\lambda| - |\rho|\ |\mu|) = M_2(\lambda) - 2H'\ |\rho| + M_2(\mu)\ |\rho|^2.$$

Whatever the value of ρ, $I(\rho)$ and $J(\rho)$ *cannot be negative*. A priori, H' may be infinite; but if so this would imply that $J(\rho)$ is negative (possibly even $-\infty$). Therefore $H' < +\infty$, so that H exists and is finite.

Let θ and ϕ denote the arguments of H and ρ, respectively; then $I(\rho)$ becomes

$$I(\rho) = M_2(\lambda) - 2\ |H|\ |\rho| \cos(\varphi - \theta) + M_2(\mu)\ |\rho|^2,$$

thus $I(\rho)$ is a minimum when $\phi = \theta$ and $|\rho| = |H|/M_2(\mu)$. $I(\rho)$ then has the value $M_2(\lambda) - |H|^2/M_2(\mu)$, and since $I(\rho)$ is never negative we have

$$M_2(\lambda) - \frac{|H|^2}{M_2(\mu)} \geqq 0, \tag{2,7,4}$$

which is (2,7,3).■

If $M_2(\mu) = 0$, μ is m-equivalent to 0 (cf. II.6), therefore so is $\lambda\bar{\mu}$, and (2,7,3) is trivial since both sides vanish.

Returning to the case $M_2(\mu) \neq 0$, note that (2,7,3) is an equality if and only if (2,7,4) is an equality, therefore if and only if there exists ρ such that $I(\rho) = M_2(\lambda - \rho\mu)$ vanishes. This means that λ is m-equivalent to $\rho\mu$; in other words, up to m-equivalence, λ and μ are proportional. Moreover, if $M_2(\mu) = 0$, i.e., if μ is m-equivalent to 0, we always can assume that μ is proportional to λ up to m-equivalence. Hence

THEOREM (2,7,4). Schwarz's inequality (2,7,3) is an equality if and only if the functions λ and μ are proportional (up to m-equivalence).

THEOREM (2,7,5). Suppose that

$$m(\mathfrak{U}) \leqq 1 \; ; \tag{2,7,5}$$

let γ and γ' be two positive numbers such that $\gamma' < \gamma$. Then for any function $\nu = \nu(u)$

$$N_{\gamma'}(\nu) \leqq N_{\gamma}(\nu). \tag{2,7,6}$$

Of course, this inequality is of interest only if its right-hand side is finite; suppose this is the case, and apply (2,7,2) with

$$\lambda = |\nu|^{\gamma'}, \; \mu = 1 \; , \; \alpha = \frac{\gamma}{\gamma'}, \; \beta = \frac{\gamma}{\gamma - \gamma'}.$$

We obtain

$$M_{\gamma'}(\nu) \leqq [N_\gamma(\nu)]^{\gamma'} \cdot \left[\int_{\mathcal{U}} 1^{\frac{\gamma}{\gamma-\gamma'}} \cdot m(du) \right]^{\frac{\gamma-\gamma'}{\gamma}} ;$$

but

$$\left[\int_{\mathcal{U}} 1^{\frac{\gamma}{\gamma-\gamma'}} \cdot m(du) \right]^{\frac{\gamma-\gamma'}{\gamma}} = [m(\mathcal{U})]^{\frac{\gamma-\gamma'}{\gamma}} \leqq 1,$$

proving (2,7,6). Theorem (2,7,1) which can easily be established directly, can also be derived from Theorem (2,7,5).

THEOREM (2,7,6) (Minkowski). Let α be any real number $\geqq 1$; then for any measure m and any "point" functions λ and μ, we have the inequality

$$N_\alpha(\lambda + \mu) \leqq N_\alpha(\lambda) + N_\alpha(\mu). \qquad (2,7,7)$$

This inequality is of interest only if both terms of the right-hand side are finite; suppose this is the case. If $\alpha = 1$, then (2,7,7) is obvious. For $\alpha > 1$, inequality (2,7,7) may either be proved directly or deduced from Theorem (2,7,3); we shall not give the proof for the general case (cf. L. Loomis [1], p. 38, and Exercise 3.11).

For $\alpha = 2$, Schwarz's inequality (2,7,3) provides a simple proof. Indeed, inequality (2,7,7) is then equivalent to

$$M_2(\lambda + \mu) = M_2(\lambda) + M_2(\mu) + 2\, N_2(\lambda)\, N_2(\mu). \qquad (2,7,8)$$

Now, in the previous notation,

$$M_2(\lambda + \mu) = M_2(\lambda) + M_2(\mu) + H + \overline{H}, \qquad (2,7,9)$$

and in view of (2,7,3) formula (2,7,9) clearly implies (2,7,8). Moreover, the two sides of (2,7,8) cannot be equal unless the two sides of (2,7,3) are equal. This is true, as we know, only if μ is m-equivalent to 0, or if there exists a constant ρ such that $\lambda = \rho.\mu$ up to m-equivalence. Then $N_2(\lambda) + N_2(\mu)$ is $(1 - |\rho|)\, N_2(\mu)$; but $|1 + \rho| = (1 + |\rho|)$ only if ρ is real and nonnegative. Hence

THEOREM (2,7,7). Let m be a measure, λ and μ two "point" functions such that

$$N_2(\lambda) < +\infty\,,\ N_2(\mu) < +\infty\,;$$

then

$$N_2(\lambda + \mu) \leqq N_2(\lambda) + N_2(\mu). \tag{2,7,10}$$

Inequality (2,7,10) is an equality if and only if λ and μ are proportional up to m-equivalence, and the proportionality factor is real and nonnegative.

8. *Types of convergence of point functions*

We have already recalled the definition of "ordinary" convergence: a sequence $\{\lambda_k\}$ of point functions (finite or not) defined over $\mathfrak{U}$ converges to the function λ if $\lim_{k\to+\infty} \lambda_k(u) = \lambda(u)$ for every $u \in \mathfrak{U}$. This definition is independent of any concept of measure. The situation is different for the new types of convergence we are about to define. In the sequel we consider a measure m, bounded or not, defined over $\mathfrak{U}$; all definitions will be *relative* to this measure.

Convergence almost everywhere

The point function $\lambda_k (k = 1, 2, 3,\dots)$ converges *almost everywhere* to the point function λ as $k \longrightarrow +\infty$ if

$$\lim_k \lambda_k(u) = \lambda(u) \text{ a.e.}\,, \tag{2,8,1}$$

i.e., if the set E of all u such that

$$\lim_{k\to+\infty} \lambda_k(u) = \lambda(u)$$

is of measure zero, $m(E) = 0$. "Ordinary" convergence obviously implies convergence a.e.

Convergence in measure

The point function λ_k converges in *measure* to the point function

λ as $k \longrightarrow +\infty$ if, for any positive ε and η, there exists a positive number $K(\varepsilon, \eta)$ such that the measure $m(E_k)$ of the set E_k of all u for which $|\lambda_k(u) - \lambda(u)| > \varepsilon$ is smaller than η for all $k > K(\varepsilon, \eta)$.

It can be proved directly (and will follow in any case from Theorem (6,6,1)) that convergence a.e. implies convergence in measure On the other hand, it is intuitively clear that λ_k may converge to λ in measure without converging to λ a.e.; for an example see VI.6, Example (6,6,2).

Convergence in α-th mean

Many other kinds of convergence may be introduced; however, unlike convergence a.e. and convergence in measure, which both depend on the measure m in an intrinsic way, they are rather artificial. Among the most useful, we mention convergence in the α-th mean: λ_k converges to λ in the α-th mean (where α is any *positive* number) if

$$\lim_{k \to +\infty} \int_{\mathcal{U}} |\lambda_k(u) - \lambda(u)|^{\alpha}\, m(du) = 0, \qquad (2,8,2)$$

i.e., if

$$\lim_{k \to \infty} M_{\alpha}(\lambda_k - \lambda) = 0.$$

The cases $\alpha = 1$ and $\alpha = 2$ are the most important in practice; convergence in the 2nd mean is known as *convergence in the quadratic mean*(1).

Convergence in the α-th mean implies convergence in measure; this may easily be deduced from Chebyshev's Theorem (2,7,2). On the other hand, λ_k may converge to λ in measure (or almost everywhere) without converging to λ in the α-th mean for any positive α, and the integral in (2,8,2) may not even exist; this is intuitively clear, and we shall cite examples in VI.6, Example (6,6,1). Nonetheless, if λ_k converges a.e. to a limit λ, and in the α-th mean to a limit μ, then λ and μ are m-equivalent.

---

(1) Abbreviation, in q.m.; German: im quadratischen Mittel; French: en moyenne quadratique.

Theorem (2,7,5) implies that if m is a bounded measure and λ_k converges to λ in the α_0-th mean ($\alpha_0 > 0$), then λ_k converges to λ in the α-th mean for any positive α smaller than α_0.

THEOREM (2,8,1). If the sequence $\{\lambda_k\}$ converges to λ in measure, there exists at least one subsequence $\{\lambda_{k_h}\}$ ($h = 1, 2, 3, \dots$) of the sequence $\{\lambda_k\}$ that converges to λ a.e.

We shall not prove this theorem; a proof can be based on Theorem (2,8,3), following the argument used below in Theorem (2,10,7).

Integrals of the type $M_\alpha(\lambda)$ yield interesting criteria for convergence a.e. Let α be a positive number, and $\{\lambda_k\}$ ($k = 1, 2, 3, \dots$) a sequence of point functions $\lambda_k(u)$, $u \in \mathfrak{U}$, such that $M_\alpha(\lambda_k) = \rho_k < +\infty$ for every k.

For every k, let ε_k be a positive number. Let e_k be the set of all $u \in \mathfrak{U}$ such that $|\lambda_k(u)| \geqq \varepsilon_k$, and put $\omega_h = \bigcup_{k>h} e_k$.

By Theorem (2,7,2),

$$m(e_k) \leqq \frac{\rho_k}{\varepsilon_k^\alpha},$$

and by (2,6,3),

$$m(\omega_h) \leqq \sum_{k>h} \frac{\rho_k}{\varepsilon_k^\alpha}. \tag{2,8,3}$$

Hence the following applications:

1) Suppose that the series ρ_k is convergent:

$$\sum_k \rho_k = \sum_k M_\alpha(\lambda_k) < +\infty. \tag{2,8,4}$$

Let e be the set of all $u \in \mathfrak{U}$ such that

$$\lim_{k\to\infty} \lambda_k(u) = 0.$$

For any positive number ε, let $e(\varepsilon)$ be the set of $u \in \mathfrak{U}$ such that

$$|\lambda_k(u)| \geqq \varepsilon$$

holds *for infinitely many* k; then, if $0 < \varepsilon' < \varepsilon$,

$$e(\varepsilon) \subset e(\varepsilon') \subset e \; ; \; \lim_{\varepsilon \to +0} e(\varepsilon) = e.$$

Consequently, by Theorem (2,6,1),

$$m(e) = \lim_{\varepsilon \to +0} m[e(\varepsilon)].$$

Now, taking $\varepsilon_k = \varepsilon$ (for all k), we have

$$e(\varepsilon) \subset \omega_h \quad \text{for any } h.$$

Together with (2,8,3), this gives

$$m[e(\varepsilon)] \leqq m(\omega_h) \leqq \frac{1}{\varepsilon^{\alpha}} \sum_{k>h} \rho_k \quad \text{for all } h.$$

It now follows from (2,8,4) that $m(e(\varepsilon)) = 0$, and consequently $m(e) = 0$. We have proved:

THEOREM (2,8,2)

In order that the λ_k converge to 0 (i.e., to the identically zero point function $\lambda(u) \equiv 0$) almost everywhere as $k \longrightarrow +\infty$, it is sufficient that the series (2,8,4) be convergent.

Note that convergence of the series (2,8,4) implies that $\lim_{k \to +\infty} \rho_k = 0$, and consequently the λ_k also converge to 0 in the α-th mean.

2) Let $\{\mu_k\}$ (k = 1, 2, 3,...) be a sequence of point functions $\mu_k(u)$, $u \in \mathfrak{U}$, such that

$$M_{\alpha}(\mu_{k+1} - \mu_k) = \rho_k < +\infty \quad \text{for every } k,$$

and suppose that there exists a sequence of positive numbers ε_k such that

$$\sum_k \varepsilon_k < +\infty , \tag{2,8,5}$$

and

$$\sum_k \frac{\rho_k}{\varepsilon_k^{\alpha}} < +\infty . \tag{2,8,6}$$

put $\lambda_1 = \mu_1$, $\lambda_k = \mu_k - \mu_{k-1}$ for $k > 1$. Let e be the set of all $u \in \mathfrak{U}$ such that

$$|\lambda_k(u)| \geqq \varepsilon_k$$

holds for *infinitely many* k. With the same notation as before, we have $e \subset \omega_h$ for every h; therefore, by (2,8,3),

$$m(e) \leqq \sum_{k>h} \frac{\rho_k}{\varepsilon_k^a} \text{ for any } h;$$

then (2,8,6) implies

$$m(e) = 0. \tag{2,8,7}$$

For any u in the complement $\check{e}$ of e, therefore, there exists K such that

$$|\mu_k(u) - \mu_{k-1}(u)| < \varepsilon_k$$

for all $k > K$ (K may of course depend on u). By (2,8,5), the series

$$\mu_1(u) + \sum_{k>1} [\mu_k(u) - \mu_{k-1}(u)] = \mu(u) \tag{2,8,8}$$

is therefore convergent, and has a sum $\mu(u)$; $\mu(u)$ is therefore a well-defined function of u on $\check{e}$. Let us define $\mu(u)$ arbitrarily on e, for instance,

$$\mu(u) = 0 \text{ if } u \in e,$$

Then the function $\mu(u)$ is defined throughout $\mathfrak{U}$. Now (2,8,8) may be written

$$\lim_{k\to+\infty} \mu_k(u) = \mu(u) \quad \text{for any} \quad u \in \check{e}.$$

Combining this with (2,8,7) we get

THEOREM (2,8,3). If conditions (2,8,5) and (2,8,6) hold, there exists a point function μ of $u \in \mathfrak{U}$ such that μ_k converges to μ almost everywhere as $k \longrightarrow +\infty$. ∎

Recall that if $\{\lambda_k\}$ is a nondecreasing sequence of real and nonnegative functions λ_k, converging (in the ordinary sense, or only almost everywhere) to a function λ, then

$$\lim_{k\to+\infty} \int_{\mathfrak{U}} \lambda_k(u)\ m(du) = \int_{\mathfrak{U}} \lambda(u)\ m(du)\ ;$$

this follows from the definition of the integral (cf. II.6). The following theorems, which the reader may easily prove as exercises, are similar. The first concerns both complex and real point functions.

THEOREM (2,8,4) (Lebesgue)

If λ_k converges to λ in measure, and there exists a nonnegative and integrable function μ such that for every k

$$|\lambda_k(u)| \leqq \mu(u) \quad \text{a.e.},$$

then λ_k converges to λ in the first mean, and, consequently,

$$\lim_{k\to+\infty} \int_{\mathfrak{U}} \lambda_k(u)\ m(du) = \int_{\mathfrak{U}} \lambda(u)\ m(du).$$

THEOREM (2,8,5) (Fatou)

Let $\{\lambda_k\}$ be a sequence of real nonnegative integrable point functions such that

$$\liminf_{k\to+\infty} \int_{\mathfrak{U}} \lambda_k(u)\ m(du) < +\infty;$$

then the function

$$\lambda(u) = \liminf_{k\to+\infty} \lambda_k(u)$$

is integrable and

$$\int_{\mathfrak{U}} \lambda(u)\ m(du) \leqq \liminf_{k\to+\infty} \int_{\mathfrak{U}} \lambda_k(u)\ m(du).$$

The study of convergence a.e., in measure, and in the quadratic mean will be taken up again and completed, in terms of Probability Theory, in VI.4, VI.5, VI.6.

9. *Totally discontinuous measures or mass distributions*

A natural question is that of the mathematical tools - which should be as easy to handle as possible - by means of which a measure or mass distribution (bounded or not) may be defined over a space $\mathcal{U}$. These tools are necessarily conditioned by the nature of $\mathcal{U}$. Later we shall study the principal methods for the case where $\mathcal{U}$ is a finite-dimensional vector space or an affine point space associated with a finite-dimensional vector space. For the moment the only general remark, applicable to any $\mathcal{U}$, that we make is the following.

A special but very simple way of constructing a mass distribution over $\mathcal{U}$ is to single out a finite or countable number of distinct elements u_k $(k = 1, 2, 3,\ldots)$ of $\mathcal{U}$ and assign a mass to those elements alone. To define the distribution it is then sufficient to specify the mass m_k assigned to u_k for each k. If E is any subset of $\mathcal{U}$ and k' denotes those k for which $u_k \in E$, it is clear that

$$m(E) = \sum_{k'} m_{k'}. \tag{2,9,1}$$

Measures or distributions of this type are called *totally discontinuous* or *discrete*.

The finite or countable set of the u_k is the support of the distribution. The masses m_k are obviously nonnegative. The right-hand side of (2,9,1) may contain infinitely many terms; it is then a series with nonnegative terms. If it is divergent, we must regard it as having a sum $+\infty$.

Note that (2,9,1) is meaningful for any subset E of $\mathcal{U}$. In general, as mentioned at the beginning of II.6, a measure may not be defined for all the subsets of $\mathcal{U}$. Totally discontinuous measures are defined, irrespective of the nature of $\mathfrak{U}$, for all subsets of $\mathfrak{U}$, and are thus an exceptional case.

Let m be a totally discontinuous measure over $\mathfrak{U}$, with support $\{u_k\}$; as above, let m_k be the mass assigned to the element u_k, and let λ be an arbitrary point function defined on $\mathfrak{U}$ and integrable with respect to m. Then the definition of the integral

$$\int_{\mathfrak{U}} \lambda(u)\, m(du)$$

in II.6 directly implies that

$$\int_{\mathfrak{U}} \lambda(u)\, m(du) = \sum_k m_k\, \lambda(u_k). \tag{2,9,2}$$

When the m_k constitute a countable set the right-hand side of (2,9,2) is a series, and so λ is integrable if and only if this series is absolutely convergent.

III. HILBERT SPACES

10. *Hilbert spaces*

Let $\mathcal{H}$ be a vector space, either real or complex; to fix ideas, we shall consider the complex case - adaptation of the argument to the real case is immediate. x, y, z,... will denote elements of $\mathcal{H}$, θ the zero of $\mathcal{H}$, and α, β, γ,... scalars, i.e., elements of the field C of complex numbers. We say that $\mathcal{H}$ is a *prehermitian-product* space if there exists a mapping which associates with any ordered pair (x, y) of elements of $\mathcal{H}$ a scalar (complex number), called the prehermitian product of x and y and denoted by x.y,, with the following properties:

1) $x.y = \overline{y.x}$, $x, y \in \mathcal{H}$;

2) $x.(y + z) = x.y + x.z$, $x, y, z \in \mathcal{H}$;

3) $(\alpha x).y = \alpha(x.y)$, $\alpha \in C$, $x, y \in \mathcal{H}$.

These properties imply that $(y + z).x = y.x + z.x$ and $x.(\alpha y) = \bar{\alpha}(x.y)$, $\alpha \in C$, $x, y \in \mathcal{H}$.

Two elements $x, y \in \mathcal{H}$ are said to be *orthogonal* if $x.y = y.x = 0$. The prehermitian product x.y is said to be a *hermitian product* if, in addition to properties 1), 2), 3) it possesses the following property:

4) $x.y = 0$ for every y implies $x = \theta$. ∎

Let us assume that $\mathcal{H}$ is a prehermitian product space; with each $x \in \mathcal{H}$ we associate the number

$$M_2(x) = x.x, \tag{2,10,1}$$

which will be called the *prehermitian form* of x; the notation $M_2(x)$, reminiscent of (2,7,1) for $\alpha = 2$, is intentional; the reason will become apparent later. Note that if $x = \theta$ then $M_2(x) = 0$ (use 3), with $y = \alpha x$, $\alpha = 0$).

By 1) the number $M_2(x)$ is *always real*. $x \in \mathcal{H}$ is said to be a *unit vector* if $M_2(x) = 1$.

It follows immediately that:

THEOREM (2,10,1). If x and y are arbitrary orthogonal elements of the prehermitian-product space $\mathcal{H}$, then

$$M_2(x + y) = M_2(x) + M_2(y). \tag{2,10,2}$$

Formula (2,10,2) is a generalization of the elementary theorem of Pythagoras. ■

In applications, $M_2(x)$ is often interpreted as equal (or at least proportional) to a quantity which is intrinsically nonnegative, such as energy or power. We are therefore particularly interested in cases where the prehermitian product has the property

$$M_2(x) \geqslant 0 \quad \text{for all } x \tag{2,10,3}$$

For any $x \in \mathcal{H}$ we can then define $N_2(x) = \sqrt{M_2(x)}$.

$N_2(x)$ will be called the *semi-norm* of x, and we shall say that the prehermitian product defines the semi-norm $N_2(x)$ in $\mathcal{H}$. The semi-norm is a real nonnegative number which vanishes when $x = \theta$; note that in the general case there may exist elements $x \neq \theta$ such that $N_2(x) = 0$.

By an almost word-for-word repetition of the argument used in II.7 for Schwarz's inequality, we prove

THEOREM (2,10,2). If $\mathcal{H}$ is a hermitian-product space with semi-norm $N_2(x)$, then, for any $x, y \in \mathcal{H}$,

$$|x \cdot y| \leqq N_2(x) \cdot N_2(y). \qquad (2,10,4$$

Inequality (2,10,4) is known as *Schwarz's inequality*; later we shall see that inequality (2,7,4) is in fact only a particular case of (2,10,4).

By the same reasoning as in Theorem (2,7,7), inequality (2,10, implies

THEOREM (2,10,3). If $\mathcal{H}$ is a prehermitian-product space with a semi norm $N_2(x)$, then, for any $x, y \in \mathcal{H}$,

$$N_2(x + y) \leqq N_2(x) + N_2(y). \qquad (2,10,5)$$

Later we shall see that (2,7,10) is a particular case of (2,10,5).

Let R be the relation defined in $\mathscr{H}$ as follows: given an ordered pair $\{x, y\}$ of elements $x, y \in \mathscr{H}$, x is in the relation R to y if

$$M_2(x-y) = 0,$$

or, equivalently,

$$N_2(x-y) = 0. \qquad (2,10,6)$$

It is clear that R is an equivalence relation, which partitions $\mathscr{H}$ into equivalence classes. An equivalence class C is determined by any $x \in \mathscr{H}$ belonging to C. For $x \in \mathscr{H}$, let C_x denote the equivalence class containing x. Let $\tilde{\mathscr{H}}$ be the set of all distinct C_x. We define addition and multiplication by a scalar in $\tilde{\mathscr{H}}$ by:

$$C_x + C_y = C_{x+y} \qquad (x, y \in \mathscr{H});$$
$$\alpha C_x = C_{\alpha x} \qquad (x \in \mathscr{H},\ \alpha \in C);$$

$\tilde{\mathscr{H}}$ is clearly a vector space, with zero element C_θ.

Define a prehermitian product $C_x.C_y$ in $\tilde{\mathscr{H}}$ by

$$C_x.C_y = x.y \ (x, y \in \mathscr{H}). \qquad (2,10,7)$$

This prehermitian product generates a prehermitian form

$$\tilde{M}_2(C_x) = x.x.$$

If $\mathscr{H}$ is semi-normed by its prehermitian product, we see that for any $C_x \in \tilde{\mathscr{H}}$

$$\tilde{M}_2(C_x) \geqq 0, \qquad (2,10,8)$$

i.e., $\mathscr{H}$ is semi-normed by the prehermitian product (2,10,7). Now it follows from (2,10,6) that the prehermitian form $\tilde{M}_2(C)$ possesses both property (2,10,8) and the property:

$M_2(C_x) = 0$ *if and only if* C_x is the zero C_θ of $\tilde{\mathcal{H}}$.

It follows that the prehermitian product (2,10,7) is in fact a hermitian product; for $C_x . C_y = 0$ for every $C_y \in \tilde{\mathcal{H}}$ implies $C_x . C_x = \tilde{M}_2(C_x) = 0$, and so $C_x = C_\theta$.

Moreover, in view of (2,10,6) it is clear that the study of $\mathcal{H}$ may always be reduced to the study of $\tilde{\mathcal{H}}$ with the hermitian product (2,10,7).

These considerations lead naturally to the following definition.

A vector space $\mathcal{H}$ with a hermitian product x.y is *normed* by its hermitian product if the hermitian form $M_2(x)$ is always nonnegative, and it vanishes *if and only if* x is the zero θ of $\mathcal{H}$.

$N_2(x) = \sqrt{M_2(x)}$ is then called the *norm* of x; it is a nonnegat real number, zero *if and only if* $x = \theta$.

Reasoning as in II.7 for Schwarz's inequality, we can prove

THEOREM (2,10,4). If $\mathcal{H}$ is normed by its hermitian product, then

1) Schwarz's inequality (2,10,4) is an equality if and only if the vectors x and y are proportional (i.e. collinear) in $\mathcal{H}$;

2) inequality (2,10,5) is an equality if and only if x and y are proportional and the proportionality factor is real and nonnegative. ∎

From now on we shall assume that $\mathcal{H}$ is normed by its hermitian product.

For any pair (x, y) of elements of $\mathcal{H}$ we define the *distance* between x and y to be $N_2(x-y)$; this defines a metric in $\mathcal{H}$. In particular, inequality (2,10,5) is simply the classical *triangle inequality* which any metric must satisfy by definition.

Another obvious property of the norm is:

For any $x \in \mathcal{H}$ and any $\rho \in C$,

$$N_2(\rho x) = |\rho| \, N_2(x). \qquad (2,10,$$

Thus $N_2(x)$ is indeed a norm in the sense of the theory of normed vector spaces.

In the sequel we shall use concepts from the general theory of metric spaces, in particular, of normed vector spaces.

Let x and y be elements of $\mathcal{H}$ such that

$$M_2(x) = M_2(y) = 1, \qquad (2,10,10)$$

i.e., x and y are unit vectors. Then

$$M_2(x + y) = (x + y) \cdot (x + y) = x \cdot x + y \cdot x + x \cdot y + y \cdot y$$
$$= 2 + y.x + x.y;$$

similarly,

$$M_2(x - y) = (x - y) \cdot (x - y) = x \cdot x - y \cdot x - x \cdot y + y \cdot y$$
$$= 2 - y.x - x.y.$$

Hence, adding these results we get

$$M_2(x + y) + M_2(x - y) = 4,$$

or

$$N_2\left(\frac{x + y}{2}\right) = \sqrt{1 - \frac{1}{4} M_2(x - y)}. \qquad (2,10,11)$$

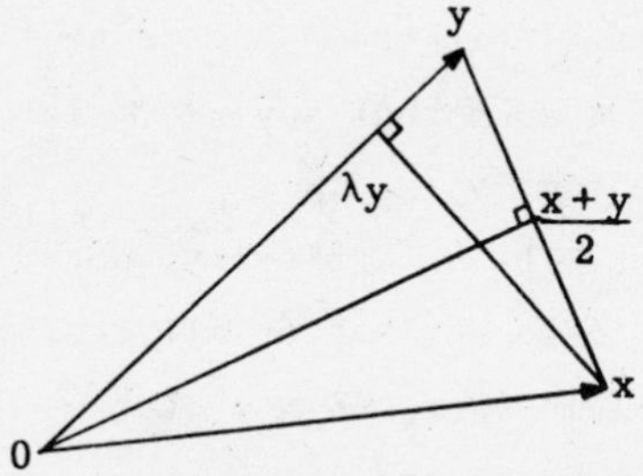

Fig. (2,10,1)

Figure (2,10,1) clarifies this computation, and provides an interpretation of (2,10,11). It is indeed intuitively clear that $N_2\left(\frac{x + y}{2}\right)$ is < 1, but the smaller $M_2(x-y)$ the closer $N_2\left(\frac{x + y}{2}\right)$ to 1. Formula (2,10,11) makes it clear that the difference between 1 and $N_2\left(\frac{x + y}{2}\right)$ does not depend on x and y themselves, but only on $M_2(x-y)$, the distance between x and y. This means that the space $\mathcal{H}$ is *uniformly convex*:

THEOREM (2,10,5). If the space $\mathcal{H}$ is normed by its hermitian product, it is uniformly convex; more precisely, equality (2,10,11) holds for any two unit vectors $x, y \in \mathcal{H}$. ▮

We shall say that a variable element $y \in \mathcal{H}$ converges to a fixed element $x \in \mathcal{H}$ if $N_2(y-x) \longrightarrow 0$. A countable sequence $\{x_k\}$ $(k = 1, 2, 3, \ldots)$ of elements of $\mathcal{H}$ is a *Cauchy sequence* if

$$\lim_{k \to +\infty} N_2(x_{k+h} - x_k) = 0, \quad \text{uniformly in} \quad h > 0. \qquad (2,10,1$$

There exists an $x \in \mathcal{H}$ such that

$$\lim_{k \to +\infty} x_k = x$$

only if $\{x_k\}$ is a Cauchy sequence.

If for any Cauchy sequence $\{x_k\}$ there exists $x \in \mathcal{H}$ such that

$$\lim_{k \to +\infty} x_k = x,$$

$\mathcal{H}$ is said to be *complete*. A vector space $\mathcal{H}$ which is normed by a hermitian product and is also complete is called a *Hilbert space*. The euclidean vector space $\mathfrak{X}_n$ of finite dimension n is a Hilbert space; examples of infinite-dimensional Hilbert spaces will be given later (cf. V.4).

A space $\mathcal{H}$ which is normed by its hermitian product but not necessarily complete is said to be *separable* if there exists a countable family $\{x_k\}$ of elements x_k of $\mathcal{H}$ such that any $x \in \mathcal{H}$ is the limit of a subsequence $\{x_{k'}\}$ of the family $\{x_k\}$. ▮

We recall the following point: let $\mathcal{H}$ be an arbitrary, not necessarily complete, metric space; let S be the set of all Cauchy sequences $u = \{x_1, x_2, \ldots, x_k, \ldots\}$ of elements x_k of $\mathcal{H}$. Let us say that two elements $u = \{x_1, x_2, \ldots, x_k, \ldots\}$ and $u' = \{x_1', x_2', \ldots, x_k, \ldots\}$ of S are equivalent, if the distance between x_k and x_k' tends to 0 as $k \longrightarrow +\infty$. This equivalence relation partitions S into equivalence classes; let $\mathcal{H}_c$ be the set of these equivalence classes.

Using the distance Δ in $\mathcal{H}$, one can define a distance Δ_c in $\mathcal{H}_c$; it turns out that, with respect to this distance, $\mathcal{H}_c$ is a complete metric space, and there is a mapping ϕ of $\mathcal{H}$ into $\mathcal{H}_c$ which preserves distances. Hence *the study of* $\mathcal{H}$ *can be replaced by that of the complete space* $\mathcal{H}_c$.

In particular, let us assume that Δ is defined by a hermitian product in $\mathcal{H}$ which norms $\mathcal{H}$. Using this hermitian product one can define a hermitian product in $\mathcal{H}_c$, which norms $\mathcal{H}_c$; it turns out that the distance defined in $\mathcal{H}_c$ by this hermitian product is identical with the distance Δ_c; hence $\mathcal{H}_c$ is a Hilbert space. Moreover, ϕ is linear and preserves hermitian products.

We are thus justified in *assuming now, once and for all, that* $\mathcal{H}$ *is a Hilbert space.* ∎

We already know that, given a countable sequence $\{x_k\}$ of elements $x_k \in \mathcal{H}$, there exists $x \in \mathcal{H}$ such that $x_k \longrightarrow x$ if and only if the sequence $\{x_k\}$ is a Cauchy sequence. An equivalent but often more convenient convergence test is given by the following theorem.

THEOREM (2,10,6). Given a countable sequence $\{x_k\}$ of elements $x_k \in \mathcal{H}$, there exists an element $x \in \mathcal{H}$ such that $\lim\limits_{k\to+\infty} x_k = x$ if and only if there is a number M such that

$$\lim_{m,n\to+\infty} x_m \cdot x_n = M,$$

independently of the way m and n tend to $+\infty$; i.e., for any $\varepsilon > 0$ there exists H such that if $m > H$ and $n > H$ then

$$|x_m \cdot x_n - M| < \varepsilon.$$

The condition is necessary, since if $\lim\limits_{k\to+\infty} x_k = x$ then, as is easily seen by using (2,10,3), $\lim\limits_{m,n\to+\infty} x_m \cdot x_n = M_2(x)$, independently of the way m and n tend to $+\infty$. Conversely, suppose that there exists a number M such that

$$\lim_{m,n\to+\infty} x_n \cdot x_m = M \qquad (2,10,13)$$

independently of the way m and n tend to + ∞; then, for h > 0,

$$M_2(x_{k+h} - x_k) = x_{k+h} \cdot x_{k+h} - x_{k+h} \cdot x_k - x_k \cdot x_{k+h} + x_k \cdot x_k.$$

Applying (2,10,13) with m = n = k or m = n = k + h we see that

$$\lim_{k \to +\infty} x_k \cdot x_k = M ;$$

$$\lim_{k \to +\infty} x_{k+h} \cdot x_{k+h} = M, \quad \text{uniformly in } h > 0;$$

putting m = k + h and n = k, or m = k, n = k + h, we have from (2,10,13)

$$\lim_{k \to +\infty} x_{k+h} \cdot x_k = \lim x_k \cdot x_{k+h} = M, \quad \text{uniformly in} \quad h > 0.$$

It follows that

$$\lim_{k \to +\infty} M_2(x_{k+h} - x_k) = 0, \quad \text{uniformly in } h > 0.$$

In other words, $\{x_k\}$ is a Cauchy sequence and consequently has a limit $x \in \mathcal{H}$. This completes the proof of Theorem (2,10,6).

The theorem may be extended without essential changes to the case of a not necessarily countable family of elements of $\mathcal{H}$. Assume, for instance, that with every real number t we associate an element x_t of $\mathcal{H}$. To say that x_t converges to a limit $x \in \mathcal{H}$ when t tends, say, to 0, means that, for any $\varepsilon > 0$, there exists $\eta > 0$ such that if $|t| < \eta$ then $N_2(x_t - x) < \varepsilon$. By an argument which is essentially the same as above it is easy to prove:

x_t has a limit as $t \longrightarrow 0$ if and only if there exists a number M such that

$$\lim_{\alpha \to 0, \beta \to 0} x_\alpha \cdot x_\beta = M$$

independently of the way α and β tend to 0; i.e., for any $\varepsilon > 0$ there exists $\eta > 0$ such that if $|\alpha| < \eta$ and $|\beta| < \eta$ then

$$|x_\alpha \cdot x_\beta - M| < \varepsilon.$$

Hilbert subspaces

Let $\mathcal{H}'$ be a subset of $\mathcal{H}$. Suppose that $\mathcal{H}'$ is a vector space with respect to the vector operations in $\mathcal{H}$, that is, $\mathcal{H}'$ is a vector subspace of $\mathcal{H}$. This implies $\theta \in \mathcal{H}'$ (consequently $\mathcal{H}'$ cannot be empty).

$\mathcal{H}'$ is normed by the hermitian product of $\mathcal{H}$. If $\mathcal{H}'$ is also complete (i.e., closed with respect to the metric in $\mathcal{H}$), it is a Hilbert space. $\mathcal{H}'$ is then said to be a Hilbert subspace of $\mathcal{H}$. This implies that $\mathcal{H}'$, like $\mathcal{H}$, is uniformly convex.

Let $\mathfrak{F}$ be any family of elements of $\mathcal{H}$, and let $\mathcal{S}(\mathfrak{F})$ denote the set of all $x \in \mathcal{H}$ for which there exist an integer n, n elements x_k ($k = 1, 2, \ldots n$) of $\mathfrak{F}$, and n scalars α^k, such that

$$x = \sum_{k=1}^{n} \alpha^k x_k .$$

$\mathcal{S}(\mathfrak{F})$ is a vector subspace of $\mathcal{H}$. Now let $\mathcal{H}(\mathfrak{F})$ denote the set of all $x \in \mathcal{H}$ which are limits of sequences of elements of $\mathcal{S}(\mathfrak{F})$; clearly, $\mathfrak{F} \subset \mathcal{S}(\mathfrak{F}) \subset \mathcal{H}(\mathfrak{F})$, and $\mathcal{H}(\mathfrak{F})$ is simply the "closure" of $\mathcal{S}(\mathfrak{F})$. It is easily seen that $\mathcal{H}(\mathfrak{F})$ is a Hilbert subspace of $\mathcal{H}$. If $\mathcal{H}'$ is a Hilbert subspace of $\mathcal{H}$ such that $\mathfrak{F} \subset \mathcal{H}'$, then necessarily $\mathcal{H}(\mathfrak{F}) \subset \mathcal{H}'$; in other words, $\mathcal{H}(\mathfrak{F})$ is the "smallest" Hilbert subspace containing $\mathfrak{F}$; we shall say that it is the Hilbert subspace *generated* by $\mathfrak{F}$. Of course, the case $\mathcal{H}(\mathfrak{F}) = \mathcal{H}$ is not excluded.

Let $\mathcal{H}'$ be a Hilbert subspace of $\mathcal{H}$, and x any fixed element of $\mathcal{H}$. Let

$$\mu = \underset{y \in \mathcal{H}'}{\text{g.l.b.}} \quad N_2(y - x) \geqq 0 ;$$

$\mu = 0$ if any only if $x \in \mathcal{H}'$. Suppose $\mu > 0$; let $\{y_k\}$ be a countable sequence of elements $y_k \in \mathcal{H}'$ such that if

$$\mu_k = N_2(y_k - x),$$

then

$$\lim_{k \to +\infty} \mu_k = \mu . \qquad (2,10,14)$$

The sequence $\{y_k\}$ is a Cauchy sequence; indeed, otherwise one can find a positive number α and arbitrary large integers m and n such that

$$N_2(y_m - y_n) > \alpha.$$

Put

$$z_k = \frac{1}{\mu_k}(x - y_k);$$

then

a) $N_2(z_m) = N_2(z_n) = 1$;

b) $N_2(z_m - z_n) = \left(\frac{1}{\mu_m} - \frac{1}{\mu_n}\right) x + \frac{1}{\mu_n} y_n - \frac{1}{\mu_m} y_m.$

Therefore, if m and n are sufficiently large,

$$N_2(z_m - z_n) > \frac{\alpha}{2\mu}.$$

Since $\mathcal{H}$ is uniformly convex (Theorem (2,10,5)), we have, for sufficiently large m and n,

$$N_2\left(\frac{z_m + z_n}{2}\right) = N_2\left[\frac{1}{2}\left(\frac{1}{\mu_m} + \frac{1}{\mu_n}\right) x - \frac{1}{2\mu_m} y_m - \frac{1}{2\mu_n} y_n\right] < 1 - \eta,$$

where η is some positive number. Putting

$$z = \frac{\mu_m \mu_n}{\mu_m + \mu_n}\left(\frac{1}{\mu_m} y_m + \frac{1}{\mu_n} y_n\right),$$

we obtain

$$N_2(x - z) < \frac{2\mu_m \mu_n}{\mu_n + \mu_n}(1 - \eta),$$

where the right-hand side is $< \mu$ if m and n are sufficiently large. Since $z \in \mathcal{H}'$, this is a contradiction; therefore $\{y_k\}$ must be a Cauchy sequence. Since $\mathcal{H}'$ is complete, there exists $y \in \mathcal{H}'$ such that

$$\lim_{k \to +\infty} y_k = y,$$

and obviously,

$$N_2(y - x) = \mu;$$

i.e., y minimizes $N_2(y - x)$ in $\mathcal{H}'$(the shortest distance from x to $\mathcal{H}'$). There is no other element $y' \in \mathcal{H}'$ such that $N_2(y' - x) = \mu$, for the existence of such a y' would contradict the uniform convexity of $\mathcal{H}$ (Theorem (2,10,15), as is easily verified. The element $y \in \mathcal{H}'$ whose existence and uniqueness have just been proved is called the *normal projection* of x on $\mathcal{H}'$; y = x if and only if $x \in \mathcal{H}'$.

GENERALIZATION (2,10,1). Let us say that a subset $\mathcal{G}$ of $\mathcal{H}$ is *convex* if for all x, $y \in \mathcal{G}$ and any nonnegative real numbers α, β such that $\alpha + \beta = 1$ we have

$$\alpha x + \beta y \quad .$$

In the previous proof the only properties of $\mathcal{H}'$used were that $\mathcal{H}'$ is *closed* and *convex;* the following generalization is immediate:

If $\mathcal{G}$ is any closed convex subset of $\mathcal{H}$, then for any $x \in \mathcal{H}$ there exists exactly one element $y \in \mathcal{G}$, called the *normal projection* of x on $\mathcal{G}$, which minimizes $N_2(x - y)$, i.e. such that

$$N_2(x - y) = \underset{z \in \mathcal{G}}{\text{g.l.b.}}\ N_2(x - z).$$

Linear mappings of a Hilbert space into a Hilbert space

Let $\mathcal{H}$ and $\mathcal{H}'$ be two Hilbert spaces over the same field (i.e., both complex or both real; to fix ideas we shall consider the complex case). Let A be a linear mapping of $\mathcal{H}$ into $\mathcal{H}'$ which maps any element $x \in \mathcal{H}$ onto a well-defined element $x' = A(x)$ of $\mathcal{H}'$. Put

$$N(A) = \underset{x \in \mathcal{H},\, x \neq \theta}{\text{l.u.b.}} \frac{N_2[A(x)]}{N_2(x)} . \qquad (2,10,15)$$

$N_2[a(x)]$ is the norm in $\mathcal{H}'$, while $N_2(x)$ is the norm in $\mathcal{H}$. $N(A)$ is a well-defined number; either $N(A) = +\infty$, or $N(A) < +\infty$. In the latter case A is said to be *bounded*. A general theorem on normed vector spaces states that A is continuous (with respect to the topologies induced in $\mathcal{H}$ and $\mathcal{H}'$ by their respective metrics or norms) if and only if A is bounded.

If there exists a one-to-one linear mapping A of $\mathcal{H}$ *onto* $\mathcal{H}'$, $\mathcal{H}$ and $\mathcal{H}'$ are said to be *isomorphic*. Suppose there exists a linear mapping U of $\mathcal{H}$ into $\mathcal{H}'$ with the following properties:

1) U is an epimorphism, i.e., $U(\mathcal{H}) = \mathcal{H}'$; in other words, fo
every $x' \in \mathcal{H}'$ there exists at least one $x \in \mathcal{H}$ such that

$$U(x) = x'.$$

2) For every x and y,

$$x \cdot y = U(x) \cdot U(y).$$

Thus U is a linear one-to-one mapping of $\mathcal{H}$ onto $\mathcal{H}'$ which preserves hermitian products and, in particular, norms and orthogonality. U is bounded, since clearly

$$N(U) = 1.$$

Then we shall say that U is *unitary*, or that it is an isomet
of $\mathcal{H}$ onto $\mathcal{H}'$. Two Hilbert spaces $\mathcal{H}$ and $\mathcal{H}'$ for which there exist an isometry of one onto the other are said to be *isometric*.

REMARK (2,10,1) *on isometries of Hilbert spaces*

Let $\mathcal{H}$ and $\mathcal{H}'$ be two Hilbert spaces and $\mathcal{Y}$ a vector subspace of $\mathcal{H}$, not necessarily complete, whose closure is $\mathcal{H}$. Let V be a linear mapping of $\mathcal{Y}$ into $\mathcal{H}'$. Put $\mathcal{Y}' = V(\mathcal{Y})$, so that V is a mappi
of $\mathcal{Y}$ *onto* $\mathcal{Y}'$, and let $\mathcal{Y}''$ be the closure of $\mathcal{Y}'$. Suppose that V preserves *hermitian products,* i.e., for any x and y in $\mathcal{Y}$ the *hermitian product* $V(x).V(y)$ of $V(x)$ and $V(y)$ in $\mathcal{H}'$ is equal to the *hermitian product* $x.y$ of x and y in $\mathcal{H}$.

V therefore preserves norms, hence is a one-to-one mapping o
onto $\mathcal{Y}'$. Consequently, V also preserves the Cauchy property: if x_k is a Cauchy sequence of elements of $\mathcal{Y}$ then $\{V(x_k)\}$ is a Cauchy sequence of elements of $\mathcal{Y}'$.

Let us define a mapping U of $\mathcal{H}$ into $\mathcal{Y}''$ as follows:

- if $x \in \mathcal{H}$ belongs to $\mathcal{Y}$, then $U(x) = V(x) \in \mathcal{Y}' \subset \mathcal{Y}''$;

- if $x \in \mathcal{H}$ does not belong to $\mathcal{Y}$, then there is a sequence $\{x_k\}$ of elements of $\mathcal{Y}$ such that $\lim\limits_{k\to+\infty} x_k = x$. $\{x_k\}$ is a Cauchy sequence, therefore so is $\{V(x_k)\}$; thus there exists an element x
of $\mathcal{Y}''$ which is the limit of $V(x_k)$ as $k \longrightarrow +\infty$. We define:

$$U(x) = \lim_{k \to +\infty} V(x_k) = x'$$

The element $U(x)$ is independent of the sequence $\{x_k\}$ used in the definition; for if $\{y_k\}$ is another sequence of elements of $\mathcal{Y}$ which converges to x, then

$$\lim_{k \to +\infty} ||y_k - x_k|| = 0.$$

Therefore, since V preserves norms,

$$\lim_{k \to +\infty} ||V(y_k - x_k)|| = \lim_{k \to +\infty} ||V(y_k) - V(x_k)|| = 0,$$

or

$$\lim_{k \to +\infty} V(y_k) = \lim_{k \to +\infty} V(y_k) = U(x).$$

Let x, y be any two elements of $\mathcal{H}$, and $\{x_k\}$, $\{y_k\}$ two sequences of elements of $\mathcal{Y}$ whose limits are x and y, respectively; then

$$x.y = \lim_{k \to +\infty} x_k.y_k,$$

$$U(x) = \lim_{k \to +\infty} V(x_k),$$

$$U(y) = \lim_{k \to +\infty} V(y_k).$$

Therefore

$$U(x).U(y) = \lim_{k \to +\infty} V(x_k).V(y_k)$$

and since V preserves hermitian products it follows that

$$V(x_k).V(y_k) = x_k.y_k.$$

Hence U is a linear one-to-one mapping of $\mathcal{H}$ onto $\mathcal{Y}''$ which preserves hermitian products (therefore also norms, orthogonality, Cauchy sequences, etc.). In other words, U is an isometry of $\mathcal{H}$ onto $\mathcal{Y}''$. If $\mathcal{Y}'' = \mathcal{H}'$, i.e., if $\mathcal{Y}'$ is dense in $\mathcal{H}'$, then U is an isometry of $\mathcal{H}$ onto $\mathcal{H}'$.

The above method is frequently used to construct isometries of Hilbert spaces. ∎

Note that the space C of complex numbers is a complex one-dimensional Hilbert space; the hermitian product of two complex numbers $z = \alpha + i\beta$ and $z' = \alpha' + i\beta'$ is by definition

$$z\,\overline{z}' = (\alpha\alpha' + \beta\beta') + i(\beta\alpha' - \alpha\beta'),$$

so that $M_2(z) = \alpha^2 + \beta^2$ and $N_2(z) = |z|$; C is normed by its hermitian product.

Let x^* be a linear mapping of $\mathcal{H}$ into C, i.e., a linear functional over $\mathcal{H}$; we shall use the notation $< x^*, x >$ to denote the number (element of C) onto which x^* maps $x \in \mathcal{H}$. The set of linear functions x^* over $\mathcal{H}$ is called the *dual* $\mathcal{H}^*$ of $\mathcal{H}$. The continuous (i.e. bounded) linear functionals form a subset $\mathcal{H}^*_f$ of $\mathcal{H}^*$; $\mathcal{H}^*_f$ is called the *strong dual* of $\mathcal{H}$. From the general theory of normed vector spaces it follows that $\mathcal{H}^*_f$ is a vector space, whose norm $N(x^*)$ is defined by (2,10,15), and it is complete with respect to this norm.

Projections

Let $\mathcal{H}'$ be a Hilbert subspace of $\mathcal{H}$; let P be the mapping of $\mathcal{H}$ into $\mathcal{H}'$ which maps $x \in \mathcal{H}$ onto its normal projection

$$y = P(x)$$

in $\mathcal{H}'$. The mapping P is called a projection; more precisely, P is the projection on $\mathcal{H}'$, or relative to $\mathcal{H}'$. The proofs of the following properties are left to the reader:

1) $x - P(x)$ is orthogonal to every element z of $\mathcal{H}'$.

This is easily proved in the particular case where $\mathcal{H}'$ is one-dimensional, using Theorem (2,10,4); the general case may then be treated, again using Theorem (2,10,4).

2) $P(x)$ is the only element of $\mathcal{H}'$ such that $x - P(x)$ is orthogonal to all elements of $\mathcal{H}'$.

3) P is a continuous (bounded) linear mapping of $\mathcal{H}$ into itsel and a continuous linear mapping of $\mathcal{H}$ onto $\mathcal{H}'$; $N(P) = 1$.

Thus, with every Hilbert subspace $\mathcal{H}'$ of $\mathcal{H}$ we associate a projection P, which is a *bounded linear mapping* and realizes a *normal and orthogonal* projection. ∎

Two subspaces of $\mathcal{H}$ are said to be orthogonal if every element of the one is orthogonal to every element of the other. Let $\mathcal{H}'$ be a Hilbert subspace of $\mathcal{H}$, and let $\mathcal{H}''$ be the set of elements of $\mathcal{H}$ othogonal to all the elements of $\mathcal{H}'$. $\mathcal{H}''$ always contains (at least) the element θ; except for the case $\mathcal{H}' = \mathcal{H}$, it contains at least one element $\neq \theta$. The reader will easily prove that

$\mathcal{H}''$ is a Hilbert subspace of $\mathcal{H}$.

Let P' and P'' be the projections on $\mathcal{H}'$ and $\mathcal{H}''$ respectively; then any $x \in \mathcal{H}$ can be expressed in the form

$$x = P'(x) + P''(x), \qquad (2,10,16)$$

where

α) $P'(x) \in \mathcal{H}'$, $P''(x) \in \mathcal{H}''$

β) $P'(x)$ and $P''(x)$ are orthogonal,

and (2,10,16) is the only such decomposition of x with properties α) and β).

This is summarized in the statement that $\mathcal{H}'$ and $\mathcal{H}''$ are *complementary.*

Canonical isomorphism

Consider an arbitrary $x \in \mathcal{H}$. Regarded as a function of $y \in \mathcal{H}$, $y.x$ is a linear functional; there therefore exists exactly one element x^* of $\mathcal{H}^*$ such that

$$\langle x^*, y \rangle \equiv y \cdot x \quad \text{for all} \quad y \in \mathcal{H}. \qquad (2,10,17)$$

Let us put

$$x^* = G(x),$$

where G denotes the mapping of $\mathcal{H}$ into $\mathcal{H}^*$ defined by (2,10,17). Schwarz's inequality (Theorem (2,10,1) immediately shows that $N[G(x)] \leqq N_2(x)$. Therefore $G(x) \in \mathcal{H}^*_f$ and G is in fact a mapping of $\mathcal{H}$ into $\mathcal{H}^*_f$.

Consider any nonzero element $x^* \in \mathcal{H}^*_f$. The set $\mathcal{H}''$ of all $y \in \mathcal{H}$ such that

$$< x^*, y > = 0,$$

is a Hilbert subspace of $\mathcal{H}$. Its complement $\mathcal{H}'$ is one-dimensional, as is easily proved. Hence we easily show that there is an element $x \in \mathcal{H}'$ such that $x^* = G(x)$. It follows that G is a one-to-one mapping of $\mathcal{H}$ into $\mathcal{H}_f^*$, with the properties:

1) $G(x + y) = G(x) + G(y)$ for x and $y \in \mathcal{H}$;

2) $G(\rho x) = \bar{\rho} G(x)$ for $x \in \mathcal{H}$, $\rho \in C$.

G is almost, though not quite (because of 2), a linear mapping; it is called the *canonical isomorphism* of $\mathcal{H}$ into $\mathcal{H}^*$.

Define the hermitian product of elements x^* and y^* in $\mathcal{H}^*$ by

$$x^* \cdot y^* = G^{-1}(y^*) \cdot G^{-1}(x^*). \tag{2,10,18}$$

One verifies that this is indeed a hermitian product, which norms $\mathcal{H}_f^*$. The resulting norm $N_2(x^*)$ of an element $x^* \in \mathcal{H}_f^*$ coincides with the norm $N(x^*)$ defined by (2,10,15), and moreover

$$N_2(x^*) = N_2[G^{-1}(x^*)].$$

Now it is known that $\mathcal{H}_f^*$ is complete with respect to the norm $N(x^*)$; therefore $\mathcal{H}_f^*$ is a Hilbert space, with the hermitian product (2,10,18). The canonical isomorphism G is a one-to-one quasi-linear mapping of $\mathcal{H}$ onto $\mathcal{H}_f^*$ which preserves norms, and even hermitian products, up to complex conjugation. Thus $\mathcal{H}$ and $\mathcal{H}_f^*$ are isometric, up to complex conjugation.

It may also be proved that $\mathcal{H}_f^*$ is separable if and only if $\mathcal{H}$ is separable.

Bases. A countable sequence $\{x_k\}$ (k = 1, 2, 3,...) of elements x_k of $\mathcal{H}$ is a *basis* if for any $x \in \mathcal{H}$ there exists a unique ordered sequence of scalars (elements of C) $\alpha^1, \alpha^2, \alpha^3, \ldots, \alpha^k, \ldots$ such that

$$x = \sum_{k=1}^{+\infty} \alpha^k x_k = \lim_{h \to +\infty} \sum_{k=1}^{h} \alpha^k x_k ; \tag{2,10,19}$$

the α^k are the *components* of x relative to the basis $\{x_k\}$.

Separability of $\mathcal{H}$ is of course a necessary condition for the existence of a basis; if $\{x_k\}$ is a basis, this implies that for any h the elements $x_1, x_2, \ldots, x_h$ are linearly independent.

A basis $\{x_k\}$ is *orthonormal* if the x_k are pairwise orthogonal unit vectors. If $\mathcal{H}$ has a basis, it must have at least one orthonormal basis. This theorem of Schmidt, well known for the case where $\mathcal{H}$ is finite-dimensional, is easily extended to the infinite-dimensional case.

Note that if $\{x_k\}$ is a countable sequence of pairwise orthogonal unit vectors $x_k \in \mathcal{H}$, then a necessary and sufficient condition for the convergence of the series $\sum_k a^k x_k$ (where $\alpha^k \in C$ for every k) is that

$$\sum_k |\alpha^k|^2 < +\infty.$$

This is a consequence of Theorem (2,10,4) and the fact that $\mathcal{H}$ is complete.

Suppose that $\mathcal{H}$ is separable. Then there exists a countable sequence $\{y_k\}$ in $\mathcal{H}$ which is dense in $\mathcal{H}$; i.e., every $x \in \mathcal{H}$ is the limit of some subsequence of $\{y_k\}$. Then it is easy to see that there exists a finite or countable sequence $\{z_k\}$ of elements $z_k \in \mathcal{H}$ such that

1) for any h, the elements $z_1, z_2, \ldots, z_h$ are linearly independent;

2) every $x \in \mathcal{H}$ is the limit of elements of the form

$$\sum_{k=1}^{h} \alpha^k z_k ,$$

where h is some positive integer, and the α^k are elements of C.

Starting from the sequence $\{z_k\}$, one employs the Schmidt process to obtain a sequence $\{x_k\}$ having the following properties:

1) the x_k are unit vectors;

2) the x_k are pairwise orthogonal;

3) every $x \in \mathcal{H}$ is the limit of elements of the form $\sum_{k=1}^{h} \alpha^k x_k$ (h finite; $\alpha^k \in C$).

Now let $\mathcal{H}'_h$ be the set of all elements of $\mathcal{H}$ of the form $\sum_{k=1}^{h} \alpha$ $\mathcal{H}'_h$ is a Hilbert subspace of $\mathcal{H}$, of finite dimension h; by consid ing the complement $\mathcal{H}''_h$ of $\mathcal{H}'_h$, the projections P'_h and P''_h on $\mathcal{H}'_h$ and and (2,10,16), it is easily proved that if we put

$$\alpha^k = x \cdot x_k \qquad (2,10,$$

then

$$x = \sum_k \alpha^k x_k ;$$

i.e., the sequence $\{x_k\}$ is an orthonormal basis of $\mathcal{H}$. Thus *ever separable Hilbert space* $\mathcal{H}$ *has at least one orthonormal basis.*

Let $\mathcal{H}$ and $\mathcal{H}'$ be two separable Hilbert spaces (both complex); let $\{x_k\}$ be an orthonormal basis of $\mathcal{H}$, and $\{x'_k\}$ an orthonormal basis of $\mathcal{H}'$. Every $x \in \mathcal{H}$ may be expressed uniquely in the form

$$x = \sum_k \alpha^k x_k,$$

where the right-hand side is a finite sum (if $\mathcal{H}$ is finite-dimensi al) or a convergent series. We have just seen that $\alpha^k = x.x_k$, an

$$\sum_k |\alpha^k|^2 < + \infty. \qquad (2,10.$$

Let us map $x \in \mathcal{H}$ onto the following element of $\mathcal{H}'$:

$$x' = U(x) = \sum_k \alpha^k x'_k ; \qquad (2,10,$$

this is meaningful, since by (2,10,21) the series (2,10,22) is co vergent. It is easy to see that the mapping U defined by (2,10,2 is an isometry: *any two separable Hilbert spaces are isometric.*

It clearly follows that Hilbert spaces have geometric properties generalizing those of Elementary Geometry, while remaining very similar to them. Moreover, if a Hilbert space is finite-dimensional, it is nothing but a euclidian or hermitian space, according to whether it is real or complex. On the other hand, there is an essential difference between an infinite-dimensional Hilbert space $\mathcal{H}$ and a finite-dimensional euclidean or hermitian space $\mathfrak{X}$, as we shall now show.

Any closed and bounded subset $e \subset \mathfrak{X}_n$ is *compact*; i.e., any infinite subset of e (if such a subset exists) contains a countable sequence $\{x_k\}$ of distinct elements which converges to some element x of e. In other words, finite-dimensional euclidian or hermitian spaces $\mathfrak{X}_n$ are *locally compact*.

However, if $\mathcal{H}$ is an infinite-dimensional Hilbert space there are subsets e of $\mathcal{H}$ which are both bounded and closed but not compact. For example, let e be the *ball* with center θ and radius 1, i.e., the set of elements $x \in \mathcal{H}$ such that

$$N_2(x) \leqq 1.$$

The set e is obviously bounded and closed (in the topology induced by the norm $N_2(x)$), but it is not compact; indeed, let $\{x_k\}$ be a countable sequence of pairwise orthogonal unit vectors of e (it follows from the preceding discussion that such sequences exist, and their terms are distinct). No element x of $\mathcal{H}$ can be a limit of the sequence $\{x_k\}$ or of any subsequence thereof. Thus:

Infinite-dimensional Hilbert spaces are not locally compact.

Example of a Hilbert space: the spaces $\mathcal{L}_2$

Let $\mathcal{U}$ be an arbitrary set of elements u, and m a measure over $\mathcal{U}$. Let $\mathcal{L}_2$ denote the set of all complex functions λ of $u \in \mathcal{U}$ such that (cf. II.7)

$$M_2(\lambda) = \int_{\mathcal{U}} |\lambda(u)|^2 \; m(du) < +\infty. \qquad (2,10,23)$$

It is easy to verify, using (among other things) Theorem (2,7,7), that

- $\mathcal{L}_2$ is a vector space.

- A hermitian product $\lambda\mu$ of elements $\lambda \in \mathcal{L}_2$ and $\mu \in \mathcal{L}_2$ is defined by

$$\lambda \cdot \mu = \int_{\mathcal{U}} \lambda(u)\, \overline{\mu(u)}\, m(du). \tag{2,10,24}$$

- $M_2(\lambda)$, defined by (2,10,23), is the hermitian norm derived from the hermitian product (2,10,24); the latter obviously defines in $\mathcal{L}_2$ the semi-norm

$$N_2(\lambda) = \sqrt{M_2(\lambda)}. \tag{2,10,25}$$

Consider the equivalence relation R in $\mathcal{L}_2$ defined by:

$\lambda \in \mathcal{L}_2$ is equivalent to $\mu \in \mathcal{L}_2$, if

$$N_2(\lambda - \mu) = \int_{\mathcal{U}} |\lambda(u) - \mu(u)|^2\, m(du) = 0,$$

and let $\tilde{\mathcal{L}}_2$ be the set of equivalence classes defined in $\mathcal{L}_2$ by R. As at the beginning of this section, $\tilde{\mathcal{L}}_2$ is seen to be *normed* by the hermitian product induced in $\tilde{\mathcal{L}}_2$ (see (2,10,6)) by the hermitian product (2,10,24) of $\mathcal{L}_2$; as before, the study of $\mathcal{L}_2$ reduces to that of $\tilde{\mathcal{L}}_2$.

We now proceed to the investigation of $\tilde{\mathcal{L}}_2$. To simplify the presentation, we shall not distinguish neither in notation nor in language between $\mathcal{L}_2$ and $\tilde{\mathcal{L}}_2$, retaining the notation $\mathcal{L}_2$. This means that two elements λ and μ of $\mathcal{L}_2$ are not considered *distinct* if $M_2(\lambda - \mu) = 0$.

$\mathcal{L}_2$ then becomes a vector space, normed by the hermitian product (2,10,24), with the norm $N_2(\lambda)$ defined by (2,10,25) and (2,10,23).

THEOREM (2,10,7) (Fischer-Riesz)

$\mathcal{L}_2$ is complete with respect to the norm (2,10,25).

Let $\{\lambda_k\}$ be a Cauchy sequence of elements of $\mathcal{L}_2$:

$$\lim_{k \to +\infty} N_2(\lambda_{k+h} - \lambda_k) = 0, \text{ uniformly in } h > 0. \qquad (2,10,26)$$

Let $\{\varepsilon_k\}$ and $\{\rho_k\}$ be two sequences of positive numbers such that

$$\sum_k \varepsilon_k < +\infty \,, \qquad (2,10,27)$$

$$\sum_k \frac{\rho_k}{\varepsilon_k^2} < +\infty \,. \qquad (2,10,28)$$

By (2,10,26), there exists an increasing sequence of integers n_k ($k = 1, 2, 3, \ldots$) such that

$$M_2(\lambda_{n_{k+h}} - \lambda_{n_k}) < \rho_k \quad \text{for any } h > 0 \,; \qquad (2,10,29)$$

in particular,

$$M_2(\lambda_{n_{k+1}} - \lambda_{n_k}) < \rho_k. \qquad (2,10,30)$$

By Theorem (2,8,3) there exists a function λ of $u \in \mathfrak{U}$ such that λ_{n_k} converges to λ almost everywhere as k tends to $+\infty$.

From (2,10,30) and the triangle inequality (2,10,4) it follows that $M_2(\lambda_{n_k})$ remains bounded as $k \longrightarrow +\infty$. By Theorem (2,8,5), it then follows that

$$M_2(\lambda) < +\infty,$$

or, in other words, $\lambda \in \mathfrak{L}_2$.

The triangle inequality implies

$$N_2(\lambda_h - \lambda) \leqq N_2(\lambda_h - \lambda_{n_k}) + N_2(\lambda_{n_k} - \lambda) \,,$$

and from (2,10,26) we deduce that

$$\lim_{h\to+\infty} N_2(\lambda_h - \lambda) = 0,$$

which proves Theorem (2,10,7). ∎

It then follows that the space $\mathfrak{L}_2$, with the hermitian product (2,10,24), is a Hilbert space.

We mention also, without proof:

THEOREM (2,10,8). If $\mathfrak{U}$ is a finite-dimensional vector or point space then $\mathfrak{L}_2$ is separable.

REMARK (2,10,3)

The space $\mathfrak{L}_2$ of point functions over $\mathfrak{U}$ satisfying (2,10,23) is often called the space of *square-integrable functions* (more precisely: absolutely square-integrable functions) over $\mathfrak{U}$.

In order that a sequence $\{\lambda_k\}$ of elements of $\mathfrak{L}_2$ converge in the norm N_2 defined by (2,10,24) and (2,10,23) to an element λ, it is necessary and sufficient that

$$\lim_{k\to+\infty} \int_{\mathfrak{U}} |\lambda_k(u) - \lambda(u)|^2 \, m(du) = 0.$$

Recalling (2,8,2), this is just convergence in the quadratic mean - already introduced for point functions in II.8: *The topology induced in the Hilbert space $\mathfrak{L}_2$ by its norm is that of convergence in the quadratic mean.*

IV. RANDOM ELEMENTS AND PROBABILITY LAWS

1. *Random elements*

Let $\mathfrak{X}$ be an arbitrary set (or space) of arbitrary elements x. An element X of $\mathfrak{X}$ whose "value" is determined by chance, i.e., by performing a trial from a certain class $\mathfrak{U}$, is called *a random element with values in* $\mathfrak{X}$[1].

Consider, for example, a point 0 of the earth's surface and some definite period of time in the future, say the year 1970 - 1971. Let $\Pi(t)$ be the atmospheric pressure at 0 at the instant t. $\Pi(t)$ cannot be forecast (only certain general characteristics of this function can be predicted); in other words, $\Pi(t)$ is random. Let $\mathfrak{X}$ be the set of all numerical real nonnegative functions of t for $1970 < t < 1971$; the "value" that chance assigns to $\Pi(t)$ is some element x of $\mathfrak{X}$, definite but actually unknown, i.e., a certain nonnegative function $\pi(t)$ of t for $1970 < t < 1971$. The function $\Pi(t)$ is a random element; more precisely, since it assumes values in a space of functions $\mathfrak{X}$, it is a *random function*[2].

A similar example is the voltage V(t) at every instant t of a given period $t_0 \leqq t \leqq t_1$ at the ends of an electric circuit containing resistances. This voltage is random owing to the random thermal agitation of the free electrons in the resistances.

The letter M on a given typewriter has a definite shape. However, if we type the letter M on a sheet of paper, the resulting image will not quite conform to the shape of the M on the typewriter (because of stains, irregularities, etc.), varying slightly with each repetition of the experiment; the image is a random element.

---

(1) Abbreviation, r.e.; German: zufällige Element; French: element aléatoire.

(2) Abbreviation: r.f.; German: zufällige Funktion; French: fonction aléatoire.

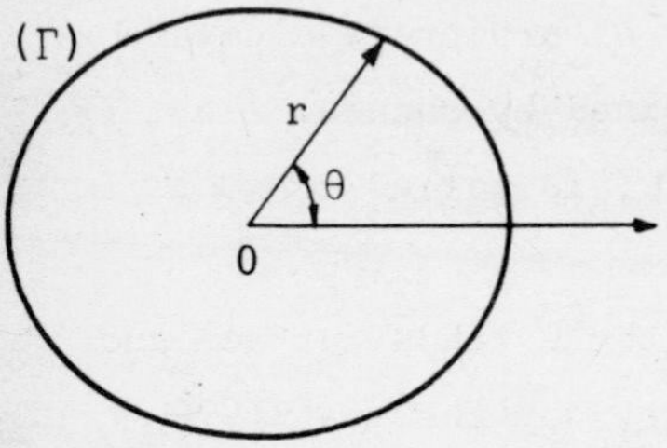

Fig. (2,11,1)

Let us select a random individual from a human population and consider his cranial contour (a plane horizontal section through a point say, half-way up the forehead); it is a closed curve Γ, as illustrate in Fig. (2,11,1), which varies according to the selected individual It may therefore be regarded as a random element. Γ may be describe in polar coordinates by an equatio

$$r = R(\theta)$$

where θ is the polar angle, r the radius-vector, and $R(\theta)$ a 2π-periodic function of θ. The random element Γ represented in this way is thus a random function $R(\theta)$. ∎

The value x assumed by a random element X with values in $\mathfrak{X}$ is determined by the actually performed trial $u \in \mathfrak{U}$. In other words, there exists a mapping

$$x = x(u) \in \mathfrak{X}, \quad u \in \mathfrak{U}$$

of $\mathfrak{U}$ into $\mathfrak{X}$ under which each $u \in \mathfrak{U}$ is assigned the value $x = x(u) \in \mathfrak{X}$ that X assumes if u is the actually performed trial. *concepts of a mapping of* $\mathfrak{U}$ *into* $\mathfrak{X}$ *and a random element* X *with values in* $\mathfrak{X}$ *are completely equivalent.*

Given a random element X with values in a space $\mathfrak{X}$, the *events associated* with X are defined as the events E whose occurrence or non-occurrence is a direct result of the value assumed by X. For instance, in the above example of atmospheric pressure $\Pi(t)$, the event E that $\Pi(t) \leqq \pi_0$ for all $1970 < t < 1971$, where π_0 is a given pressure, is associated with the random element $\Pi(t)$. It suffices to know the value $\pi(t)$ assumed by $\Pi(t)$ in order to know whether E has occurred, i.e., whether

$$\pi(t) \leqq \pi_0 \text{ for all } 1970 < t < 1971.$$

Probability law of a random element

Given an event E associated with the random element X, the elements x of $\mathfrak{X}$ may be divided into two sets (complementary with respect to $\mathfrak{X}$) e and $\check{e}$: e is the set of all $x \in \mathfrak{X}$ such that if X = x, E occurs, while $\check{e}$ is the set of all $x \in \mathfrak{X}$ such that if X = x, E does not occur. The set e and the event E are equivalent; any subset $e \subset \mathfrak{X}$ defines an event E associated with X, viz., the event

$$X \in e.$$

Now, as an event, E may be identified with a subset of $\mathfrak{U}$; this is obviously the subset

$$\{u \in \mathfrak{U} \mid x = x(u) \in e\} = x^{-1}(e),$$

so that

$$\Pr(E) = \Pr(X \in e) = \Pr[u \in x^{-1}(e)]. \qquad (2,11,1)$$

We shall employ the term *intrinsic study* of a random element X with values in $\mathfrak{X}$ for the study of its associated events. From the standpoint of its intrinsic study, a random element X with values in a space $\mathfrak{X}$ is equivalent to a game consisting of the random selection of an element from the space $\mathfrak{X}$. The set of the probabilities of all events E associated with X is called the *probability law*[1] of X (more rigorously, we should say: the set of the probabilities of all *probabilized* events E associated with X).

As before, let us introduce the set e:

$$e = \{x \in \mathfrak{X} \mid \text{if } X = x, \text{ E occurs}\}$$

and put

$$n(e) = \Pr(E) = \Pr(X \in e).$$

(1) Probability law or probability distribution: German: Wahrscheinlichkeits-Verteilung; French: loi de probabilité.

Note that if two subsets e_1 and e_2 of $\mathfrak{X}$ are disjoint, the corresponding subsets $x^{-1}(e_1)$ and $x^{-1}(e_2)$ of $\mathfrak{U}$ are disjoint. Generally speaking, the family of events associated with $\mathfrak{X}$ is a σ-algebra of subsets of $\mathfrak{U}$.

In view of Axiom II and (2,5,3) it is immediate that $n(e) \geqq 0$ $n(e) \leqq n(\mathfrak{X}) = 1$.

Axiom III (Total Probability) implies that if $e_1, e_2, \ldots, e_k$, are disjoint subsets of $\mathfrak{X}$ (finite or countable in number) then

$$n\left(\bigcup_k e_k\right) = \sum_k n(e_k).$$

The intrinsic study of X *is equivalent to that of its probability law.* We now see that the latter is determined by the set function $n(e)$, which is a bounded *measure* over $\mathfrak{X}$. Alternatively, it is given by the distribution of a bounded unit mass over $\mathfrak{X}$. *The probability law of a r.e. with values in $\mathfrak{X}$ and the distribution of a unit mass over $\mathfrak{X}$ are equivalent concepts.*

Induced measures

The preceding discussion may be formulated in more general terms. Let $\mathfrak{U}$ be an arbitrary set or space of elements u, and $m(E)$ $(E \subset \mathfrak{U})$ a measure over $\mathfrak{U}$. Let $\mathfrak{X}$ be an arbitrary space of elements x, and $x = x(u)$ any mapping of $\mathfrak{U}$ into $\mathfrak{X}$. For any subset $e \subset \mathfrak{X}$, define

$$n(e) = m\,[x^{-1}(e)].$$

This defines a function n over subsets e of $\mathfrak{X}$, obviously real and nonnegative. Now if e_1 and e_2 are any two disjoint subsets of $\mathfrak{X}$, then $x^{-1}(e_1)$ and $x^{-1}(e_2)$ are necessarily disjoint subsets of $\mathfrak{U}$; it follows that the complete additivity of m over $\mathfrak{U}$ implies the complete additivity of n over $\mathfrak{X}$. Consequently, n is a *measure* (over $\mathfrak{X}$); we shall call it the *measure* (corresponding to the measure m over) induced over $\mathfrak{X}$ by the mapping $x = x(u)$ of $\mathfrak{U}$ into $\mathfrak{X}$.

REMARK (2,11,1)

Thus, if X is the random element with values in $\mathfrak{X}$ defined by the mapping $x = x(u)$ of the space $\mathfrak{U}$ of trials into the space $\mathfrak{X}$, the probability law n(e) of X is just the measure, corresponding to the probability Pr(E) over $\mathcal{U}$, induced over $\mathfrak{X}$ by the mapping $x = x(u)$.

Change of variable in an integral

Let m(E) be a measure over a space $\mathcal{U}$ of elements u, and n(e) the measure corresponding to m, induced over $\mathfrak{X}$ by the mapping $x = x(u) \in \mathfrak{X}$. Let $\lambda(x)$ be a point function defined over $\mathfrak{X}$, integrable with respect to the measure n, i.e., the integral

$$\int_{\mathfrak{X}} \lambda(x)\, n(dx) \tag{2,11,2}$$

exists. Then:

THEOREM (2,11,1). The existence of the integral (2,11,2) implies that the point function $\lambda[x(u)]$ is integrable with respect to m, and

$$\int_{\mathfrak{X}} \lambda(x)\, n(dx) = \int_{\mathcal{U}} \lambda[x(u)]\; m(du). \tag{2,11,3}$$

The proof of this theorem, whose importance will become apparent later, proceeds as follows. First note that if $\lambda(x)$ is a simple function (cf. II.6) of x over $\mathfrak{X}$ then $\lambda[x(u)]$ is a simple function of u over $\mathcal{U}$, and under these conditions (2,11,3) is obvious. Moreover if, $\{\lambda_k(x)\}$ is a nondecreasing sequence of simple functions of x over $\mathfrak{X}$ converging to $\lambda(x)$, then $\lambda_k[x(u)]$ is a nondecreasing sequence of simple functions of u over $\mathcal{U}$ converging to $\lambda[x(u)]$. To prove Theorem (2,11,1) it then suffices to refer to the definition of the integral in II.6.

Random elements which are functions of others

Let X be a random element with values in a space $\mathfrak{X}$, and λ a mapping of $\mathfrak{X}$ into a space $\mathfrak{X}'$ which maps every $x \in \mathfrak{X}$ onto a well-defined element $x' = \lambda(x)$ of $\mathfrak{X}'$. Let X' be the random element with values in $\mathfrak{X}'$ which assumes the value $x' = \lambda(x) \in \mathfrak{X}'$ if x is the value assumed by X. We shall say that X' is a function of X, using the notation $X' = \lambda(X)$ for the relation whereby X' is derived from X. Note that X' is derived from X by a mapping of $\mathfrak{X}$ into $\mathfrak{X}'$ just as X was obtained above by a mapping of $\mathfrak{U}$ into $\mathfrak{X}$.

Let $m(e) = Pr(X \in e)$, where e is any subset of $\mathfrak{X}$, and $m'(e') = Pr(X' \in e')$ where e' is any subset of $\mathfrak{X}'$. The set of all $m'(e')$, where e' runs through all subsets of $\mathfrak{X}'$, is the probability law of X'. The following analogue of (2,11,1) is obvious:

$$m'(e') = Pr[X \in \lambda^{-1}(e')] = m[\lambda^{-1}(e')]. \qquad (2,11,4)$$

In other words, m' is the measure (corresponding to the measure m over $\mathfrak{X}$) induced over $\mathfrak{X}'$ by the mapping λ of $\mathfrak{X}$ into $\mathfrak{X}'$.

It is clear that the study of the random elements X' which are functions of the random element X is one aspect of the intrinsic study of X.

Random variables

A simple example of a random element is a random element X with values in the space R of real numbers, which we call a random variable.(1) A random variable is a real number whose value is determined by chance. In mathematical terms, it is a mapping $x = x(u)$ of the class of trials $\mathfrak{U}$ into R, i.e., a real function over $\mathfrak{U}$. The importance of random variables is clearly fundamental Among the various characteristics of a trial, we are particularly interested in those which are quantitative, or at least characterizable in numerical terms; a number X measuring (or characterizing) a quantity of this kind is a random variable.

EXAMPLE (2,11,1). In tossing a coin under the usual conditions, with Pr(head) = Pr(tail) = 1/2, we win 1 franc if the coin falls heads, nothing otherwise. Our gain X is a random variable with possible values 0 or 1, each with probability 1/2.

EXAMPLE (2,11,2). In rolling two dice, let i denote the number on the first die, and j the number on the second; then the sum $X = i + j$ is a random variable.

---

(1) Abbreviation: r.v.; German: zufällige Grösse, zufällige Veränderliche; French: variable aléatoire.

EXAMPLE (2,11,3). Referring to the beginning of this section, the value of the atmospheric pressure $\Pi(t)$ or the voltage $V(t)$ at any given instant t is a random variable.

EXAMPLE (2,11,4). Let X be a random variable and $x' = \lambda(x)$ a real function of $x \in R$, i.e., a mapping of R into R. Then $X' = \lambda(X)$ is a r.v., a function of the r.v. X; for instance, the square X^2 of X is a r.v. which is a function of X ($\lambda(x) = x^2$).

n-*dimensional random vectors and variables*

The space R is a one-dimensional euclidean vector space. A natural generalization of the previous notion of random variable is that of a random element X (or $\vec{X}$) with values in a real vector space $\mathfrak{X}_n$ of finite dimension n, where n is an arbitrary positive integer: note, by the way, that there is no loss of generality in assuming $\mathfrak{X}_n$ to be euclidean.

If we agree to call any element of a given vector space such as $\mathfrak{X}_n$ a vector then we may clearly say that X (or $\vec{X}$) is an *n-dimensional random vector*.

Let E_n be an affine point space associated with the vector space $\mathfrak{X}_n$, therefore also of dimension n. If 0 is an arbitrary fixed point of E_n, the equality $\overrightarrow{ON} = \vec{x}$ establishes a one-to-one correspondence between the points N of E_n and the vector x(or $\vec{x}$) of $\mathfrak{X}_n$. The study of the random *vector* X with values in $\mathfrak{X}_n$, and that of the random *point* N with values in E_n defined by $\overrightarrow{ON} = \vec{X}$, are therefore equivalent; in fact, except for linguistic details, there is almost no difference between them. In this subsection, at least, we shall continue to speak of the random *vector* X with values in the *vector* space $\mathfrak{X}_n$.

Let **b** be an arbitrary fixed basis in $\mathfrak{X}_n$ and let $X_1, X_2, \ldots, X_n$ (in this order) be the components of the vector X relative to this basis **b**. For each j ($1 \leqq j \leqq n$), X_j is a real random number; the study of X is therefore equivalent to that of the ordered n-tuple of random variables $\{X_1, X_2, \ldots, X_n\}$. Conversely, let

$\{X_1, X_2, \ldots, X_n\}$ be an ordered n-tuple of random variables. In thi order, $X_1, X_2, \ldots, X_n$ may always be interpreted as the components a random vector X in a real n-dimensional vector space $\mathfrak{X}_n$ relative to an arbitrary fixed basis $\mathbf{b}$, and the study of the n-tuple $\{X_1, X_2, \ldots, X_n\}$ of random variables an *n-dimensional random variat* Thus:

1) The random variables with values in R considered above are just one-dimensional random variables; throughout the sequel, unles otherwise stated, *the term "random variable" will always mean "one-dimensional random variable."*

2) The study of an n-dimensional random vector $\vec{X}$ is equivalent to that of the n-dimensional r.v. formed by its components relative to any fixed basis, and vice versa.

3) The concept of n-dimensional random variable (or vector) is natural and convenient whenever we wish to study n random variables $X_1, X_2, \ldots, X_n$ *simultaneously*.

EXAMPLE (2,11,5). The atmospheric pressure Π at a point 0 of the earth's surface at a definite future instant can be regarded, as before, as a random variable; the temperature θ can be also regarded as a random variable. Now Π and θ may be studied separately but, say, in order to determine whether there is some relation between these two quantities, we may wish to study Π and θ simultaneously, i.e., the two-dimensional random variable $\{\Pi, \theta\}$.

EXAMPLE (2,11,6). Let $\{X_1, X_2, \ldots, X_n\}$ be an n-dimensional r.v.; then the sum

$$Y = X_1 + X_2 + \cdots + X_n$$

of the X_j is a one-dimensional r.v., a function of the n-dimensiona r.v. $\{X_1, X_2, \ldots, X_n\}$. It corresponds to the mapping of $\mathfrak{X}_n$ into R defined by

$$y = x_1 + x_2 + \cdots + x_n .$$

REMARK (2,11,2)

Referring to a previous remark, note that, given an ordered n-tuple $\{X_1, X_2, \ldots, X_n\}$ of r.v., the X_j may also be interpreted as the *coordinates* of a random *point* N in the affine point space E_n associated with the vector space $\mathfrak{X}_n$ relative to the reference system $(0; \mathbf{b})$. The interpretation of the ordered n-tuple $\{X_1, X_2, \ldots, X_n\}$ of r.v. as a random vector $X \in \mathfrak{X}_n$ and its interpretation as a random point $N \in E_n$ are entirely equivalent; we may choose whichever is more convenient.

Lebesgue measure, absolutely continuous measures or mass distributions over $\mathfrak{X}_n$

We have remarked that without loss of generality $\mathfrak{X}_n$ may be assumed euclidean; we therefore adopt this assumption.

Let $\mathcal{B}$ be the smallest σ-algebra of subsets of $\mathfrak{X}_n$ containing all open or closed subsets of $\mathfrak{X}_n$; any set $e \subset \mathfrak{X}_n$ belonging to $\mathcal{B}$ will be called a *Borel set*. It is clear that $\mathcal{B}$ contains all polyhedra, all unions of finitely countably many disjoint polyhedra. We define:

1) a σ-algebra $\mathcal{L}$ of subsets of $\mathfrak{X}_n$, called the σ-algebra of *Lebesgue sets*;

2) a measure v(e) defined over $\{\mathfrak{X}_n, \mathcal{L}\}$, called *Lebesgue measure*.

$\mathcal{L}$ and the measure v have the following characteristic properties:

a) $\mathcal{B} \subset \mathcal{L}$;

b) if e is a polyhedron, then v(e) is the *volume* of e in the sense of Elementary Geometry;

c) if $e \in \mathcal{L}$ and $v(e) = 0$, then $e' \subset e$ implies $e' \in \mathcal{L}$ and consequently $v(e') = 0$; any measure with this property is called a *complete* measure;

d) $\mathcal{L}$ is the smallest σ-algebra (of subsets of $\mathfrak{X}_n$) possessing properties a), b), c).

It follows from the properties of the volume of a polyhedron that if $e_1, e_2 \subset \mathfrak{X}_n$ are congruent in the sense of Elementary Geomet then $e_1 \in \mathcal{L}$ implies $e_2 \in \mathcal{L}$, and $v(e_1) = v(e_2)$.

Lebesgue measure coincides with the elementary notion of volume. Like volume, Lebesgue measure is defined only up to a real positive proportionality factor, corresponding to the arbitrary choice of unit volume. Lebesgue measure is σ-bounded, but not bounded. ■

Let $\rho(x)$ be a real-valued *nonnegative* function of $x \in \mathfrak{X}_n$; for any subset e of $\mathfrak{X}_n$, consider the integral

$$\int_e \rho(x)\ v(dx) \qquad (2,11,$$

defined by the definitions of II.6 applied to the measure v(e) ove $\mathfrak{X}_n$. This integral is obviously nonnegative.

In the general case, this is what is called an L-integral or Lebesgue integral; it reduces to the classical multiple Riemann integral if e is sufficiently simple and $\rho(x)$ sufficiently regular say continuous.

Now put

$$m(e) = \int_e \rho(x)\ v(dx) \qquad (2,11,$$

This set function m(e) is obviously a measure or mass distribution over $\mathfrak{X}_n$; $\rho(x)$ is said to be its *density*(1).

Measures over $\mathfrak{X}_n$ defined in this way by (2,11,6) as integrals of a density are called *absolutely continuous*.

Let x be any fixed element of $\mathfrak{X}_n$, and e_α a family of sets depending on the parameter $\alpha > 0$, such that $x \in e_\alpha$ for any α and

$$\lim_{\alpha \to 0} v(e_\alpha) = 0.$$

Define $m(e_\alpha)$ by (2,11,6):

$$m(e_\alpha) = \int_{e_\alpha} \rho(y)\ v(dy).$$

---

(1) Abbreviation, d.; German: Dichte; French: densité.

It is known that, except perhaps for certain exceptional values of x, we have

$$\lim_{\alpha \to 0} \frac{m(e_\alpha)}{v(e_\alpha)} = \rho(x) , \tag{2,11,7}$$

provided the e_α satisfy certain conditions, which are however quite weak. For instance, (2,11,7) holds if the e_α are (n-dimensional) spheres with center x and radius α, or (n-dimensional) cubes with center x and side α, etc.

REMARK (2,11,3)

Consider an orthonormal basis of $\mathfrak{X}_n$; any element x of $\mathfrak{X}_n$ is determined by its components $x_1, x_2, \ldots, x_n$ relative to the basis. The function $\rho(x)$ is in fact a function $\rho(x_1, x_2, \ldots, x_n)$ of the n real variables $x_1, x_2, \ldots, x_n$ and the volume element $v(dx)$ is simply the product $dx_1, dx_2, \ldots, dx_n$ of the differentials of the x_j. Consequently, the integral (2,11,6) coincides with the multiple integral

$$m(e) = \underbrace{\int \cdots \int_e}_{n} \rho(x_1, x_2, \ldots, x_n) dx_1\, dx_2 \ldots dx_n . \tag{2,11,8}$$

Now let $\lambda(x) = \lambda(x_1, x_2, \ldots, x_n)$ be a numerical function of $x \in \mathfrak{X}_n$, i.e., a mapping of $\mathfrak{X}_n$ into the space R of real numbers (or the space C of complex numbers). Assume that λ is integrable over $\mathfrak{X}_n$ with respect to the measure m, and consider the integral

$$\int_{\mathfrak{X}_n} \lambda(x)\ m(dx).$$

By the definition (cf. II.6) of this integral, it is easy to see that

$$\int_{\mathfrak{X}_n} \lambda(x)\ m(dx) = \int_{\mathfrak{X}_n} \lambda(x)\ \rho(x)\ v(dx) , \tag{2,11,9}$$

where the integral on the right-hand side is the multiple integral

$$\underbrace{\int \cdots \int_{\mathfrak{X}_n}}_{n} \lambda(x_1, x_2, \ldots, x_n)\ \rho(x_1, x_2, \ldots, x_n)\ dx_1\, dx_2 \ldots dx_n .$$

This is a Lebesgue integral, though if $\lambda(x)$ and $\rho(x)$ are sufficiently regular (say continuous) it reduces to a classical multiple Riemann integral.

Let $\mathfrak{X}'_n$ be an n-dimensional vector space and λ a mapping of $\mathfrak{X}_n$ into $\mathfrak{X}'_n$ which maps every $x \in \mathfrak{X}_n$ onto a well-defined element $x' = \lambda(x)$ of $\mathfrak{X}'_n$. For any subset $e' \subset \mathfrak{X}'_n$, put

$$m'(e') = m[\lambda^{-1}(e')], \qquad (2,11,$$

so that m' is the measure over $\mathfrak{X}'_n$, corresponding to m, induced by the mapping λ of $\mathfrak{X}_n$ into $\mathfrak{X}'_n$.

Let us assume that λ is one-to-one; then, if the distribution $m(e)$ is absolutely continuous with density $\rho(x)$ in $\mathfrak{X}_n$ and λ is differentiable, $m'(e')$ is absolutely continuous in $\mathfrak{X}'_n$. Let $\rho'(x')$ be its density; to evaluate it, take orthonormal bases in $\mathfrak{X}_n$ and $\mathfrak{X}'_n$, and denote the components of x and x' by $x_1, x_2, \ldots,$ and $x'_1, x'_2, \ldots, x'_n$ respectively. The relation $x' = \lambda(x)$ is expressed analytically by a system of n equalities

$$x'_j = \lambda_j(x_1, \ldots, x_n) \quad (j = 1, 2, \ldots, n) \qquad (2,11,1$$

expressing each x'_j as a function of the x_k $(k = 1, 2, \ldots, n)$. Let

$$\Delta(x) = \Delta(x_1, \ldots, x_n)$$

denote the Jacobian with respect to the x_k of the functions

$$x'_j = \lambda_j(x_1, \ldots, x_n).$$

$\rho(x)$ and $\rho'(x')$ are functions of $x_1, \ldots, x_n$ and $x'_1, \ldots, x'_n$, respectively; in view of (2,11,8) and the fact that λ is one-to-one, formula (2,11,10) may be expressed by

$$\underbrace{\int \cdots \int_{\lambda(e)}}_{n} \rho'(x'_1, \ldots, x'_n)\, dx'_1 \ldots dx'_n = \underbrace{\int \cdots \int_e}_{n} \rho(x_1, \ldots, x_n)\, dx_1 \ldots dx_n \qquad (2,11,12)$$

Performing the change of variables defined by (2,11,11) in the left-hand side of (2,11,12) we obtain

$$\underbrace{\int \cdots \int_e}_{n} \rho'[\lambda_1(x_1, \ldots, x_n), \ldots, \lambda_n(x_1, \ldots, x_n)]\, |\Delta(x_1, \ldots, x_n)|\, dx_1 \ldots dx_n$$

Since equality (2,11,12) holds *for any* $e \subset \mathfrak{X}_n$, we obtain, provided the functions ρ, ρ', Δ are sufficiently regular (say continuous),

$\rho(x_1,\ldots, x_n) = \rho'[\lambda_1(x_1,\ldots, x_n),\ldots, \lambda_n(x_1,\ldots, x_n)]\,|\Delta(x ,\ldots, x_n)|$,

which we express in symbolic form:

$$\rho(x) = \rho'[\lambda(x)] \; |\Delta(x)| . \qquad (2,11,13)$$

Uniform measures or mass distributions

Let $A \subset \mathfrak{X}_n$ be any subset of $\mathfrak{X}_n$, and ρ a *constant*; let us define:

$$\rho(x) = \begin{cases} \rho & \text{if } x \in A, \\ 0 & \text{if } x \notin A ; \end{cases} \qquad (2,11,14)$$

the absolutely continuous measure $m(e)$ whose density $\rho(x)$ is given by (2,11,14) is said to be *uniform over* A.

Obviously,

$$m(A) = \rho\, v(A) ; \qquad (2,11,15)$$

$$m(e) = m(e \cap A) \leqq m(A) \quad \text{for any} \quad e \subset \mathfrak{X}_n ; \qquad (2,11,16)$$

in particular,

$$m(\mathfrak{X}_n) = m(A), \qquad (2,11,17)$$

and thus a measure uniform on A is bounded if and only if $v(A) < +\infty$.

REMARK (2,11,4)

Returning to the above notion of an absolutely continuous measure over $\mathfrak{X}_n$, defined by (2,11,6), we see that the property of absolute continuity of a measure $m(e)$ over $\mathfrak{X}_n$ is relative to another measure, here the measure $v(e)$. Part a) of Theorem (2,6,7) shows that if $m(e)$ is absolutely continuous, then $m(e) = 0$ for any set e such that $v(e) = 0$. These remarks are the motive for important generalizations.

Let m(e) and n(e) be two measures over an arbitrary space $\mathcal{U}$ of elements u, where e denotes any subset of $\mathcal{U}$. The measure n is said to be *absolutely continuous* with respect to the measure m if n(e) = 0 for any set e such that m(e) = 0. One can prove:

THEOREM (2,11,2) (Radon-Nikodym). If the measure n is absolutely continuous with respect to the measure m, then there exists a "point" function $\lambda(u)$ such that

$$n(e) = \int_e \lambda(u)\ m(du) \tag{2,11,18}$$

for any set e. ∎

λ is called the *density* of n with respect to m. The usual form of the Radon-Nikodym Theorem is rather more general than Theorem (2,11,2), in that the completely additive set functions m(e) and n(e) need not be measures. We shall prove neither the general Radom-Nikodym Theorem nor Theorem (2,11,2), confining ourselves to the following remarks.

1) Any point function μ which is m-equivalent to λ may clearly be substituted for λ in (2,11,18); conversely a function μ such that

$$\int_e \mu(u)\ m(du) = \int_e \lambda(u)\ m(du) \quad \text{for any } e$$

is necessarily equivalent to λ; in other words, n has a unique density with respect to m, up to m-equivalence.

2) Since $n(e) \geqq 0$ for any e, it follows that $\lambda(u) \geqq 0$ almost everywhere with respect to m.

Convergence of a sequence of measures

Let m be a measure, and consider a countable sequence of measures m_k (k = 1, 2, 3,...) defined over the same space $\mathcal{U}$ of elements u; under what conditions may it be said that the measure m_k converges to the measure m as $k \longrightarrow +\infty$? This question is very important, particularly in Probability Theory, as we shall see in Chapter VI (cf. VI.10 - 15).

The most natural definition, at first sight, is: m_k converges to m if, for any $e \subset \mathfrak{U}$,

$$\lim_{k \to +\infty} m_k(e) = m(e). \tag{2,11,19}$$

Unfortunately, as can be shown by examples, the condition that (2,11,19) hold for *any* $e \subset \mathfrak{U}$ is extremely restrictive, except in certain particular cases; we must be content to require (2,11,19) for sets e of a certain class $\mathcal{E}$. The problem then arises of defining and specifying $\mathcal{E}$, etc.; in short, the suggested definition is clearly complicated and difficult to handle.

We shall adopt another, less direct definition, which will prove more convenient. It seems natural to require that if $\lambda = \lambda(u)$ is a point function (real or complex), the convergence of m_k to m should imply

$$\lim_{k \to +\infty} \int_{\mathfrak{U}} \lambda(u)\, m_k(du) = \int_{\mathfrak{U}} \lambda(u)\, m(du). \tag{2,11,20}$$

This motivates the following formulation of the definition. Let $\mathfrak{F}$ be a class of point functions $\lambda = \lambda(u)$ over $\mathfrak{U}$; we shall say that m_k converges to m if (2,11,20) holds for any $\lambda \in \mathfrak{F}$. Of course, the class $\mathfrak{F}$ must be appropriately chosen. For instance, $\mathfrak{F}$ cannot contain all point functions. Indeed, a minimum requirement is that the integrals in both sides of (2,11,20) exist for any $\lambda \in \mathfrak{F}$. On the other hand, an overly restricted class $\mathfrak{F}$ is not desirable. Needless to say, different choices of $\mathfrak{F}$ may lead to equivalent definitions for the convergence of m_k to m.

We cannot discuss these questions thoroughly; having given the reader an idea of their nature, we shall limit ourselves to the definition in the following particular case.

Let $\mathfrak{U}$ be a finite-dimensional euclidean vector space or point space with the natural topology. Let $\mathfrak{F}$ be the family of point functions λ defined and continuous over $\mathfrak{U}$ which vanish outside a bounded set, and $\mathfrak{G}$ the family of point functions λ defined, continuous, and bounded over $\mathfrak{U}$; obviously, $\mathfrak{F} \subset \mathfrak{G}$.

Suppose that the measures m_k and m are bounded; we shall say that

1) m_k converges *weakly* to m as $k \longrightarrow +\infty$ if for any function $\lambda \in \mathcal{F}$

$$\lim_{k \to +\infty} \int_{\mathcal{U}} \lambda(u)\, m_k(du) = \int_{\mathcal{U}} \lambda(u)\, m(du), \quad \lambda \in \mathcal{F}.$$

2) m_k converges *completely* to m as $k \longrightarrow +\infty$ if for any function $\lambda \in \mathcal{G}$

$$\lim_{k \to +\infty} \int_{\mathcal{U}} \lambda(u)\, m_k(du) = \int_{\mathcal{U}} \lambda(u)\, m(du), \quad \lambda \in \mathcal{G}.$$

Complete convergence clearly implies weak convergence; on the other hand, m_k may converge weakly to m but not completely (cf. III.4 and Exercise 3.4). It is quite easy to show that m_k converges to m completely if and only if m_k converges to m weakly and in addition

$$\lim_{k \to +\infty} m_k(\mathcal{U}) = m(\mathcal{U}).$$

If the function λ is not bounded, it may happen that the integral $\int_{\mathcal{U}} \lambda(u) m_k(du)$ exists for every k and m_k converges to m completely, but nevertheless the integral $\int_{\mathcal{U}} \lambda(u) m(du)$ does not exist. Even if the integral does exist, we may have

$$\lim_{k \to +\infty} \int_{\mathcal{U}} \lambda(u)\, m_k(du) \neq \int_{\mathcal{U}} \lambda(u)\, m(du).$$

Suppose that, for every k, m_k is absolutely continuous with density ρ_k, and m_k converges to m completely; this does not imply that m is absolutely continuous. Even if m is absolutely continuous with density ρ, we may have

$$\lim_{k \to +\infty} \rho_k \neq \rho$$

[cf. Theorem (6,14,2)].

The reader may investigate for himself how to extend the above definitions of convergence to the case of unbounded measures ($\mathcal{U}$ being always finite-dimensional).

12. *Impossible and almost impossible events. Certain and almost certain events*

Since the probability law of an n-dimensional r.v. is just the distribution of a unit mass over $\mathfrak{X}_n$, this distribution or probability law may be absolutely continuous.

EXAMPLE (2,12,1). Consider a particle in Brownian motion in a plane Oxy. Its velocity $\vec{V}$ at a fixed instant t in the future is random and unpredictable. In particular, the angle Θ that $\vec{V}$ forms with Ox ($0 \leqq \Theta < 2\pi$) is random; it may assume any value in the interval $[0, 2\pi[$ and no one value is more probable than any other. It is thus reasonable to suppose that, for instance, $\Pr(\theta_1 \leqq \Theta < \theta_2)$ (where θ_1 and θ_2, $\theta_1 < \theta_2$, are two numbers in $[0, 2\pi]$) is proportional to $(\theta_2 - \theta_1)$. Since $\Pr(0 \leqq \Theta < 2\pi) = 1$, it follows that

$$\Pr(\theta_1 \leqslant \Theta < \theta_2) = \frac{\theta_2 - \theta_1}{2\pi}.$$

Thus Θ is a r.v. whose probability law is absolutely continuous, with the following density $\rho(\theta)$:

$$\rho(\theta) = \begin{cases} 0 & \text{if } \theta \notin [0, 2\pi[\\ 1/2\pi & \text{if } \theta \in [0, 2\pi[. \end{cases}$$

In other words, the probability law of Θ is the uniform law on $[0, 2\pi[$. ∎

Now let X be a r.v. (one-dimensional, to fix ideas) having an absolutely continuous probability law, with density f(x); let x be any fixed number, and Δx a positive number. Then for any $\Delta x > 0$ it follows from (2,11,8) that

$$\Pr(X = x) \leqq \Pr(x - \Delta x < X < x + \Delta x) = \int_{x-\Delta x}^{x+\Delta x} f(y)\, dy.$$

Hence, letting Δx tend to $+0$, we obtain

$$\Pr(X = x) = 0.$$

Thus the event $X = x$ has probability zero. We know (cf. II.5, result γ, p. 63) that any impossible event has probability zero. Now the event $X = x$ may indeed be impossible, but it may also be possible; thus, in Example (2,12,1) above, the event $\Theta = \theta$ always has probability zero, but if $\theta \notin [0, 2\pi[$ it is a *possible event*.

This remark justifies the following definition: Any event with probability zero will be called *almost impossible*. Any impossible event is almost impossible but an almost impossible event may be possible.

The distinction between impossible and almost impossible is therefore necessary in order to avoid logical contradictions. Moreover, we know that probability is a measure; *the concept of an almost impossible event is just the concept, already defined in II.6, of a set of measure zero.*

Nevertheless, for Probability Theory, which deals only with probabilities, the distinction between an event which is almost impossible and yet possible and an event which is actually impossible is immaterial. Neither should we make such a distinction from a practical point of view. Referring again to Example (2,12,1), though no purely logical considerations prevent Θ from assuming a preassigned value θ ($\theta \in [0, 2\pi[$), we do not expect this event to happen; moreover, no physical measurement allows us to verify that Θ has assumed the exact value θ.

The following dual definition is natural: Any event of probability one will be called *almost certain*[1] (or *almost sure*). Any certain (sure) event is therefore almost certain, but an almost certain event may not be certain.

---

(1) Almost certain, almost certainly (or: almost sure, almost surely): abbreviation, a.c. (or: a.s.); German: fast sicher; French: presque-sûr (or presque-certain).

By (2,5,9), the complementary event $\check{A}$ of an event A is almost impossible if A is almost certain, almost certain if A is almost impossible.

If A is a sure event and E an arbitrary event, we have $A \cap E = E$, and of course $\Pr(A \cap E) = \Pr(E)$. If A is almost certain, it may happen that $A \cap E \neq E$, but we always have $\Pr(A \cap E) = \Pr(E)$. Indeed, the complementary event $\check{A}$ of A is almost impossible, i.e. $\Pr(\check{A}) = 0$; since $\check{A} \cap E \subset \check{A}$, it follows from (2,5,6) that $\Pr(\check{A} \cap E) = 0$. Since A and $\check{A}$ are mutually exclusive, and $A \cup \check{A}$ is certain, the Axiom of Total Probability implies

$$\Pr(E) = \Pr(A \cap E) + \Pr(\check{A} \cap E) = \Pr(A \cap E).$$

Thus:

THEOREM (2,12,1). If A is an almost certain event, i.e. its complementary event $\check{A}$ is almost impossible, that is to say, if $\Pr(A) = 1$ and $\Pr(\check{A}) = 0$, then for any event E

$$\Pr(\check{A} \cap E) = 0, \quad \Pr(A \cap E) = \Pr(E).$$

REMARK (2,12,1)

It follows from Axiom III (cf. in particular (2,5,8) and (2,6,3)) that the union S of a finite or countable number of almost impossible events E is almost impossible (even if the events E are not pairwise mutually exclusive). ∎

Let us return to the case of a r.v. X with an absolutely continuous probability law and density $f(x)$. Let E_x denote the event $X = x$; the events E_x corresponding to different values of x are pairwise mutually exclusive, and each of them is almost impossible, while their union is obviously certain: it is the event that X assumes some value, and this is a logical consequence of the definition of X. Thus a union S of almost-impossible events may not be almost impossible; but this cannot happen unless S is the union of an *uncountable* set of almost impossible events.

We shall often use the adverb *almost certainly (almost surely)* to express the fact that an event is almost certain (almost sure);

these expressions are the translation, in the terminology of Probability Theory, of the expression "almost everywhere" used in Measure Theory (cf. II.6).

Equivalent random elements

The preceding discussion motivates the following definition, which coincides with that of m-equivalent mappings (cf. II.6) when the measure m is a probability law.

Two random elements X and Y are said to be equivalent if

$$\Pr(X = Y) = 1, \tag{2,12,}$$

i.e., if X and Y are almost certainly equal.

Two equivalent r.e. may not be identical from a strictly logical point of view, but in Probability Theory and in making decisions concerning our behavior, a r.e. may be replaced by any other equivalent r.e.

If two r.e. X and Y are equivalent, then there exists a space $\mathfrak{Z}$ such that X and Y assume the same value in $\mathfrak{Z}$ almost certainly; consequently, it is almost certainly legitimate to regard X and Y as two r.e. with values in the same space $\mathfrak{Z}$. Regarded as r.e. with values in $\mathfrak{Z}$, they have the same probability laws.

A non-random real number c is often called a *certain number*. If a r.v. X is such that P(X = c) = 1, we shall say that X is an *almost certain number*, or that X reduces to the certain number c, etc. In general, a nonrandom element c of a space $\mathfrak{X}$ will be calle a *certain element*, and if a r.e. X with values in $\mathfrak{X}$ is such that Pr(X = c) = 1 we shall say that it is an *almost certain element*, o that it reduces to the certain element c, or that it is equivalent to the certain element c, etc.

REMARK (2,12,2)

Recall that in defining random variables in II.11 as random elements with values in R, we meant the space R of real *finite* numbers: the numbers + ∞ and - ∞ were not regarded as elements of

R. Since a random variable X is by definition a r.e. with values in R, this implies that the event $X = \pm\infty$ is impossible.

Now let us imagine a real random number X for which the event $X = \pm\infty$ is *possible, though almost impossible*. It is quite legitimate to call such a random number a random variable, since in any event X is equivalent to a r.e. with values in R. There exist random numbers Y for which the event $Y = \pm\infty$ is not only possible, but of positive probability. In such cases we shall also say that Y is a random variable, but we shall call it a *degenerate* random variable. Consider, for instance, the r.v. X of Example (2,11,1), which assumes the value 0 with probability 1/2, and suppose that for some reason we are interested in its reciprocal $Y = 1/X$. There is a positive probability 1/2 that Y will be equal to ∞; it is thus a degenerate random variable.

Similarly, when we considered the n-dimensional vector space $\mathfrak{X}_n$ (or the point space E_n associated with $\mathfrak{X}_n$) with regard to r.e. with values in $\mathfrak{X}_n$ and measures (or mass distributions) and probability laws over $\mathfrak{X}_n$, it was understood that $\mathfrak{X}_n$ contains only *finite* vectors.

One consequence of this should be noted. Let $e_1, e_2, \ldots, e_k, \ldots$ be a countable sequence of subsets of $\mathfrak{X}_n$, such that $e_k \subset e_{k+1}$ and any vector x of $\mathfrak{X}_n$ (x is thus a *finite* vector) belongs to at least one, say e_{k_0}, of the e_k. This implies that x belongs to all e_k with $k \geqq k_0$, and moreover $\bigcup_k e_k = \mathfrak{X}_n$. Then, if m(e) is a measure (bounded or not) over $\mathfrak{X}_n$, we have, applying Theorem (2,6,1),

$$m(\mathfrak{X}_n) = \lim_{h \to +\infty} m(e_h). \qquad (2,12,2)$$

An n-dimensional random vector which is infinite with positive probability (or a random point which is at infinity with positive probability) will be called *degenerate*. The term n-dimensional random vector alone will imply that the random vector in question is infinite with probability zero. The same terminology will be used for n-dimensional r.v.

Random numbers usually represent the size of a physical magnitude and normally such magnitudes cannot be infinite. Therefore throughout the rest of this book we shall limit ourselves, unless otherwise stated, to the study of one-dimensional or n-dimensional random variables. On this basis, it should be quite easy to extend the discussion to degenerate r.v.

BIBLIOGRAPHIC NOTES FOR CHAPTER II

Regarding the philosophical problems of the foundations of Probability Theory and the resulting mathematical problems, a concise survey of the various points of view may be found in R. Fortet [2]; the reader may also consult J.M. Keynes [1] (which contains a large bibliography), E. Borel [3], A. Cournot [1], B. de Finetti [1], R. de Mises [1], H. Reichenbach [1], E. Tornier [1], J. Ville [1], R. Carnap [1], B.O. Koopman [1], [2], [3] (not including publications already listed in the bibliography of Keynes [1]), Actualités Scientifiques [1], [2] where several papers on the question are collected, and Actualités Scientifiques [3] where these papers are discussed, Actualités Scientifiques [4] (Proceedings of the International Congress on the Philosophy of Science (Probability Section), Paris, 1949). A modern subjectivist point of view may be found in L.J. Savage [1].

A relatively elementary treatment of the foundations and general elements of Probability Theory may be found in L. Bachelier [1], J. Bertrand [1], E. Borel [1], [2], E. Borel and R. Deltheil [1], W. Feller [3], B.V. Gnedenko and A. Khinchin [1], J. Bass [1]; a deeper mathematical treatment is given in J. Dubourdieu [1], M. Fréchet [1], A. Kolmogorov [1], P. Lévy [1], M. Loève [2], E. Parzen [1], A. Renyi [4], A. Tortrat [1].

A rigorous mathematical formulation of the Axioms of Probability may be found in A. Kolmogorov [1], M. Loève [2]; for a formulation of these axioms in terms of Boolean Algebra, see A. Kolmogorov [3]; an extended axiomatic system is proposed by A. Renyi [1].

Probabilities associated with systems of arbitrary, not necessarily mutually exclusive, events are investigated in M. Fréchet [2], M. Loève [1], and A. Renyi [2]; some considerations on the general concept of random elements and probability laws are given in Chapter III of A. Blanc-Lapierre and R. Fortet [1]; see also M. Fréchet [3], E. Mourier [1].

For Set Theory, Measure and Integration, the reader may consult C. de la Vallée - Poussin [1], Saks [1], P. Halmos [1], where the construction of the integral is based, as in our exposition, on the concept of measure; N. Bourbaki [1], [2], where the integral is defined by its properties as a linear functional over a function space; J. Neveu [1], where the algebraic characteristics of the family of measurable sets play a fundamental role.

For the theory of Hilbert spaces, the reader may consult the elementary book of P. Halmos [2], and the following more detailed books: M.H. Stone [1], N.I. Achieser and I.M. Glassmann [1], F. Riesz and B. Sz.-Nagy [1].

EXERCISES

2.1. Let $\mathfrak{F}$ be any family of events A; show (cf. II.1) that:

α) $\bigcup_{\mathfrak{F}} \check{A}$ is the event complementary to $\bigcap_{\mathfrak{F}} A$;

β) $\bigcap_{\mathfrak{F}} \check{A}$ is the event complementary to $\bigcup_{\mathfrak{F}} A$.

2.2. Let $A_1, A_2, \ldots, A_n$ be n events; set

$$p_j = \Pr(A_j) ;$$

$$p_{j_1 j_2} = \Pr\left(A_{j_1} \cap A_{j_2}\right) \qquad (j_1 < j_2 \leqq n) ;$$

. . . .

$$p_{j_1 j_2 \ldots j_s} = \Pr\left(\bigcap_{k=1}^{s} A_{j_k}\right) \qquad (j_1 < j_2 < \ldots < j_s \leqq n) ;$$

. . . .

P = Pr(exactly one of the A_j occurs).

Show that:

α) $$\Pr\left(\bigcup_{j=1}^{n} A_j\right) = \sum_j p_j - \sum_{j_1<j_2} p_{j_1 j_2} + \ldots$$

$$+ (-1)^{s-1} \sum_{j_1<j_2<\cdots<j_s} p_{j_1 j_2 \ldots j_s} + \ldots + (-1)^{n-1} p_{123\ldots n}$$

(Hint: for n = 2, this is formula (2,5,7); continue by induction on n).

β) $P = \sum_{j} p_j - 2 \sum_{j_1<j_2} p_{j_1 j_2} + \cdots + (-1)^{s-1} s \sum_{j_1<j_2<\cdots<j_s} p_{j_1 j_2 \cdots j_s}$

$+ \cdots + (-1)^{n-1} n \, p_{123\ldots n}$;

γ) $\Pr\left(\bigcap_{j=1}^{n} A_j\right) \geqq 1 - \sum_{j=1}^{n} \Pr(\check{A}_j)$.

2.3. Discuss the extension of parts α), β) of the preceding exercise to the case of a countable number of events $A_1, A_2, \ldots, A_n, \ldots$

2.4. *The Theorem of Borel-Cantelli*

Let $E_1, E_2, \ldots, E_k, \ldots$ be an infinite countable sequence of events, and A the event that infinitely many E_k occur. Show that A is the upper limit of the sequence $\{E_k\}$, defined in II.4 as

$$A = \overline{L} = \bigcap_{h} \left(\bigcup_{k>h} E_k\right).$$

Show that if $\sum_{k} \Pr(E_k) < +\infty$, then $\Pr(A) = 0$, i.e., almost certainly, at most finitely many E_k occur. [*Hint*: put $B_h = \bigcup_{k>h} E_k$; then $\Pr(B_h) \leqq \Pr(E_k)$ by (2,6,3); $A = \bigcap_h B_h$, consequently $\Pr(A) \leqq \Pr(B_h)$ for any h].

2.5. Let α and β be two positive real numbers such that $1/\alpha + 1/\beta = 1$. Let n be an arbitrary positive integer and $s_1, s_2, \ldots, s_n$ and $a_1, a_2, \ldots, a_n$ given positive real numbers.

Let $b_1, b_2, \ldots, b_n$ be n variables satisfying the conditions

$$\left(\sum_{k=1}^{n} s_k b_k^{\beta}\right)^{1/\beta} = B, \quad b_k \geqslant 0 \ (k = 1, 2, \ldots, n), \qquad (2.5 - 1)$$

where B is a given positive number.

Put

$$H = \sum_{k=1}^{n} s_k a_k b_k ;$$

which values of the b_k satisfying (2.5 - 1) maximize H? What is the value of the corresponding maximum of H?

[*Answer*: $b_k = B\left(\sum_{k=1}^{n} s_k a_k^{\alpha}\right)^{-1/\beta} a_k^{1/\beta - 1}$; maximum of

$H = \left(\sum_{k=1}^{n} s_k a_k^{\alpha}\right)^{1/\alpha} \cdot B$].

Hence derive a proof of Hölder's inequality (2,7,2) in the case where λ and μ are simple functions (cf. II.6); then, using Lemma (2,6,1), prove (2,7,2) in the general case.

2.6. Let α be a real number > 1; let n be an arbitrary positive number, $s_1, s_2, \ldots, s_n$ and $a_1, a_2, \ldots, a_n$ given positive real numbers, $b_1, b_2, \ldots, b_n$ n real variables satisfying the conditions

$$\left(\sum_{k=1}^{n} s_k b_k^{\alpha}\right)^{1/\alpha} = B,\ b_k \geqslant 0\ (k = 1, 2, 3, \ldots, n), \qquad (2.6 - 1)$$

where B is a given positive number. Put

$$S = \left[\sum_{k=1}^{n} s_k (a_k + b_k)^{\alpha}\right]^{1/\alpha};$$

which values of the b_k satisfying (2.6 - 1) maximize S?

What is the value of the corresponding maximum of S?

[*Answer*: $b_k = B\left(\sum_{h=1}^{n} s_h a_h^{\alpha}\right)^{-1/\alpha} a_k$; maximum of

$S = \left(\sum_{k=1}^{n} s_k a_k^{\alpha}\right)^{1/\alpha} + B$].

Hence derive a proof of Minkowski's inequality (2,7,7) in the case where λ and μ are simple functions (cf. II.6); then, using Lemma (2,6,1), prove (2,7,7) in the general case.

CHAPTER III

DISTRIBUTION FUNCTIONS

CHAPTER III

DISTRIBUTION FUNCTIONS

1. *Distribution functions*

In view of the preceding discussion, a natural starting-point for Probability Theory is the intrinsic study of one-dimensional random variables, followed by random vectors or n-dimensional random variables for any n.

The intrinsic study of an n-dimensional random vector X with values in a real n-dimensional vector space $\mathfrak{X}_n$ is that of its probability law, which is equivalent to a bounded measure or mass distribution over $\mathfrak{X}_n$. We are already acquainted (cf. II.9 and II.11) with the mathematical description of a measure, bounded or not, in two particular cases:

1) totally discontinuous measure;

2) absolutely continuous measure.

However, there exist mass distributions which are neither totally discontinuous nor absolutely continuous (cf. III.2, singular distribution functions). Moreover, we may have to introduce a mass distribution without knowing in advance whether it is totally discontinuous, absolutely continuous, or of some other type. We therefore need an analytical description which is applicable to any mass distribution over $\mathfrak{X}_n$, irrespective of its specific nature.

A description of this kind will now be introduced - distribution functions[1]. Before further discussion, it must be emphasized that *as a rule* (with occasional exceptions) we shall not make

---

[1] Abbreviation, d.f.; German; Verteilungs-Funktion; French; fonction de répartition.

explicit mention of the σ-algebras of sets over which the measures or mass distributions under discussion are defined. Every set figuring in the discussion will be assumed to belong to the appropriate σ-algebra; this may follow either from a tacit assumption or from a proof which is omitted.

Now let $\{x_1, x_2, \ldots, x_n\}$ be an n-tuple of real variables, each assuming values in $(-\infty, +\infty)$. An *n-dimensional distribution function* of the variables $x_1, x_2, \ldots, x_n$ is any function $F(x_1, x_2, \ldots, x_n)$ of these variables which is *finite, real, and monotone nondecreasing* in each of the variables x_j; i.e., for any j $(j = 1,2,\ldots,n)$, and any fixed $x_1, \ldots, x_{j-1}, x_{j+1}, \ldots, x_n$ the function $F(x_1, \ldots, x_{j-1}, x_j, x_{j+1}, \ldots, x_n)$ is monotone nondecreasing in x_j over $(-\infty < x_j < +\infty)$.

The introduction of distribution functions is motivated by the following remark. With any measure or mass distribution m(e) over $\mathfrak{X}_n (e \subset \mathfrak{X}_n)$ we can associate an n-dimensional d.f.; moreover, this may be done in more than one way. Let us prove this statement, assuming first that the measure is *bounded.*

Let $\mathfrak{b}$ be any fixed basis of $\mathfrak{X}_n$, and let $\{y_1, y_2, \ldots, y_n\}$ denote the components relative to this basis of an arbitrary vector y of $\mathfrak{X}_n$. With any n-tuple $x_1, x_2, \ldots, x_n$ we associate the subset $e(x_1, x_2, \ldots, x_n)$ of $\mathfrak{X}_n$ defined by $e(x_1, x_2, \ldots, x_n) = \{y \in \mathfrak{X}_n | y_1 < x_1, y_2 < x_2, \ldots, y_n < x_n\}$. Now put

$$F(x_1, x_2, \ldots, x_n) = m\,[e(x_1, x_2, \ldots, x_n)], \qquad (3,1,1)$$

where the right hand side is the mass assigned by the mass distribution m to the subset $e(x_1, x_2, \ldots, x_n)$ of $\mathfrak{X}_n$.

It follows immediately [cf. (2,5,4)] that $F(x_1, x_2, \ldots, x_n)$ is finite, real, and a monotone nondecreasing function of each of the variables x_j. It is therefore an n-dimensional d.f. of the variables x_j. The d.f. defined by (3,1,1) will be called *the distribution function associated with the (bounded) measure m.*

REMARK (3,1,1)

There are other ways of associating a d.f. with a measure. First, the choice of the basis $\mathfrak{b}$ in $\mathfrak{X}_n$ is arbitrary; but even when the basis has been selected once and for all there are still several ways of constructing a d.f. based on m. For instance, let $\bar{e}(x_1,x_2,\ldots,x_n)$ denote the subset of $\mathfrak{X}_n$ defined by

$$\bar{e}(x_1, x_2, \ldots, x_n) = \{y \in \mathfrak{X}_n \mid y_1 \leqq x_1,\ y_2 \leqq x_2, \ldots, y_n \leqq x_n\},$$

and define

$$G(x_1, x_2, \ldots, x_n) = m\,[\bar{e}(x_1, x_2, \ldots, x_n)]. \tag{3,1,2}$$

Like $F(x_1,x_2,\ldots,x_n)$, $G(x_1,x_2,\ldots,x_n)$ is an n-dimensional d.f. However, F and G are generally distinct. Nevertheless we intuitively feel (and this is in fact correct) that they provide equivalent descriptions of the measure m; this remark will lead to the concept of equivalent d.f. later in this section.

A d.f. may be associated with a measure m even if m is not bounded; it is sufficient to require that $m(e) < +\infty$ for any bounded subset $e \subset \mathfrak{X}_n$. In fact, all we require is a slight modification of the procedure defined by (3,1,1). We demonstrate this in the simple case n = 1; $\mathfrak{X}_1$ may then be identified with the space R of real numbers. For $x \in R$, we define

$$F(x) = \begin{cases} m([0,x[) & \text{if } x > 0, \\ 0 & \text{if } x = 0, \\ -m([x,0[) & \text{if } x < 0. \end{cases}$$

The assumption that $m(e) < +\infty$ for any bounded subset $e \subset R$ implies that F(x) is a d.f.

REMARK (3,1,2)

It may already be conjectured (and this will be verified later that the d.f. $F(x_1,x_2,\ldots,x_n)$ constructed on the basis of a measure m(e) over $\mathfrak{X}_n$ by (3,1,1) (or a similar procedure) is a convenient tool for defining and investigating the measure m(e) itself. However, this construction presupposes the choice of a definite basis **b** in $\mathfrak{X}_n$, and this choice is essentially arbitrary. In other words $F(x_1,x_2,\ldots,x_n)$ is not an *intrinsic* tool for defining and investigating m(e); a more appropriate designation would be "the d.f. base on m(e) (e.g. associated with m(e) by (3,1,1)) *relative to* **b**." Thi non-intrinsic character, which sometimes proves inconvenient, must always be kept in mind. In certain cases, among all possible bases in $\mathfrak{X}_n$ one specific **b** may be particularly suitable, e.g., because of the concrete meaning of the corresponding components x_j. It may then be especially expedient and interesting to utilize the d.f. $F(x_1, x_2, \ldots, x_n)$ based on m(e) relative to this basis **b**.

Basic properties of distribution functions

The monotonicity of d.f. immediately implies the existence of certain limits. We begin with the case n = 1; let F(x) be a one-dimensional d.f. of the variable x, i.e., a monotone nondecreasing function of x over $(-\infty, +\infty)$:

$$F(x') \leqq F(x'') \text{ if } x' < x''. \tag{3,1,3}$$

Hence the existence of the following limits for every x:

$$\lim_{\varepsilon \to +0} F(x + \varepsilon) = F(x + 0) ;$$

$$\lim_{\varepsilon \to +0} F(x - \varepsilon) = F(x - 0) ;$$

$$\lim_{\varepsilon \to +\infty} F(x) = F(+\infty) ;$$

$$\lim_{x \to -\infty} F(x) = F(-\infty) .$$

Moreover, for every x,

$$-\infty \leqq F(-\infty) \leqq F(x - 0) \leqq F(x) \leqq F(x + 0) \leqq F(+\infty) \leqq +\infty . \tag{3,1,4}$$

The d.f. F(x) remains bounded over any bounded interval (a,b). We shall say that it is

bounded below if $F(-\infty) > -\infty$,

bounded above if $F(+\infty) < +\infty$,

bounded if it is bounded above and below, i.e., if

$$-\infty < F(-\infty) \leqq F(+\infty) < +\infty. \tag{3,1,5}$$

The difference $F(+\infty) - F(-\infty)$ is known as the *total variation* of F.

These statements and definitions can be extended to any n. If $F(x_1, x_2, \ldots, x_n)$ is an n-dimensional d.f., the limits $F(-\infty, x_2, \ldots$ $\ldots, x_n)$, $F(+\infty, x_2, \ldots, x_n)$, $F(-\infty, -\infty, x_3, \ldots, x_n)$, $F(-\infty, +\infty,$

$x_3,\ldots,x_n),\ldots,F(-\infty,-\infty,\ldots,-\infty)$, $F(+\infty,+\infty,\ldots,+\infty)$ exist; F is said to be

bounded below if

$$F(-\infty,\ -\infty,\ldots,\ -\infty) > -\infty;$$

bounded above if

$$F(+\infty,\ +\infty,\ldots,\ +\infty) < +\infty;$$

bounded if it is bounded both below and above.

The limits $F(x_1-0,\ x_2,\ldots,\ x_n)$, $F(x_1+0,\ x_2,\ldots,\ x_n)$, $F(x_1+0,\ x_2-x_3,\ldots,\ x_n)$, etc., also exist.

Case of a distribution function associated with a measure

Let m(e) be a bounded measure over $\mathfrak{X}_n$, let $F(x_1,x_2,\ldots,x_n)$ be the n-dimensional d.f. associated with it by (3,1,1); for any integer $k > 0$ put

$$e_k = e(x_1 - 1/k,\ x_2 - 1/k,\ldots,\ x_n - 1/k).$$

Obviously,

$$e_k \subset e_{k+1} \quad (k = 1,\ 2,\ 3,\ldots)\ ;\ \bigcup_{k=1}^{+\infty} e_k = e(x_1,\ x_2,\ldots,\ x_n).$$

It suffices to apply Theorems (2,6,1), (2,6,2) and Remark (2,9,2) in order to establish:

THEOREM (3,1,1). If m(e) is a bounded measure or mass distribution over $\mathfrak{X}_n$ and $F(x_1,x_2,\ldots,x_n)$ is the distribution function associated with it by (3,1,1), then

1)
$$\lim_{\varepsilon_1\to+0,\varepsilon_2\to+0,\ldots,\varepsilon_n\to+0} F(x_1-\varepsilon_1,\ x_2-\varepsilon_2,\ldots,\ x_n-\varepsilon_n)$$
$$= F(x_1-0,\ x_2-0,\ldots,\ x_n-0) = F(x_1,\ x_2,\ldots,\ x_n) \qquad (3,1,6)$$

for any finite $x_1,x_2,\ldots,x_n$. This remains true if some of the x_j are $+\infty$, for instance:

$$\lim_{\varepsilon_2\to+0,\ldots,\varepsilon_n\to+0} F(+\infty,\ x_2-\varepsilon_2,\ldots,\ x_n-\varepsilon_n)$$
$$= F(+\infty,\ x_2-0,\ldots,\ x_n-0) = F(+\infty,\ x_2,\ldots,\ x_n).$$

2) $F(+\infty,+\infty,\ldots,+\infty) = m(\mathfrak{X}_n)$. (3,1,7)

3) If any one of the x_j tends to $-\infty$, then $F(x_1,x_2,\ldots,x_n)$

tends to 0; it follows, for instance, that

$$F(-\infty, x_2, \ldots, x_n) = 0, \; F(-\infty, +\infty, x_3, \ldots, x_n) = 0, \text{ etc.} \qquad (3,1,8)$$

Measure or mass distribution defined by a distribution function

Let $G(x_1, x_2, \ldots, x_n)$ be any given n-dimensional d.f.; let $\mathfrak{X}_n$ be any real n-dimensional vector space, with a fixed but arbitrary basis **b**. We shall now see that, in a sense to be made precise, the function G defines a measure n(e) over $\mathfrak{X}_n$, which is uniquely determined once b has been chosen. To simplify the notation we limit ourselves to n = 2, but the results are clearly valid for any n.

Thus, let $\mathfrak{X}_2$ be a real 2-dimensional vector space with some given basis **b**; it will be convenient to use the notation x, y for the two components (relative to **b**) of any vector of $\mathfrak{X}_2$. A *rectangular domain* is any subset of $\mathfrak{X}_2$ of the form

$$\alpha \leqq x < \beta, \; \gamma \leqq y < \delta, \qquad (3,1,9)$$

where $\alpha, \beta, \gamma, \delta$, $(\alpha < \beta, \gamma < \delta)$ are arbitrary numbers (possibly infinite).

Let e be the rectangular domain defined by (3,1,9), and let e_1, e_2, e_3, e_4 be the sets defined by

$$\begin{aligned} e_1&: \; x < \beta, \; y < \delta; \\ e_2&: \; x < \beta, \; y < \gamma; \\ e_3&: \; x < \alpha, \; y < \delta; \\ e_4&: \; x < \alpha, \; y < \gamma. \end{aligned}$$

Then:

$$e_1 = e \cup e_2 \cup (e_3 - e_4),$$

and $e, e_2, (e_3 - e_4)$ are pairwise disjoint.

Then if m(e) is a measure over $\mathfrak{X}_2$ such that $m(e) < +\infty$ for any bounded subset e of $\mathfrak{X}_2$, and F(x,y) is the 2-dimensional d.f. associated with it by (3,1,1), the complete additivity of m and Theorem (3,1,1) imply:

$$m(e) = F(\beta - 0, \delta - 0) - F(\alpha - 0, \delta - 0) - F(\beta - 0, \gamma - 0) + F(\alpha - 0, \gamma - 0);$$

This suggests the following argument. If $e \subset \mathfrak{X}_2$ is a rectangular domain of type (3,1,9), and

$$\Delta(e) = G(\beta - 0, \delta - 0) - G(\alpha - 0, \delta - 0) - G(\beta - 0, \gamma - 0) + G(\alpha - 0, \gamma - 0), \qquad (3,1,10)$$

then the right-hand side is necessarily $\geqq 0$, as is immediately verified.

Now let $\mathcal{B}$ denote the σ-algebra of Borel subsets of $\mathfrak{X}_2$.

THEOREM (3,1,2). There exists a unique measure n(e) over $\mathcal{B}$ such that

$$n(e) = \Delta(e) \text{ for any rectangular domain } e. \qquad (3,1,11)$$

The measure n(e) whose existence and uniqueness are established by Theorem (3,1,2) will be called the *measure defined by* G.

We shall accept the first part of Theorem (3,1,2) without proof, namely, we assume the existence of *at least one* measure n(e) satisfying (3,1,11) for any rectangular domain e. To prove uniqueness, suppose there exist two measures n(e) and n'(e) satisfying (3,1,11) for any rectangular domain e. We must show that n(e) = n'(e) for any $e \in \mathcal{B}$.

1) By definition, $n(e) = n'(e) = \Delta(e)$ for any e which is a rectangular domain of the type (3,1,9).

2) Assume that e is the union of finitely or countably many pairwise disjoint rectangular domains e_k of the type (3,1,9); by virtue of the complete additivity of n and n' we have

$$n(e) = \sum_k n(e_k) = \sum_k \Delta(e_k) ;$$

$$n'(e) = \sum_k n'(e_k) = \sum_k \Delta(e_k) ;$$

therefore n'(e) = n(e).

This is true in particular for any open set e, since any open set e is the union of finitely or countably many pairwise disjoint rectangular domains e_k.

3) If e is a set whose complement $\check{e}$ is the countable union of

pairwise disjoint rectangular domains, then n(e) = n'(e); indeed, by complete additivity,

$$n(e) = n(\mathfrak{X}_2) - n(\check{e}), \quad n'(e) = n'(\mathfrak{X}_2) - n'(\check{e}) \;;$$

but $n(\mathfrak{X}_2) = n'(\mathfrak{X}_2)$ by 1), and $n(\check{e}) = n'(\check{e})$ by 2). If $n(\mathfrak{X}_2) = n'(\mathfrak{X}_2) < +\infty$, this proves that n(e) = n'(e); but the equality n(e) = n'(e) may also be proved without serious difficulty if $n(\mathfrak{X}_2) = n'(\mathfrak{X}_2) = +\infty$.

Now this applies in particular to closed sets e, since the complement $\check{e}$ of a closed set is open.

4) Finally, one proves the equality n(e) = n'(e) for arbitrary e, where e is neither a rectangular domain, a countable union of disjoint rectangular domains, nor the complement of such a set, but belongs nevertheless to $\mathcal{B}$. The proof, however, is more subtle, and we shall omit it. We only hint that it utilizes, among other things, the quite natural idea of "approaching" e via simple combinations of open or closed sets.

REMARK (3,1,3)

Note that from the foregoing argument one can evolve a proof that there is at least one measure n(e) over $\mathfrak{X}_2$ satisfying (3,1,11) for any rectangular domain e.

REMARK (3,1,4)

Let $\mathfrak{F}$ be a σ-algebra of subsets of $\mathfrak{X}_2$ such that $\mathfrak{F} \supset \mathcal{B}$. There may exist a measure n'(e) over $\mathfrak{F}$ which also satisfies

$$n'(e) = \Delta(e)$$

for any rectangular domain e, therefore $n'(e) \equiv n(e)$ for any $e \in \mathcal{B}$. However, for an $e \in \mathfrak{F}$ and $\notin \mathcal{B}$, n'(e) is defined, while the above construction defines n(e) only over $\mathcal{B}$. If $\mathfrak{F}$ is not given, the d.f. G is not sufficient to determine it. Thus, if a measure m is given and a d.f. G constructed with a view to its description, the two concepts are not completely equivalent - their equivalence is limited to subsets $\in \mathcal{B}$. This restriction will be of no consequence for the sequel; we shall in fact use the d.f. only for problems concerning $\mathcal{B}$.

REMARK (3,1,5)

By (3,1,11) and (3,1,10)

$$n(\mathfrak{X}_2) = G(+\infty, +\infty) - G(-\infty, +\infty) - G(+\infty, -\infty) + G(-\infty, -\infty). \quad (3,1,12)$$

Thus the measure n(e) defined by the d.f. G is bounded if and only if the right-hand side of (3,1,12) is finite, i.e., as is easily seen, if and only if G is bounded.

An immediate corollary of Theorem (3,1,2) is:

THEOREM (3,1,3). A bounded measure m(e) over $\mathfrak{X}_2$ is uniquely determined (over $\mathcal{B}$) by the distribution function F associated with it by (3,1,1): it is precisely the measure "defined by F" in the above sense.

Equivalent distribution functions

Remark (3,1,1) explains and justifies the following definition. Two 2-dimensional d.f. F and G are said to be *equivalent* if they define the same measure over $\mathfrak{X}_2$ ($\mathfrak{X}_2$ is of course considered with respect to the same fixed basis $\mathfrak{b}$ for F and G).

THEOREM (3,1,4). If a bounded measure m(e) over $\mathfrak{X}_2$ is defined by a distribution function G, then G is equivalent to the distribution function F associated with m(e).

Equivalence of d.f. is obviously an equivalence relation (in the algebraic sense), since

1) any d.f. F is equivalent to itself;

2) if the d.f. F is equivalent to the d.f. G, then G is equivalent to F;

3) if both the d.f. F and G are equivalent to the same d.f. H, then F and G are equivalent to each other.

Consequently, if the d.f. F and G are not equivalent and the d.f. H is equivalent to F, it cannot be equivalent to G. Thus the d.f. are divided into *disjoint* equivalence classes such that

a) if the d.f. F and G belong to the same class, they are

equivalent;

b) if the d.f. F and G belong to two distinct classes, they are not equivalent.

Any equivalence class C is uniquely determined by any d.f. F belonging to it (since C is just the set of d.f. equivalent to F), or equivalently by the (unique) measure defined by the d.f. that belong to it.

Totally discontinuous distribution functions

A 2-dimensional d.f. F(x,y) is said to be *totally discontinuous* if the measure m(e) that it defines over $\mathfrak{X}_2$ is totally discontinuous (cf.II.9).

Absolutely continuous distribution functions

A 2-dimensional d.f. is said to be *absolutely continuous* if the measure m(e) that it defines over $\mathfrak{X}_2$ is absolutely continuous (cf.II.11). Fix an *orthonormal* basis in $\mathfrak{X}_2$, and let $\rho(x,y)$ be the density of m(e); then (2,8,7), (3,1,10) and (3,1,11) easily imply the following assertion:

With the possible exception of certain elements (x,y) of $\mathfrak{X}_2$, the partial derivative

$$\frac{\partial^2}{\partial x\, \partial y} F(x,y)$$

exists and is equal to $\rho(x,y)$:

$$\frac{\partial^2}{\partial x\, \partial y} F(x,y) = \rho(x,y). \tag{3,1,13}$$

The density ρ of m will also be called the density of the absolutely continuous d.f. F.

Integral of a point function with respect to a distribution function

Again we assume n = 2 to simplify the notation.

Let F(x,y) be a given 2-dimensional d.f., m the measure over $\mathfrak{X}_2$ that it defines, $A \subset \mathfrak{X}_2$ any subset of $\mathfrak{X}_2$. Let u denote the element of $\mathfrak{X}_2$ whose components are (x,y). Finally, let $\lambda(u) = \lambda(x,y)$ be a point function (real or complex) defined over $\mathfrak{X}_2$, or at least over A. The integral of the function λ over A with respect to the d.f. F, denoted by

$$\int_A \lambda(x, y)\, dF(x, y), \tag{3,1,14}$$

is defined as the integral

$$\int_A \cdot \lambda(u)\, m(du) \tag{3,1,15}$$

of λ over A with respect to the measure m; in particular:

$$\int_A 1 \cdot dF(x, y) = \int_A dF(x, y) = m(A). \tag{3,1,16}$$

The integral (3,1,14) is therefore merely another notation for (3,1,15); its properties are those of (3,1,15), described in II.6.

Now that we have established the particular importance of distribution functions for Probability Theory, we shall devote the remainder of this chapter to a fairly detailed study of their properties. This will minimize the need for mathematical digressions in the coming chapters. If he so wishes, the reader may proceed directly to Chapter IV, returning to the present chapter for definitions and proofs only as the need arises.

2. *One-dimensional distribution functions*

We shall first treat the case n = 1, i.e., one-dimensional distribution functions. Throughout this book, unless otherwise stated, the unqualified term "distribution function" will mean a one-dimensional distribution function depending on one variable.

It should now be clear that the properties of a d.f. derive from those of the measure or mass distribution that it defines over a one-dimensional vector space $\mathfrak{X}_1$ (or, equivalently, the x-axis or the space R of real numbers). However, in the following sections we shall generally study d.f. directly, i.e., without explicit appeal to the corresponding measures. This direct investigation is in fact easy and not without interest.

Recall then that a *distribution function* of a real variable x, or a distribution function over the x-axis is any real *finite* function F(x), defined and monotone nondecreasing for $-\infty < x < +\infty$; we have already mentioned some direct consequences of this definition.

If $F(x) = F(x - 0)$, F is said to be *continuous from the left* at the point x; if $F(x) = F(x + 0)$, F is said to be *continuous from the right* at the point x; F is *continuous* at the point x if it is continuous both from the left and from the right. Let us put

$$s(x) = F(x + 0) - F(x - 0) \geqslant 0 ;$$

if $s(x) = 0$, the distribution function F is continuous at the point x; if $s(x) > 0$, the distribution function has a discontinuity of the "first kind" or a jump of positive size s(x) at the point x.

Suppose that in some bounded open interval $]a,b[$ $(a < b)$ the function F(x) has jumps at distinct points $x_1, x_2, \ldots, x_n, \ldots$ constituting a finite or countable set; then necessarily

$$\sum_n s(x_n) \leqslant F(b) - F(a). \tag{3,2,1}$$

If the x_n form a countable set, the series $\sum_n s(x_n)$ must be convergent. Let α be an arbitrary positive number and N a positive integer such that

$$N\alpha > F(b) - F(a). \tag{3,2,2}$$

Were N or more of the x_n to correspond to jumps of size greater than α, we would have, by (3,2,2),

$$\sum_n s(x_n) > F(b) - F(a),$$

contradicting (3,2,1). Hence follows easily:

THEOREM (3,2,1). A distribution function F(x) can have only discontinuities of the first kind, i.e., jumps; the set of points of discontinuity of F(x) is finite or at most countable.

Equivalent distribution functions

At the end of III.1 we formulated a general definition, valid for any n, for the equivalence of n-dimensional distribution functions. The reader can easily verify that for n = 1 this definition has the following equivalent form:

A distribution function G(x) is *equivalent* to a d.f. F(x) if there exists a constant h such that

$$G(x) = h + F(x) \tag{3,2,3}$$

for any x that is not a discontinuity point of F; thus, for any x,

$$G(x+0) = h + F(x+0) \;;\; G(x-0) = h + F(x-0) \;;$$

$$F(x+0) - F(x-0) = G(x+0) - G(x-0).$$

Therefore, if x is a discontinuity point of F, it is also a discontinuity point of G, and F and G have jumps of the same size at x.

REMARK (3,2,1)

If G is any d.f., and we define F(x) = G(x - 0), then the function F is a d.f. which is *continuous from the left* and equivalent to G.

REMARK (3,2,2)

In general, let (a,b) denote an arbitrary interval (open or closed), not necessarily the interval (- ∞, + ∞). One can then define a d.f. over (a,b) to be a function F(x) defined and finite for $x \in (a,b)$, and monotone nondecreasing throughout (a,b).

Let F(x) be a d.f. over (a,b), say, to fix ideas, over [a,b]; consider then d.f. over (- ∞, + ∞) defined by

$$G(x) = \begin{cases} F(x) & \text{if } x \in [a,b], \\ F(a) & \text{if } x < a, \\ F(b) & \text{if } x > b. \end{cases}$$

It is clear that the study of G(x) is equivalent to that of F(x). From now on we shall therefore limit ourselves to studying d.f. over (- ∞, + ∞), since the latter include d.f. over any interval (a,b).

In III.10, however, for reasons to be specified then, we shall study the special case of the interval $(0, +\infty)$. With this exception, we shall limit ourselves to d.f. over $(-\infty, +\infty)$; moreover, since these are the only ones of interest in Probability Theory, we shall only deal with d.f. over $(-\infty, +\infty)$ which are *bounded below*; the extension to d.f. which are not bounded below, if necessary, is immediate. We shall use the term *normalized distribution function* for any d.f. $F(x)$, *continuous from the left and bounded below*, such that $F(-\infty) = 0$.

Note that if one d.f. of an equivalence class C is bounded below, then all the d.f. in C are bounded below. In view of Remark (3,2,1) it is clear that an equivalence class C of d.f. which are bounded below contains a unique normalized d.f., which therefore determines C. Thus, up to equivalence, there is no loss of generality in limiting the discussion to normalized d.f.

Normalized totally discontinuous distribution functions

The notion of a totally discontinuous n-dimensional d.f. F, defined in III.1 for arbitrary n, can be defined directly for $n = 1$, without appeal to the measure m defined by F.

Let F be a normalized d.f., and $x_1, x_2, \ldots, x_n, \ldots$ its discontinuity points; let

$$\begin{aligned} s(x_n) &= F(x_n + 0) - F(x_n) \\ &= F(x_n + 0) - F(x_n - 0) \end{aligned}$$

be the sizes of the corresponding jumps; it is clear that the function $S(x)$ defined by

$$S(x) = \sum_{\{n \mid x_n < x\}} s(x_n) \qquad (3,2,4)$$

is a normalized d.f. This might be called the "jump function" associated with F. Note that the sum (3,2,4) may contain an infinite (countable) number of terms, but the assumption that F is normalized implies that it must converge. F is said to be *totally discontinuous* if F(x) is *identical* with S(x). In other words, a (normalized) totally discontinuous d.f. F varies only by jumps: if a and b ($a < b$) are two consecutive discontinuity points of F (i.e., there is no point c, $a < c < b$, such that F has a jump at c), then F(x) remains constant and equal to F(a) throughout [a,b[.

Dirac distribution functions

The simplest example of a one-dimensional totally discontinuous normalized d.f. is the Dirac d.f. $\Delta(x)$ defined by

$$\Delta(x) = \begin{cases} 0 & \text{if } x \leqq 0, \\ 1 & \text{if } x > 0. \end{cases} \qquad (3,2,5)$$

This is the d.f. associated by (3,1,1) with the mass distribution consisting of a single mass 1 at the origin 0. The d.f. corresponding to the mass distribution consisting of a single mass m at the point a is $m\Delta(x - a)$.

Absolutely continuous normalized distribution functions

The general concept of n-dimensional absolutely continuous d.f. introduced at the end of III.1 may also be defined directly for $n = 1$:

A normalized d.f. F is *absolutely continuous* if there exists a function f(x), nonnegative and Lebesgue-integrable over $(-\infty, a)$ for any $a < +\infty$, such that

$$F(x) = \int_{-\infty}^{x} f(y)\, dy \quad \text{for all } x. \qquad (3,2,6)$$

In general the integral on the right-hand side of (3,2,6) is a Lebesgue integral; however, if f(x) is Riemann-integrable (for instance, if f(x) is continuous, as often happens in practice), it may be evaluated as a classical Riemann integral.

We shall say that f(x) is the *density* of F(x).

Formula (3,2,6) implies that F(x) is continuous; it has a derivative f(x), except perhaps for a set of values of x of (Lebesgue) measure *zero*; for the meaning of this term, we refer the reader to II.6 and II.11. However, neither here not in the sequel shall we go into details on this point. At any rate, if f(x) is continuous then F(x) is differentiable with derivative f(x) for any x, i.e., there are no "exceptional" values of x.

Singular distribution functions

A d.f. F(x) is singular if it has the following properties:

1) it is continuous;

2) except for a set of values of x of Lebesgue measure zero, it has a derivative equal to 0;

3) it is not a constant.

Singular distribution functions are complicated in structure and cannot be easily visualized; in fact, they almost never appear in applications, but examples may be constructed (cf. Exercise 3.1).

We have now defined three special classes of d.f.: totally discontinuous, absolutely continuous, and singular d.f. The reason for this classification is the following theorem.

THEOREM (3,2,2). Any normalized distribution function F(x) may be expressed uniquely as the sum of three distribution functions S(x), I(x), R(x):

$$F(x) = S(x) + I(x) + R(x), \tag{3,2,7}$$

where

1) S(x) is a totally discontinuous normalized distribution function;

2) I(x) is an absolutely continuous normalized distribution function, with density f(x);

3) R(x) is a singular normalized distribution function.

To avoid lengthening our presentation, we shall not give the rather lengthy proof of this theorem (cf. M. Fréchet [1], p.315). Clearly, S(x) is just the jump function associated with F(x); it obviously follows from (3,2,6) that, apart from exceptional values of x, F(x) has a derivative equal to that of I(x), i.e., f(x).

Of course, one or two of the three terms on the right-hand side of (3,2,7) may be absent, i.e., vanish identically, in particular, the singular term R(x) is generally absent.

Moreover:

THEOREM (3,2,3). Any continuous bounded distribution function is uniformly continuous over the whole interval $(-\infty, +\infty)$.

The easy proof of this theorem is left to the reader.

3. *Riemann-Stieltjes integrals*

Let F(x) be a d.f., m the measure that it defines over the space R of real numbers (or the x-axis), $\lambda(x)$ a real or complex function of x (point function), and e a subset of R. At the end of III.1 we defined the integral

$$I = \int_e \lambda(x)\, dF(x) \tag{3,3,1}$$

as the integral

$$I = \int_e \lambda(x)\, m(dx),$$

whose properties are already known (cf. II.6). In particular:

1) If $\lambda(x)$ is real and nonnegative, the integral

$$\int_e \lambda(x)\, dF(x)$$

is real and nonnegative.

2)

$$\left|\int_e \lambda(x)\, dF(x)\right| \leqq \int_e |\lambda(x)|\, m(dx) \leqq M \cdot m(e), \tag{3,3,2}$$

where M is any upper bound of $|\lambda(x)|$ over e.

3) For fixed e, $\int_e \lambda(x)\, dF(x)$ is a linear functional of λ.

4) If G is a d.f. equivalent to F, then

$$\int_e \lambda(x)\, dG(x) = \int_e \lambda(x)\, dF(x).$$

5) If F(x) is totally discontinuous, it follows, returning to the notation of (3,2,4), that

$$\int_e \lambda(x)\, dF(x) = \sum_{\{n | x_n \in e\}} \lambda(x_n)\, s(x_n). \tag{3,3,3}$$

The right-hand side of (3,3,3) is either a finite sum or a series, which converges if and only if the integral on the left-hand side exists.

6) If $F(x)$ is absolutely continuous with density $f(x) = \frac{d}{dx}F(x)$, i.e., m is absolutely continuous with density $f(x)$, then

$$\int_e \lambda(x)\, dF(x) = \int_e \lambda(x)\, f(x)\, dx. \tag{3,3,4}$$

In other words, the integral on the left-hand side reduces to a Lebesgue integral, or even a Riemann integral if $f(x)$ and $\lambda(x)$ are sufficiently regular, e.g. continuous.

We note in passing that the integral of λ over e with respect to F, as defined above, includes as particular cases the concept of series (at least - absolutely convergent series; cf. (3,3,3)), and also the Riemann and even Lebesgue integral. However, in general it is an essentially new concept, for it is applicable even when $F(x)$ is singular, in which case it does not reduce to any more elementary integral.

If e is an interval (a,b), a direct and quite elementary definition, of independent interest, is available for the integral (3,3,1). This definition is applicable to a certain class of functions defined over (a,b), which we shall call the RS-class. The integral itself is then called the Riemann-Stieltjes (RS) integral of λ with respect to the d.f. F over (a,b).

We first consider the case where (a,b) is finite and $a < b$. Then (a,b) may be open or closed at a, open or closed at b. We shall see later that if a and b are points of continuity of F, there is no need to distinguish these different cases; this is not so if a or b is a point of discontinuity of F. To fix ideas, we shall assume that $(a,b) = [a,b]$ is closed.

Let us partition $[a,b]$ into n subintervals by $n + 1$ distinct points $x_0, x_1, \ldots, x_n$ in order of increasing magnitude, where $x_0 = a$, $x_n = b$; let us call the number μ defined by

$$\mu = \underset{j}{\text{l.u.b.}} \quad (x_j - x_{j-1})$$

the *modulus* of the partition. Now put

$$\Delta_j F = F(x_j) - F(x_{j-1}) \geqslant 0.$$

In each subinterval (x_{j-1}, x_j) we select an *arbitrary* abscissa ξ_j and form the sum

$$i = \sum_{j=1}^{n} \lambda(\xi_j)\, \Delta_j F, \tag{3,3,5}$$

which we shall call the Riemann-Stieltjes sum associated with the partition.

Suppose there exists a number I with the following property: For any given $\varepsilon > 0$, there exists a number $\eta > 0$ such that

$$|i - I| < \varepsilon$$

for any Riemann-Stieltjes sum i associated with a partition of modulus $\mu < \eta$; then we shall say that 1) λ belongs to class RS over [a,b] with respect to F, or that it is integrable in the Riemann-Stieltjes sense (or RS-integrable) over [a,b] with respect to the d.f. F, and 2) I is the Riemann-Stieltjes integral (or RS-integral) of λ over [a,b] with respect to F. We denote this integral I by

$$I = \int_a^b \lambda(x)\, dF(x), \tag{3,3,6}$$

or, more explicitly,

$$I = \int_a^b \lambda(x)\, dF(x) \quad \text{(RS)}. \tag{3,3,7}$$

The theory of the RS-integral may be developed in complete analogy with that of the classical Riemann integral. In particular, the following theorem is easily proved.

THEOREM (3,3,1). If $\lambda(x)$ is continuous over a bounded interval [a,b], it is RS-integrable over [a,b] with respect to any distribution function.

It is clear that when e is the interval (a,b) and λ is of class RS over e the *RS-integral* (3,3,7) coincides with the *integral* (3,3,1). The basic properties of the RS-integral are of course those of the integral (3,3,1) in general; however, we mention the useful rule of integration by parts for the RS-integral (3,3,7).

Integration by parts

If $\lambda(x)$ has a sufficiently regular derivative $\dot{\lambda}(x)$ (e.g. continuous) and a and b are continuity points of F, then the rule of integration by parts applies to the RS-integral, in the form:

$$\int_a^b \lambda(x)\, dF(x) = [\lambda(b)\, F(b) - \lambda(a)\, F(a)] - \int_a^b F(x)\, \dot{\lambda}(x)\, dx, \quad (3,3,8)$$

where the integral on the right-hand side is an ordinary Riemann integral. If $\lambda(x)$ itself is a d.f. which is continuous at a and b, then

$$\int_a^b \lambda(x)\, dF(x) = [\lambda(b)\, F(b) - \lambda(a)\, F(a)] - \int_a^b F(x)\, d\lambda(x). \quad (3,3,9)$$

Extensions

The RS-integral over an interval [a,b] where $b < a$ is of course defined by

$$\int_a^b \lambda(x)\, dF(x) = - \int_b^a \lambda(x)\, dF(x).$$

The extension to the case of infinite (a,b) follows the same lines as the analogous extension in the theory of the classical Riemann integral.

To fix ideas, let $(a,b) = (-\infty, +\infty)$; we shall say that a function $\lambda(x)$ defined over $(-\infty, +\infty)$ is RS-summable over $(-\infty, +\infty)$ with respect to F if

1) $\lambda(x)$ and $|\lambda(x)|$ are RS-integrable with respect to F over any finite interval (a,b) $(a < b)$; this is true if, for instance, $\lambda(x)$ is continuous over $(-\infty, +\infty)$;

2) $\int_a^b |\lambda(x)|\, dF(x)$ is bounded above when a and b vary arbitrarily.

It is easy to prove that if $\lambda(x)$ is RS-summable over $(-\infty, +\infty)$ there exists a number I such that

$$\lim_{\substack{a \to -\infty \\ b \to +\infty}} \int_a^b \lambda(x)\, dF(x) = I \quad (3,3,10)$$

independently of the way a and b tend to $-\infty$ and $+\infty$, respectively. We then define:

$\lambda(x)$ is RS-integrable with respect to F over $(-\infty, +\infty)$ if and only if it is RS-summable over $(-\infty, +\infty)$; by definition, its RS-integral over $(-\infty, +\infty)$ is the number I defined by (3,3,10), i.e.,

$$\int_{-\infty}^{+\infty} \lambda(x)\, dF(x) = \lim_{\substack{a \to -\infty \\ b \to +\infty}} \int_a^b \lambda(x)\, dF(x). \tag{3,3,11}$$

If $\lambda(x)$ is RS-integrable over $(-\infty, +\infty)$, the same is true for $|\lambda(x)|$ (and vice versa), and we have

$$\left| \int_{-\infty}^{+\infty} \lambda(x)\, dF(x) \right| \leqq \int_{-\infty}^{+\infty} |\lambda(x)|\, dF(x). \tag{3,3,12}$$

It is quite obvious that even this extension of the RS-integral is merely a particular case of the more general integral (3,3,1).

4. *Convergence of distribution functions*

Given a countable sequence of d.f. $F_1, F_2, \ldots, F_n, \ldots$ and a d.f. F, we shall say that F_n converges to F ($F_n \to F$) as $n \to +\infty$, if the sequence $F_n(x)$ converges to the number $F(x)$ for any x that is not a discontinuity point of F. If $F_n \to F$ and G is any d.f. such that $G(x) = F(x)$ for any x that is not a discontinuity point of F (so that G is equivalent to F), then F_n also converges to G.

If $F_n \to F$ and $F_n(-\infty) = 0$ for any n, then $F(-\infty) \geqq 0$, but it may happen that $F(-\infty) > 0$. In other words, the limit of a sequence of normalized d.f. is not necessarily normalized. Similarly, if $F_n \to F$ and $F_n(+\infty) = 1$ for any n, then $F(+\infty) \leqq 1$, but it may happen that $F(+\infty) < 1$ (cf. Exercise 3.4).

Compactness Theorem (3,4,1)

Let $\mathfrak{F}$ be an infinite family of distribution functions, with the following property: for any a and b $(a < b)$, there exist two numbers $m(a)$ and $M(b)$ such that, for any distribution function $G \in \mathfrak{F}$ and any $x \in (a,b)$,

$$m(a) < G(x) < M(b). \tag{3,4,1}$$

Then there exist a distribution function F (which may or may not belong to $\mathfrak{F}$) and a countable sequence $F_1, F_2, \ldots, F_n, \ldots$ of distribution functions belonging to $\mathfrak{F}$ such that F_n converges to F as n tends to $+\infty$.

We shall only outline the proof of this theorem, leaving the details to the reader.

Let (a,b) be any fixed finite interval, $a < b$. Let $x_1, x_2, \ldots, x_n, \ldots$ be a countable set of points of (a,b), which is *dense* in (a,b); for instance, the x_k may be the rational numbers in (a,b). By (3,4,1), the set of all numbers $G(x_1)$ where G is an arbitrary function in $\mathfrak{F}$, is a bounded set. There therefore exists a countable sequence S_1 of d.f. $G_{1,1}, G_{1,2}, \ldots, G_{1,h}, \ldots$ belonging to $\mathfrak{F}$ such that

$$y_1 = \lim_{h \to +\infty} G_{1,h}(x_1)$$

exists. By (3,4,1), the numbers $G_{1,h}(x_2)$ constitute a bounded set, and there exists a countable sequence S_2 of d.f. $G_{2,1}, G_{2,2}, G_{2,3}, \ldots, G_{2,h}, \ldots$ *belonging to* S_1 such that

1) $G_{2,1} = G_{1,1}$;

2) $y_2 = \lim_{h \to -\infty} G_{2,h}(x_2)$ exists.

Since the numbers $G_{2,h}(x_3)$ still constitute a bounded set, there exists a countable sequence S_3 of d.f. $G_{3,1}, G_{3,2}, G_{3,3}, \ldots, G_{3,h}, \ldots$ *belonging to* S_2 such that

1) $G_{3,1} = G_{2,1}$, $G_{3,2} = G_{2,2}$;

2) $y_3 = \lim_{h \to +\infty} G_{3,h}(x_3)$ exists,

and so on. In general, we find a sequence $S_n (n = 1,2,3,\ldots)$ of d.f. $G_{n,1}, G_{n,2}, G_{n,3}, \ldots, G_{n,h}, \ldots$ belonging to S_{n-1} such that

1) $G_{n,j} = G_{n-1,j}$ for $j = 1, 2, \ldots, n-1$;

2) $y_n = \lim_{h \to +\infty} G_{n,h}(x_n)$ exists

Now define $F_n = G_{n,n}$, so that $F_n \in S_j$ for any j, and consequently

$$\lim_{n \to +\infty} F_n(x_k) = y_k. \qquad (3,4,2)$$

Define a function F(x) over (a,b) by

$$F(x) = y_k \text{ if } x = x_k;$$

$$F(x) = \underset{\{j|x_j<x\}}{\text{l.u.b.}} \; y_j \text{ if } x \text{ is not one of the } x_k.$$

This construction of F(x), known as the "diagonal procedure," appears in many other branches of mathematics.

It is easily verified that F(x) is a bounded and monotone non-decreasing function over (a,b). For any $x \in (a,b)$ which is one of the x_k, we have, by (3,4,2),

$$\lim_{n\to+\infty} F_n(x) = F(x). \tag{3,4,3}$$

We must prove (3,4,3) for any $x \in (a,b)$ that is not a discontinuity point of F(x). Let x be such a point, and let x' and x'' be two points of the sequence of x_k such that

$$x' < x < x''.$$

It follows from the monotonicity of F_n and F that

$$F_n(x') < F_n(x) < F_n(x'') \text{ for any } n,$$
$$F(x') < F(x) < F(x'').$$

Since

$$\lim_{n\to+\infty} F_n(x') = F(x'), \quad \lim_{n\to+\infty} F_n(x'') = F(x''),$$

it follows that for fixed $\varepsilon > 0$ and sufficiently large n

$$F(x') - \varepsilon \leqslant F_n(x) \leqslant F(x'') + \varepsilon. \tag{3,4,4}$$

Now x is a continuity point of F; since the sequence of the x_k is dense in (a,b), x' and x'' may be chosen arbitrarily close to x, in particular, so that

$$F(x') \geqslant F(x) - \varepsilon, \quad F(x'') \leqslant F(x) + \varepsilon; \tag{3,4,5}$$

then (3,4,4) and (3,4,5) imply

$$F(x) - 2\varepsilon \leqslant F_n(x) \leqslant F(x) + 2\varepsilon$$

for sufficiently large n. Since ε is arbitrarily small, we get (3,4,3).

This construction may be applied for a = – h, b = h, where h is a positive integer. Letting h assume the values 1,2,3,... it is not difficult to prove Theorem (3,4,1) as formulated above.

Consider a sequence of d.f. $F_1, F_2, \ldots, F_n, \ldots$ and a d.f. F. Suppose that $F_n \to F$ as $n \to +\infty$. Let $\lambda(x)$ be a given function, and assume that for all n the integral

$$I_n = \int_{-\infty}^{+\infty} \lambda(x)\, dF_n(x)$$

exists. Does it follow that the integral

$$I = \int_{-\infty}^{+\infty} \lambda(x)\, dF(x)$$

exists and $\lim_{n\to+\infty} I_n = I$?

It is easy to find examples showing that I may not exist, and that even if it exists it may not be true that $\lim_{n\to+\infty} I_n = I$. However, some interesting results of this type may be established, for instance:

THEOREM (3,4,2). Let $F_1, F_2, \ldots, F_n, \ldots$ be a uniformly bounded countable sequence of distribution functions, i.e., there exist two numbers m and M such that

$$m < F_n(x) < M \text{ for any n and any x.}$$

Let F be a distribution function. Then, in order that F_n converge to F as n tends to $+\infty$, it is necessary that for any continuous function $\lambda(x)$ which vanishes outside a bounded set, the RS-integrals

$$I_n = \int_{-\infty}^{+\infty} \lambda(x)\, dF_n(x)$$

converge to the RS-integral

$$I = \int_{-\infty}^{+\infty} \lambda(x)\, dF(x).$$

This condition is also sufficient if $F_n(-\infty) = F(-\infty)$ for any n (in other words, the condition is sufficient "up to equivalence" of distribution functions).

The proof of Theorem (3,4,2) is left to the reader.

REMARK (3,4,1)

Let $m_n (n = 1,2,3,...)$ and m be the measures over the space R of real numbers defined by the d.f. $F_n (n = 1,2,3,...)$ and F. It is natural to ask: what are the relations between the convergence of F_n to F as defined above and the convergence of m_n to m in the sense of convergence of measures, defined at the end of II.11?

Theorem (3,4,2) shows that the convergence of F_n to F implies the *weak* convergence of m_n to m; conversely, weak convergence of m_n to m implies convergence of F_n to F, *up to equivalence of d.f.*

Distance between two distribution functions

Let F and G be two d.f.; it is natural (cf. M. Fréchet [1], p.331) to call the number

$$\Delta(F,G) = \underset{x}{\text{l.u.b.}} |F(x) - G(x)|$$

the *distance* $\Delta(F,G)$ between the two d.f. Unfortunately, the number $\Delta(F,G)$ is generally $+\infty$. On the other hand, if F and G are bounded it is finite, and the definition of distance is then acceptable; in the sequel we therefore limit ourselves to bounded d.f.

Let $F_1, F_2, \ldots, F_n, \ldots$ be a sequence of bounded d.f. and F a bounded d.f.

Let m_n, m denote the measures defined (over R) by F_n, F, respectively.

1) Suppose that F_n converges to F as $n \to +\infty$; this does not necessarily imply that $\Delta(F_n,F)$ converges to 0 (cf. Exercise 3,4); $\Delta(F_n,F)$ may not converge to 0, in particular, owing to discontinuities of F.

2) Suppose that $\lim_{n\to+\infty} \Delta(F_n,F) = 0$; it follows that for any x the number $F_n(x)$ converges to the number $F(x)$, and therefore F_n converges to F. Moreover, the convergence of $F_n(x)$ to $F(x)$ is also uniform in $x \in (-\infty, +\infty)$; in particular, $\lim_{n\to+\infty} F_n(-\infty) = F(-\infty)$, $\lim_{n\to+\infty} F(+\infty) = F(+\infty)$. This amounts to saying that the measure m_n converges to m *completely* (cf. II,11).

It is easy to prove:

If F *is continuous*, then in order that F_n converge to F and $\lim_{n\to\infty} [F_n(+\infty) - F_n(-\infty)] = F(+\infty) - F(-\infty)$ (i.e., in order that m_n converge to m completely) it is necessary and sufficient that $\Delta(F_n,F)$ converge to 0.

Moreover, the definition of the distance $\Delta(F,G)$ may be modified in such a way that this result remains valid when F (though bounded) has discontinuities (cf. M. Fréchet [1], p.331).

REMARK (3,4,2)

Suppose that the d.f. F_n converges to F as $n \to +\infty$, and that each F_n is absolutely continuous, with density f_n. This does not imply that F is necessarily absolutely continuous; even if F is absolutely continuous with density f, f_n need not converge to f.

Nevertheless, assume that

α) the d.f. F_n is absolutely continuous with density f_n, and $F_n(-\infty) = 0$ for any n;

β) the d.f. F is absolutely continuous with density f, and $F(-\infty) = 0$;

γ) f_n converges to f as $n \to +\infty$;

δ) the functions $f_n(x)$ are uniformly bounded in x and n.

Then F_n converges to F as $n \to +\infty$.

5. *Review of matrix theory*

In the sequel we shall need certain matrix concepts and notations. Since these are classical definitions and properties, we shall limit ourselves to a brief review, beginning with real matrices, i.e., matrices with real elements. Now any n-tuple $x = \{x_1,x_2,\dots,x_n\}$ of real numbers $x_1,x_2,\dots,x_n$ may be regarded as an $n \times 1$ matrix x [column matrix], or an element (vector) of the vector space R^n (n^{th} cartesian power of the space R of real numbers; R^n is of dimension n).

Let A be any $n \times p$ matrix of real or complex elements (a_{jk}). We shall use the notation $°A$ for the $p \times n$ matrix (a_{kj}), i.e., the transpose of A. The matrix A is symmetric if and only if

$$A = {}^{\circ}A,$$

implying p = n; in other words,

$$a_{jk} = a_{kj} \quad (j, k = 1, 2, \ldots, n).$$

Note that a 1×1 matrix is always symmetric.

Let A be a real symmetric matrix (p = n); the quadratic form associated with A is the function

$$\sum_{jk} a_{jk} x_j x_k = {}^{\circ}x\ A\ x = {}^{\circ}x\ {}^{\circ}A\ x \qquad (3,5,1)$$

of n real variables $\{x_1, x_2, \ldots, x_n\}$, the latter regarded as the elements of the $n \times 1$ matrix x. It is a homogeneous quadratic polynomial in the x_j. Conversely, any homogeneous quadratic polynomial in the x_j may be expressed uniquely in the form

$$\sum_{jk} a_{jk} x_j x_k, \qquad (3,5,2)$$

with coefficients $a_{jk}(j,k = 1,2,\ldots,n)$ satisfying the condition

$$a_{jk} = a_{kj} \quad (j, k = 1, 2, \ldots, n).$$

Thus the a_{jk} are the elements of a symmetric $n \times n$ matrix A, so that we can write (3,5,2) in the form (3,5,1) and say that if (3,5,2) is the quadratic form associated with A, then A is the symmetric matrix associated with the form (3,5,2).

An $n \times n$ matrix D of elements (d_{jk}) is *diagonal* if $d_{jk} = 0$ for $j \neq k$. Let D be a real diagonal matrix and let α be the number of positive elements along its principal diagonal, β the number of negative elements and γ the number of zero elements ($\alpha \geq 0$, $\beta \geq 0$, $\gamma \geq 0$, $\alpha + \beta + \gamma = n$; $\gamma > 0$ if and only if D is singular, i.e., has determinant zero). α may be called the positive weight of D, β its negative weight, γ its zero weight. Since it is diagonal, D is symmetric and its associated quadratic form ${}^{\circ}x\ D\ x$ has the simple "reduced" form

$${}^{\circ}x\ D\ x = \sum_{h=1}^{n} d_{hh} x_h^2, \qquad (3,5,3)$$

i.e. it is a sum of squares weighted by coefficients d_{hh}, of which α are positive, β negative and γ zero.

If the real $n \times n$ matrix A is symmetric and T is any real $n \times n$ matrix, then the matrix $°TAT$ is symmetric. Its associated quadratic form, as a function of n variables $\{y_1, y_2, \ldots, y_n\}$ (interpreted as the elements of an $n \times 1$ matrix y) is

$$°y \; °T \; A \; T \; y \, .$$

This form may be derived from (3,5,1) by performing the change of variables $x = Ty$. This change of variables is particularly interesting when T is nonsingular (and therefore has an inverse T^{-1}), since then the correspondence between x and y is one-to-one ($x = Ty \Longleftrightarrow y = T^{-1}x$).

Now there exists a nonempty family $\mathcal{D}_A$ of real nonsingular $n \times n$ matrices T, which we call diagonalizing matrices, such that the matrix

$$D = °T \; A \; T$$

is not only symmetric but also diagonal. Consequently, by the change of variables $x = Ty$, $T \in \mathcal{D}_A$, the form (3,5,1) is transformed into a reduced form (3,5,3). It may also be proved that for any $T \in \mathcal{D}_A$, the positive, negative, and zero weights α, β, γ of $D = °T \; A \; T$ are the same; they are therefore intrinsic properties of A, which we shall call the positive, negative, and zero weights of A. A is said to be *positive-definite* if its negative weight β is zero, which is equivalent to

$$°x \; A \; x \geq 0 \text{ for any } x.$$

A positive-definite matrix A is said to be *degenerate* if its zero weight γ is positive. If A is a degenerate positive-definite matrix, there exist nonzero real $n \times 1$ matrices x such that $°x \; A \; x = 0$, while if A is nondegenerate positive-definite $°x \; A \; x = 0$ only if x is zero. A can be degenerate only if it is singular (its determinant is zero); more precisely, if A is a positive-definite symmetric matrix, i.e., of zero negative weight, its positive weight is equal to its rank.

If the real symmetric matrix A is positive-definite, it is easy to see that for any real nonsingular matrix T the matrix $°T\ A\ T$ is not only symmetric, but also positive-definite; it is degenerate if and only if A itself is degenerate.

A real $n \times n$ matrix T with elements (t_{jk}) is *orthogonal* if $°TT = I$, where I denotes the unit matrix (the diagonal matrix in which all the elements on the principal diagonal are 1). Then the determinant of T is ± 1, and $°T = T^{-1}$; the elements t_{jk} satisfy the (necessary and sufficient) conditions

$$\sum_{h=1}^{n} t_{jh}^2 = 1 \quad (j = 1, 2, \ldots, n), \tag{3,5,4}$$

$$\sum_{h=1}^{n} t_{jh}\, t_{kh} = 0 \quad (j \neq k\ ;\ j, k = 1, 2, \ldots, n). \tag{3,5,5}$$

If T_k denotes the kth column of T, i.e., the $n \times 1$ matrix with elements t_{hk} $(h = 1,2,\ldots,n)$, then following the classical geometric interpretation, we say that as vectors of R^n the T_k are pairwise orthogonal unit vectors [by (3,5,4) and (3,5,5)].

If $T \in \mathcal{D}_A$ and T is orthogonal, then the elements $d_{kk} = s_k$ on the principal diagonal of $D = °T\ A\ T$ are the n characteristic roots (distinct or not, but real) of the symmetric matrix A, and T_k is a characteristic vector of A associated with the characteristic root s_k.

Conversely, if A is a real symmetric matrix, let us denote its n characteristic roots, distinct or not, by $s_1, s_2, \ldots, s_n$. For each $k(k = 1,2,\ldots,n)$ let T_k be an $n \times 1$ matrix which is a characteristic vector of A associated with the characteristic root s_k, such that the elements t_{hk} $(h = 1,2,\ldots,n)$ satisfy conditions (3,5,4) and (3,5,5); such vectors are known to exist. Then the $n \times n$ matrix T whose k^{th} column is T_k $(k = 1,2,\ldots,n)$ is orthogonal and belongs to $\mathcal{D}_A$. Note that the positive, negative and zero weights of A are the numbers of its positive, negative and zero characteristic roots, respectively.

Complex matrices

An ordered n-tuple $x = \{x_1, x_2, \ldots, x_n\}$ of complex numbers $x_1, x_2, \ldots, x_n$ may be regarded as an $n \times 1$ complex matrix x, or as an element (vector) of the vector space C^n (the nth Cartesian power of the space C of complex numbers; C^n is of dimension n).

A complex matrix is a matrix whose elements are complex numbers. Let A be any complex $n \times p$ matrix with elements (a_{jk}); we use the notation $\overline{A}$ for its conjugate matrix, i.e., the $n \times p$ matrix $\overline{A}$ whose elements are the conjugates $\overline{a_{jk}}$ of the elements a_{jk} of A. The matrix A is said to be *hermitian* if

$$A = {}^\circ\overline{A}; \tag{3,5,6}$$

in other words $p = n$, and

$$a_{jk} = \overline{a_{kj}} \quad (j, k = 1, 2, \ldots, n),$$

which implies that the elements $a_{hh}(h = 1, 2, \ldots, n)$ on the principal diagonal are real.

A hermitian form of an n-tuple $x = \{x_1, x_2, \ldots, x_n\}$ of complex variables x_j is any expression of the form:

$$\sum_{jk} a_{jk}\, x_k\, \overline{x_k}$$

where the constant (complex) coefficients (a_{jk}) satisfy the conditions

$$a_{jk} = \overline{a_{kj}}\,.$$

A hermitian form always assumes real values.

If A is a complex $n \times n$ hermitian matrix with elements (a_{jk}), then A and the hermitian form

$$\sum_{jk} a_{jk}\, x_j\, \overline{x_k} \tag{3,5,7}$$

are said to be *associated*. Regarding x as an $n \times 1$ matrix with elements x_j, the form (3,5,7) may be expressed in matrix notation:

$$\sum_{jk} a_{jk} x_j \overline{x_k} = {}^\circ\overline{x}\, A\, x = {}^\circ\overline{x}\, {}^\circ\overline{A}\, x. \tag{3,5,8}$$

An $n \times n$ complex matrix T with elements (t_{jk}) is said to be *unitary* if $^\circ\overline{T}T = I$, i.e., if

$$\sum_h |t_{jh}|^2 = 1 \quad (j = 1, 2, \ldots, n), \tag{3,5,9}$$

$$\sum_h t_{jh} \overline{t_{kh}} = 0 \quad (j \neq k,\ j,\ k = 1, 2, \ldots, n). \tag{3,5,10}$$

It is well known that all the above properties of real matrices extend to complex matrices, by substituting hermitian matrix for symmetric matrix, hermitian form for quadratic form, and unitary matrix for orthogonal matrix.

The reduced form analogous to (3,5,3) is then of the type

$$^\circ\overline{x}\ D\ x = \sum_{jk} d_{hh}\ |x_h|^2, \tag{3,5,11}$$

where the coefficients d_{hh} are real, since a complex diagonal matrix D is hermitian only if the elements d_{hh} on its principal diagonal are real.

It is clear on the other hand that since real numbers are complex numbers which are equal to their conjugates all definitions and results concerning complex matrices include as particular cases those concerning real matrices.

Positive-definite matrices

A square $n \times n$ matrix A with complex elements a_{jk} $(j,k = 1,2,\ldots,n$; in particular, the a_{jk} may be real) is said to be *positive-definite* if, for any n complex numbers $\{x_1,x_2,\ldots,x_n\}$, the associated form

$$\sum_{jk} a_{jk}\ x_j\overline{x}_k \tag{3,5,12}$$

is real and nonnegative. All the characteristic roots of A are then real and nonnegative. A is degenerate positive-definite if there exsit $x_1,x_2,\ldots,x_n$, not all zero, such that the form (3,5,12) vanishes. At least one of the characteristic roots of A is then zero. The terms "positive-definite" and "degenerate (or nondegenerate) positive-definite" may apply to the form (3,5,12) or the associated matrix A.

The following properties of positive-definite matrices are easily verified:

1) If A is a positive-definite $n \times n$ matrix with elements a_{jk}, then

a) $A = {}^\circ\overline{A}$, i.e., $a_{kj} = \overline{a}_{jk}$ for all j, k; in other words, A is hermitian (or symmetric if the a_{jk} are real); the matrix $\overline{A}$ is positive-definite.

b) The product ρA of A and a nonnegative real number ρ is a positive-definite matrix.

2) If A and B are two positive-definite $n \times n$ matrices, the matrix $A + B$ is positive-definite.

3) All the principal minors of A are real and nonnegative; in particular, for any j, k:

$$\begin{aligned} &a_{jj} \text{ is real and nonnegative,} \\ &|a_{jk}|^2 \leq a_{jj}\, a_{kk}; \end{aligned} \tag{3,5,13}$$

conversely, if all the principal minors of a square matrix A are real and nonnegative then A is positive-definite.

Let A and B be two $n \times n$ matrices with elements a_{jk} and b_{jk}, respectively; the *Hadamard product* of A by B, denoted by $A \oplus B$, is $n \times n$ matrix C whose elements are $c_{jk} = a_{jk}\, b_{jk}$. Clearly, if A and B are hermitian, $A \oplus B$ is hermitian; but we can say even more:

LEMMA (3,5,1)

If A and B are positive-definite, then $A \oplus B$ is positive-definite.

To prove this, note that if the $n \times n$ matrix A with elements a_{jk} is hermitian, there exists a unitary $n \times n$ matrix T with elements t_{jk} such that

$${}^\circ\overline{T}\, A\, T = D, \tag{3,5,14}$$

where D is a diagonal matrix whose diagonal elements are the (real) characteristic roots $s_1, s_2, \ldots, s_n$ of A. Since T is unitary, ${}^\circ\overline{T} = T^{-1}$, and from (3,5,14) we obtain

$$A = T\,D\,{}^{\circ}\overline{T},$$

or, since D is diagonal,

$$a_{jk} = \sum_h s_h\,\overline{t}_{hj}\,t_{hk}\ (j,k = 1,2,\ldots,n). \qquad (3,5,15)$$

Similarly, if the $n \times n$ matrix B with elements b_{jk} is hermitian and we denote its (real) characteristic roots by $\sigma_\ell(\ell = 1,2,\ldots,n)$ and the elements of a suitable unitary matrix by u_{jk} $(j,k = 1,2,\ldots,n)$, then

$$b_{jk} = \sum_\ell \sigma_\ell \overline{u}_{\ell j} u_{\ell k}\ (j,k = 1,2,\ldots,n). \qquad (3,5,16)$$

The elements c_{jk} of $A \oplus B$ are then

$$c_{jk} = a_{jk}\,b_{jk} = \sum_{h,\ell} s_h\,\sigma_\ell\,\overline{t}_{hj}\,\overline{u}_{\ell j}\,t_{hk}\,u_{\ell k}. \qquad (3,5,17)$$

Consider the hermitian form $\sum_{j,k} c_{jk}\,x_j\overline{x}_k$ associated with $A \oplus B$; by (3,5,17) it follows that

$$\sum_{j,k} c_{jk}\,x_j\overline{x}_k = \sum_{j,k,h,\ell} s_h\,\sigma_\ell\,\overline{t}_{hj}\,\overline{u}_{\ell j}\,t_{hk}\,u_{\ell k}\,x_j\overline{x}_k$$

$$= \sum_{h,\ell} s_h\,\sigma_\ell\,(\sum_{j,k} \overline{t}_{hj}\,\overline{u}_{\ell j}\,x_j\,t_{hk}\,u_{\ell k}\,\overline{x}_k)$$

$$= \sum_{h,\ell} s_h\,\sigma_\ell\,|\sum_m \overline{t}_{hm}\,\overline{u}_{\ell m}\,x_m|^2.$$

This proves Lemma (3,5,1).

6. *The special role of exponential functions*

In the following sections, and again in Chapters IV, V, VI, a special role will be assigned to exponential functions; it is important to understand the reason for the privileged role of exponential functions in certain domains.

Let t be a real variable in $(-\infty, +\infty)$; it is always possible, and convenient, to interpret t as time. The effect of a translation on the time axis is to change t to t + h, where h is arbitrary. The set of all translations over the time axis is a group $\mathfrak{G}$. Now consider the set $\mathfrak{F}$ of all real or complex functions of t; $\mathfrak{F}$ is a vector space. With the translation $t \to t + h$ we can associate a linear operator T_h over $\mathfrak{F}$ which maps any function f of $\mathfrak{F}$ onto the function $g = T_h F$ defined by

$$g(t) = f(t + h).$$

The study of $\mathfrak{G}$ is equivalent to that of the family of operators T_h $(-\infty < h < +\infty)$, and it is of special interest to consider the functions $f \in \mathfrak{F}$, not identically zero, which have the following property For any h, $T_h f = \rho f$, where ρ is a constant that may depend on h; the main point is that, *for any t and h,*

$$f(t + h) = \rho(h) f(t) \qquad (3,6,1)$$

where $\rho(h)$ may be any function depending only on h. Formula (3,6,1) may be interpreted as the statement that $\rho(h)$ is a characteristic root of the linear operator T_h, and f a characteristic vector of T_h associated with the characteristic root $\rho(h)$. The functions f satisfying this condition are therefore characteristic vectors of T *for any* h: they are called the *characters* of the translation group $\mathfrak{G}$. More precisely, the term "characters" is customarily applied to nonzero solutions f of (3,6,1) which are also *continuous* and *bounded*. However, there is no need here to go into details on this point. In the sequel, (in particular, Remark (3,6,2)) we shall hint at the reasons for the particular importance of those solutions of (3,6,1) which are continuous and bounded.

Note that if f is not the identically zero solution of (3,6,1), then f(t) cannot vanish for any value of t. Moreover, if f satisfies (3,6,1), then so does any function proportional to f; taking this into account, let us impose on f the additional condition

$$f(0) = 1. \qquad (3,6,2)$$

Putting t = 0 in (3,6,1), it follows that

$$\rho(h) = f(h),$$

and finally (3,6,1) becomes

$$f(t + h) = f(t)f(h). \qquad (3,6,3)$$

Any exponential function $e^{i\lambda t}$, where λ is a complex constant, satisfies (3,6,3) and (3,6,2), and it is known that the exponential functions $e^{i\lambda t}$ are the only *continuous* functions satisfying (3,6,3) and (3,6,2).

These considerations may be given a physical interpretation. Many devices operate as follows: when excited by a certain function s of t, called the *input signal*, their response is a function r of t, called the *output signal*. Let us make the following assumptions, which are generally valid.

1) The set $\mathfrak{F}_s$ of input signals accepted by the device is a vector space of functions of t, which are sufficiently regular; there is no need here to specify this regularity, which may vary from one case to another. In any case $\mathfrak{F}_s$ is a vector subspace of $\mathfrak{F}$.

2) If $s \in \mathfrak{F}_s$, then $T_h s \in \mathfrak{F}_s$ for any h.

3) The exponential functions $e^{i\lambda t}$ belong to $\mathfrak{F}_s$, at least for real λ. If λ has a nonzero imaginary part, then $e^{i\lambda t}$ does not remain bounded over $(-\infty, +\infty)$ and then it is physically impossible that $e^{i\lambda t} \in \mathfrak{F}_s$. In particular, if the imaginary part of λ is positive, then $\lim_{t\to-\infty} |e^{i\lambda t}| = +\infty$, and this would mean that the device received an infinitely great signal at an infinitely remote past ($t = -\infty$) - a situation with no physical interpretation.

To say that r is the response of the device to the input s means that if t is an arbitrary fixed instant, the value of r(t) is determined (by the device) as a function of the set of values $s(\tau)$ at different instants τ ($-\infty < \tau < +\infty$). I.e., for each t there exists a function Ψ_t defined over the various $s(\tau)$ ($-\infty < \tau < +\infty$) whose value is r(t):

$$r(t) = \Psi_t\{s(\tau)\} \text{ for } -\infty < \tau < +\infty. \qquad (3,6,4)$$

But in many cases (electrical circuits such as quadrupoles, amplifiers, etc.) the device has two additional properties:

1) It is *linear*, i.e.,

α) if s is the sum of two functions s_1 and s_2 (of $\mathfrak{F}_s$), the corresponding response r is the sum of the responses r_1 and r_2 of the device to s_1 or s_2 alone;

β) if r is the response corresponding to an input signal s, and ρ any constant, the response corresponding to the input signal ρ.s is ρ.r.

2) It is *stationary*, i.e., its properties are invariant with respect to time. In other words, if r is the response corresponding to the input signal s, then the response to the input signal $T_h s$ is $T_h r$, and this holds for any $s \in \mathfrak{F}_s$ and $h \in (-\infty, +\infty)$.

The linearity property means that for each t the mapping Ψ_t is linear. Let r_λ be the response in the case where s is the exponential function $e^{i\lambda t}$. By β), the response to $e^{i\lambda(t+h)} = e^{i\lambda h}.e^{i\lambda t}$, where h is an arbitrary constant, is $e^{i\lambda h}\, r_\lambda$. Now by virtue of stationarity, since $e^{i\lambda(t+h)}$ is the image under T_h of $e^{i\lambda t}$ the response to $e^{i\lambda(t+h)}$ must be $T_h r_\lambda$. Therefore

$$T_h r_\lambda = e^{i\lambda h}\, r_\lambda,$$

i.e., for any t and h,

$$r_\lambda\,(t + h) = e^{i\lambda h}\, r_\lambda\,(t). \tag{3,6,5}$$

Putting t = 0 in (3,6,5), we easily see that

$$r_\lambda(t) = G(\lambda)e^{i\lambda t}$$

where G(λ) may depend on λ, but not on t.

Now suppose that the input signal s is a linear combination of exponential functions $e^{i\lambda_1 t}, e^{i\lambda_2 t}, \ldots, e^{i\lambda_k t}$:

$$s(t) = \sum_k \ell_k\, e^{i\lambda_k t}$$

where the coefficients ℓ_k are *constants* (independent of t), and the λ_k various constants; by virtue of linearity, the corresponding response r will be

$$r(t) = \sum_k \ell_k \, G(\lambda_k) \, e^{i\lambda_k t}.$$

This extends in a natural way to the case where s is a linear combination of a countable set or a continuum of functions $e^{i\lambda_k t}$. In general, all the functions of $\mathfrak{F}_s$ are linear combinations of exponential functions with constant coefficients. This demonstrates, from a new angle, the particular importance of exponentials in a certain type of problem.

It is clear that a device of the above type may also be defined by the function $G(\lambda)$, which is called its *gain*, and also that it may generally be represented mathematically by a linear mapping Φ of $\mathfrak{F}_s$ into $\mathfrak{F}$:

$$r = \Phi(s),$$

whose defining property is that it transforms the exponential function $e^{i\lambda t}$ into $G(\lambda).e^{i\lambda t}$. This may also be expressed by the statement that $G(\lambda)$ is a characteristic root of Φ, with characteristic function $e^{i\lambda t}$.

EXAMPLE (3,6,1)

Consider two contacts a and b connected by an electrical circuit comprising capacitance C, resistance R, and self-induction L, all constant. If a voltage s(t) is applied at the contacts (a,b), the charge r(t) of the capacitance at time t is related to s(t) $(-\infty < t < +\infty)$ by the following differential equation with constant coefficients:

$$L \, \ddot{r}(t) + R \, \dot{r}(t) + \frac{r(t)}{C} = s(t) \qquad (3,6,6)$$

Let $\mathfrak{F}_s$ be a class of functions of t having properties 1), 2), 3) of page 170, which are also bounded.

For given s(t), equation (3,6,6) has infinitely many solutions r, but if we note that the solutions of the algebraic equation

$$L\rho^2 + R\rho + \frac{1}{C} = 0 \qquad (3,6,7)$$

have negative real parts, we see that if s is bounded then there is only one solution of (3,6,6), say

$$r = \Phi(s),$$

which is bounded for $-\infty < t < +\infty$. This solution corresponds to the actual operation of the circuit, assuming it in operation from $t = -\infty$. Then, interpreting s as an input signal and $r = \Phi(s)$ as the corresponding response, the circuit is a device of the above type, i.e., linear and stationary.

Its gain $G(\lambda)$ is

$$G(\lambda) = 1/(-L\lambda^2 + iR\lambda + \frac{1}{C}),$$

since the bounded solution of (3,6,6) for $s(t) = e^{i\lambda t}$ is

$$\frac{e^{i\lambda t}}{-L\lambda^2 + iR\lambda + \frac{1}{C}}.$$

This gain $G(\lambda)$ is therefore infinite for values of $i\lambda = \rho$ that are solutions of (3,6,7), but then $s(t) = e^{i\lambda t}$ is not finite, which we have assumed impossible.

REMARK (3,6,1)

It is obvious that no actual physical device can foretell the future, i.e., its response cannot take into account any signal not yet received. In mathematical terms this means that, given t, the function Ψ_t in (3,6,4) does not depend on the signals $s(\tau)$ for all τ between $-\infty$ and $+\infty$, only on those for which $\tau \leq t$. There is no reason to limit ourselves to this particular case here, but this remark is clearly very important in many applications.

REMARK (3,6,2)

We have already noted that exponential functions with real λ are characterized by the fact that they remain bounded for any *real* t. Consequently, from certain points of view particular interest attaches to linear combinations of exponential functions $e^{2\pi i\nu t}$ with real ν, which leads to the study of Fourier transforms. In Probability Theory, mainly in connection with the addition of independent random variables (cf. Chapter VI), it is often very convenient to study d.f. by way of their Fourier transforms.

We shall therefore devote a few sections to Fourier transforms. However, we do not intend to study the subject in general; we shall limit ourselves to classes of functions for which elementary procedures are adequate, and which suffice to yield the results of interest for Probability Theory.

Functions of classes (S_1) *and* (S_2) *and their Fourier transforms*

We begin with a lemma.

LEMMA (3,6,1)

Let [a,b] be a bounded closed interval and f(x) a real or complex function of the real variable x, defined and continuous over [a,b]; then

$$\lim_{\alpha\to+\infty} \int_b^a \sin \alpha x\ f(x)\ dx = 0,$$

where α is real.

It is clearly sufficient to prove the lemma for real f(x). Let h be the greatest nonnegative integer such that

$$a + \frac{2h\pi}{\alpha} \leqq b ,$$

and put

$$x_k = a + \frac{2k\pi}{\alpha} \qquad (k = 0, 1, 2, \ldots, h) ;$$

$$I_k = \int_{x_k}^{x_{k+1}} \sin \alpha x\ f(x)\ dx \qquad (k = 0, 1, 2, \ldots, h-1);$$

$$J = \int_{x_h}^{b} \sin \alpha x\ f(x)\ dx .$$

It is immediate that

$$\lim_{\alpha\to+\infty} J = 0 .$$

Now put $x = x_k + y$; owing to the periodicity of $\sin\alpha x$ we have

$$I_k = \int_0^{2\pi/\alpha} \sin\alpha(x_k + y)\ f(x_k + y)\ dy$$

$$= \int_0^{2\pi/\alpha} \sin\alpha(a + y)\ f(x_k + y)\ dy.$$

Any function which is continuous on a bounded closed interval is uniformly continuous there. Fix $\varepsilon > 0$ and put

$$f(x_k + y) = f(x_k) + \omega_k(y); \tag{3,6,8}$$

then, if α is sufficiently large,

$$|\omega_k(y)| < \varepsilon \text{ for any } k\ (k = 0, 1, 2, \ldots, h - 1)$$
$$\text{and any } y,\ 0 \leqslant y \leqslant 2\pi/\alpha. \tag{3,6,9}$$

The integral

$$\int_0^{2\pi/\alpha} \sin\alpha(a + y)\ dy$$

vanishes, so that by (3,6,8) and (3,6,9) we have

$$|I_k| = \left| \int_0^{2\pi/\alpha} \sin\alpha(a + y)\ \omega_k(y)\ dy \right| < \frac{2\pi}{\alpha}\ \varepsilon\ ;$$

$$\sum_{k=0}^{h-1} |I_k| < \frac{2h\pi}{\alpha}\ \varepsilon < (b - a)\ \varepsilon\ .$$

Since ε is arbitrarily small, this completes the proof of Lemma (3,6,1). ∎

Now let x denote a real variable over $(-\infty, +\infty)$; we shall say that a real or complex function $f(x)$ of x [over $(-\infty, +\infty)$] is of class (S_1) if it is Lebesgue-measurable and

$$\int_{-\infty}^{+\infty} |f(x)|\ dx < +\infty.$$

We shall say that $f(x)$ is of class (S_2) if it is of class (S_1) and in addition:

1) f(x) is continuous and has a continuous first derivative $\dot{f}(x) = \frac{d}{dx} f(x)$ and second derivative $\ddot{f}(x) = \frac{d^2}{dx^2} f(x)$;

2) the integrals

$$\int_{-\infty}^{+\infty} |\dot{f}(x)| \, dx \quad \text{and} \quad \int_{-\infty}^{+\infty} |\ddot{f}(x)| \, dx$$

are *finite*;

3) $\lim_{x \to \pm\infty} f(x) = \lim_{x \to \pm\infty} \dot{f}(x) = 0$;

4) $\int_{-\infty}^{\infty} |f(x)|^2 \, dx < + \infty$.

Now let r(t) be a complex-valued function of a real variable $t \in (- \infty, + \infty)$; the *Fourier transform* of this function is the complex-valued function $R(\nu)$ of the real variable $\nu \in (- \infty, + \infty)$ defined by

$$R(\nu) = \int_{-\infty}^{+\infty} e^{-2\pi i\nu t} \, r(t) \, dt, \qquad (3,6,10)$$

which we represent by the symbol

$$R = \mathfrak{F} [r]. \qquad (3,6,11)$$

Note that if t is interpreted as time, then ν may be interpreted as *frequency*; $\omega = 2\pi\nu$ may then be interpreted as *angular frequency*.

The problem of the existence of the integral (3,6,10) must be considered, and if it exists, its properties have to be studied.

The following theorem deals with this question.

THEOREM (3,6,1). a) If a complex-valued function r(t) of the real variable $t \in (- \infty, + \infty)$ is of class (S_1), then its Fourier transform $R(\nu)$ i.e., the function of the real variable $\nu \in (- \infty, + \infty)$ defined by

$$R(\nu) = \int_{-\infty}^{+\infty} e^{-2\pi i\nu t} \, r(t) \, dt, \text{ or } R = \mathfrak{F}[r], \qquad (3,6,12)$$

exists, and is bounded and uniformly continuous over $(-\infty, +\infty)$.

b) If *in addition* $r(t)$ is of class (S_2), then as a function of $\nu \in (-\infty, +\infty)$ $R(\nu)$ is of class (S_1), and the integral

$$\int_{-\infty}^{+\infty} |R(\nu)|^2 \, d\nu$$

is finite; finally we have the *reciprocity formula*:

$$r(t) = \int_{-\infty}^{+\infty} e^{2\pi i t\nu} R(\nu)\, d\nu, \text{ or } r = \mathfrak{F}^{-1}[R]. \qquad (3,6,13)$$

REMARK (3,6,3)

Note that by (3,6,12) and (3,6,13) $\mathfrak{F}$ and its inverse $\mathfrak{F}^{-1}$ are of the same form, except that i is replaced by -i; consequently, theorems valid for $\mathfrak{F}$ remain valid for $\mathfrak{F}^{-1}$, interchanging the roles of t and ν, $r(t)$ and $R(\nu)$.

The proof of a) is immediate. Let us proceed to b). In view of properties 1), 2), 3) of class (S_2), repeated integration by parts in (3,6,12) yields (for $\nu \neq 0$)

$$R(\nu) = \frac{1}{4\pi^2\nu^2} \int_{-\infty}^{+\infty} e^{-2\pi i\nu t}\, \ddot{r}(t)\, dt. \qquad (3,6,14)$$

If we put

$$a(\nu) = \int_{-\infty}^{+\infty} e^{2\pi i\nu t}\, \ddot{r}(t)\, dt,$$

then $a(\nu)$ is a bounded function of ν, uniformly continuous over $(-\infty, +\infty)$.

It follows that the function $s(t)$ defined by

$$s(t) = \int_{-\infty}^{+\infty} e^{2\pi i\nu t} R(\nu)\, d\nu$$

exists, and is bounded and uniformly continuous in t over $(-\infty, +\infty)$. Letting α denote a positive real number, we may write

$$s(t) = \lim_{\alpha\to+\infty} \int_{\alpha}^{+\alpha} e^{2\pi i\nu t} R(\nu)\, d\nu.$$

Now note that if β denotes a positive real number, then

$$\lim_{\beta\to+\infty} \int_{-\beta}^{+\beta} e^{-2\pi i\nu t}\, r(t)\, dt = R(\nu)$$

uniformly in ν over $(-\infty, +\infty)$. Thus, in view of (3,6,14),

$$s(t) = \lim_{\alpha\to+\infty} \lim_{\beta\to+\infty} \int_{-\alpha}^{+\alpha} \int_{-\beta}^{+\beta} e^{-2\pi i\nu(\tau-t)}\, r(\tau)\, d\nu\, d\tau.$$

Set

$$A(\alpha,\beta) = \int_{-\alpha}^{+\alpha} \int_{-\beta}^{+\beta} e^{-2\pi i\nu(\tau-t)}\, r(\tau)\, d\nu\, d\tau.$$

Integrating first with respect to ν, we get

$$A(\alpha,\beta) = \frac{1}{\pi} \int_{-\beta}^{+\beta} \frac{\sin 2\pi\alpha(\tau-t)}{\tau-t}\, r(\tau)\, d\tau. \qquad (3,6,15)$$

Let ε and a denote fixed numbers such that $0 < \varepsilon < a < \beta$, and put

$$B(\alpha) = \frac{1}{\pi} \int_{|\tau-t|\le\varepsilon} \frac{\sin 2\pi\alpha(\tau-t)}{\tau-t}\, r(\tau)\, d\tau, \qquad (3,6,16)$$

$$C(\alpha) = \frac{1}{\pi} \int_{\varepsilon<|\tau-t|\le a} \sin 2\pi\alpha(\tau-t)\, \frac{r(\tau)}{\tau-t}\, d\tau, \qquad (3,6,17)$$

$$D(\alpha,\beta) = \frac{1}{\pi} \int_{a<|\tau-t|\le\beta} \sin 2\pi\alpha(\tau-t)\, \frac{r(\tau)}{\tau-t}\, d\tau. \qquad (3,6,18)$$

Clearly

$$A(\alpha,\beta) = B(\alpha) + C(\alpha) + D(\alpha,\beta),$$

and

$$|D(\alpha,\beta)| \le \frac{1}{\pi a} \int_{a}^{+\infty} |r(\tau)|\, d\tau.$$

Thus if a is sufficiently large $D(\alpha,\beta)$ is arbitrarily small, uniformly in α and β if $\beta > a$.

By Lemma (3,6,1), for any fixed ε and a,

$$\lim_{\alpha \to +\infty} C(\alpha) = 0.$$

In the interval $[t - \varepsilon, t + \varepsilon]$, put $\tau = t + x$. By the mean-value theorem for this interval,

$$r(\tau) = r(t) + \dot{r}(t + \theta x)x,$$

where θ is some number between 0 and 1. Thus

$$B(\alpha) = r(t)\frac{1}{\pi}\int_{|x|\le\varepsilon} \frac{\sin 2\pi\alpha x}{x}\,dx + \frac{1}{\pi}\int_{|x|\le\varepsilon} \sin 2\pi\alpha x\,\dot{r}(t + \theta x)\,dx.$$

Since $\dot{r}$ is continuous, for any fixed $\varepsilon_0 > 0$ there exists a number $M > 0$ such that

$$|\dot{r}(t + \theta x)| \le M$$

for any x such that $|x| \le \varepsilon_0$. Consequently, the integral

$$\frac{1}{\pi}\int_{|x|\le\varepsilon} \sin 2\pi\alpha x\,\dot{r}(t + \theta x)\,dx$$

is arbitrarily small, if ε is sufficiently small.

On the other hand, it is known (cf. E. Goursat [1], Vol. 1, p. 235) that for any fixed $\varepsilon > 0$

$$\lim_{\alpha \to +\infty} \int_{|y|<\alpha\varepsilon} \frac{\sin y}{y}\,dy = \int_{-\infty}^{+\infty} \frac{\sin y}{y}\,dy = \pi. \qquad (3,6,19)$$

Since a may be chosen arbitrarily large and ε arbitrarily small, one easily sees that

$$s(t) = r(t),$$

which establishes (3,6,13) and part b) of Theorem (3,6,1).

Formula (3,6,13) is the *Fourier representation* of $r(t)$, i.e., its representation as a linear combination of exponential functions with purely imaginary exponents.

EXAMPLE (3,6,2)

Consider the function

$$r(t) = e^{-t^2/2} \qquad (3,6,20)$$

of $t \in (-\infty, +\infty)$, which is obviously of class (S_2). Let us find its Fourier representation. To this end, we determine its Fourier transform $R(\nu)$ according to (3,6,10):

$$R(\nu) = \int_{-\infty}^{+\infty} e^{-2\pi i\nu t - t^2/2}\, dt. \qquad (3,6,21)$$

Introduce the complex variable $z = t + 2\pi i\nu$, and the function

$$f(z) = e^{-z^2/2}.$$

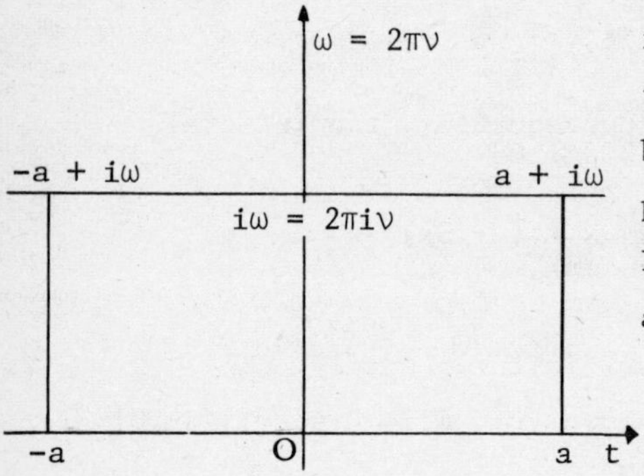

Fig. (3,6,1)

$f(z)$ is holomorphic throughout the complex z-plane. Let $J(\omega)$ be its integral in the complex plane along the straight line $\text{Im}z = 2\pi\nu$ parallel to the real axis [see Figure (3,6,1)]:

$$J(\omega) = \int_{t=-\infty}^{t=+\infty} e^{-z^2/2}\, dz.$$

Let a denote any positive real number. The integral of $f(z)$ along the closed contour $(a, a + i\omega, -a + i\omega, -a, a)$ must vanish. One easily shows that for any real ω, $J(\omega) = J(0)$, i.e., $J(\omega)$ is equal to the real integral:

$$\int_{-\infty}^{+\infty} e^{-t^2/2}\, dt.$$

It is known (cf. E. Goursat [1], Vol. 1, p. 331) that

$$\int_{-\infty}^{+\infty} e^{-t^2/2}\, dt = \sqrt{2\pi}, \qquad (3,6,22)$$

which reduces to (1,1,6) if we substitute $u = t^2/2$.

Now perform the change of variable $z = t + 2\pi i\nu = t + i\omega$ in (3,5,21); then

$$R(\nu) = e^{-\omega^2/2}\, J(\omega),$$

and consequently

$$r(t) = e^{-t^2/2} = \int_{-\infty}^{+\infty} e^{it\omega} \cdot \frac{1}{\sqrt{2\pi}}\, e^{-\omega^2/2}\, d\omega. \blacksquare \qquad (3,6,23)$$

More generally, if σ denotes any positive constant, we see that the Fourier representation of the function

$$e^{-(\sigma^2/2)t^2}$$

is

$$e^{-\sigma^2 t^2/2} = \int_{-\infty}^{+\infty} e^{i\omega t} \cdot \frac{1}{\sqrt{2\pi}\sigma}\, e^{-\omega^2/2\sigma^2}\, d\omega. \qquad (3,6,24)$$

Let $r(t)$, $s(t)$, $\phi(t)$ be three functions of $t \in (-\infty, +\infty)$, of class (S_2), and $R(\nu)$, $S(\nu)$, $f(\nu)$ their Fourier transforms. Consider the double integral

$$I = \int_{-\infty}^{+\infty} \int_{-\infty}^{+\infty} \phi(v - u)\, r(u)\, \overline{s(v)}\, du\, dv;$$

it can be verified that this integral exists. Replacing $\phi(v - u)$ by its value according to (3,6,13),

$$\phi(v - u) = \int_{-\infty}^{+\infty} e^{2\pi i(v-u)\nu}\, f(\nu)\, d\nu,$$

we may write

$$I = \int_{-\infty}^{+\infty} \int_{-\infty}^{+\infty} \int_{-\infty}^{+\infty} e^{2\pi i(v-u)\nu}\ f(\nu)\ r(u)\ \overline{s(v)}\ du\ dv\ d\nu.$$

In view of the properties of the class (S_2), the order of integration may be changed, and we get

$$I = \int_{-\infty}^{+\infty} \left(\int_{-\infty}^{+\infty} e^{-2\pi i\nu u}\ r(u)\ du \right) \left(\int_{-\infty}^{+\infty} e^{2\pi i\nu v}\ \overline{s(v)}\ dv \right) f(\nu)\ d\nu.$$

Combining this with (3,6,12) and Remark (3,6,3), we obtain the following lemma:

LEMMA (3,6,2)

Let $r(t)$, $s(t)$, $\phi(t)$ be three functions and $R(\nu)$, $S(\nu)$, $f(\nu)$ their Fourier transforms; if $r(t)$, $s(t)$, $\phi(t)$ are of class (S_2), or if $R(\nu)$, $S(\nu)$, $f(\nu)$ are of class (S_2), then

$$\int_{-\infty}^{+\infty} \int_{-\infty}^{+\infty} \phi(v-u)\ r(u)\ \overline{s(v)}\ du\ dv = \int_{-\infty}^{+\infty} f(\nu)\ R(\nu)\ \overline{S(\nu)}\ d\nu. \tag{3,6,25}$$

We shall now use Lemma (3,6,23) with $f(\nu) = e^{-\sigma^2\nu^2/2}$, where σ is an arbitrary positive constant. It follows immediately from (3,6,24) that

$$\phi(t) = \frac{\sqrt{2\pi}}{\sigma}\ e^{-2\pi^2 t^2/\sigma^2}.$$

Now let σ tend to $+0$; clearly, the right-hand side of (3,6,25) tends to

$$\int_{-\infty}^{+\infty} R(\nu)\ \overline{S(\nu)}\ d\nu.$$

On the other hand, let ε be an arbitrary fixed positive number, and put

$$J = \iint_{|v-u|>\varepsilon} \frac{\sqrt{2\pi}}{\sigma}\ e^{-2\pi^2(v-u)^2/\sigma^2}\ r(u)\ s(v)\ du\ dv,$$

$$H = \iint\limits_{|v-u| \leq \varepsilon} \frac{\sqrt{2\pi}}{\sigma} e^{-2\pi^2 (v-u)^2/\sigma^2} \, r(u) \, s(v) \, du \, dv.$$

The left-hand side of (3,6,25) is J + H. Clearly, for any fixed $\varepsilon > 0$,

$$\lim_{\sigma \to 0} J = 0.$$

As for H, putting v = u + w and

$$K(u) = \frac{\sqrt{2\pi}}{\sigma} \int_{-\varepsilon}^{+\varepsilon} e^{-2\pi^2 \omega^2/\sigma^2} \, \overline{s(u+w)} \, dw,$$

we get

$$H = \int_{-\infty}^{+\infty} r(u) \, K(u) \, du.$$

Now let

$$s(u+w) = s(u) + \Omega(w);$$

then, owing to the uniform continuity of the function s(u), $|\Omega(w)|$ has an upper bound over $[-\varepsilon, +\varepsilon]$ which is independent of u and infinitely small if $\varepsilon \to 0$. Using (3,6,22) and putting

$$K(u) = \overline{s(u)} + L(u),$$

we see that L(u) tends to 0 as $\varepsilon \to +0$, uniformly in u and σ. It follows that as $\sigma \to +0$, the left-hand side of (3,6,25) tends to

$$\int_{-\infty}^{+\infty} r(u) \, \overline{s(u)} \, du.$$

Combining this with Remark (3,6,3), we have:

THEOREM (3,6,2). Let r(t) and s(t) be two functions of $t \in (-\infty, +\infty)$ and $R(\nu)$ and $S(\nu)$ their respective Fourier transforms $[\nu \in (-\infty, +\infty)]$; suppose r(t) and s(t) are of class (S_2) (or $R(\nu)$ and $S(\nu)$ of class (S_2)).

Then

$$\int_{-\infty}^{+\infty} r(t)\ \overline{s(t)}\ dt = \int_{-\infty}^{+\infty} R(\nu)\ \overline{S(\nu)}\ d\nu. \qquad (3,6,26)$$

CONVOLUTION

Let $r(t)$ and $s(t)$ be two functions of $t \in (-\infty, +\infty)$ of class (S_1). The convolution[1] of r by s, denoted by $s * r$, is the function of $t \in (-\infty, +\infty)$ defined by

$$\int_{-\infty}^{+\infty} s(t-\tau)\ r(\tau)\ d\tau. \qquad (3,6,27)$$

It is immediate that $s * r$ is of class (S_1), $s * r = r * s$ (commutativity), and that convolution is associative.

Let $R(\nu)$ and $S(\nu)$ be the Fourier transforms of $r(t)$ and $s(t)$. Then

$$\int_{-\infty}^{+\infty} e^{-2\pi i\nu t} \left(\int_{-\infty}^{+\infty} s(t-\tau)\ r(\tau)\ d\tau\right) dt$$

$$= \int_{-\infty}^{+\infty}\int_{-\infty}^{+\infty} e^{-2\pi i\nu t}\ s(t-\tau)\ r(\tau)\ dt\ d\tau$$

$$= \int_{-\infty}^{+\infty}\int_{-\infty}^{+\infty} e^{-2\pi i\nu(t-\tau)}\ s(t-\tau).\ e^{-2\pi i\nu\tau}\ r(\tau)\ dt\ d\tau$$

$$= \int_{-\infty}^{+\infty}.\int_{-\infty}^{+\infty} e^{-2\pi i\nu x}\ s(x).\ e^{-2\pi i\nu\tau}\ r(\tau)\ dx\ d\tau$$

(by the change of variable $t = \tau + x$). This result may also be obtained by applying Theorem (3,6,2) and Theorem (3,6,1).

---

[1] French: produit de convolution (or convolution, produit de composition); German: Faltung.

Consequently

THEOREM (3,6,3). The convolution $s * r = r * s$ of two functions $r(t)$ and $s(t)$ of class (S_1) over $t \in (-\infty, +\infty)$, is a function of $t \in (-\infty, +\infty)$, of class (S_1), whose Fourier transform is the product of the Fourier transforms of $r(t)$ and $s(t)$.

Fourier transform in $\mathcal{L}_2$

Let $\mathcal{L}_2$ be the space of Lebesgue-measurable functions $r(t)$ of $t \in (-\infty, +\infty)$ such that

$$\int_{-\infty}^{+\infty} |r(t)|^2 \, dt < +\infty. \tag{3,6,28}$$

It was shown in II.10 that $\mathcal{L}_2$ is a separable Hilbert space. The hermitian product r.s of an element $r \in \mathcal{L}_2$ and an element $s \in \mathcal{L}_2$ is defined by

$$r.s = \int_{-\infty}^{+\infty} r(t) \, \overline{s(t)} \, dt. \tag{3,6,29}$$

A variable element $r \in \mathcal{L}_2$ tends to a fixed element $s \in \mathcal{L}_2$ if

$$\lim \int_{-\infty}^{+\infty} |r(t) - s(t)|^2 \, dt = 0.$$

By property 4) of the class (S_2) (note that this is its first application), the functions of class (S_2) form a vector subspace $\mathcal{L}_2'$ of $\mathcal{L}_2$. $\mathcal{L}_2'$ is not closed, but it is well known that $\mathcal{L}_2'$ is dense in $\mathcal{L}_2$. Now Theorems (3,6,1) and (3,6,3) prove that the Fourier transform $\mathfrak{F}$ is a linear mapping of $\mathcal{L}_2'$ into $\mathcal{L}_2$ which preserves hermitian products. By the procedure indicated in Remark (2,10,1), the mapping $\mathfrak{F}$ may be extended to $\mathcal{L}_2$ as an *isometry* $\mathfrak{F}_c$ of $\mathcal{L}_2$ onto $\mathcal{L}_2$.

Thus the natural domain of definition of the Fourier transform is the space $\mathcal{L}_2$ defined by (3,6,28). For any $r \in \mathcal{L}_2$, we define its Fourier transform $R \in \mathcal{L}_2$ by

$$R = \mathfrak{F}_c[r].$$

In view of the fact that $\mathcal{F}_c$ is an isometry, formula (3,6,28) may be written:

$$\int_{-\infty}^{+\infty} |r(t)|^2 \, dt = \int_{-\infty}^{+\infty} |R(\nu)|^2 \, d\nu < +\infty.$$

THEOREM (3,6,4) (Uniqueness). 1) If r(t) is of class (S_2), its Fourier transform is identically zero if and only if r(t) is identically zero.

2) If r(t) belongs to $\mathcal{L}_2$, its Fourier transform $R(\nu) = \mathcal{F}_c[r]$ vanishes almost everywhere (with respect to Lebesgue measure over $\nu \in (-\infty, +\infty)$) if and only if r(t) vanishes almost everywhere (with respect to Lebesgue measure over $t \in (-\infty, +\infty)$).

3) If r(t) is of class (S_1), its Fourier transform $\mathcal{F}_c[r]$ is identically zero if and only if r(t) is zero almost everywhere [with respect to Lebesgue measure over $t \in (-\infty, +\infty)$].

Part 1) is an immediate consequence of (3,6,13); part 2) results from the fact that $\mathcal{F}_c$ is an isometry of $\mathcal{L}_2$ onto itself; we leave it to the reader to prove part 3 by using 2).

THEOREM (3,6,5) (Inversion). If r(t) and its Fourier transform R(ν) are of class (S_1), then

$$r(t) = \int_{-\infty}^{+\infty} e^{2\pi i t\nu} R(\nu) \, d\nu$$

for any value of t for which r(t) is continuous.

The reader may find a proof of Theorem (3,6,5) in R. Goldberg [1].

7. *Positive-definite functions*

Let u be a real variable over $(-\infty, +\infty)$, and $\phi(u)$ a complex-valued function of u. $\phi(u)$ is said to be positive-definite[1] if for any positive integer n, any n-tuple $\{u_1, u_2, \ldots, u_n\}$ of values u, and n-tuple $\{\lambda_1, \lambda_2, \ldots, \lambda_n\}$ of complex numbers λ_j, the number

(1) German: positiv bestimmte; French: définie positive.

$$\sum_{j,k} \phi(u_k - u_j)\, \lambda_j \bar{\lambda}_k \qquad (3,7,1)$$

is always real and nonnegative:

$$\sum_{j,k} \phi(u_k - u_j)\, \lambda_j \bar{\lambda}_k \geq 0. \qquad (3,7,2)$$

In other words, the $n \times n$ matrix with elements $\phi(u_k - u_j)$ is always positive-definite.

The properties of positive-definite matrices reviewed in III.5 immediately yield the following results.

1) If $\phi(u)$ is a positive-definite function of $u \in (-\infty, +\infty)$, then

a) its product $\rho\phi(u)$ by a nonnegative real constant ρ is positive-definite;

b) $\phi(u)$ is "hermitian," i.e.,

$$\phi(-u) = \overline{\phi(u)}, \qquad (3,7,3)$$

so that (3,7,1) is a hermitian form, which is moreover positive-definite. $\phi(0)$, which is therefore real, must also be nonnegative, and for any $u \in (-\infty, +\infty)$ we have

$$|\phi(u)| \leq \phi(0). \qquad (3,7,4)$$

Inequality (3,7,4) shows that any positive-definite function is bounded; it may be proved by putting $n = 2$, $u_1 = 0$, $u_2 = u$ in (3,7,2).

c) $\overline{\phi(u)}$ is positive-definite.

2) If $\phi_1(u)$ and $\phi_2(u)$ are two positive-definite functions of $u \in (-\infty, +\infty)$, then their sum $\phi_1(u) + \phi_2(u)$ and product $\phi_1(u)\phi_2(u)$ are positive-definite [cf. Lemma (3,5,1)]. In particular, if $\phi(u)$ is a positive-definite function of $u \in (-\infty, +\infty)$ its real part $\frac{1}{2}\{\phi(u) + \overline{\phi(u)}\}$ and its squared modulus $|\phi(u)|^2 = \phi(u)\overline{\phi(u)}$ are positive-definite.

3) If a function $\phi(u)$ is the limit of positive-definite functions over $u \in (-\infty, +\infty)$, then $\phi(u)$ is positive-definite.

4) If $\phi(u)$ is positive-definite over $u \in (-\infty, +\infty)$, then for

any u and $\Delta u \in (-\infty, +\infty)$ we have

$$|\varphi(u + \Delta u) - \varphi(u)|^2 \leqslant (\varphi(0)^2 - |\varphi(\Delta u)|^2) + 2\,\varphi(0)\,|\varphi(0) - \varphi(\Delta u)|\,. \quad (3,7,5)$$

Inequality (3,7,5) is proved by setting $n = 3$, $u_1 = 0$, $u_2 = u$, $u_3 = u + \Delta u$ in (3,7,2); the matrix

$$\begin{pmatrix} \varphi(0) & \overline{\varphi(u)} & \overline{\varphi(u+\Delta u)} \\ \varphi(u) & \varphi(0) & \overline{\varphi(\Delta u)} \\ \varphi(u+\Delta u) & \varphi(\Delta u) & \varphi(0) \end{pmatrix}$$

must be positive-definite, and consequently its determinant must be real and nonnegative. Using (3,7,4), this implies (3,7,5).

It follows from (3,7,5) that if $\phi(u)$ is positive-definite, it is continuous over $u \in (-\infty, +\infty)$ if and only if it is continuous at $u = 0$; it is then *uniformly* continuous over $(-\infty, +\infty)$.

As a consequence of properties 1) and 2) it is clear that any nonnegative integral power of a positive-definite function is positive definite; any finite linear combination of positive-definite functions with real nonnegative coefficients is positive-definite; this also holds for infinite linear combinations, provided they are convergent.

LEMMA (3,7,1)

If $\phi(u)$ is a continuous function of $u \in (-\infty, +\infty)$, it is positive-definite (over $(-\infty, +\infty)$) if and only if for any function $A(u)$ of $u \in (-\infty, +\infty)$ the double Riemann integral

$$\int_{-\infty}^{+\infty} \int_{-\infty}^{+\infty} \varphi(v - u)\, A(u)\, \overline{A(v)}\, du\, dv \qquad (3,7,6)$$

is real and nonnegative, provided it exists and is absolutely convergent.

Lemma (3,7,1) results from the fact that the Riemann integral (3,7,6) is the limit of "Riemann sums" which are double sums of the type (3,7,1).

The integral (3,7,6) may be real and nonnegative for functions

A(u) for which (3,7,6) is not a Riemann integral but only a Lebesgue integral; we leave the study of such extensions to the reader.

REMARK (3,7,1)

If $\phi(u)$ is a positive-definite function of $u \in (-\infty, +\infty)$ and h is a nonzero real constant, then $\phi(h + u)$ is not a positive-definite function of $u \in (-\infty, +\infty)$ unless $\phi(u)$ is a constant (necessarily real and nonnegative).

Fourier transforms of bounded distribution functions

Let F(x) be any *bounded* distribution function over $(-\infty < x < +\infty)$. It is natural to define the Fourier transform of F(x) as the function of $v \in (-\infty, +\infty)$ defined by the RS-integral

$$\int_{-\infty}^{+\infty} e^{-2\pi ivx}\, dF(x).$$

To simplify the notation (and to conform with accepted usage) we shall replace the variable v by the variable $u = -2\pi v$.

We shall therefore define the Fourier transform of the bounded d.f. F(x) as the function $\phi(u)$ of $u \in (-\infty, +\infty)$ defined by the RS-integral

$$\varphi(u) = \int_{-\infty}^{+\infty} e^{iux}\, dF(x). \qquad (3,7,7)$$

If F(x) is absolutely continuous with density f(x), (3,7,7) is a Lebesgue integral:

$$\varphi(u) = \int_{-\infty}^{+\infty} e^{iux}\, f(x)\, dx, \qquad (3,7,8)$$

which is even a Riemann integral if f(x) is sufficiently regular, e.g., continuous. If F(x) is totally discontinuous, then, in the notation of (3,3,3), the integral (3,7,7) becomes a series:

$$\varphi(u) = \sum_{n} e^{iux_n}\, s(x_n). \qquad (3,7,9)$$

F(x) occurs in (3,7,7) only up to equivalence: equivalent

bounded d.f. have the same Fourier transform. Thus, if necessary, when using (3,7,7) we can assume without loss of generality that F(x) is normalized.

It is immediate that $\phi(u)$ possesses the following properties:

1) $\phi(0) = F(+\infty) - F(-\infty)$, i.e., $\phi(0)$ is equal to the total variation (necessarily nonnegative) of F(x);

2) $\phi(u)$ is a continuous function, even uniformly continuous over $(-\infty < u < +\infty)$.

Even more:

LEMMA (3,7,2)

The Fourier transform $\phi(u)$ of a bounded distribution function F(x) is continuous and positive-definite.

Indeed, note first that for any $x \in (-\infty, +\infty)$, any positive integer n, any n real numbers (distinct or not) $u_1, u_2, \ldots, u_n$, and any n complex numbers $a_1, a_2, \ldots, a_n$, the expression

$$\sum_{j,k=1}^{n} e^{i(u_k - u_j)x} a_j \overline{a_k}$$

is always real and nonnegative, since

$$\sum_{j,k=1}^{n} e^{i(u_k-u_j)x} a_j \overline{a_k} = \left(\sum_{j=1}^{n} e^{-iu_j x} a_j\right)\left(\sum_{k=1}^{n} e^{iu_k x} \overline{a_k}\right)$$

$$= \left(\sum_{j=1}^{n} e^{-iu_j x} a_j\right)\overline{\left(\sum_{k=1}^{n} e^{-iu_k x} a_k\right)}$$

$$= \left|\sum_{j=1}^{n} e^{-iu_j x} a_j\right|^2 \geqslant 0.$$

In other words, for any $x \in (-\infty, +\infty)$ the exponential function e^{iux} is a positive-definite function of $u \in (-\infty, +\infty)$.

Now

$$\sum_{j,k=1}^{n} \varphi(u_k - u_j)\, a_j \overline{a_k} = \sum_{j,k=1}^{n} \left[\int_{-\infty}^{+\infty} e^{i(u_k-u_j)x}\, dF(x)\right] a_j \overline{a_k}$$

$$= \int_{-\infty}^{+\infty} \left[\sum_{j,k=1}^{n} e^{i(u_k-u_j)x} a_j \overline{a_k}\right] dF(x)$$

and, by the previous remark, this expression must be nonnegative.

EXAMPLE (3,7,1)

The function

$$f(x) = \frac{e^{-\frac{x^2}{2\sigma^2}}}{\sqrt{2\pi}\,\sigma} \qquad (\sigma \text{ a positive constant}),$$

which is of class (S_2), is real and nonnegative; the integral

$$F(x) = \int_{-\infty}^{x} \frac{e^{-\frac{y^2}{2\sigma^2}}}{\sqrt{2\pi}\,\sigma}\, dy$$

is a bounded, absolutely continuous d.f., whose Fourier transform is, by Example (3,6,2),

$$\varphi(u) = e^{-\frac{\sigma^2 u^2}{2}} .$$

The function $e^{-\frac{\sigma^2 u^2}{2}}$ is therefore positive-definite. ∎

Several questions arise here; in particular:

a) Is the correspondence between bounded d.f. and their Fourier transforms one-to-one - of course, up to equivalence of d.f.? If it is not, the idea of associating a bounded d.f. with its Fourier transform is of only limited interest.

b) Assuming that this correspondence is one-to-one, its significant interest will depend on its specific properties, which must therefore be investigated. In particular, is the correspondence continuous in some sense?

c) We have just seen that the Fourier transform of a bounded d.f. is continuous and positive-definite; is the converse true, i.e. is any positive-definite continuous function $\phi(u)$ the Fourier transform of a bounded d.f.?

The rest of this section will be devoted to answering these questions. We first prove:

THEOREM (3,7,1). If a countable sequence of bounded normalized d.f. $F_k(k = 1,2,3,\ldots)$ converges to a bounded d.f. F as $k \to +\infty$ and in addition $F_k(+\infty)$ converges to $F(+\infty)$, then the Fourier transform $\phi_k(u)$ of F_k converges to the Fourier transform $\phi(u)$ of F uniformly in u over $(-\infty, +\infty)$.

Let a and b be two positive numbers such that -a and +b are continuity points of F, so large that

$$\int_{-\infty}^{-a} dF(x) + \int_{b}^{+\infty} dF(x) = [F(-a) - F(-\infty)] + [F(+\infty) - F(b)] < \varepsilon,$$

where ε is an arbitrary fixed positive number. Now

$$\left|\int_{-\infty}^{-a} e^{iux}\, dF_k(x) + \int_{b}^{+\infty} e^{iux}\, dF_k(x)\right| \leq [F_k(-a) - F_k(-\infty)] + [F_k(+\infty) - F_k(b)] \quad ; \tag{3,7,10}$$

therefore for large enough k the left-hand side of (3,7,10) is bounded by 2ε, uniformly in u. Now choose n + 1 points in the interval (-a, b):

$$x_0 = -a < x_1 < x_2 < \dots < x_j < x_{j+1} < \dots < x_n = b,$$

all continuity points of F. e^{iux} is a continuous function of x over (-a, +b), uniformly in u. It follows that the integrals

$$\int_{-a}^{b} e^{iux}\, dF_k(x), \quad \int_{-a}^{b} e^{iux}\, dF(x)$$

may be approximated arbitrarily closely, uniformly in u, by the Riemann-Stieltjes sums

$$\sum_j e^{iux_j} [F_k(x_{j+1}) - F_k(x_j)], \quad \sum_j e^{iux_j} [F(x_{j+1}) - F(x_j)].$$

But it is clear that as $k \to +\infty$ the first of these sums converges to the second, uniformly in u. Theorem (3,7,1) follows. ∎

Suppose that a bounded d.f. F(x) has density f(x) of class (S_2). The Fourier transform $\phi(u)$ of F(x) is given by (3,7,8), and by the inversion formula (3,6,13) it follows that

$$f(x) = \frac{1}{2\pi} \int_{-\infty}^{+\infty} e^{-iux}\, \varphi(u)\, du. \tag{3,7,11}$$

Thus, in the particular case where f exists and is of class (S_2) the correspondence between F and ϕ is indeed one-to-one, and

(3,7,11) provides a method for determining F up to equivalence, given ϕ.

More precisely, still assuming $f(x) \in (S_2)$, we have

$$f(x) = \lim_{a \to +\infty} \frac{1}{2\pi} \int_{-a}^{+a} e^{-iux} \varphi(u)\, du,$$

uniformly in x. Thus for any α and β $(\alpha < \beta)$,

$$F(\beta) - F(\alpha) = \int_{\alpha}^{\beta} f(x)\, dx = \lim_{a \to +\infty} \frac{1}{2\pi} \int_{-a}^{+a} \int_{\alpha}^{\beta} e^{-iux} \varphi(u)\, du\, dx$$

$$= \lim_{a \to +\infty} \frac{i}{2\pi} \int_{-a}^{+a} \frac{e^{-i\beta u} - e^{-i\alpha u}}{u} \varphi(u)\, du. \quad (3,7,12)$$

Now consider the case of an arbitrary bounded d.f. F(x), where α and β $(\alpha < \beta)$ are any two continuity points of F(x); consider the integral

$$A(a) = \frac{i}{2\pi} \int_{-a}^{+a} \frac{e^{-i\beta u} - e^{-i\alpha u}}{u} \varphi(u)\, du,$$

where a is an arbitrary positive number. Putting

$$\Lambda(y) = \int_{0}^{y} \frac{\sin v}{v}\, dv,$$

and replacing $\phi(u)$ by its expression (3,7,7), we have

$$A(a) = \frac{1}{\pi} \int_{-\infty}^{+\infty} \{\Lambda[(\beta - x)a] - \Lambda[(\alpha - x)a]\}\, dF(x).$$

By (3,6,19), $\Lambda(y)$ tends to $\pi/2$ as $y \to +\infty$, while $\Lambda(y)$ tends to $-\pi/2$ as $y \to -\infty$. Hence,

$$\lim_{a \to +\infty} A(a) = F(\beta) - F(\alpha),$$

i.e., (3,7,12) remains true:

THEOREM (3,7,2). A bounded distribution function F(x) is uniquely determined, up to equivalence, by its Fourier transform $\phi(u)$; in fact, for any pair (α,β) of continuity points of F(x) $(\alpha < \beta)$ we have

$$F(\beta) - F(\alpha) = \lim_{a \to +\infty} \frac{i}{2\pi} \int_{-a}^{+a} \frac{e^{-i\beta u} - e^{-i\alpha u}}{u} \varphi(u)\, du. \quad (3,7,13)$$

REMARK (3,7,2)

Formula (3,7,13) is inconvenient in that it expresses $F(\beta) - F(\alpha)$ as a limit; a modified form of (3,7,13) avoids this shortcoming.

$\mathcal{R}[\phi(u)]$ is an even function, while $\mathcal{I}[\phi(u)]$ is odd; consequently,

$$\int_{-a}^{+a} \mathcal{I}\left[\frac{e^{-i\beta u} - e^{-i\alpha u}}{u} \phi(u)\right] du = 0 \text{ for any } a;$$

If α and β $(\alpha < \beta)$ again denote arbitrary continuity points of $F(x)$, formula (3,7,13) is easily expressed in the *finite* form

$$F(\beta) - F(\alpha) = \frac{i}{2\pi} \int_{-\infty}^{+\infty} \mathcal{R}\left[\frac{e^{-i\beta u} - e^{i\alpha u}}{u} \phi(u)\right] du. \quad (3,7,14)$$

Using (3,7,13) we can prove

THEOREM (3,7,3). If the Fourier transform $\phi(u)$ of a bounded distribution function $F(x)$ is of class (S_1), i.e.,

$$\int_{-\infty}^{+\infty} |\phi(u)| \, du < +\infty,$$

then $F(x)$ has a density $f(x)$ which is given by

$$f(x) = \frac{i}{2\pi} \int_{-\infty}^{+\infty} e^{-iux} \phi(u) \, du. \quad (3,7,15)$$

Note that (3,7,15) is not an immediate consequence of the Inversion Theorem (3,6,5); a proof of (3,7,15) is in fact included in that of Theorem (3,7,4), which we now formulate.

THEOREM (3,7,4) (Bochner). A complex function $\phi(u)$ of $u \in (-\infty, +\infty)$ is the Fourier transform of a bounded distribution function if and only if it is continuous and positive-definite.

Necessity is proved in Lemma (3,7,2); we shall prove that the condition is sufficient. Let $\phi(u)$ be a continuous and positive-definite function over $(-\infty, +\infty)$.

If $\phi(u)$ is of class (S_2), sufficiency is easily established by using the inversion formula (3,6,13). However, since $\phi(u)$ is not necessarily of class (S_2), the general proof is more complicated.

If σ is any positive number, the continuous function

$$e^{-\sigma^2 u^2/2}$$

is positive-definite (Example (3,7,1)). By property 2) of positive-definite functions, the function

$$\phi_\sigma(u) = \phi(u)\, e^{-\sigma^2 u^2/2}$$

is positive-definite and continuous. Moreover,

$$|\phi_\sigma(u)| \le \phi(0)\, e^{-\sigma^2 u^2/2}.$$

It follows that the function

$$f_\sigma(x) = \frac{1}{2\pi}\int_{-\infty}^{+\infty} e^{-iux}\, \phi_\sigma(u)\, du \qquad (3,7,16)$$

exists and is bounded and uniformly continuous in x. We claim that

$$\phi_\sigma(u) = \int_{-\infty}^{+\infty} e^{iux}\, f_\sigma(x)\, dx. \qquad (3,7,17)$$

Let α be any positive number, and put

$$\phi_{\sigma,\alpha}(u) = \int_{-\infty}^{+\infty} \frac{1}{\sqrt{2\pi}\alpha}\, e^{-(u-v)^2/2\alpha^2}\, \phi_\sigma(v)\, dv = \int_{-\infty}^{+\infty} \frac{1}{\sqrt{2\pi}\alpha}\, e^{-v^2/2\alpha^2}\, \phi_\sigma(u-v)\, dv$$

Note that

$$\lim_{\alpha\to+0} \phi_{\sigma,\alpha}(u) = \phi_\sigma(u)$$

uniformly in u, moreover $\phi_{\sigma,\alpha}(u)$ is of class (S_2). It can be verified that

$$\frac{1}{2\pi}\int_{-\infty}^{+\infty} e^{-iux}\, \phi_{\sigma,\alpha}(u)\, du = e^{-\alpha^2 x^2/2}\, f_\sigma(x),$$

and since $\phi_{\sigma,\alpha}(u)$ is of class (S_2) it follows from (3,6,13) that

$$\phi_{\sigma,\alpha}(u) = \int_{-\infty}^{+\infty} e^{iux}\, e^{-\alpha^2 x^2/2}\, f_\sigma(x)\, dx. \tag{3,7,18}$$

Finally, let β be any positive number, and put

$$R(x) = \frac{1}{\sqrt{2\pi}\beta}\, e^{-(x-x_0)^2/2\beta^2}.$$

The Fourier transform of this function is

$$r(u) = \int_{-\infty}^{+\infty} e^{iux}\, R(x)\, dx = e^{-(\beta^2/2)u^2 + ix_0 u}.$$

Further, put

$$H_\sigma = \int_{-\infty}^{+\infty}\int_{-\infty}^{+\infty} \phi_\sigma(v-u)\, r(u)\, \overline{r(v)}\, du\, dv,$$

$$H_{\sigma,\alpha} = \int_{-\infty}^{+\infty}\int_{-\infty}^{+\infty} \phi_{\sigma,\alpha}(v-u)\, r(u)\, \overline{r(v)}\, du\, dv.$$

The integrals H_σ, $H_{\sigma,\alpha}$ exist, and

$$\lim_{\alpha\to+0} H_{\sigma,\alpha} = H_\sigma.$$

Moreover, Lemma (3,7,1) implies that H_σ is real and

$$H_\sigma \geq 0. \tag{3,7,19}$$

But by Lemma (3,6,2)

$$H_{\sigma,\alpha} = 2\pi \int_{-\infty}^{+\infty} \frac{1}{\beta^2} e^{-(x-x_0)^2/\beta^2}\, e^{-\alpha^2 x^2/2}\, f_\sigma(x)\, dx.$$

Hence

$$H_\sigma = \lim_{\alpha\to+0} H_{\sigma,\alpha} = 2\pi \int_{-\infty}^{+\infty} \frac{1}{\beta^2}\, e^{-(x-x_0)^2/\beta^2}\, f_\sigma(x)\, dx,$$

and consequently, by (3,7,19),

$\frac{1}{\sqrt{\pi}\beta} \int_{-\infty}^{+\infty} e^{-(x-x_0)^2/\beta^2}\, f_\sigma(x)\, dx$ is real and nonnegative. But

$$\lim_{\beta\to+0} \frac{1}{\sqrt{\pi}\beta} \int_{-\infty}^{+\infty} e^{-(x-x_0)^2/\beta^2}\, f_\sigma(x)\, dx = f_\sigma(x_0);$$

therefore f(x) is a *nonnegative real function.*

In particular, it follows from (3,7,18) that

$$\phi_{\sigma,\alpha}(0) = \int_{-\infty}^{+\infty} e^{-\alpha^2x^2/2} f_\sigma(x)\, dx.$$

Since

$$\lim_{\alpha\to+0} \phi_{\sigma,\alpha}(0) = \phi_\sigma(0) = \phi(0),$$

it follows that the integral

$$\int_{-\infty}^{+\infty} f_\sigma(x)\, dx = \phi(0)$$

is finite. Now consider the bounded d.f. F defined (up to equivalence) by

$$dF_\sigma(x) = f_\sigma(x)\, dx.$$

Theorem (3,4,1) is applicable to the family, indexed by σ, of functions F_σ; then there exist a bounded d.f. F and a sequence $\{\sigma_k\}$ of values of σ such that

$$\sigma_k > 0, \quad \lim_{k\to+\infty} \sigma_k = 0, \quad \lim_{k\to+\infty} F_{\sigma_k} = F.$$

It is clear that

$$\phi(u) = \lim_{k\to+\infty} \phi_{\sigma_k}(u) = \lim_{k\to+\infty} \int_{-\infty}^{+\infty} e^{iux}\, dF_{\sigma_k}(x) = \int_{-\infty}^{+\infty} e^{iux}\, dF(x),$$

and this completes the proof of Theorem (3,7,4).

8. *Converse of Theorem* (3,7,1)

Let $F_k (k = 1,2,\ldots)$ be a countable sequence of bounded normalized d.f. with Fourier transforms ϕ_k. Suppose that the $\phi_k(u)$ converge to a limit $\phi(u)$ as $k \to +\infty$. We know that $\phi(u)$ is positive-definite; moreover, the fact that $\phi_k(0) = F_k(+\infty)$ converges to $\phi(0)$ implies that the F_k form a family of d.f. uniformly bounded in k. Let $\mathfrak{F}$ be the family of all d.f. F such that there exists a subsequence of indices k' such that

$$\lim_{k'\to+\infty} F_{k'} = F.$$

By Theorem (3,4,1) $\mathfrak{F}$ is not empty.

Let $F \in \mathfrak{F}$, and denote the Fourier transform of F by $\psi(u)$.

If

$$\varphi(0) = \lim_{k' \to +\infty} F_{k'}(+\infty) = F(+\infty), \tag{3,8,1}$$

it follows from Theorem (3,7,1) that

$$\varphi(u) = \lim_{k' \to +\infty} \varphi_{k'}(u) = \psi(u). \tag{3,8,2}$$

However, (3,8,1) need not be true (cf. Exercise 3.4).

Let $\alpha > 0$ be arbitrary, and consider the integral

$$\int_{-a}^{+a} \varphi_k(u)\, du = \int_{-a}^{+a} \int_{-\infty}^{+\infty} e^{iux}\, du\, dF_k(x)$$

$$= \int_{-\infty}^{+\infty} \left[\int_{-a}^{+a} e^{iux}\, du \right] dF_k(x)$$

$$= 2 \int_{-\infty}^{+\infty} \frac{\sin \alpha x}{x}\, dF_k(x).$$

If a is an arbitrary positive number, then

$$\left| \int_{|x|>a} \frac{\sin \alpha x}{x}\, dF_k(x) \right| \leqslant \frac{1}{a}\,[F_k(+\infty) - F(a \pm 0)] + \frac{1}{a}\,[F(-a \pm 0) - F(-\infty)]$$

$$\leqslant \frac{2\,\varphi(0)}{a};$$

in other words, the integrals $\int_{-\infty}^{+\infty} \frac{\sin \alpha x}{x}\, dF_k(x)$ converge uniformly in k and α. Thus, for any α

$$\lim_{k' \to +\infty} \int_{-a}^{+a} \varphi_{k'}(u)\, du = \int_{-a}^{+a} \psi(u)\, du.$$

Now suppose that ϕ_k converges to ϕ uniformly in u for $|u| < \eta$, where η is any fixed positive number; for all $\alpha < \eta$, we have

$$\lim_{k' \to +\infty} \int_{-a}^{+a} \varphi_{k'}(u)\, du = \int_{-a}^{+a} \varphi(u)\, du;$$

therefore:

$$\int_{-a}^{+a} \varphi(u)\, du = \int_{-a}^{+a} \psi(u)\, du. \tag{3,8,3}$$

Since (3,8,3) holds for any $\alpha < \eta$, it follows that

$$\psi(u) = \phi(u) \text{ for } |u| < \eta;$$

in particular, for u = 0,

$$F(+\infty) = \psi(0) = \varphi(0) = \lim_{k'\to+\infty} F_{k'}(+\infty).$$

Thus (3,8,1) holds, and this implies (3,8,2): since F is an arbitrary element of $\mathfrak{F}$, it follows that $\mathfrak{F}$ can contain *only one* element F, whose Fourier transform is $\phi(u)$. Thus, as $k \to +\infty$ the sequence F_k converges to a unique element F such that $\lim_{k\to+\infty} F_k(+\infty) = F(+\infty)$. Thus:

THEOREM (3,8,1). Let F_k be a countable sequence of bounded normalized distribution functions with Fourier transforms $\phi_k (k = 1,2, 3,...)$. Suppose that for any $u \in (-\infty, +\infty)$ the function $\phi_k(u)$ converges to a limit $\phi(u)$ as $k \to +\infty$, and that there exists a positive number η, such that $\phi_k(u)$ converges to $\phi(u)$ uniformly in u over the interval $(-\eta, +\eta)$. Then there exists a bounded normalized distribution function F such that

1) $\lim_{k\to+\infty} F_k = F$, $\lim_{k\to+\infty} F_k(+\infty) = F(+\infty)$;

2) ϕ is the Fourier transform of F.

REMARK (3,8,1)

If m_k, m denote the measures defined by the d.f. F_k, F of Theorem (3,8,1), this means that under the above conditions m_k converges *completely* to m as $k \to +\infty$.

Convolution of two bounded distribution functions

Let F(x), G(x), be two bounded d.f. such that $F(-\infty) = G(-\infty) = 0$; the *convolution* $G * F$ of F by G is the function H defined by

$$H(x) = \int_{-\infty}^{+\infty} G(x - y)\, dF(y). \qquad (3,8,4)$$

The integral (3,8,4) is an integral of type (3,3,1); it is a RS-integral if, e.g., G(x) is continuous.

The integral (3,8,4) exists for any x, and the function H(x) is itself a bounded d.f. such that $H(-\infty) = 0$. If G' is a d.f. equivalent to G, $G'(-\infty) = 0$, then $G' * F$ is equivalent to $G * F$. In this connection, note that if G is normalized, i.e. left-continuous, then H is also left-continuous and therefore normalized.

Let M denote the set of equivalence classes of bounded d.f. which vanish when $x = -\infty$. The above remarks show that the convolution may be regarded as an (internal) operation in M, or (see III, 1), an operation in the set of bounded measures over R (the space of real numbers). The latter interpretation is clearly the most meaningful; *we adopt it from now on* throughout this book; in other words, the d.f. occurring in convolutions will be considered only up to *equivalence*.

The following assertions are now evident.

1) Convolution is commutative; indeed, under a transformation of type (3,3,8) the right-hand side of (3,8,4) becomes

$$\int_{-\infty}^{+\infty} F(x - y)\, dG(y). \tag{3,8,5}$$

In other words, the convolution $F * G$ of G by F coincides with the convolution $G * F$ of F by G; we shall therefore say that the function H defined by (3,8,4) or (3,8,5) is the convolution of F *and* G.

2) M contains an identity element with respect to the convolution operation - the Dirac d.f. $\Delta(x)$ (cf. III.2); indeed, it is immediate that for any d.f. G

$$\int_{-\infty}^{+\infty} G(x - y)\, d\Delta(y) = G(x) \quad \text{(up to equivalence)}.$$

3) By (3,8,4), if G is absolutely continuous with density g, then $H = G * F$ is absolutely continuous with density h given by

$$h(x) = \int_{-\infty}^{+\infty} g(x - y)\, dF(y)\,. \tag{3,8,6}$$

Thus, if F is also absolutely continuous with density f(x) then

$$h(x) = \int_{-\infty}^{+\infty} g(x - y)\, f(y)\, dy, \tag{3,8,7}$$

where (3,8,7) is an ordinary Lebesgue or even Riemann integral if f and g are sufficiently regular. h is then the convolution of the functions f and g as defined in III.6; obviously, f and g are of class (S_1).

A literal repetition of the proof on p. 175 yields

THEOREM (3,8,2). If F and G are two bounded distribution functions such that $F(-\infty) = G(-\infty) = 0$, with Fourier transforms $\phi(u)$ and $\psi(u)$, respectively, then the Fourier transform of their convolution $F * G$ is the product $\phi(u).\psi(u)$.

Theorem (3,8,2) is the main reason for the application of Fourier transforms in studying the addition of independent random variables (cf. Chapter VI).

REMARK (3,8,2)

In (3,8,4) or (3,8,5) we assumed that the d.f. F and G are bounded and $F(-\infty) = G(-\infty) = 0$. However, in certain cases these integrals exist even when F and G are not bounded; in such cases, therefore, the convolution of two unbounded d.f. may be defined; this will be done, for instance, in III.10.

REMARK (3,8,3)

The group of translations over the x-axis (i.e. the additive group R of real numbers) is the source of the fundamental notions of III.6, III.7, III.8: the characters of this group give rise to the exponential functions; transformations based on these characters lead in our case to exponential and (in particular) Fourier transforms, the notion of positive-definite functions, Bochner's Theorem, etc. It is intuitively clear (and can in fact be proved) that analogous notions with the same properties may be associated with any abelian group, provided that it is locally compact. Examples are the group of rotations on a circle, the additive group of a euclidean vector space $\mathfrak{X}_n$ of arbitrary finite dimension n. This situation will be useful in the study of n-dimensional d.f. (c.f. III.11) and n-dimensional random variables (cf. Chapter V).

9. *Absolute and algebraic moments of a bounded distribution function*

Given a d.f. F(x) over $(-\infty, +\infty)$, let α be any fixed non-negative number and a any fixed abscissa. The following integral is called the *absolute moment of order* α *about a*, denoted by ${}_a m^*_\alpha$, if it exists:

$$ {}_{a}m_{\alpha}^{*} = \int_{-\infty}^{+\infty} |x - a|^{\alpha} \, dF(x). \tag{3,9,1} $$

If α *is an integer,* the *algebraic moment of order* α *about* a, denoted by ${}_{a}m_{\alpha}$, is the following integral, if it exists:

$$ {}_{a}m_{\alpha} = \int_{-\infty}^{+\infty} (x - a)^{\alpha} \, dF(x). \tag{3,9,2} $$

Our first task is to consider the existence of the integrals (3,9,1) and (3,9,2); since (3,9,2) exists if and only if (3,9,1) exists, it will suffice to consider (3,9,1).

Now (3,9,1) cannot exist unless $F(x)$ is bounded: thus *the notion of moment is relevant only for bounded d.f.,* and in the sequel the d.f. are therefore assumed bounded.

The following remarks are obvious.

1) For given $\alpha \geq 0$, if (3,9,1) exists for $a = a_0$, then (3,9,1) exists for any value of a;

2) If (3,9,1) exists for $\alpha = \beta$, then a fortiori (3,9,1) exists for any $\alpha < \beta$. Thus there exists a number $\alpha_0 \geq 0$ such that for any $\alpha < \alpha_0$ $(\alpha \geq 0)$, (3,9,1) exists, while for any $\alpha > \alpha_0$, (3,9,1) does not exist. The cases $\alpha_0 = 0$ or $\alpha_0 = +\infty$ are also possible.

For any integer α such that (3,9,1) exists, and for any a,

$$ |{}_{a}m_{\alpha}| \leqslant {}_{a}m_{\alpha}^{*}. \tag{3,9,3} $$

Moreover, if $\alpha = 2k$ is an even integer, the algebraic moment coincides with the absolute moment:

$$ {}_{a}m_{2k} = {}_{a}m_{2k}^{*}. $$

Naturally, the reference point a is often chosen as $a = 0$; we use the simplified notation m_{α}^{*} (m_{α}) for ${}_{0}m_{\alpha}^{*}$ $({}_{0}m_{\alpha})$, the absolute (algebraic) moment of order α about the origin 0.

Let n be the measure over $(-\infty, +\infty)$ - i.e., the space R of real numbers - defined by the bounded d.f. $F(x)$; suppose that

$$ n(R) = F(+\infty) - F(-\infty) \leq 1. \tag{3,9,4} $$

Then

$$m_\alpha^* = \int_{-\infty}^{+\infty} |x|^\alpha \, dF(x) = \int_{-\infty}^{+\infty} |x|^\alpha \, n(dx) \quad (\alpha \geq 0).$$

As remarked above, there exists a real number $\alpha_0 \geq 0$ such that m_α^* exists if $\alpha < \alpha_0$ and does not exist if $\alpha > \alpha_0$. Thus, if $\alpha_0 > 0$, $(m_\alpha^*)^{1/\alpha}$ is a function of α defined over $[0, \alpha_0[$. In the notation of II.7, m_α^* may thus be interpreted as $m_\alpha^* = M_\alpha(\nu)$, where ν is the point function defined over R by $\nu(x) = x$. Thus

$$(m_\alpha^*)^{1/\alpha} = N_\alpha(\nu).$$

Now by (3,9,4) the condition (2,7,5) is satisfied. An application of Theorem (2,7,5) yields:

THEOREM (3,9,1). If β is a positive number such that m_β^* exists, then $(m_\alpha^*)^{1/\alpha}$ is a non-decreasing function of α in the interval $[0, \beta]$; in particular, if $\beta \geq 2$, then

$$|m_1| = \left|\int_{-\infty}^{+\infty} x \, dF(x)\right| \leq m_1^* < \sqrt{m_2^*}. \qquad (3,9,5)$$

Relations between moments and Fourier transforms

Consider the Fourier transform $\phi(u)$ of a bounded d.f. $F(x)$; assume that (3,9,1) is convergent for $\alpha = 1$, i.e. m_1^* exists. It follows from (3,7,7) by differentiation under the integral sign that

1) $\phi(u)$ is differentiable and its derivative $d/du \, \phi(u)$ is

$$\frac{d}{du}\varphi(u) = i \int_{-\infty}^{+\infty} e^{iux} \, x \, dF(x) \; ; \qquad (3,9,6)$$

2) $d/du \, \phi(u)$ is a uniformly continuous function of u over $(-\infty, +\infty)$;

3) the value of the derivative at $u = 0$ is im_1.

More generally, let $\varphi^{(k)}(u)$ denote the kth derivative of $\phi(u)$. In the same way we see that if m_k^* exists, then $\varphi^{(k)}(u)$ exists and is a uniformly continuous function of u over $(-\infty, +\infty)$, and

$$\varphi^{(k)}(u) = i^k \int_{-\infty}^{+\infty} e^{iux} \, x^k \, dF(x). \qquad (3,9,7)$$

Hence

$$|\varphi^{(k)}(u)| \leqq m_k^* \tag{3,9,8}$$

and

$$\varphi^{(k)}(0) = i^k \; m_k \; . \tag{3,9,9}$$

Conversely, suppose that $\phi(u)$ has n derivatives $(n \geq 1)$. It may be proved (cf. R. Fortet [2]) that if n is even, then m_n^* exists; if n is odd, then m_{n-1}^* exists, but m_n^* may not exist.

Suppose that F(x) has moments of orders up to and including n, where n is an arbitrary positive integer; the real and imaginary parts A(u) and B(u) of $\phi(u)$ are

$$A(u) = \int_{-\infty}^{+\infty} \cos ux \; dF(x), \quad B(u) = \int_{-\infty}^{+\infty} \sin ux \; dF(x).$$

The existence of the moments of F of orders up to n implies that the *real* functions A(u) and B(u) have n derivatives, given by the formulas:

$$\frac{d^k A(u)}{du^k} = \int_{-\infty}^{+\infty} \mathcal{R}(i^k \; e^{iux}) \; x^k \; dF(x) \; ;$$

$$\frac{d^k B(u)}{du^k} = \int_{-\infty}^{+\infty} \mathcal{J}(i^k \; e^{iux}) \; x^k \; dF(x) \qquad (k = 1, 2, \ldots, n).$$

In particular, for u = 0,

$$\left(\frac{d^k A(u)}{du^k}\right)_{u=0} = \mathcal{R}(i^k) \; m_k,$$

$$\left(\frac{d^k B(u)}{du^k}\right)_{u=0} = \mathcal{J}(i^k) \; m_k \qquad (k = 1, 2, \ldots, n).$$

For k = n and any u, we have

$$\frac{d^n A(u)}{du^n} = \int_{-\infty}^{+\infty} \cos\left(ux + \frac{n\pi}{2}\right) x^n \; dF(x)$$

$$\frac{d^n B(u)}{du^n} = \int_{-\infty}^{+\infty} \sin\left(ux + \frac{n\pi}{2}\right) x^n \; dF(x).$$

Thus not only do $\frac{d^n A(u)}{du^n}$ and $\frac{d^n B(u)}{du^n}$ exist, but they are uniformly continuous functions of u over $(-\infty, +\infty)$. We may therefore write:

$$\frac{d^n A(u)}{du^n} = \left(\frac{d^n A(u)}{du^n}\right)_{u=0} + \alpha(u),$$

$$\frac{d^n B(u)}{du^n} = \left(\frac{d^n B(u)}{du^n}\right)_{u=0} + \beta(u),$$

where the continuous functions $\alpha(u)$ and $\beta(u)$ have the property

$$\lim_{u\to 0} \alpha(u) = \lim_{u\to 0} \beta(u) = 0. \qquad (3,9,10)$$

Now apply the classical Maclaurin formula to A(u) and B(u), putting $m_0 = 1$; we find that

$$\varphi(u) = A(u) + i\,B(u) = \sum_{k=0}^{n} \frac{i^k m_k}{k!} u^k + \omega(u)\, u^n, \qquad (3,9,11)$$

where

$$\omega(u) = \frac{1}{n!}\,[\alpha(u_1) + \beta(u_2)]$$

with $|u_1| \leq |u|$, $|u_2| \leq |u|$.
Thus, by (3,9,9),

$$\lim_{u\to 0} \omega(u) = 0. \qquad (3,9,12)$$

Now

$$\alpha(u) = \int_{-\infty}^{+\infty} \left[\cos\left(ux + \frac{n\pi}{2}\right) - \cos n\,\frac{\pi}{2}\right] x^n\, dF(x). \qquad (3,9,13)$$

If we assume that F also has a moment of order n + 1, formula (3,9,13) may be differentiated under the integral sign, and we get

$$\left|\frac{d}{du}\alpha(u)\right| = \left|\int_{-\infty}^{+\infty} \sin\left(ux + \frac{n\pi}{2}\right) x^{n+1}\, dF(x)\right| \leqq m^*_{n+1}, \qquad (3,9,14)$$

and similarly,

$$\left|\frac{d}{du}\beta(u)\right| \leqq m^*_{n+1}. \qquad (3,9,15)$$

Hence, obviously,

$$|\alpha(u)| \leqq m^*_{n+1}\,|u|, \quad |\beta(u)| \leqslant m^*_{n+1}\,|u|,$$

and we can make (3,9,12) more precise:

$$|\omega(u)| \leqq 2 \frac{m^*_{n+1}}{n!} |u| \text{ for any } u. \tag{3,9,16}$$

Now assume that F(x) has moments of all orders; it is then natural to ask whether

$$\varphi(u) = \sum_{k=0}^{+\infty} \frac{i^k m_k}{k!} u^k , \tag{3,9,17}$$

provided, of course, that the power series (3,9,17) has a nonzero radius of convergence ρ. The following assertion is easily proved.

If the radius of convergence ρ of the power series (3,9,17), where u is regarded as a complex variable, is positive, then there exists a function of u holomorphic in the domain $|\mathfrak{I}(u)| < \rho$ whose Taylor expansion about 0 is the series (3,9,17) and which coincides with $\phi(u)$ for real u.

It follows that the d.f. F is uniquely determined (up to equivalence) by its algebraic moments m_k, since the latter determin $\phi(u)$ by (3,9,17), and $\phi(u)$ in turn determines F by (3,7,13). Conversely, the Taylor expansion of $\phi(u)$ about 0 often provides a convenient method for determining the m_k.

REMARK (3,9,1)

Let F be a d.f. and $F_1, F_2, \ldots, F_n, \ldots$ a countable sequence of d.f. Let m_k, $m_k(n)$ denote the algebraic moments of order $k(k = 0, 1,2,\ldots)$ (about the origin, to fix ideas) of F and F_n, respectively if they exist.

As observed in III.4, before Theorem (3,5,2), if $\lambda(x)$ is a given function the assumption that $F_n \rightarrow F$ and the integrals

$$I_n = \int_{-\infty}^{+\infty} \lambda(x)\, dF_n(x)$$

exist need not imply that I_n converges to the integral

$$I = \int_{-\infty}^{+\infty} \lambda(x)\, dF(x),$$

nor even that the integral I exists. In fact, by Theorem (3,4,2), I exists and I_n converges to I only if $\lambda(x)$ is continuous and vanishes outside a bounded set; in general this is not true if $\lambda(x)$ is arbitrary.

Considering the case $\lambda(x) = x^k$ or $\lambda(x) = |x|^k$ (these are not bounded functions), it is clear that in general the fact that $F_n \to F$ and $m_k(n)$ exists does not imply that $m_k(n) \to m_k$, nor even that m_k exists [cf. Example (6,11,3)].

Now consider the converse problem. Suppose that

1) for any $k (k = 0,1,2,...)$ and any n, m_k and $m_k(n)$ exist;

2) for any k, $\lim_{n\to+\infty} m_k(n) = m_k$.

Do conditions 1) and 2) imply that $F_n \to F$ as $n \to +\infty$? The answer is negative. However, if the following condition is added:

3) the power series

$$\sum_{k=0}^{+\infty} \frac{i^k m_k}{k!} u^k$$

has a positive radius of convergence ρ, then the following frequently useful result may be proved: Conditions 1), 2), and 3) imply that $F_n \to F$ as $n \to +\infty$.

REMARK (3,9,2)

Naturally, the existence and values of the moments depend essentially on the behavior of $F(x)$ as $x \to \pm\infty$, and it is clear from (3,9,10), or again from (3,9,16) if this formula is valid, that the moments determine the behavior of $\phi(u)$ as $u \to 0$. Qualitatively speaking, we may therefore conclude that the behavior of $F(x)$ as $x \to \pm\infty$ and that of $\phi(u)$ as $u \to 0$ are related and determine each other.

10. *Distribution functions over* $[0, +\infty)$ *and their Laplace transforms*

In III.6 and III.7 we introduced the Fourier transform of a d.f. over $(-\infty, +\infty)$, which involves exponential functions e^{iux} with a pure imaginary exponent (u and x are real). However, the source of the properties of the Fourier transform is essentially the condition (3,6,3), which is satisfied by all exponential functions, not only by those with pure imaginary exponent. In principle, therefore, it should be equally interesting to use exponential functions with any exponent; one might even prefer real exponents in order to avoid introducing imaginary elements in questions involving essentially real elements.

Now the advantage of exponential functions with pure imaginary exponents is that under the sole assumption that a d.f. F(x) is bounded, the RS-integral (3,7,7) always converges, while if λ has a nonzero real part then the RS-integral

$$\int_{-\infty}^{+\infty} e^{\lambda x}\, dF(x) \tag{3,10,1}$$

may diverge even for bounded F, since $|e^{\lambda x}|$ is not bounded over $(-\infty, +\infty)$. Nevertheless the integral (3,10,1) may indeed converge if we stipulate that F, whether bounded or not, belong to a convenient class of d.f. This is the case for d.f. over $[0, +\infty)$.

D.f. over $[0, +\infty)$ are introduced whenever a mass distribution or probability law is restricted to the nonnegative half of the real axis. This is the case when the discussion concerns a random quantity or variable which is essentially nonnegative by nature, such as a period of *time*, e.g., the life of an individual (animate or not). This indicates the importance of this class of d.f. If F(x) is a d.f. over $[0, +\infty)$, it may also be regarded as a d.f. over $(-\infty, +\infty)$, defining it for $x < 0$ by $F(x) = F(0)$. The entire theory of d.f. over $(-\infty, +\infty)$ may then be applied, but it is also possible, and generally more convenient, to study d.f. over $[0, +\infty)$ by specific methods. This is the object of this section.

Any d.f. over $[0, +\infty)$ figuring in this section will always be assumed zero and continuous at $x = 0$; i.e., if F is such a d.f. then $F(0) = F(+0) = 0$. The extension to the general case presents no difficulties, but the reader must be careful in applying the results of this section to d.f. F such that $F(+0) - F(0) > 0$; for instance, in such cases one must specify in integrals of the type (3,10,2), (3,10,22) whether the integration interval $(0, +\infty)$, $(0,x)$ is open or closed, etc.

Let F(x) be a d.f., bounded or *not*, over $[0, +\infty)$. The Laplace transform $\phi(u)$ of F is the function of the *real* variable u defined by

$$\varphi(u) = \int_0^{+\infty} e^{-ux} \, dF(x). \qquad (3,10,2)$$

It is clear that two equivalent d.f. over $[0, +\infty)$ have the same Laplace transform. The question of the existence of the RS-integral (3,10,2) must now be examined. It is immediate that if (3,10,2) exists for some value u_0 of u, it exists for all $u > u_0$. This motivates the following definition: For any fixed real number a, we shall say that a d.f. F(x) over $[0, +\infty)$ is of class $\mathfrak{F}(a)$ if $F(0) = F(+0) = 0$ and the integral (3,10,2) is convergent for any $u > a$. Thus the Laplace transform $\phi(u)$ defined by (3,10,2) exists and is a continuous, monotone, nonincreasing, nonnegative, real function over $]a, +\infty)$.

If the d.f. F over $[0, +\infty)$ belongs to $\mathfrak{F}(a)$, it belongs a fortiori to $\mathfrak{F}(b)$ for any $b > a$; if F is bounded, it obviously belongs to $\mathfrak{F}(0)$. Whenever we employ Laplace transforms, we must therefore limit the discussion to d.f. of some class $\mathfrak{F}(a)$ over $[0, +\infty)$ (where a may depend on the specific d.f.). Luckily, the d.f. encountered in applications generally satisfy this condition. A useful remark is that if $F \in \mathfrak{F}(a)$ then for any $u > a$ the d.f. $G_u(x)$ defined by

$$G_u(x) = \int_0^{x} e^{-uy} \, dF(y) \qquad (3,10,3)$$

is bounded.

Let us introduce the complex variable $z = u - i\omega$, where u and ω are real, and consider the function $\psi(z)$ defined by

$$\psi(z) = \int_0^{+\infty} e^{-zx} \, dF(x) = \int_0^{+\infty} e^{(-u+i\omega)x} \, dF(x). \qquad (3,10,4)$$

If $F \in \mathfrak{F}(a)$, $\psi(z)$ is clearly holomorphic in the half-plane $\mathcal{R}(z) = u > a$. Now $\psi(z)$ and $\phi(u)$ coincide for real z, i.e., for $\omega = 0$; it follows that the values of $\phi(u)$ in $]a, +\infty)$ uniquely determine $\psi(z)$ in the entire half-plane $\mathcal{R}(z) > a$. In particular, for any fixed real number $u_0 > a$, the function $\lambda(\omega)$ defined by

$$\lambda(\omega) = \psi(u_\circ - i\omega)$$

is well defined. In view of (3,10,4) and (3,10,3), $\lambda(\omega)$ is precisely the Fourier transform of the *bounded* d.f. $G_{u_0}(x)$. By Theorem (3,7,2), both $G_{u_0}(x)$ and $F(x)$ (in view of (3,10,3)) are completely determined up to equivalence:

THEOREM (3,10,1). A distribution function $F(x)$ of class $\mathfrak{F}(a)$ over $[0, +\infty)$ is uniquely determined (up to equivalence) by its Laplace transform $\phi(u)$ over $]a, +\infty)$.

EXAMPLE (3,10,1)

Consider the following d.f. $F(x)$ over $[0, +\infty)$:

$$F(x) = 1 - e^{-x} ; \tag{3,10,5}$$

it is obviously of class $\mathfrak{F}(a)$ for any $a \geq -1$; its Laplace transform is

$$\varphi(u) = \int_0^{+\infty} e^{-ux}\, d(1 - e^{-x}) = \int_0^{+\infty} e^{-(u+1)x}\, dx = \frac{1}{u+1}, \tag{3,10,6}$$

so that $\psi(z)$, obtained in this case by simply substituting z for u in (3,10,6), is

$$\psi(z) = \frac{1}{z+1} = \frac{1}{u+1-i\omega}.$$

It may be verified that $\psi(z)$ is holomorphic in the half-plane $\mathcal{R}(z) > -1$. For arbitrary $u_0 > -1$,

$$\lambda(\omega) = 1/(u_\circ + 1 - i\omega)$$

which is clearly the Fourier transform of the normalized, absolutely continuous d.f. $G_{u_0}(x)$ over $(-\infty, +\infty)$ whose density is

$$g_{u_0}(x) = \begin{cases} 0 & \text{if } x < 0, \\ e^{-(u_0+1)x} & \text{if } x > 0. \end{cases}$$

Applying (3,10,3) with $u = u_0$ to this d.f. $G_{u_0}(x)$ does indeed yield the d.f. $F(x)$ of (3,10,5), up to equivalence. ∎

Bochner's Theorem (3,7,4) proves that any continuous and positive-definite function is the Fourier transform of a bounded d.f. over $(-\infty, +\infty)$, and conversely; is there a corresponding property for Laplace transforms of d.f. of class $\mathfrak{F}(a)$ over $[0, +\infty)$?

In the sequel f(x) denotes a function (not necessarily real) of the real variable x, and (as usual) $f^{(k)}(x)$ denotes f(x) itself if k = 0, and the k-th derivative of f if $k \geq 1$.

We introduce the following definition:

Absolutely monotone functions

A real function $\phi(u)$ of the real variable u over an arbitrary fixed open interval]a,b[(a < b) (possibly infinite), is said to be absolutely monotone over]a,b[if

1) it is infinitely differentiable over]a,b[;

2) for any $u \in$]a,b,[, and any integer $k \geq 0$,

$$(-1)^k \varphi^{(k)}(u) \geqq 0 \quad (k = 0, 1, 2, \ldots). \qquad (3,10,7)$$

The point of this definition is immediately apparent:

LEMMA (3,10,1)

If the distribution function F(x) over $[0, +\infty)$ is of class $\mathfrak{F}(a)$ and $\phi(u)$ is its Laplace transform, then

1) $\phi(u)$ is absolutely monotone over $]a, +\infty)$;

2) $\lim_{u \to +\infty} \varphi(u) = 0$.

We leave the proof of this lemma, which is not difficult, to the reader. The condition F(0) = F(+ 0) = 0 in the definition of the class $\mathfrak{F}(a)$ is required only for 2). The converse of Lemma (3,10,1), which we shall now prove, is much more difficult.

We first note several obvious properties of absolutely monotone functions.

1) The sum $\phi_1(u) + \phi_2(u)$ of two absolutely monotone functions over]a,b[, is absolutely monotone over]a,b[.

2) If the function $\phi(u)$ is absolutely monotone over]a,b[and c is a nonnegative real constant, then $c\phi(u)$ is absolutely monotone over]a,b[.

3) If the functions $\phi_1(u)$ and $\phi_2(u)$ are absolutely monotone over]a,b[, then the product $\phi_1(u).\phi_2(u)$ is absolutely monotone over]a,b[.

4) Let $\{\phi_n\}$ be a sequence of functions which are absolutely monotone over]a,b[, and $\phi(u)$ a function defined and infinitely differentiable over]a,b[; if

$$\lim_{n\to+\infty} \varphi_n^{(k)}(u) = \varphi^{(k)}(u) \quad \text{for all} \quad k = 0, 1, 2, \ldots,$$

then $\phi(u)$ is absolutely monotone over]a,b[.

5) If the function $\phi(u)$ is absolutely monotone over]a,b[, and h is any constant, then the function $\psi(u) = \phi(u + h)$ is absolutely monotone over]a - h, b - h[.

We now prove two lemmas.

LEMMA (3,10,2)

Let]a,b[be an open (possibly infinite) interval $(a < b)$, g(x) a function of $x \in\ ]a,b[$ which is n + 1 times continuously differentiable over]a,b[. For any u, u_0 in]a,b[and k = 0,1,2,...,n set

$$R_k(u) = \int_{u_0}^{u} \frac{(u-x)^k}{k!} g^{(k+1)}(x)\,dx \quad (k = 0, 1, 2, \ldots, n); \qquad (3,10,8)$$

then:

$$R_n(u) = g(u) - \sum_{k=0}^{n} \frac{g^{(k)}(u_0)}{k!}(u-u_0)^k. \qquad (3,10,9)$$

In fact, integrating by parts in (3,10,8), we get

$$R_k(u) = R_{k-1}(u) - \frac{g^{(k)}(u_0)}{k!}(u-u_0)^k \quad (k = 1, 2, \ldots, n),$$

and since

$$R_0(u) = g(u) - g(u_0),$$

this gives (3,10,9).

LEMMA (3,10,3)

Let $\phi(u)$ be a real function of the real variable u which is absolutely monotone over]a,b[$(a < b)$, and u_0 any fixed number in

]a,b[. Then for any $u \in]a,b[$ such that $|u - u_0| < |u_0 - a|$ we have

$$\varphi(u) = \sum_{k=0}^{+\infty} \frac{\varphi^{(k)}(u_o)}{k\,!} (u - u_o)^k, \tag{3,10,10}$$

i.e. $\phi(u)$ may be expanded in a Taylor series about u_0; more precisely, if we set

$$\rho_n(u) = \sum_{k=n+1}^{+\infty} \frac{\varphi^{(k)}(u_o)}{k\,!} (u - u_o)^k$$

(the "tail" of the series (3,10,10)) and let α denote an arbitrary number between a and u_0 and Δ the positive difference $u_0 - \alpha$, then for $u < \alpha$:

$$|\rho_n(u)| \leqq \varphi(\alpha) \left|\frac{u - u_o}{\Delta}\right|^{n+1}. \tag{3,10,11}$$

In fact, applying (3,10,9) to $\phi(u)$, we have

$$\varphi(u) = \sum_{k=0}^{n} \frac{\varphi^{(k)}(u_o)}{k\,!} (u - u_o)^k + R_n(u) \tag{3,10,12}$$

where

$$R_n(u) = \int_{u_o}^{u} \frac{(u - y)^n}{n\,!} \varphi^{(n+1)}(y)\, dy.$$

The substitution $y = u_0 + (u - u_0)x$ and (3,10,7) immediately show that

$$|R_n(u)| < |u - u_o|^{n+1} \int_0^1 \frac{(1-x)^n}{n!} (-1)^{n+1} \varphi^{(n+1)}[u_o+(u - u_o)x]\, dx. \tag{3,10,13}$$

If it is assumed that

$$u > \alpha,$$

then for $x \geq 0$

$$u_o + (u - u_o)x \geqslant u_o - \Delta x. \tag{3,10,14}$$

Now $(-1)^{n+1} \varphi^{(n+1)}(u)$ is a nonincreasing function of u, since by (3,10,7) its derivative $(-1)^{n+1} \varphi^{(n+2)}(u)$ is not positive; therefore, by (3,10,14),

$$(-1)^{n+1} \varphi^{(n+1)}[u_o + (u - u_o)x] \leqslant (-1)^{n+1} \varphi^{(n+1)}[u_o - \Delta\mathbf{x}]$$

and consequently

$$|R_n(u)| \leqslant |u - u_o|^{n+1} \int_0^1 \frac{(1 - x)^n}{n\,!} (-1)^{n+1} \varphi^{(n+1)}(u_o - \Delta x)\, dx. \quad (3,10,15)$$

Setting

$$g(x) = \varphi(u_o - \Delta x),$$

we have

$$g^{(k)}(x) = \Delta^k (-1)^k \varphi^{(k)}(u_o - \Delta x) \quad (k = 0, 1, 2, \ldots). \quad (3,10,16)$$

Therefore, again by (3,10,14)

$$|R_n(u)| \leqslant \left|\frac{u - u_o}{\Delta}\right|^{n+1} \cdot \int_0^1 \frac{(1 - x)^n}{n\,!} g^{(n+1)}(x)\, dx.$$

Application of Lemma (3,10,2) to the function g(x) (with $u_0 = 0$, $u = 1$) gives

$$\int_0^1 \frac{(1 - x)^n}{n\,!} g^{(n+1)}(x) = g(1) - \sum_{k=0}^{n} \frac{g^{(k)}(0)}{k\,!} \ ;$$

substituting g(1) and all $g^{(k)}(0)$ from (3,10,16) and taking (3,10,7) into account, we obtain

$$\int_0^1 \frac{(1 - x)^n}{n\,!} g^{(n+1)}(x) \leqslant \varphi(\alpha),$$

$$|R_n(u)] \leqslant \varphi(\alpha) \left|\frac{u - u_o}{\Delta}\right|^{n+1}, \quad (3,10,17)$$

which implies that if $u > \alpha$ then the remainder $\rho_n(u)$ coincides with $R_n(u)$; this proves Lemma (3,10,3). ▮

Let $\phi(u)$ be a real function of the real variable u which is absolutely monotone over $]0, +\infty)$ and such that $\lim_{u \to +\infty} \varphi(u) = 0$; let λ be a positive real parameter and consider $\varphi(\lambda - \lambda e^{-u/\lambda})$ for arbitrary fixed positive u. By (3,10,10),

$$\varphi(\lambda - \lambda e^{-u/\lambda}) = \sum_{k=0}^{+\infty} (-1)^k \frac{\varphi^{(k)}(\lambda)}{k\,!} \lambda^k e^{-ku/\lambda}. \quad (3,10,18)$$

Let $F_\lambda(x)$ be the normalized totally discontinuous d.f. with points of discontinuity

$$x_k = \frac{k}{\lambda} \quad (k = 0, 1, 2, \ldots),$$

where the jump σ_k at x_k is

$$\sigma_k = (-1)^k \frac{\varphi^{(k)}(\lambda)}{k!} \lambda^k.$$

$F_\lambda(x)$ vanishes identically for $x \leq 0$, and may be regarded as a d.f. over $[0, +\infty)$. Using (3,3,9), we rewrite (3,10,18) as

$$\varphi(\lambda - \lambda e^{-u/\lambda}) = \int_0^{+\infty} e^{-ux} \, dF_\lambda(x). \qquad (3,10,19)$$

Let $y > 0$ be an arbitrary fixed number, n the greatest integer such that

$$\frac{n}{\lambda} < y,$$

and $\alpha > 0$ arbitrary but fixed. By (3,10,10)

$$\varphi(\alpha) = \sum_{k=0}^{+\infty} \frac{\varphi^{(k)}(\lambda)}{k!} (\alpha - \lambda)^k.$$

By (3,10,7) the terms of this series are nonnegative if $\lambda > \alpha$. Thus

$$\varphi(\alpha) \geqslant \sum_{k=0}^{n} \frac{\varphi^{(k)}(\lambda)}{k!} (\alpha - \lambda)^k = \sum_{k=0}^{n} (-1)^k \frac{\varphi^{(k)}(\lambda)}{k!} \lambda^k \left(1 - \frac{\alpha}{\lambda}\right)^k$$

$$\geqslant \sum_{k=0}^{n} \sigma_k (1 - \alpha/\lambda)^k \geqslant (1 - \alpha/\lambda)^{\lambda y} \cdot \sum_{k=0}^{n} \sigma_k = (1 - \alpha/\lambda)^{\lambda y} F_\lambda(y).$$

But $\lim_{\lambda \to +\infty} \left(1 - \frac{\alpha}{\lambda}\right)^\lambda = e^{-\alpha}$; thus $F_\lambda(x)$ is bounded uniformly in λ over any finite interval $[0,y)$ as $\lambda \to +\infty$. From Theorem (3,4,1) it follows that there exist a d.f. over $[0, +\infty)$ and a countable sequence $\{\lambda_j\}$ of values of λ such that

$$\lim_{j \to +\infty} \lambda_j = +\infty \; ; \; \lim_{j \to +\infty} F_{\lambda_j} = F \text{ over } [0, +\infty).$$

Moreover, it is clear that

$$F(y) \leqq \varphi(\alpha) e^{\alpha y},$$

so that

$$F(+0) \leqq \varphi(\alpha)$$

for all $\alpha > 0$. Since $\lim_{\alpha \to +\infty} \varphi(\alpha) = 0$, it follows that

$$F(+0) = 0. \tag{3,10,20}$$

Let us put

$$\rho_n(u) = \sum_{k>n} (-1)^k \frac{\varphi^{(k)}(\lambda)}{k!} \lambda^k e^{-ku/\lambda}.$$

Now $\lambda e^{-u/\lambda}$ is asymptotically equivalent to $\lambda - u$ as $\lambda \to +\infty$. Assuming $\alpha < u$ and taking λ sufficiently large, we can apply (3,10,11) (replacing u_0 by λ, u by $\lambda - \lambda e^{-u/\lambda}$), and we get

$$|\rho_n(u)| \leqslant \varphi(\alpha) \left| \frac{\lambda e^{-u/\lambda}}{\lambda - \alpha} \right|^{n+1}. \tag{3,10,21}$$

Inspection of the right-hand side of (3,10,21) shows that there are two positive numbers λ_0 and η, independent of y, such that

$$|\rho_n(u)| \leqslant \varphi(\alpha) e^{-\eta y} \quad \text{for all } x > 0 \text{ and all } \lambda > \lambda_o,$$

$$\rho_n(u) = \varphi(\lambda - \lambda e^{-u/\lambda}) - \int_0^{y-0} e^{-ux} \, dF_\lambda(x).$$

Setting $\lambda = \lambda_j$ and letting $j \to +\infty$, we obtain

$$\left| \varphi(u) - \int_0^{y-0} e^{-ux} \, dF(x) \right| \leqslant \varphi(\alpha) e^{-\eta y}.$$

If we now let $y \to +\infty$, it is clear that

1) the integral

$$\int_0^{+\infty} e^{-ux} \, dF(x)$$

exists for any $u > 0$, which together with (3,10,20) implies that F is of class $\mathfrak{F}(0)$;

2)

$$\varphi(u) = \int_0^{+\infty} e^{-ux} \, dF(x),$$

i.e., $\phi(u)$ is the Laplace transform of F(x).

Now consider a function $\phi(u)$ which is absolutely monotone over an interval $]a, +\infty)$, where a is an arbitrary finite number; then the function $\phi_1(u) = \phi(a + u)$ is absolutely monotone over $]0, +\infty)$ and the foregoing discussion may be applied. The result is

THEOREM (3,10,2). A real function $\phi(u)$ of the real variable u is the Laplace transform of a distribution function of class $\mathfrak{F}(a)$ if and only if it is defined and absolutely monotone over $]a, +\infty)$, and $\lim_{u \to +\infty} \varphi(u) = 0$. ■

Using moments (as in Remark (3,9,1)) one can throw light on the connection between the behavior of a d.f. F(x) over $[0, +\infty)$ as $x \to +\infty$ and that of its Laplace transform as $u \to 0$; however, one can prove a more precise result, which is also stronger in that it holds for certain unbounded d.f.

THEOREM (3,10,3) (Karamata). Let F(x) be a distribution function of class $\mathfrak{F}(0)$ over $[0, +\infty)$ with Laplace transform $\phi(u)$. Suppose there exist constants A and $\rho \geq 0$ such that $\phi(u)$ is asymptotically equivalent to A/u^{ρ} as $u \to +0$; then F(x) is asymptotically equivalent to $Ax^{\rho}/\Gamma(\rho + 1)$ as $x \to +\infty$.

The reader can easily prove this theorem in the case $\rho = 0$, when it reduces to the following assertion: If $\phi(u)$ tends to a limit A as $u \to 0$, then F(x) tends to the limit A as $x \to +\infty$. In the case $\rho > 0$ the proof is lengthy and delicate and is therefore omitted (cf. Exercise 3.12).

CONVOLUTION

If F(x) and G(x) are two d.f. over $[0, +\infty)$, the integral (3,8,4) becomes

$$H(x) = \int_0^x G(x - y)\, dF(y). \tag{3,10,22}$$

This integral exists for all $x \in [0, +\infty)$ and may be interpreted as defining the convolution $G * F$ of F by G, which is again a d.f. H(x) over $[0, +\infty)$. By a transformation similar to (3,3,8) it can be verified that

$$F * G = G * F.$$

It is easily verified that if F is of class $\mathfrak{F}(a)$ and G of class $\mathfrak{F}(b)$, then $G * F$ is of the class $\mathfrak{F}(c)$ where $c = \max(a,b)$. Moreover, the Laplace transform of $G * F$ over $]c, +\infty)$ is the product $\phi(u).\psi(u)$ of the Laplace transform $\phi(u)$ of F and the Laplace transform $\psi(u)$ of G.

Here, as in III.10, we have assumed that the d.f. F and G are zero and continuous at $x = 0$; we leave it to the reader to make (3,10,22) rigorous and to study it in the case when the above condition is not satisfied.

Generating functions

If we put $e^{\lambda} = z$ in an exponential function $e^{\lambda x}$, we get z^x. Exponential transformations of type (3,10,1), in particular, Fourier and Laplace transforms, may be translated into this new notation. Thus, with any d.f. we associate a function $g(z)$, called its *generating function*. This device is useful only in the very particular case (though frequently encountered in practice) of a totally discontinuous d.f. $F(x)$ with points of discontinuity at nonnegative integers (so that F is a d.f. over $[0, +\infty)$, but we do not assume here that $F(+0) - F(0) = s = 0$). For a d.f. of this type, put

$$F(k + 0) - F(k - 0) = s_k \quad \text{for } k = 0, 1, 2, \ldots$$

where the s_k are nonnegative (possibly zero); then by (3,3,9) the generating function

$$g(z) = \int_{-\infty}^{+\infty} z^x \, dF(x) \qquad (3,10,23)$$

becomes

$$g(z) = \sum_{k=0}^{+\infty} s_k z^k, \qquad (3,10,24)$$

a power series in z. In particular, if all the s_k vanish for k greater than a fixed integer n,

$$g(z) = \sum_{k=0}^{n} s_k z^k \qquad (3,10,25)$$

- a polynomial of degree at most n.

Subject to appropriate existence conditions, the generating function possesses properties corresponding to those of the Fourier or Laplace transform. Thus, the formula (3,10,24) (or (3,10,25)) yields algebraic relations with integer coefficients between the algebraic moments m_k (about the origin 0) of F(x) and the derivatives $g^{(k)}(1)$ of g(z) at z = 1; for instance:

$$g^{(1)}(1) = m_1$$

$$g^{(2)}(1) = m_2 - m_1;$$

etc. As before, it is easy to prove the following assertion:

If F_1 and F_2 are two d.f. over $[0, +\infty)$ with generating functions $g_1(z)$ and $g_2(z)$, then the generating function of the convolution $G * F$ is the product $g_1(z).g_2(z)$.

11. *n-dimensional distribution functions*

Let $x_1, x_2, \ldots, x_n$ be an n-tuple of real variables. Recall that the x_j may be regarded as the n components of a vector x of an n-dimensional vector space $\mathfrak{X}_n$ relative to an arbitrary fixed basis **b**; $\mathfrak{X}_n$ may even be assumed euclidean and **b** orthonormal. The x_j may also be interpreted as the n coordinates of a point N of the affine point space E_n associated with $\mathfrak{X}_n$, relative to the reference system defined by **b**. These interpretations are essentially equivalent; to fix ideas, we shall adopt the former.

The n-tuple $\{x_1, x_2, \ldots, x_n\}$, i.e. the vector x, may vary over any subset of $\mathfrak{X}_n$; we shall limit ourselves to the case where it varies over the entire space $\mathfrak{X}_n$. No loss of generality is involved, since if a mass is distributed over a subset e of $\mathfrak{X}_n$, we can always regard it as distributed over all of $\mathfrak{X}_n$, with zero mass assigned to $\mathfrak{X}_n$ - e.

Recall (cf. III.1) that a finite real function $F(x_1, x_2, \ldots, x_n)$ is an *n-dimensional distribution function* of $\{x_1, x_2, \ldots, x_n\}$ or a distribution function of the n (ordered) variables $x_1, x_2, \ldots, x_n$ if for any $k (k = 1,2,\ldots,n)$ and any fixed $x_1, x_2, \ldots, x_{k-1}, x_{k+1}, \ldots, x_n$, $F(x_1, x_2, \ldots, x_n)$ is a monotone nondecreasing function of x_k (i.e. a one-dimensional d.f. of x_k).

The one-dimensional d.f. studied above are therefore particular cases (n = 1) of n-dimensional d.f. We reiterate that throughout this book the term distribution function will mean a one-dimensional d.f., unless otherwise stated.

In III.1 we stated and proved several properties of n-dimensional d.f.; most of the notions and properties of one-dimensional d.f. extend to n-dimensional d.f.

Note first that a d.f. $F(x_1,x_2,\ldots,x_n)$ is bounded over any bounded set of $\mathfrak{X}_n$; $F(x_1,x_2,\ldots,x_n)$ may have point discontinuities analogous to the jumps of one-dimensional d.f., and as in the latter case the set of these discontinuities is at most countably infinite.

However, when $n > 1$, $F(x_1,x_2,\ldots,x_n)$ may also have discontinuities of a different type. Recall that F defines a measure or mass distribution, and imagine, for instance, a mass distribution over a plane E_2 ($n = 2$); let Γ be a curve in this plane; the mass may be distributed only over Γ, no mass at all being assigned to the remaining part of the plane. Thus Γ is simply a plane material wire, ideally without width. Over Γ, the distribution may be continuous, even absolutely continuous; but when it is regarded as a distribution over the plane, Γ is obviously a line of singularity, and, apart from exceptions, there is no surface density at any of its points. We shall not attempt to analyze questions of this kind; they play no essential role in the sequel.

Henceforth *all our discussion of n-dimensional d.f. will refer to the case $n = 2$.* In other words, we shall consider d.f. $F(x,y)$ of two variables x and y, which may be regarded as the components of a vector $u \in \mathfrak{X}_2$. The choice of $n = 2$ will simplify the notation considerably, and on the other hand *extension to arbitrary n is immediate*.

n-dimensional Riemann-Stieltjes integrals

Let $F(x,y)$ be a 2-dimensional d.f., m the measure that it defines over $\mathfrak{X}_2$, $\lambda(x,y)$ or $\lambda(u)$ a point function, i.e. a real or complex function of the pair $u = (x,y)$ defined over a subset $e \subset \mathfrak{X}_2$. At the end of III.1 we defined the integral

$$I = \int_e \lambda(x,y)\, dF(x,y) \tag{3,11,1}$$

as the integral

$$I = \int_e \lambda(u)\ m(du). \qquad (3,11,2)$$

The properties of (3,11,1) are therefore those of (3,11,2), as described in II.6; the main properties were studied in III.3 for the case n = 1, and there is no need to go into details for arbitrary n. We confine ourselves to the following remarks.

1) The integral (3,11,1) is usually represented by the symbol

$$I = \underbrace{\int\int_e}_{2} \lambda(x\ ,\ y)\ dF(x\ ,\ y), \qquad (3,11,3)$$

the number of integral signs $\int$ being equal to the number of the variables (in this case 2).

2) If G(x,y) is a d.f. equivalent to F(x,y), then

$$\int\int_e \lambda(x\ ,\ y)\ dG(x\ ,\ y) = \int\int_e \lambda(x\ ,\ y)\ dF(x\ ,\ y).$$

3) Let F(x,y) be totally discontinuous, i.e. let m be totally discontinuous; suppose the support of the mass distribution over $\mathfrak{X}_2$ defined by m is some (finite or countable) set of elements $u_k = (x_k, y_k)$ of $\mathfrak{X}_2$, and let m_k be the mass assigned to u_k. Then

$$\int\int_e \lambda(x\ ,\ y)\ dF(x\ ,\ y) = \sum_{\{k|u_k \in e\}} \lambda(x_k\ ,\ y_k)\ m_k. \qquad (3,11,4)$$

4) If F(x,y) is absolutely continuous, with density

$$f(x\ ,\ y) = \frac{\partial^2}{\partial x\ \partial y} F(x\ ,\ y), \qquad (3,11,5)$$

i.e. m is absolutely continuous with density f, then

$$\int\int_e \lambda(x,\ y)\ dF(x\ ,\ y) = \int\int_e \lambda(x,\ y)\ f(x,\ y)\ dx\ dy. \qquad (3,11,6)$$

The integral on the left-hand side therefore reduces to a Lebesgue integral - even a Riemann integral if λ and f are sufficiently regular, e.g. continuous.

5) The case $e = \mathfrak{X}_2$ is frequently encountered. In this case the integral (3,11,3) may be written

$$I = \int_{-\infty}^{+\infty} \int_{-\infty}^{+\infty} \lambda(x, y)\, dF(x, y), \qquad (3,11,7)$$

or

$$I = \iint_{\mathfrak{X}_2} \lambda(x, y)\, dF(x, y).$$

Moreover (cf. II.6), the integral (3,11,3) for arbitrary e may always be expressed in the form (3,11,7); in fact let $\phi(e; x,y)$ be the indicator of the set e, i.e. the function

$$\varphi(e\,;\, x, y) = \begin{cases} 1 \text{ if } (x, y) \in e, \\ 0 \text{ if } (x, y) \notin e. \end{cases}$$

Set $\mu(x,y) = \lambda(x,y)\phi(e; x,y)$; then the integral (3,11,3) coincides with the integral

$$\int_{-\infty}^{+\infty} \int_{-\infty}^{+\infty} \mu(x, y)\, dF(x, y).$$

Recall that (3,11,7) exists if and only if

$$\int_{-\infty}^{+\infty} \int_{-\infty}^{+\infty} |\lambda(x, y)|\, dF(x, y) < +\infty,$$

and then

$$\left| \int_{-\infty}^{+\infty} \int_{-\infty}^{+\infty} \lambda(x, y)\, dF(x, y) \right| \leqq \int_{-\infty}^{+\infty} \int_{-\infty}^{+\infty} |\lambda(x, y)|\, dF(x, y). \quad (3,11,8)$$

As in the case n = 1, if e satisfies certain conditions and λ belongs to what we have called the Riemann-Stieltjes class (RS-class) over e with respect to F, we can give a direct and simple definition of the Riemann-Stieltjes integral or RS-integral (of λ over e with respect to F) which obviously coincides with the integral (3,11,3). It is enough to imitate the case n = 1, which now proceed to do.

We begin with the case of a "rectangular domain" e defined by

$$a \leqq x < b\,,\ c \leqq y < d. \qquad (3,11,9)$$

In this case we replace (3,11,6) by the notation

$$\int_{a-0}^{b-0} \int_{c-0}^{d-0} \lambda(x, y)\, dF(x, y). \qquad (3,11,10)$$

Consider the case $\lambda(x,y) = 1$ (on e); by the definition of (3,11,6), i.e. (3,11,2), formula (3,1,11) implies

$$\int_{a-0}^{b-0}\int_{c-0}^{d-0} 1 \cdot dF(x,y) = m(e)$$

$$= F(b-o, d-o) - F(a-o, d-o) - F(b-o, c-o) + F(a-o, c-o).$$

With this in mind, let us return to the case of arbitrary λ. Divide the interval (a,b) into subintervals by points $x_0 = a, x_1, \ldots, x_j, \ldots, x_m = b$, $x_j < x_{j+1}$ ($j = 0,1,\ldots,m-1$), and the interval (c,d) by points $y_0 = c, y_1, \ldots, y_k, \ldots, y_n = d$, $y_k < y_{k+1}$ ($k = 0,1,\ldots,n-1$); Let e_{jk} denote the rectangular domain

$$x_j \leqslant x < x_{j+1}, \quad y_k \leqslant y < y_{k+1},$$

then e is divided into mn disjoint rectangular domains e_{jk} ($j = 0,1,\ldots,m-1$; $k = 0,1,\ldots,n-1$). Put

$$\Delta^2_{jk}F = F(x_{j+1} - 0, y_{k+1} - 0) - F(x_{j+1} - 0, y_k - 0)$$

$$- F(x_j - 0, y_{k+1} - 0) + F(x_j - 0, y_k - 0).$$

In each e_{jk} choose an arbitrary element

$$x = \alpha_{jk}, \quad y = \beta_{jk}, \quad (x_j \leqslant \alpha_{jk} < x_{j+1}\,;\ y_k \leqslant \beta_{jk} < y_{k+1}),$$

and form the Riemann-Stieltjes sum:

$$\sum_{jk} \lambda(\alpha_{jk}, \beta_{jk})\, \Delta^2_{jk}F. \tag{3,11,11}$$

We define the RS-integral

$$\int_{a-0}^{b-0}\int_{c-0}^{d-0} \lambda(x,y)\, dF(x,y) \quad (RS) \tag{3,11,12}$$

as the limit, if it exists, of the RS-sums (3,11,11) as the partitions of (a,b) and (c,d) are made uniformly infinitely fine. The analogy with the classical theory of multiple Riemann integrals is complete. The set of functions λ for which the limit exists is defined as the RS-class (for the rectangular domain e and the d.f. F).

As in the classical theory of multiple Riemann integrals, we proceed to the case of arbitrary e, i.e. not necessarily rectangular, even unbounded. The set e must satisfy certain conditions; we shall not consider them in detail, since they are quite weak and almost always satisfied by the sets e encountered in practice.

We conclude our discussion of multiple RS-integrals with the following remarks.

1) If e is compact and λ is continuous over e, then λ is RS-integrable over e with respect to any 2-dimensional d.f. F.

2) Consider, in particular, the case e = $\mathfrak{X}_2$; here e is not bounded. The corresponding RS-integral may be denoted

$$\int_{-\infty}^{+\infty} \int_{-\infty}^{+\infty} \lambda(x, y)\ dF(x, y) \quad (RS), \qquad (3,11,13)$$

or

$$\int\int_{\mathfrak{X}_2} \lambda(x, y)\ dF(x, y) \quad (RS). \qquad (3,11,14)$$

It exists if and only if $|\lambda(x,y)|$ is RS-integrable over $\mathfrak{X}_2$ and

$$\int_{-\infty}^{+\infty} \int_{-\infty}^{+\infty} |\lambda(x, y)|\ dF(x, y) \quad (RS) < +\infty , \qquad (3,11,15)$$

so that (3,11,8) is applicable.

3) If the integral (3,11,3) happens to be an RS-integral, i.e. λ belongs to the RS-class, and we wish to emphasize this, then as for (3,11,12), (3,11,13), (3,11,14) and (3,11,15) we use the notation:

$$\int\int_{e} \lambda(x, y)\ dF(x, y) \quad (RS).$$

In general this is not necessary, and we shall usually employ the notation (3,11,3) without the symbol (RS), even if λ is of the RS-class.

REMARK (3,11,1)

For any fixed y the 2-dimensional d.f. F(x,y) is a one-dimensional d.f. of x; on occasion, we shall consider the one-dimensional integral (e.g., over $(-\infty, +\infty)$), of a function $\lambda(x)$ of x with respect to $F(x,y)$ for fixed y; to avoid ambiguity, we shall represent this integral by the notation

$$\int_{-\infty}^{+\infty} \lambda(x)\, F(dx, y).$$

Similarly, we can represent (3,11,7) by the notation

$$\int_{-\infty}^{+\infty} \int_{-\infty}^{+\infty} \lambda(x, y)\, F(dx, dy) \qquad (3,11,16)$$

which will be used instead of (3,11,7) when no confusion may arise.

REMARK (3,11,2)

Consider the integral

$$\int \int_{e} \lambda(x, y)\, dF(x, y) \qquad (3,11,17)$$

where e is any given subset of $\mathfrak{X}_2$. Let M denote the set of x for which there is at least one pair (x',y') in e such that $x' = x$, and N the set of y for which there is at least one pair (x',y') in e such that $y' = y$. For each $x \in M$, let e_x be the set of y such that $(x,y) \in e$; for each $y \in N$, let e_y be the set of x such that $(x,y) \in e$.

Suppose first that $F(x,y)$ is absolutely continuous with density $f(x,y)$; then (3,11,17) is the Lebesgue integral

$$\int \int_{e} \lambda(x, y)\, f(x, y)\, dx\, dy. \qquad (3,11,18)$$

It may be proved that (3,11,18) is equal to either of two iterated integrals:

$$\int\int_{e} \lambda(x, y)\, f(x, y)\, dx\, dy = \int_{M} \left[\int_{e_x} \lambda(x, y)\, f(x, y)\, dy \right] dx =$$
$$\int_{N} \left[\int_{e_y} \lambda(x, y)\, f(x, y)\, dx \right] dy. \qquad (3,11,19)$$

Formula (3,11,19) is a particular case of the well-known Fubini Theorem; we omit a general statement of this theorem, and prove neither the general formulation nor even the particular case (3,11,19). Note that if λ, f and e are sufficiently regular, so that the integrals involved in (3,11,19) are Riemann integrals, then (3,11,19) is an elementary theorem.

An analogue of (3,11,19) is valid in the general case, when the d.f. F(x,y) is not necessarily absolutely continuous; we shall not deal with this generalization, confining ourselves to a few remarks.

a) The following assertions are obvious:

1) One simple case of (3,11,19) is that in which e is the cartesian product $M \times N$ of M and N, i.e. $e_x = N$ does not depend on x and $e_y = M$ does not depend on y; formula (3,11,19) is then

$$\iint_{M\times N} \lambda(x, y)\, f(x, y)\, dx\, dy = \int_M \left[\int_N \lambda(x, y)\, f(x, y)\, dy \right] dx$$

$$= \int_N \left[\int_M \lambda(x, y)\, f(x, y)\, dx \right] dy \; ; \quad (3,11,20)$$

2)

$$\int_{-\infty}^{+\infty}\int_{-\infty}^{+\infty} \lambda(x)\; F(dx, dy) = \int_{-\infty}^{+\infty} \lambda(x)\; [F(dx, +\infty) - F(dx, -\infty)]$$

$$= \int_{-\infty}^{+\infty} \lambda(x)\; F(dx, +\infty) - \int_{-\infty}^{+\infty} \lambda(x)\, F(dx, -\infty). \quad (3,11,21)$$

b) *Separation of variables*

Let A(x) be a one-dimensional d.f. of x, and B(y) a one-dimensional d.f. of y; their product

$$F(x, y) = A(x)\; B(y) \qquad (3,11,22)$$

is clearly a 2-dimensional d.f. When an n-dimensional d.f. may be expressed as the product of n one-dimensional d.f. of n different variables, we say that the variables are *separated*. This is a particular but very important case, especially in Probability Theory, where it is connected with the property of *independence* (cf. Chap. IV). When F has the form (3,11,22), the integral

$$\int_{-\infty}^{+\infty} \lambda(x)\; F(dx, y)$$

of Remark (3,11,1) becomes

$$\int_{-\infty}^{+\infty} \lambda(x)\ F(dx, y) = \int_{-\infty}^{+\infty} \lambda(x)\ dA(x)\ B(y) = \left[\int_{-\infty}^{+\infty} \lambda(x)\ dA(x)\right] . B(y), \quad (3,11,23)$$

while (3,11,21) may be written

$$\int_{-\infty}^{+\infty}\int_{-\infty}^{+\infty} \lambda(x)\ F(dx, dy) = \left[\int_{-\infty}^{+\infty} \lambda(x)\ dA(x)\right] \cdot \left[B(+\infty) - B(-\infty)\right]. \quad (3,11,24)$$

More generally, if $\mu(x)$ and $\nu(y)$ are any two functions, $\mu(x)$ defined for $x \in M$, and $\nu(y)$ for $y \in N$, then

$$\int\int_{M\times N} \mu(x)\ \nu(y)\ dF(x, y) = \int\int_{M\times N} \mu(x)\ \nu(y)\ dA(x)\ dB(y)$$

$$= \left[\int_M \mu(x)\ dA(x)\right]\left[\int_N \nu(y)\, dB(y)\right]. \quad (3,11,25)$$

In particular, if $M = N = (-\infty, +\infty)$, then

$$\int_{-\infty}^{+\infty}\int_{-\infty}^{+\infty} \mu(x)\ \nu(y)\ dF(x, y) = \left[\int_{-\infty}^{+\infty} \mu(x)\ dA(x)\right]\left[\int_{-\infty}^{+\infty} \nu(y)\, dB(y)\right]. \quad (3,11,26)$$

If $\lambda(x,y)$ is any function of (x,y) over the cartesian product $e = M \times N$, then:

$$\int\int_{M\times N} \lambda(x, y)\ dF(x, y) = \int_M \left[\int_N \lambda(x, y)\ dB(y)\right] dA(x)$$

$$= \int_N \left[\int_M \lambda(x, y)\ dA(x)\right] dB(y). \quad (3,11,27)$$

Formula (3,11,27) is the analogue of (3,11,20) for the case of a function F with separated variables.

In particular, if $M = N = (-\infty, +\infty)$, formula (3,11,27) gives

$$\int_{-\infty}^{+\infty}\int_{-\infty}^{+\infty} \lambda(x, y)\ dF(x, y) = \int_{-\infty}^{+\infty}\left[\int_{-\infty}^{+\infty} \lambda(x, y)\ dB(y)\right] dA(x)$$

$$= \int_{-\infty}^{+\infty}\left[\int_{-\infty}^{+\infty} \lambda(x, y)\ dA(x)\right] dB(y). \quad (3,11,28)$$

F-equivalent functions

Let F(x,y) be a 2-dimensional d.f., and m the measure that it defines over $\mathfrak{X}_2$. Let $\lambda = \lambda(x,y,)$, $\mu = \mu(x,y)$ be two point functions. In II.6 we defined the meaning of the statements: λ and μ are m-equivalent, λ is m-equivalent to 0, etc. Since m is defined by F, we shall also say that λ and μ are F-equivalent, λ is F-equivalent to 0, etc. Clearly, if G is a d.f. equivalent to F and λ and μ are F-equivalent, then they are also G-equivalent.

12. *Convergence of a sequence of n-dimensional distribution functions*

Let F and F_k (k = 1,2,3,...) be an n-dimensional d.f. and a sequence of n-dimensional d.f. of n real variables $x_1, x_2, \ldots, x_n$. As before, we regard the x_j as the components of a vector of an n-dimensional vector space $\mathfrak{X}_n$. Let m and m_k (k = 1,2,3,...) be the measures over $\mathfrak{X}_n$ defined by F and F_k, respectively; suppose that m and the m_k are bounded uniformly in k. We shall say that F_k converges to F *up to equivalence* as $k \to +\infty$ if m_k converges to m weakly in the sense of the definition at the end of II.11; i.e., if for any point function $\lambda(x_1, x_2, \ldots, x_n)$ which is continuous and vanishes outside a compact subset of $\mathfrak{X}_n$,

$$\lim_{k\to+\infty} \underbrace{\int_{-\infty}^{+\infty} \cdots \int_{-\infty}^{+\infty}}_{n} \lambda(x_1, x_2, \ldots, x_n)\, dF_k(x_1, x_2, \ldots, x_n)$$

$$= \underbrace{\int_{-\infty}^{+\infty} \cdots \int_{-\infty}^{+\infty}}_{n} \lambda(x_1, x_2, \ldots, x_n)\, dF(x_1, x_2, \ldots, x_n).$$

If

$$F_k(-\infty, -\infty, \ldots, -\infty) = F(-\infty, -\infty, \ldots, -\infty)$$

for any k and $F(x_1, x_2, \ldots, x_n)$ is absolutely continuous, then convergence of F_k to F implies that

$$\lim_{k\to+\infty} F_k(x_1, x_2, \ldots, x_n) = F(x_1, x_2, \ldots, x_n) \qquad (3,12,1)$$

for any x_j.

In III.4 we gave a direct (but equivalent, in view of Remark

(3,4,1)) definition of the convergence of a sequence of d.f. to a limit d.f. for $n = 1$; this direct definition has no natural extension to the case $n > 1$.

Fourier transform of an n-dimensional bounded distribution function

To simplify the notation, let us return to the case $n = 2$, with the notation of III.11. Let $F(x,y)$ be a *bounded* 2-dimensional d.f., and u and v two real variables over $(-\infty, +\infty)$. The 2-*dimensional Fourier transform* of F is the function $\phi(u,v)$ of two real variables u and v $(-\infty < u, v < +\infty)$ defined by

$$\varphi(u, v) = \int_{-\infty}^{+\infty}\int_{-\infty}^{+\infty} e^{i(ux+vy)}\, dF(x, y). \qquad (3,12,2)$$

The theory of 2-dimensional Fourier transforms may be developed exactly as in III.7 and III.8 for the case $n = 1$ (cf. Remark (3,8,3)); we only indicate the main points, leaving it to the reader to provide the proofs.

A complex function $\phi(u,v)$ of two real variables u and v over $(-\infty, +\infty)$ is said to be *positive-definite* if it has the following property. For any positive integer h, h-tuples of real numbers $u_1, u_2, \ldots, u_h$ (not necessarily different) and $v_1, v_2, \ldots, v_h$ (not necessarily different), and h complex numbers $a_1, a_2, \ldots, a_h$, the number

$$\sum_{j,k=1}^{h} \varphi(u_k - u_j, v_k - v_j)\, a_j\, \bar{a}_k \qquad (3,12,3)$$

is real and nonnegative.

A function $\phi(u,v)$ is the Fourier transform of a bounded 2-dimensional d.f. $F(x,y)$ if and only if $\phi(u,v)$ is continuous and positive-definite.

Positive-definite functions of two variables have properties analogous to the properties 1), 2), 3), 4) of positive-definite functions of one variable (cf. III.7).

A 2-dimensional bounded d.f. is uniquely determined (up to equivalence) by its 2-dimensional Fourier transform.

Consider a countable sequence of 2-dimensional d.f. $F_k(x,y)$ of two variables x,y uniformly bounded in k (k = 1,2,3,...); let $\phi_k(u,v)$ denote the 2-dimensional Fourier transform of F_k and m_k the measure over $\mathfrak{X}_2$ defined by F_k (k = 1,2,3,...).

THEOREM (3,12,1). 1) If there exists a 2-dimensional d.f. F(x,y) such that F_k converges to F as $k \to +\infty$, then F is bounded and consequently has a 2-dimensional Fourier transform.

2) If F_k converges to a d.f. F as $k \to +\infty$, and in addition

$$\lim_{k\to+\infty} [F_k(+\infty, +\infty) - F_k(-\infty, -\infty)] = F(+\infty, +\infty) - F(-\infty, -\infty),$$

i.e. m_k converges *completely* to the measure m defined by F, then ϕ_k converges to the Fourier transform ϕ of F uniformly in u,v.

3) If ϕ_k converges to a limit ϕ as $k \to +\infty$, uniformly in u,v in the region $|u| \leq \eta$, $|v| \leq \eta$ for any fixed positive number η, then there exists a bounded d.f. F(x,y) with Fourier transform ϕ such that F_k converges to F as $k \to +\infty$, and moreover

$$\lim_{k\to+\infty} [F_k(+\infty, +\infty) - F_k(-\infty, -\infty)] = F(+\infty, +\infty) - F(-\infty, -\infty),$$

i.e. m_k converges *completely* to the measure m defined by F.

CONVOLUTION

Given two 2-dimensional bounded d.f. F(x,y) and G(x,y) such that $F(-\infty, -\infty) = G(-\infty, -\infty) = 0$, their convolution $G * F = H(x,y)$ is defined by

$$H(x, y) = \int_{-\infty}^{+\infty} \int_{-\infty}^{+\infty} G(x - \alpha, y - \beta)\, dF(\alpha, \beta). \qquad (3,12,4)$$

The 2-dimensional Fourier transform of $G * F$ is the product of the 2-dimensional Fourier transforms of F and G.

Moments of a bounded n-dimensional distribution function

Given a 2-dimensional bounded d.f. F(x,y), let α and β be two nonnegative numbers, and a and b any two fixed real numbers. The *absolute moment* of order (α,β) about a and b, denoted by ${}_{a,b}m^*_{\alpha,\beta}$, is the RS-integral

$$ {}_{a,b}m^*_{\alpha,\beta} = \int_{-\infty}^{+\infty}\int_{-\infty}^{+\infty} |x - a|^{\alpha}\, |y - b|^{\beta}\, dF(x\, ,\, y). \qquad (3,12,5) $$

If α and β are integers, the *algebraic moment* of order (α,β) about a and b, denoted by ${}_{a,b}m_{\alpha,\beta}$, is the RS-integral

$$ {}_{a,b}m_{\alpha,\beta} = \int_{-\infty}^{+\infty}\int_{-\infty}^{+\infty} (x - a)^{\alpha}\, (y - b)^{\beta}\, dF(x\, ,\, y). \qquad (3,12,6) $$

Of course, these moments exist if and only if the corresponding RS-integrals are meaningful, i.e., absolutely convergent. $\alpha + \beta$ is called the *total order* of the moment. When $a = b = 0$, ${}_{o,o}m^*_{\alpha,\beta}$, ${}_{o,o}m_{\alpha,\beta}$ will be denoted briefly by $m^*_{\alpha,\beta}$, $m_{\alpha,\beta}$.

The 2-dimensional Fourier transform $\phi(u,v)$ and the algebraic moments of F satisfy relations generalizing those of III.9. Thus, assuming α and β integers, let

$\dfrac{\partial^{\alpha+\beta}}{\partial u^{\alpha}\ \partial v^{\beta}}\varphi(0\, ,\, 0)$ denote the derivative $\dfrac{\partial^{\alpha+\beta}}{\partial u^{\alpha}\ \partial v^{\beta}}\varphi(u\, ,\, v)$

at $u = 0$, $v = 0$. Then

$$ \frac{\partial^{\alpha+\beta}}{\partial u^{\alpha}\ \partial v^{\beta}}\varphi(0\, ,\, 0) = i^{\alpha+\beta}\, m_{\alpha,\beta}. \qquad (3,12,7) $$

Analogues of (3,9,10) and (3,9,16) also hold; for instance, corresponding to (3,9,16) we have

$$ \varphi(u\, ,\, v) = \sum_{\alpha,\beta=0}^{+\infty} i^{\alpha+\beta}\, \frac{(\alpha + \beta)!}{\alpha!\ \beta!}\, m_{\alpha,\beta}\, u^{\alpha}\, v^{\beta}. \qquad (3,12,8) $$

Interpretation of the moments of order 1

We have seen how algebraic moments derive in a natural way from the Fourier transform, to be precise, from its differentiability properties. However, the first two algebraic moments are elementary concepts in mechanics, in the study of "material systems," i.e. mass distributions.

Consider a *bounded* measure or mass distribution m(e) over the affine point space E_2 associated with the vector space $\mathfrak{X}_2$, and let $m(E_2) = M$. m(e) may be interpreted as a material system (a plane sheet, since E_2 is 2-dimensional) of total mass M. Fix some reference system $[0; \mathfrak{b}]$ in E_2 (cf. II.11), and let $F(x,y)$ be a 2-dimensional d.f. defining $m(e)$.

We first consider the case where $m(e)$ is a totally discontinuous measure which assigns masses m_k to finitely or countably many fixed points N_k with coordinates (x_k, y_k) $(k = 1,2,3,\ldots)$ (cf. II.9). The *center of gravity* of the plane region or the mass distribution $m(e)$ is the point $G \in E_2$ with coordinates

$$\bar{a} = \frac{1}{M} \sum_k x_k m_k, \quad \bar{b} = \frac{1}{M} \sum_k y_k m_k ; \tag{3,12,9}$$

when the right-hand sides of (3,12,9) are series, the point G exists if and only if they are absolutely convergent. The existence of the center of gravity G and its position do not depend on the choice of the reference system $[0; \mathfrak{b}]$.

Now let $m(e)$ be an absolutely continuous measure with density $\rho(x,y)$ (cf. II.11); then the center of gravity G is defined as the point whose coordinates $(\bar{a}, \bar{b})$ are given (assuming the reference system $[0; \mathfrak{b}]$ orthonormal) by the formulas

$$\bar{a} = \frac{1}{M} \int_{-\infty}^{+\infty} \int_{-\infty}^{+\infty} x\, \rho(x, y)\, dx\, dy, \quad \bar{b} = \frac{1}{M} \int_{-\infty}^{+\infty} \int_{-\infty}^{+\infty} y\, \rho(x, y)\, dx\, dy . \tag{3,12,10}$$

The point G exists if and only if the integrals (3,12,10) are absolutely convergent, and its existence and position are again independent of the choice of the reference system $[0; \mathfrak{b}]$.

Now in view of (3,11,4) and (3,11,6) the formulas (3,12,9) and (3,12,10) may both be expressed in the form

$$\bar{a} = \frac{1}{M} \int_{-\infty}^{+\infty} \int_{-\infty}^{+\infty} x\, dF(x, y), \quad \bar{b} = \frac{1}{M} \int_{-\infty}^{+\infty} \int_{-\infty}^{+\infty} y\, dF(x, y). \tag{3,12,11}$$

The right-hand sides of (3,12,11) are clearly the moments $m_{1,0}$ and $m_{0,1}$, i.e. the algebraic moments of total order 1 of the d.f. F about the origin 0. Formula (3,12,11) may therefore be rewritten as

$$\bar{a} = \frac{1}{M} m_{1,0}, \quad \bar{b} = \frac{1}{M} m_{0,1} . \tag{3,12,12}$$

In mechanics, the mass distributions usually considered are

either totally discontinuous, absolutely continuous, or a combination of these two types. However, the foregoing reasoning leads naturally to the following generalization: For any bounded mass distribution m(e) over E_2, the *center of gravity is defined as the point* G *with coordinates* $(\bar{a},\bar{b})$ *given by (3,12,12).* The center of gravity exists if and only if the moments $m_{1,0}$ and $m_{0,1}$ exist.

In the general case the center of gravity has the same properties as in classical mechanics. In particular we shall see below that its existence and position are independent of the choice of the reference system in E_2. The import of (3,12,12) is the light it throws on the connection between the moments of total order 1 and the well-known notion of the center of gravity.

Now let m(e) be a bounded mass distribution, not over the point space E_2, but over the associated vector space $\mathfrak{X}_2$. The quantities $\bar{a}$ and $\bar{b}$ defined by (3,12,12) are no longer interpreted as the coordinates of a point G, but as the components of a vector $\vec{g} \in \mathfrak{X}_2$, in other words, the vector $\overrightarrow{OG}$. We shall call the vector $\vec{g}$ defined by (3,12,12) the *mean vector* of the mass distribution or the measure m(e) over $\mathfrak{X}_2$. It exists if and only if $m_{1,0}$ and $m_{0,1}$ exist; its existence and value do not depend on the basis **b** chosen in $\mathfrak{X}_2$. Of course, the notions of mean vector and center of gravity are essentially identical.

Central moments

Given a 2-dimensional d.f. F(x,y), the *central moment* of order α and β, denoted by $\mu^{\bullet}_{\alpha,\beta}$ (absolute moment of order (α,β)) or $\mu_{\alpha,\beta}$ (algebraic moment of order (α,β)) is the absolute or algebraic moment of order (α,β) about $a = \bar{a}$ and $b = \bar{b}$, where $\bar{a}$ and $\bar{b}$ are defined by (3,12,12). It is clear that

$$\mu_{1,0} = \mu_{0,1} = 0.$$

Moments of order 2 and moments of inertia

In mechanics, the description of a material system involves both its center of gravity and its *moments of inertia.* As before, let us consider a bounded mass distribution m(e) over E_2 relative to a reference system [0; **b**], in other words, a plane E_2 relative to coordinate axes Oxy, which we assume to be orthonormal. $F(x,y)$ denotes a 2-dimensional d.f. defining m(e).

Motivated as above by the simple case of a totally discontinuous or absolutely continuous measure, we arrive at the following conclusions.

Let Q be a point with coordinates (a,b); the moment (both absolute and algebraic)

$$_{a,b}m_{2,o} = \int_{-\infty}^{+\infty}\int_{-\infty}^{+\infty} (x - a)^2 \, dF(x , y)$$

may be interpreted as the moment of inertia of the mass distribution with respect to an axis Qx' parallel to Ox through y = b. The moment

$$_{a,b}m_{1,1} = \int_{-\infty}^{+\infty}\int_{-\infty}^{+\infty} (x - a)(y - b) \, dF(x , y),$$

is the so-called "inertia product."

Thus $_{a,b}m_{2,o}$, $_{a,b}m_{1,1}$, $_{a,b}m_{o,2}$ are the coefficients of the equation

$$_{a,b}m_{o,2}(x - a)^2 - 2 \, _{a,b}m_{1,1}(x - a)(y - b) + {}_{a,b}m_{2,o}(y - b)^2 = 1$$

of the *inertia ellipse* at the point Q.

$_{a,b}m_{1,1} = 0$ means that Qx', Qy' are the axes of symmetry of this ellipse.

The reader will easily prove the following well-known theorem.

THEOREM (3,12,2). As above, let $(\overline{a},\overline{b})$ denote the coordinates of the center of gravity and M the total mass $M = m(E_2)$; then

$$_{a,b}m_{2,o} = \mu_{2,o} + M(a - \overline{a})^2,$$

$$_{a,b}m_{o,2} = \mu_{o,2} + M(b - \overline{b})^2.$$

Application of the Hölder inequalities to moments

The inequalities proved in II.7 yield useful information about moments. In particular, they provide upper bounds for the moments - an example is Theorem (3,9,1). Still confining ourselves to the case n = 2, let us examine some consequences of the Hölder inequalities (2,7,2).

Let $m(e)$ be the measure over E_2 defined by a d.f. $F(x,y)$, and $\lambda = \lambda(u) = \lambda(x,y)$ a function of $u = (x,y) \in E_2$ defined over E_2. Recalling the notation of II.7, we put

$$M_\alpha(\lambda) = \int_{E_2} |\lambda(u)|^\alpha \, m(du) = \int_{-\infty}^{+\infty} \int_{-\infty}^{+\infty} |\lambda(x\,,\,y)|^\alpha \, dF(x\,,\,y)$$

for $\alpha > 0$, and

$$N_\alpha(\lambda) = [M_\alpha(\lambda)]^{\frac{1}{\alpha}}.$$

To say, for instance, that the absolute moment

$${}_{o,o}m^*_{\alpha,o} = \int_{-\infty}^{+\infty} \int_{-\infty}^{+\infty} |x|^\alpha \, dF(x\,,\,y)$$

exists, is the same as saying that the function $\lambda(x,y) = x$ has the property

$$M_\alpha[\lambda] < +\infty\,.$$

If β is a positive number and the absolute moment

$${}_{o,o}m^*_{o,\beta} = \int_{-\infty}^{+\infty} \int_{-\infty}^{+\infty} |y|^\beta \, dF(x\,,\,y)$$

exists, this is the same as saying that the function $\mu(x,y) = y$ has the property

$$M_\beta(\mu) < +\infty\,.$$

If in addition $\frac{1}{\alpha} + \frac{1}{\beta} = 1$, then (2,7,2) shows that

1) the absolute and algebraic moments

$${}_{o,o}m^*_{1,1} = \int_{-\infty}^{+\infty} \int_{-\infty}^{+\infty} |x|\,|y| \, dF(x,\ y)\,,\quad {}_{o,o}m_{1,1} = \int_{-\infty}^{+\infty} \int_{-\infty}^{+\infty} xy \, dF(x,\ y)$$

exist;

2)

$$0 \leqq |{}_{o,o}m_{1,1}| \leqq {}_{o,o}m^*_{1,1} \leqq [{}_{o,o}m^*_{\alpha,o}]^{\frac{1}{\alpha}} \cdot [{}_{o,o}m^*_{o,\beta}]^{\frac{1}{\beta}}. \qquad (3,12,13)$$

For instance, for the case $\alpha = \beta = 2$, i.e. the Schwarz inequality (2,7,3), we get the following proposition:

If the (algebraic and absolute) moments of order (2,0) and

(0,2) exist, then the moments of order (1,0) and (0,1) exist [this was already proved in (3,9,5)]; moreover, the algebraic and absolute moments of order (1,1) exist, and

$$0 \leq ({}_{o,o}m_{1,1})^2 \leq ({}_{o,o}m^*_{1,1})^2 \leq {}_{o,o}m_{2,o} \cdot {}_{o,o}m_{o,2}. \tag{3,12,14}$$

If we put

$${}_{o,o}m_{1,1} = \sqrt{{}_{o,o}m_{2,o} \cdot {}_{o,o}m_{o,2}} \cdot r\ , \tag{3,12,15}$$

it follows from (3,12,14) that

$$|r| \leq 1. \tag{3,12,16}$$

Thus the coefficient r defined by (3,12,15), which is obviously dimensionless, is always between -1 and +1. Moreover, we know from Theorem (2,7,4) that $|r| = 1$ if and only if the functions $\lambda(x,y) = x$ and $\mu(x,y) = y$ are proportional, up to m-equivalence. This is the case if and only if there exists a straight line Δ in E_2 through the origin 0 such that the mass distributed over E_2 by the distribution m(e) is concentrated entirely on Δ; i.e. m(e) = 0 for any subset e of E_2 disjoint from Δ.

On the other hand, it is clear that r = 0 if and only if the axe of symmetry of the inertia ellipse with respect to 0 coincide with the coordinate axes Oxy of the reference system chosen in E_2.

Of course, since the point 0 is completely arbitrary with respect to the mass distribution m(e), these results apply to any point of E_2, in particular to the center of gravity G (the inertia ellipse with respect to G is often called the central inertia ellipse).

Case of separated variables

The reader will easily ascertain what happens to some of the previous results when the variables can be separated, i.e. (cf. end of III.11) when the 2-dimensional d.f. F(x,y) is the product of two one-dimensional d.f. A(x) and B(y):

$$F(x, y) = A(x)\ B(y). \tag{3,12,17}$$

We restrict ourselves to the following remark. If F has the

form (3,12,17), then by (3,12,2) its Fourier transform is

$$\varphi(u, v) = a(u)\, b(v), \tag{3,12,18}$$

where a(u) is the one-dimensional Fourier transform of A(x) and b(v) that of B(y); indeed, in view of (3,11,26),

$$\varphi(u, v) = \int_{-\infty}^{+\infty}\int_{-\infty}^{+\infty} e^{i(ux+vy)}\, dF(x, y) = \int_{-\infty}^{+\infty}\int_{-\infty}^{+\infty} e^{iux} \cdot e^{ivy}\, dA(x)\, dB(y)$$

$$= \left[\int_{-\infty}^{+\infty} e^{iux}\, dA(x)\right] \cdot \left[\int_{-\infty}^{+\infty} e^{ivy}\, dB(y)\right].$$

We know that a(u) and b(v) are positive-definite functions.

Conversely, assume that $\phi(u,v)$ is a positive-definite function of the pair (u,v); we know that then

$$0 \leqq |\varphi(u, v)| \leqq \varphi(0, 0).$$

Disregarding the trivial case $\phi(0,0) = 0$, which implies $\phi(u,v) \equiv 0$, we assume $\phi(0,0) > 0$.

Assume that $\phi(u,v)$ has the form (3,12,18); this decomposition is not unique, since

$$\varphi(u, v) = [\rho \cdot a(u)] \cdot \left[\frac{b(v)}{\rho}\right]$$

for any nonzero ρ. ρ can be chosen so that $\rho.a(0)$ and $\frac{b(0)}{\rho}$ are real and positive; i.e. there is no loss of generality in assuming that a(0) and b(0) are real and positive.

Now this implies that a(u) and b(v) are positive-definite functions. Indeed, consider an arbitrary positive integer h, h arbitrary pairs (u_j, v_j) $(j = 1,2,3,\ldots,h)$, and h arbitrary complex numbers $\tau_1, \tau_2, \ldots, \tau_h$. By assumption, the number

$$\sum_{jk} \varphi(u_j - u_k, v_j - v_k)\, \tau_j\, \overline{\tau}_k$$

is real and nonnegative; taking $v_1 = v_2 = \ldots = v_h$, we obtain

$$b(0) \sum_{jk} a(u_j - u_k)\, \tau_j\, \overline{\tau}_k,$$

so that a(u) is positive-definite. A similar argument proves the statement for b(v).

Conversely, let a(u) be a positive-definite function of u and b(v) a positive-definite function of v; their product

$$\varphi(u, v) = a(u)\, b(v)$$

is a positive-definite function of (u,v); this is an immediate consequence of Lemma (3,5,1).

THEOREM (3,12,3). If a complex function $\phi(u,v)$ of two real variables u and v $(-\infty < u, v < +\infty)$ is the product of a function a(u) of u alone and a function b(v) of v alone such that a(0) and b(0) are nonnegative, then $\phi(u,v)$ is a positive- definite function of (u,v) if and only if a(u) is a positive-definite function of u and b(v) a positive-definite function of v.

Intrinsic study

We have already indicated (Remark (3,1,2)) that a measure m(e) over $\mathfrak{X}_2$ may always be studied by fixing a basis $\mathfrak{b}$ in $\mathfrak{X}_2$ and representing m(e) by a d.f. F(x,y) that defines it relative to $\mathfrak{b}$, say the d.f. (3,1,1). However, the fact that this depends on an arbitrary basis $\mathfrak{b}$ is a disadvantage. It is therefore useful, even necessary, to study m(e) with regard to its intrinsic properties, i.e. those independent of any reference system in $\mathfrak{X}_2$.

Let $\mathfrak{X}$ be a vector space; the set of all linear functionals over $\mathfrak{X}$ is a vector space $\mathfrak{X}^*$, called the dual of $\mathfrak{X}$.

NOTATION

We shall always use the notation

$$< z^*, z >, \tag{3,12,19}$$

already introduced in II.10, for the result of applying a linear functional z* (an element of $\mathfrak{X}^*$) to an element z of $\mathfrak{X}$. Recall that if z is not the zero θ of $\mathfrak{X}$, then $\mathfrak{X}^*$ contains at least one linear functional z* such that $< z^*, z > \neq 0$.

It is known that when $\mathfrak{X} = \mathfrak{X}_n$ is of finite dimension n, $\mathfrak{X}^* = \mathfrak{X}_n^*$ is also n-dimensional.

We shall consider the case n = 2, but the definitions and results can be immediately extended to any finite n.

Thus, let $\mathfrak{X}_2$ be a (real) 2-dimensional vector space and m(e) a measure or mass distribution over $\mathfrak{X}_2$; as before, we denote M = $m(\mathfrak{X}_2)$, and confine ourselves to the case of a *bounded* measure, $M < +\infty$.

DEFINITION

A mean vector of the distribution m *is an element* g *(or* $\vec{g}$*) of* $\mathfrak{X}_2$ *such that*

$$M < z^*, g > = \int_{\mathfrak{X}_2} < z^*, z > \cdot m(dz) \qquad (3,12,20)$$

for any $z^* \in \mathfrak{X}_2^*$. Note first that if such an element g exists it is unique; for were there another mean vector g', $g - g' \neq \theta$, there would be a linear functional, say z_0^*, such that

$$< z_0^*, g - g' > \neq 0,$$

or

$$< z_0^*, g > \neq < z_0^*, g' >.$$

But then it would follow from (3,12,20) that $< z^*, g > = < z^*, g' >$ for any $z^* \in \mathfrak{X}_2^*$, including $z^* = z_0^*$. The intrinsic character of the mean vector g defined by (3,12,20), if it exists, is thus evident.

To show that the mean vector does indeed exist, consider some basis **b** in $\mathfrak{X}_2$ and let **b*** be the dual basis in $\mathfrak{X}_2^*$; let (x,y) and (u,v) be the components of $z \in \mathfrak{X}_2$ relative to **b** and of $z^* \in \mathfrak{X}_2^*$ relative to **b***, respectively. It is known that for any choice of **b**

$$< z^*, z > = ux + vy.$$

On the other hand, let F(x,y) be a 2-dimensional distribution function defining m(e) with respect to **b**; then

$$\int_{\mathfrak{X}_2} < z^*, z > m(dz) = \int_{-\infty}^{+\infty} \int_{-\infty}^{+\infty} (ux + vy)\, dF(x, y)$$

$$= M(u\bar{a} + v\bar{b}),$$

where $\bar{a}$ and $\bar{b}$ are precisely the numbers defined by (3,12,11). It follows that there is a vector g (and only one) satisfying (3,12,20), namely the vector with components $(\bar{a},\bar{b})$ relative to **b**.

The mean vector g is an important characteristic of the distribution m but it is obviously not sufficient to define m; the problem of an entirely *intrinsic* definition of m is as yet unsolved. Now consider the function $\xi(z^*)$ of $z^* \in \mathfrak{X}_2^*$ defined by

$$\xi(z^*) = \int_{\mathfrak{X}_2} e^{i\langle z^*, z\rangle} m(dz). \qquad (3,12,21)$$

This complex function is obviously of intrinsic significance; we shall call it the Fourier functional transform of the measure m. It defines m uniquely, for, considering the bases **b** and **b*** in $\mathfrak{X}_2$ and $\mathfrak{X}_2^*$, respectively, and using the same notation, we see that

$$\xi(z^*) = \int_{-\infty}^{+\infty}\int_{-\infty}^{+\infty} e^{i(ux+vy)}\, dF(x, y) = \phi(u, v), \qquad (3,12,22)$$

which is none other than the Fourier transform $\phi(u,v)$ of the 2-dimensional d.f. F; we already know that $\phi(u,v)$ determines F uniquely, up to equivalence.

If $\mathcal{Y}$ is an arbitrary *real* vector space of elements y, and $\xi(y)$ a complex-valued function of $y \in \mathcal{Y}$, then, recalling that $\mathcal{Y}$ is an abelian group with respect to addition, we call $\xi(y)$ a *positive-definite function* of $y \in \mathcal{Y}$ if for any integer $h > 0$, any h-tuple of elements $y_1, y_2, \ldots, y_h$ of $\mathcal{Y}$, and any h-tuple of complex numbers $a_1, a_2, \ldots, a_h$, the number

$$\sum_{jk} \xi(y_j - y_k)\, a_j \bar{a}_k$$

is real and nonnegative.

Thus $\phi(u,v)$ is a positive-definite function of the 2 real variables u and v if and only if $\xi(z^*)$ is a positive-definite function of $z^* \in \mathfrak{X}_2^*$.

BIBLIOGRAPHIC NOTES FOR CHAPTER III

A study of one-dimensional d.f. may be found in M. Fréchet [1].

E. C. Titchmarsh [1] presents the general theory of Fourier transforms; S. Bochner [1], [2], E. Lukacs [1] are more specialized and consider Fourier transforms of bounded d.f.; see also P. Lévy [1], S. Bochner and K. Chandrasekharan [1].

For a very complete treatment of the theory of Laplace transforms see G. Doetsch [1].

For n-dimensional d.f., see H. Cramer [4].

On the theory of groups and the general theory of Fourier transforms, see A. G. Kurosh [1], A. Weil [1], N. Bourbaki [2], L. Schwartz [1], [2], L. M. Loomis [1], Bachman [1]; for multidimensional d.f., see E. Lukacs [2].

EXERCISES

3.1 (E. Borel [1], p. 12). For any real number x such that $0 \leq x < 1$, there is exactly one countable sequence $\{\alpha_k\}$ of integers α_k such that

a) $0 \leqslant \alpha_k \leqslant 2$;

b) $$\sum_{k=1}^{+\infty} \frac{\alpha_k}{3^k} = x . \tag{3-1,1}$$

The series (3-1,1) is the *ternary expansion* of x.

Let $\mathfrak{A}$ be the set of all x such that

1) $0 \leqslant x < 1$;

2) in the ternary expansion of x at least one α_k is 1, but only finitely many α_k are 1.

For $x \in \mathfrak{A}$, let n be the first k such that $\alpha_k = 1$, and put

$$y(x) = \sum_{k=1}^{n} \frac{\alpha_k}{3^k} . \tag{3-1,2}$$

Let z(x) be the number obtained by substituting 1 in the right-hand side of (3-1,2) for any $\alpha_k = 2$ (if such α_k exist); z(x) is a monotone nondecreasing function of x. Show that there exists exactly one d.f. F(x) such that

1) F(x) is continuous from the left;

2) $F(x) \equiv 0$ if $x < 0$;

3) $F(x) \equiv 1$ if $x \geq 1$;

4) $F(x) = z(x)$ if $x \in \mathfrak{A}$.

Show that $F(x)$ is continuous. Show that there is a countable set of pairwise disjoint sub-intervals of $[0,1]$ such that $F(x)$ is constant over each of them. Hence show that $F(x)$ is a singular d.f.

3.2 Let $F(x)$ be any given d.f. Show that there is a countable sequence $\{F_k\}$ $(k = 1,2,3,\ldots)$ of d.f. $F_k(x)$ such that

a) for any k, F_k is totally discontinuous;

b) $\lim_{k\to+\infty} F_k = F$.

3.3 Let $F(x)$ be any given d.f.; show that there is a countable sequence $\{F_k\}$ $(k = 1,2,3,\ldots)$ of d.f. $F_k(x)$ such that:

a) for any k, F_k is absolutely continuous;

b) $\lim_{k\to+\infty} F_k = F$.

3.4 Let $F_m(x)$ be the absolutely continuous d.f. defined by

$$F_m(x) = \frac{1}{\sqrt{2\pi}} \int_{-\infty}^{x} e^{-(y-m)^2/2}\, dy,$$

where m is an arbitrary real constant, and $G_\sigma(x)$ the absolutely continuous d.f. defined by

$$G_\sigma(x) = \frac{1}{\sqrt{2\pi}\sigma} \int_{-\infty}^{x} e^{-y^2/2\sigma^2}\, dy,$$

where σ is a positive constant.

$F_m(-\infty) = G_\sigma(-\infty) = 0$, and by (3,6,22) $F_m(+\infty) = G_\sigma(+\infty) = 1$. Find the Fourier transforms $\phi_m(u)$ and $\psi_\sigma(u)$ of F_m and G_σ respectively [cf. Example (3,6,2)]. Show that there exists a d.f. such that $\lim_{m\to+\infty} F_m = F$, and $F(+\infty) \neq 1$; how does ϕ_m behave as $m \to +\infty$? How do G_σ and ψ_σ behave as $\sigma \to +\infty$? [Relate this to Theorems (3,7,1) and (3,8,1).] How do the distances $\Delta(F_m,F)$, $\Delta(G_\sigma,G)$ (cf. III.4) behave as $m \to +\infty$, $\sigma \to +\infty$, respectively?

3.5 Verify that the function $R(x) = \frac{1}{\pi}\frac{1}{1+x^2}$ belongs to class (S_2). Find its Fourier transform $r(u)$ [the method of residues may be used]; explain why $r(u)$ is positive-definite (*Answer*: $r(u) = e^{-|u|}$).

3.6 Let $F(x)$ be a d.f. and x_0 any given value of x; $F(x)$ is said to be symmetric with respect to x_0 if for any $h > 0$

$$F(x_0 + h + 0) - F(x_0 + 0) = F(x_0 - 0) - F(x_0 - h - 0) ;$$

$F(x)$ is said to be symmetric if it is symmetric with respect to $x_0 = 0$. What can be said about the density $f(x)$ of an absolutely continuous symmetric d.f.?

Show that if $F(x)$ is bounded, it is symmetric if and only if its Fourier transform $\phi(u)$ is real (for any real u) [use (3,6,10)].

3.7 If α is a positive number and $F(x)$ a d.f. over $[0, +\infty)$ such that

$$\int_0^{+\infty} x^\alpha \, dF(x) = m_\alpha^* < +\infty, \quad F(+0) = 0,$$

show that

$$\lim_{x \to +\infty} x^\alpha [1 - F(x)] = 0$$

and

$$\alpha \int_0^{+\infty} x^{\alpha-1} [1 - F(x)] \, dx = m_\alpha^*$$

[Show that $x^\alpha[1 - F(x)] \leqq \int_x^{+\infty} y^\alpha \, dF(y)$; then carry out integration by parts using (3,3,8)].

3.8 Let $F(x_1, x_2, \ldots, x_n)$ be an n-dimensional d.f. Show that there exists a sequence $\{F_k\}$ of n-dimensional d.f. $F_k(x_1, x_2, \ldots, x_n)$ such that

1) for any k, F_k is totally discontinuous;
2) $\lim_{k \to +\infty} F_k = F$.

3.9 Prove Karamata's Theorem (3,9,3).

In the case $\rho > 0$, for $y > 0$, put

$$F\left(\frac{1}{y}\right) = \int_{-\infty}^{+\infty} e^{-yx} g(e^{-yx}) \, dF(x),$$

where $g(v)$ denotes the function defined over $0 \leqq v \leqq 1$ by

$$g(v) = \begin{cases} \frac{1}{v} \text{ if } \frac{1}{e} \leqq v \leqq 1 \, ; \\ 0 \text{ if } 0 \leqq v \leqq \frac{1}{e} \, . \end{cases}$$

Study the behavior of the integral

$$\int_{-\infty}^{+\infty} e^{-yx} \, p(e^{-yx}) \, dF(x),$$

as $y \to +0$, where $p(v)$ is a polynomial in v over $[0,1]$; then approximate $g(v)$ over $[0,1]$ in a suitable manner by polynomials in v.

CHAPTER IV

RANDOM VARIABLES

AXIOM OF CONDITIONAL PROBABILITY

I. RANDOM VARIABLES

1. *Distribution function of a random variable*

Recall that throughout this book, unless otherwise stated, the term "random variable" always refers to a one-dimensional nondegenerate r.v. [cf. Remark (2,12,2)]. Section I of this chapter will be devoted to the intrinsic study (cf. II.11) of random variables. The intrinsic study of a r.v. is that of its probability law, or of the corresponding measure m(e) or distribution of unit mass over the space R of real numbers (or over the x-axis). The condition

$$\Pr(X = \pm\infty) = 0 \qquad (4,1,1)$$

means that X is nondegenerate. The *distribution function* F(x) of X is the function of the real variable x defined for all $x \in (-\infty, \infty)$ by

$$F(x) = \Pr(X < x). \qquad (4,1,2)$$

F(x) is none other than the d.f. associated by (3,1,1) with the measure m(e), i.e. with the probability law of X.

It follows from Theorem (3,1,1) that

1) $F(x - 0) = F(x)$; the d.f. of a r.v. is *continuous from the left*. (4,1,3)

2) $F(+\infty) = 1$. (4,1,4)

3) $F(-\infty) = 0$. (4,1,5)

It is clear that F(x) is a *normalized* d.f. (cf. III.2).

When no ambiguity is possible, we shall frequently use the term "distribution function" for the d.f. of a r.v., i.e. a normalized d.f. F such that $F(+\infty) = 1$.

On the other hand, we know from Theorem (3,1,2) that F(x) uniquely determines m(e), i.e. the probability law of X. In particular, formula (3,1,10) (in the one-dimensional case) shows that for any two numbers a and b (not necessarily finite) such that $a <$ if e denotes the set of all x such that $a \leqq x < b$ and E the event $a \leqq X < b$, then

$$m(e) = \Pr(E) = \Pr(a \leqq X < b) = F(b) - F(a), \qquad (4,1,6)$$

whence, by Theorem (2,6,2),

$$\begin{aligned} \Pr(X = x) &= F(x+0) - F(x-0) \\ &= F(x+0) - F(x). \end{aligned} \tag{4,1,7}$$

REMARK (4,1,1). We know that the probability law of X may be defined by any d.f. equivalent to F(x) (cf. III. 1 and III.2). However, it is convenient to choose one of these equivalent d.f. as *the* distribution function of X. The universally adopted convention is arbitrary, and [cf. Remark (3,1,1)] the function

$$G(x) = F(x+0) = \Pr(X \leqslant x)$$

is another possibility. In the latter case the d.f. of X would be continuous from the right instead of from the left (cf. III.2).

Discrete random variables. In III.2 we divided normalized d.f. into three categories: totally discontinuous d.f., absolutely continuous d.f., and singular d.f. By Theorem (3,2,2) any normalized d.f. F(x) may be expressed uniquely as the sum of three normalized d.f. (some of which may be identically zero), one totally discontinuous, the second absolutely continuous and the third singular.

We shall say that a r.v. X is *discrete* if its (normalized) d.f. F(x) is totally discontinuous (cf. III.2), i.e. if its probability law m(e) is totally discontinuous (cf. II.9). In view of (2,9,1), therefore, a discrete r.v. is a r.v. X having only finitely or countably many possible values $x_1, x_2, \ldots x_k, \ldots$ Thus, if e is a subset of R which contains none of the x_k, the event $X \in e$ is impossible, or at least almost impossible:

$$m(e) = \Pr(X \in e) = 0. \tag{4,1,8}$$

Note that for some values of k the event $X = x_k$ may also be impossible or almost impossible.

The x_k are not necessarily indexed in order of magnitude; moreover, it may well be impossible to index them in this way, for instance if some x_k converge to a limit λ as $k \to +\infty$, while other x_k converge to a limit $\mu \neq \lambda$ (Example: $x_k = (-1)^k + 1/k$).

On the other hand, there is no loss of generality in assuming that all the x_k are distinct, and we shall do so. Let us put

$$p_k = \Pr(X = x_k).$$

By (4,1,8) and the Axioms of II.5, it is obvious that

$$p_k \geqslant 0 \; ; \; \sum_k p_k = 1. \qquad (4,1,9)$$

The case $p_k = 0$ is not excluded; if the x_k form a countable set, formula (4,1,9) implies that the series $\sum_k p_k$ is convergent. The p_k determine $F(x)$ uniquely; indeed, if $K(x)$ is the set of all k such that $x_k < x$, then the Total Probability Axiom implies

$$F(x) = \sum_{k \in K(x)} p_k . \qquad (4,1,10)$$

Conversely, the p_k may be derived from $F(x)$, since by (4,1,7)

$$p_k = F(x_k + 0) - F(x_k). \qquad (4,1,11)$$

Indicator. As an example of a discrete random variable, let us consider an indicator. The *indicator* of an event E is the variable X which is 1 if E occurs, 0 otherwise. This notion is in fact identical with that of the indicator of a set, introduced in II.6, p. 68 . *Using indicators, we can reduce any problem of probability to the study of random variables*; the importance of random variables is thus re-emphasized.

Thus, if we put

$$p = \Pr(E), \; q = 1 - p = 1 - \Pr(E) = \Pr(\check{E}),$$

the d.f. $F(x)$ of the indicator X of the event E is clearly

$$F(x) = \begin{cases} 0 & \text{for } x \leqslant 0 \; ; \\ q & \text{for } 0 < x \leqslant 1 \; ; \\ q + p = 1 & \text{for } 1 < x. \end{cases}$$

The graph of $F(x)$ is illustrated in Figure (4,1,1)

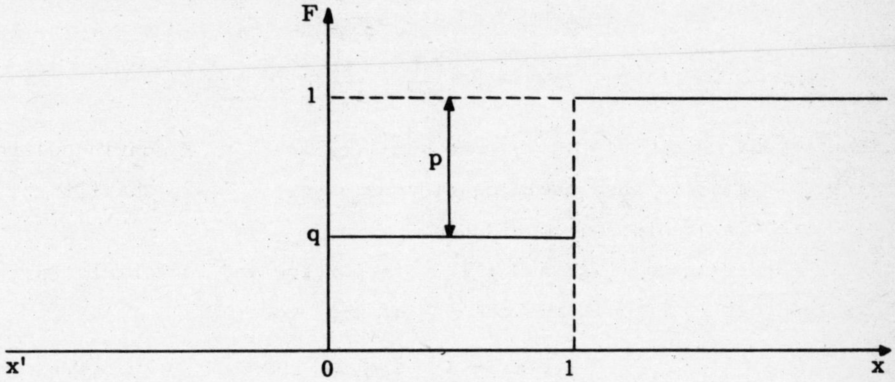

Fig. (4,1,1)

The general property (4,1,11), by which $F(x)$ has a jump discontinuity of size p_k when X passes through the value x_k (if $p_k \neq 0$), is evident in this particular case.

EXAMPLE (4,1,1) The random variable $X = i + j$ of Example (2,11,2) takes only integer values 2,3,4, ..., 12; it is therefore a discrete r.v.

Let $F(x)$ be its d.f.; obviously,

$$F(x) = 0 \text{ for } x \leqslant 2 \text{ ; } F(x) = 1 \text{ for } x > 12.$$

Let us evaluate $F(4)$. $X < 4$ if and only if $X = i + j = 2$ or 3. This is the case if and only if one of the following three events occurs:

E_1 = the event $i = 1$, $j = 1$;

E_2 = the event $i = 2$, $j = 1$;

E_3 = the event $i = 1$, $j = 2$.

But E_1, E_2, E_3 are mutually exclusive, therefore

$$F(4) = Pr(E_1) + Pr(E_2) + Pr(E_3).$$

Let C_{ij} be the event: the first die shows i and the second j ($i = 1,2,\ldots,6$; $j = 1,2,\ldots,6$); we saw in Example (2,2,1) that

$$Pr(C_{ij}) = \frac{1}{36} \text{ ;}$$

but

$$E_1 = C_{1,1}\ ,\ E_2 = C_{2,1}\ ,\ E_3 = C_{1,2}\ ,$$

and so

$$F(4) = 3 \times \frac{1}{36} = \frac{1}{12}\ .$$

Continuous random variables. We shall call a r.v. X continuous if its d.f. F(x) is absolutely continuous, i.e. if its probability law m(e) is absolutely continuous.

Referring to II.11 and III.2, in particular, to formula (3,2,5), we see that if f(x) is the density of m(e) then

$$F(x) = \int_{-\infty}^{x} f(y)\,dy, \tag{4,1,12}$$

and, apart perhaps from exceptional values of x,

$$f(x) = \frac{d}{dx} F(x). \tag{4,1,13}$$

f(x) is called the *probability density*[1] of the (continuous) r.v. X; in view of (4,1,12) and (4,1,13), it clearly determines the probability law of X uniquely.

By the monotonicity of F(x) and formulas (4,1,4) and (4,1,5),

$$f(x) \geq 0, \int_{-\infty}^{+\infty} f(x)\,dx = 1. \tag{4,1,14}$$

Uniformly distributed random variables. Let A be an arbitrary subset of the set R of real numbers. A continuous r.v. X is said to have a *uniform law over A*, or to be *uniformly distributed over A*, if its probability law is uniform over A (cf. II.11); by (2,11,14), (2,11,15) and (2,11,17), the p.d. f(x) of X is

$$f(x) = \begin{cases} 0 \text{ if } x \notin A, \\ \\ \dfrac{1}{v(A)} \text{ if } x \in A. \end{cases} \tag{4,1,15}$$

Since R is of dimension 1, the "volume" v(A) must be interpreted here as the *length* of A; for instance, if A is an interval (a,b) (a < b) then v(A) = b - a.

---

(1) Probability density; abbreviation: p.d.; in German: Wahrscheinlichkeitsdichte; in French: densité de probabilité.

REMARK (4,1,2). Referring again to the classification of d.f. described in III.2, one might call a random variable X singular if its d.f. is singular; a r.v. X such that the singular component R(x) of its d.f. F(x) in the representation (3,2,6) is identically zero would then be called mixed (partly discrete and partly continuous), and so on. However, singular, mixed, etc. r.v. are much rarer in applications than discrete and continuous r.v.

Certain number, almost certain number. As already mentioned in II.12, a non-random number a is often called a certain number, in contradistinction to a random variable. It may always be regarded as a r.v. X which assumes the value a certainly; we can say that *it reduces to the certain number* a. Of course, given a r.v. X and a number a, if $Pr(X = a) = 1$, it does not follow (unless more information is available) that X is certainly equal to a, but only *almost certainly* (or almost surely) equal to a (cf. II.12). In this case we can say that a is an almost certain number, and that X reduces to an almost certain number.

If X is a r.v. which is certainly or almost certainly equal to a number a, then X is discrete and its probability law is described by a unit mass 1 at the point a of the x-axis; its d.f. is $\Delta(x - a)$ (cf. III.2).

Types of random variables and distribution functions

In measuring a directed quantity, the origin and the unit of measurement (also often the direction, though we shall disregard this possibility in order to simplify the discussion) are arbitrary, apart from questions of practical convenience. It is thus convenient to associate with every r.v. X the set of all r.v. Y which are linear functions of X (for the general notion of functions of an r.e., see II.11):

$$Y = aX + b, \tag{4,1,16}$$

where a and b are constants (non-random; $a > 0$ since the direction is preserved). We shall say that Y is of the same *type* as X. It is clear from (4,1,16) that the fact that two r.v. X and Y are of the same type is an equivalence relation; thus types of r.v. are disjoint equivalence classes of r.v.

If the d.f. of X is F(x) and Y is related to X by (4,1,16), then its d.f. G(y) may be derived from F(x) as follows:

$$G(y) = \Pr(Y < y) = \Pr(aX + b < y)$$

$$= \Pr\left(X < \frac{y - b}{a}\right) = F\left(\frac{y - b}{a}\right) . \qquad (4,1,17)$$

If F(x) is the d.f. of a r.v., any d.f. $F(\alpha x + \beta)$, where α and β are arbitrary (but positive), is said to be of the same type as F(x). Note that if F(x) has density f(x), then $F(\alpha x + \beta)$ has density $\alpha f(\alpha x + \beta)$, the derivative of $F(\alpha x + \beta)$ with respect to x. The densities f(x), $\alpha f(\alpha x + \beta)$ are also said to be of the same type.

Symmetric random variables and distribution functions

An r.v. X and its d.f. F(x) are said to be *symmetric with respect to* x_0 if for any $h > 0$

$$\Pr(X < x_o - h) = \Pr(X > x_o + h),$$

i.e., if

$$F(x_o - h) = 1 - F(x_o + h + 0) \qquad (4,1,18)$$

[cf. Exercise 3.6]. X and F(x) are said to be symmetric if they are symmetric with respect to $x_0 = 0$.

If F(x) has density f(x), then (4,1,18) is equivalent to

$$f(x_o - h) = f(x_o + h), \qquad (4,1,19)$$

i.e. $f(x_0 + h)$ is an *even* function of h.

2. *Description of a random variable*

For both practical and theoretical reasons, one of the main objects of Probability Theory is to learn how to "describe" probability laws. To describe a probability law means to express its general behavior and main characteristics in terms of a limited number of well-chosen parameters or data. In this section we shall deal with the description of probability laws of random variables.

Suppose that we are recording the heights of N individuals of the same sex - say males, and the same age (say twenty to twenty-one), in an almost homogeneous and fairly large population. Let us regard the height X of the individuals of this population as a r.v. with density f(x). On a graph, plot the heights (expressed, say, in centimeters) along the x-axis, and let y = N(x) be the number of individuals whose observed height is x.

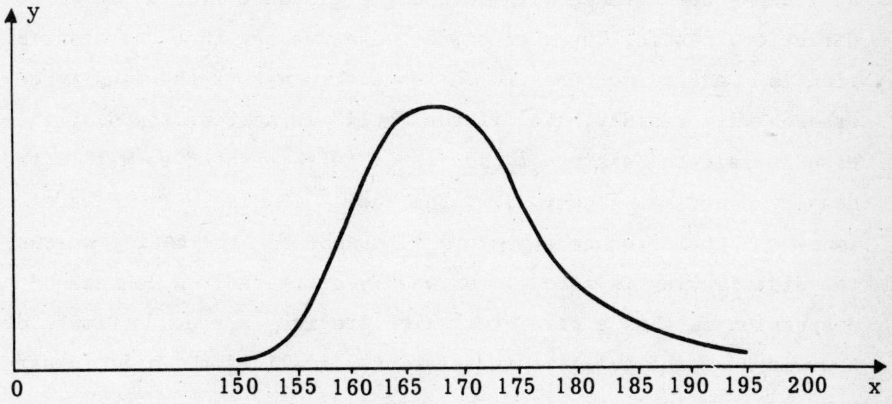

Fig. (4,2,1)

One expects the curve to have the shape of Fig. (4,2,1), and this is indeed the case. This is also the type of curve obtained in most biometric measurements concerning a *homogeneous* population; it is the classical bell-shaped curve, whose essential feature is that it displays one (and only one) central zone of relatively high density f(x), the density decreasing more or less rapidly and finally becoming negligible with increasing distance from the central zone in either direction.

Clearly, in order to describe a probability law or distribution of this type it is natural, even necessary, to introduce two parameters:

a central value;

a measure of dispersion.

Before explaining what is meant by these two expressions, we remark that these two parameters are often sufficient for a description, perhaps concise, but accurate, of the distribution.

A *central value* is any abscissa x_0 situated more or less at the center of the high-density zone; specification of a central value is therefore an approximate indication of the location of this zone.

Once a central value x_0 has been fixed, we see that the density tends to decrease with increasing distance from x_0 in either direction. As for the rate of decrease, we say that the distribution is *weakly scattered* (with respect to x_0) if the density decreases very rapidly, i.e. if the "bell" is narrow; the distribution is said to be *strongly scattered* (with respect to x_0) if the density decreases slowly, i.e. the "bell" is wide. It is very important to introduce a *measure of dispersion* indicating whether the distribution is strongly or weakly scattered. A measure of dispersion is thus a parameter which provides a rough estimate of the width of the bell; it is relatively small if the bell is narrow, relatively large if the bell is wide.

Clearly, the notions of central value and measure of dispersion are essentially qualitative; for instance, in the case of Figure (4,2,1) one might say that any number x_0 between 164 and 170 is admissible as a central value; there is no reason to choose $x_0 = 165$ rather than $x_0 = 166$ (or vice versa), insofar as our aim is only to "describe" the distribution in the sense specified at the beginning of this section. Nonetheless, one must adopt some definite rule for the choice of a central value. Any such rule R will yield a well-defined central value x_0, but it will always be possible to formulate another rule R' which yields a central value x_0' other than x_0, though of equal qualitative significance. Of course analogous arguments hold true for measures of dispersion.

Mathematical expectation of a random variable

The probability law of a r.v. X with d.f. F(x) is simply the distribution of a unit mass over the x-axis. Physicists tackled the problem of describing a material system long ago, and developed two notions: center of gravity and moment of inertia. It is therefore natural that probability-theoreticians have appealed to the same ideas.

The *mathematical expectation*[1] of an r.v. X, denoted by E(X) or E[X], is defined as the abscissa of the center of gravity of the mass distribution over the x-axis described by its d.f. F(x). Adapting III.12 to the one-dimensional case, E(X) is defined by the formula

$$E(X) = \int_{-\infty}^{+\infty} x \, dF(x). \tag{4,2,1}$$

If X is a discrete r.v., then, in the notation of IV.1, formula (4,2,1) is equivalent to

$$E(X) = \sum_k x_k p_k , \tag{4,2,2}$$

and if X is a continuous r.v. with density f(x), then (4,2,1) is equivalent to

$$E(X) = \int_{-\infty}^{+\infty} x \, f(x) \, dx. \tag{4,2,3}$$

It is essential to note that E(X) *exists if and only if the*

---

(1) Expectation: abbreviation: m.e. German: matematische Erwartung; French: espérance mathématique. Equivalent terms in English are mean value, mean, average; in German: Mittelwert, gewogenes Mittel; in French: valeur moyenne or simply moyenne. We do not recommend these terms, which should be reserved for other purposes. Neither shall we use the notation $\overline{X}$ often used by physicists for E(X); in this book $\overline{X}$ denotes the complex conjugate of a complex-valued r.v.

Riemann-Stieltjes integral (4,2,1) - possibly in the form (4,2,2) or (4,2,3) - *is absolutely convergent*. This is not always the case, i.e. there are r.v. with no m.e. However, the r.v. encountered in applications generally have a m.e. In particular, any bounded r.v. X, i.e. such that a number M exists for which $|X| < M$ almost certainly, has a m.e.

REMARK (4,2,1)

1) If a r.v. X reduces to the almost certain number a, then clearly $E(X) = a$.

2) If a r.v. X can assume only nonnegative values, i.e. if $X \geq 0$ almost certainly, then its m.e., if it exists, cannot be negative; moreover, it vanishes if and only if X is almost certainly zero.

EXAMPLE (4,2,1). Let us compute the m.e. E(X) of the indicator X of an event E with probability $Pr(E) = p$; it is immediate from (4,2,2) that

$$E(X) = 0 \times [1 - Pr(E)] + 1 \times Pr(E) = p.$$

Thus the probability of an event may always be interpreted as the m.e. of its indicator.

Computation of mathematical expectations

We can compute E(X) by formulas (4,2,1), (4,2,2) or (4,2,3) when F(x), p_k, or f(x) are known. Frequently, however, F(x) is not given directly.

We first recall how a r.v. X is associated with a class of trials. Let $\mathfrak{U}$ denote the class of trials u; let $p(E) = Pr(E)$, where E is an arbitrary event or subset of $\mathfrak{U}$, be the probability over $\mathfrak{U}$, and m(e), as above, the probability law of X. We saw in II.11 that X is equivalent to a mapping $x = \lambda(u)$ of $\mathfrak{U}$ into R or into the x-axis, and if e is any subset of R then

$$m(e) = p[\lambda^{-1}(e)], \tag{4,2,4}$$

i.e. m(e) is the measure induced over R by the mapping λ, corresponding to the measure p(E) over $\mathfrak{U}$. Let e be the set e(x) of all real numbers < x [cf. (3,1,1)]; then (4,2,4) gives

$$m[e(x)] = F(x) = p\{\lambda^{-1}[e(x)]\}, \qquad (4,2,5)$$

which shows how the d.f. F(x) can be determined if p and λ are known.

Now the integral (4,2,1) defining the m.e. is (cf. III.3)

$$\int_R x\, m(dx).$$

Applying Theorem (2,11,1), we get

$$E(X) = \int_{\mathfrak{U}} \lambda(u)\, p(du), \qquad (4,2,6)$$

which expresses E(X) in terms of p and λ rather than the d.f. F(x).

The r.v. X is often introduced in the following indirect way. Let S be an exhaustive system of pairwise mutually exclusive events C whose union is certain (cf. II.1). Let $x = \mu(C)$ be a mapping of S into R, which maps each $C \in S$ onto a number $x = \mu(C)$. By assumption, exactly one event C of S occurs; it is therefore meaningful to define a r.v. X by stipulating that it assumes the value $\mu(C)$ when the event C occurs. Any event ω defined by S, i.e. any subset of S, is the union of certain elements C of S [cf. Example (2,5,1)]; if we put $n(\omega) = \Pr(\omega)$, then $n(\omega)$ is a measure over S, $n\left(\bigcup_S C\right) = 1$.

First consider the case when S contains only finitely or countably many events C: $C_1, C_2, \ldots, C_h, \ldots$; it is clear that the measure $n(\omega)$ is totally discontinuous. The possible values of X are the numbers $\mu(C_h)$; since they are finite or at most countable in number, X must be a discrete r.v. Let x_k $(k = 1,2,3,\ldots)$ be the possible *distinct* values of X, and put

$$p_k = \Pr(X = x_k).$$

Let H_k denote the set of all h such that $\mu(C_h) = x_k$; H_k contains at least one element, but may contain more, even infinitely many. It is clear from the Total Probability Axiom that

$$p_k = \Pr\left(\bigcup_{h \in H_k} C_h\right) = \sum_{h \in H_k} \Pr(C_h).$$

This determines F(x) [cf. (4,1,10)]. Moreover, by (4,2,2) we have

$$E(X) = \sum_k x_k\, p_k = \sum_k x_k \Big(\sum_{h \in H_k} Pr(C_h) \Big)$$

$$= \sum_k \Big(\sum_{h \in H_k} x_k\, Pr(C_h) \Big)$$

$$= \sum_k \Big(\sum_{h \in H_k} \mu(C_h)\, Pr(C_h) \Big). \tag{4,2,7}$$

Since the sets H_k (k = 1,2,3,...) are pairwise disjoint and their union contains all the h,

$$E(X) = \sum_k x_k\, p_k = \sum_h \mu(C_h)\, Pr(C_h). \tag{4,2,8}$$

Note that if the sums involved in (4,2,8) are infinite the terms may be rearranged as in (4,2,7) or (4,2,8) only if the series $\sum_k x_k p_k$ is absolutely convergent, i.e. if E(X) exists. Formula (4,2,8) then means that if either of the series $\sum_k x_k p_k$ or $\sum_h \mu(C_h)\, Pr(C_h)$ is absolutely convergent then so is the other, and the two series have the same sum.

EXAMPLE (4,2,2). Let us return to Example (4,1,1), retaining the same notation. The set S of 36 events C_{ij} (i,j = 1,2,3,4,5,6) is an exhaustive system; with the event C_{ij} we associate the number

$$\mu(C_{ij}) = i + j.$$

Since $Pr(C_{ij}) = 1/36$, the m.e. E(X) of the r.v. X = i + j is (by formula (4,2,8))

$$E(X) = \sum_{i,j} (i + j)\, \frac{1}{36} = \frac{1}{36} \sum_i \Big[\sum_j (i + j) \Big]$$

$$= \frac{1}{36} \sum_i (6i + 21) = \frac{12 \times 21}{36} = 7. \;\blacksquare \tag{4,2,9}$$

Now consider the general case, in which the exhaustive system S is arbitrary. Note that the probability law m(e) of X is simply the measure induced over R (or the x-axis) by the mapping μ of S into R, corresponding to the measure $n(\omega)$ over S (cf. II.11). The analogues of (4,2,4) and (4,2,5) are therefore

$$m(e) = n[\mu^{-1}(e)] \qquad (4,2,10)$$

$$m[e(x)] = F(x) = n\{\mu^{-1}[e(x)]\}. \qquad (4,2,11)$$

E(X) is defined by

$$E(X) = \int_{-\infty}^{+\infty} x\, dF(x) = \int_R x\, m(dx),$$

so that by Theorem (2,11,1)

$$E(X) = \int_S \mu(C)\, n(dC). \qquad (4,2,12)$$

Formula (4,2,12) includes (4,2,8) as a particular case; moreover, the reasoning used to establish (4,2,8) is a proof of Theorem (2,11,1) for the particular case of a totally discontinuous measure. Formula (4,2,12) yields a direct computation of E(X), without resort to F(x), if n and μ are known. It is also clear that (4,2,12) is essentially the same as (4,2,6).

The foregoing may be presented in the following equivalent manner.

Let Y be a random element with values in an arbitrary space $\mathcal{Y}$ of elements y, whose probability law $n(\omega)$ is known or assumed known, where ω denotes a subset of $\mathcal{Y}$ and $n(\omega)$ is a measure over $\mathcal{Y}$ such that $n(\mathcal{Y}) = 1$. Let $x = \mu(y)$ be a given mapping of $\mathcal{Y}$ into R, and let X be the r.v. which assumes the value $x = \mu(y)$ if Y assumes the value y, i.e. the r.v. X is the function $X = \mu(Y)$ of the r.e. Y (cf. II.11).

Since the different events Y = y ($y \in \mathcal{Y}$) form an exhaustive system, it is clear that the situation is the same as above: m(e) is the measure induced by the mapping μ of $\mathcal{Y}$ into R, corresponding to the measure $n(\omega)$ over $\mathcal{Y}$, so that

$$m(e) = n[\mu^{-1}(e)], \qquad (4,2,13)$$

$$m[e(x)] = F(x) = n\{\mu^{-1}[e(x)]\}, \qquad (4,2,14)$$

$$E(X) = \int_{\mathcal{Y}} \mu(y)\, n(dy). \qquad (4,2,15)$$

EXAMPLE (4,2,3). Let $Z = \lambda(X)$ be a r.v. which is a function of a r.v. X, where $\lambda(x)$ denotes a real function of the real variable x R, i.e. λ is a mapping of R into R. It follows immediately from (4,2,15) that

$$E(Z) = E[\lambda(X)] = \int_R \lambda(x)\, m(dx)$$

$$= \int_{-\infty}^{+\infty} \lambda(x)\, dF(x). \qquad (4,2,16)$$

In particular, if $\lambda(X) = aX + b$, where a and b are any two certain numbers, so that Z is of the same type as X (if $a > 0$), then

$$E(aX + b) = a\, E(X) + b. \qquad (4,2,17)$$

It follows that by a suitable change of the origin, without changing either unit or direction, we can always make $E(X) = 0$.

REMARK (4,2,2). For the moment, the only significance that the m.e. has for us is that of a *central value* for the probability law of a r.v. In Chapter VI, however, the theory of addition of r.v. will show us that the m.e. has a considerably more precise and important concrete and probability-theoretical meaning. This is related to an important property of the m.e., which we can already establish.

Let X and Y be two r.v. Their sum S is also a r.v. If $x = \lambda(u)$ and $y = \mu(u)$ are the mappings of the class of trials $\mathfrak{U}$ into R defining X and Y, then S is defined by the mapping $s(u) = \lambda(u) + \mu(u)$ of $\mathfrak{U}$ into R; then it follows from (4,2,6) and Theorem (2,6,4) that

$$E(S) = E(X + Y) = \int_{\mathfrak{U}} [\lambda(u) + \mu(u)]\, p(du)$$

$$= \int_{\mathfrak{U}} \lambda(u)\, p(du) + \int_{\mathfrak{U}} \mu(u)\, p(du) = E(X) + E(Y).$$

THEOREM (4,2,1). If two random variables X and Y have mathematical expectations E(X), E(Y), the mathematical expectation of their sum X + Y is the sum of E(X) and E(Y):

$$E(X + Y) = E(X) + E(Y). \qquad (4,2,18)$$

EXAMPLE (4,2,4). Let us return to Example (4,2,2). To distinguish between the two dice, let I be the r.v. whose value is the number i on the first die, J the r.v. whose value is the number j on the second die, so that X = I + J; since the six faces of each die are equiprobable, formula (4,2,2) gives

$$E(I) = E(J) = \frac{1}{6}(1 + 2 + 3 + 4 + 5 + 6) = 3.5,$$

and by (4,2,18)

$$E(X) = E(I) + E(J) = 7,$$

agreeing with (4,2,9).

Second moments and dispersion. In III.12 we saw that the moment of inertia about a of the probability law or mass distribution of a r.v. X with d.f. F(x) coincides with the second moment of F about a:

$${}_a m_2 = \int_R (x - a)^2 \, m(dx) = \int_{-\infty}^{+\infty} (x - a)^2 \, dF(x). \qquad (4,2,19)$$

Now by (4,2,16) we can write

$${}_a m_2 = E[(X - a)^2]. \qquad (4,2,20)$$

The analogy with mechanics suggests that we use the second moment to estimate the dispersion of X (or of its probability law) with respect to a. But we must first justify this idea, and determine its precise import. One obvious requirement is that the second moment must exist, i.e. the integral (4,2,19) must converge. Let us assume that this is the case; recall that if (4,2,19) is convergent for one value of a, then it is convergent for any a.

Let λ be any positive number, and A the set of x such that $|x - a| > \lambda$. Then Chebyshev's Theorem (2,7,2), applied to the measure m(e) and the function $(x - a)^2$, implies that

$$\Pr(|X - a| > \lambda) = m(A) \leqslant \frac{{}_a m_2}{\lambda^2}. \qquad (4,2,21)$$

This inequality gives an upper bound for the probability that the difference between X and a exceeds λ in absolute value, for any positive λ, in particular, for large values of λ. The upper bound provided by (4,2,21) is often much greater than the exact value of Pr(X - a); however, as is easily seen, formula (4,2,21) is the best possible estimate for $\Pr(|X - a| > \lambda)$ if the only available information on the probability law of X is the existence and value of ${}_a m_2$.

Thus, the use of second moments as estimates of dispersion appears to be justified. ${}_a m_2$ is of course of dimension 2 relative to X; we shall therefore often use $({}_a m_2)^{1/2}$ instead of ${}_a m_2$: $({}_a m_2)^{1/2}$ is of dimension 1 and is known as the *root-mean-square deviation*[1] (about a) of the r.v.

Of course, in principle we wish to use (4,2,21) when it is most exact, i.e. for a value of a such that ${}_a m_2$ is a minimum. Now

$${}_a m_2 = E[(X - a)^2] = E(X^2 - 2aX + a^2),$$

or, by Theorem (4,2,1),

$${}_a m_2 = E(X^2) - 2E(X)\,a + a^2 ;$$

hence ${}_a m_2$ is a quadratic polynomial in a, and it is a *minimum* if

$$a = E(X)$$

[1] Root-mean-square deviation: abbreviation: r.m.s.d.; German: Mittleres Abweichungsquadrat; in French: écart moyen quadratique.

(this also follows from Theorem (3,12,2)), and thus, for any a,

$$ {}_a m_2 = E[(X - a)^2] = [a - E(X)]^2 + E\{[X - E(X)]^2\}. \qquad (4,2,22) $$

If we put a = E(X), then ${}_a m_2 = E\{[X - E(X)]^2\}$ is the second *central* moment (cf. III.12), which is called the *variance*[1] of X, often denoted by $\mathfrak{V}(X)$. The square root of the variance is the *root-mean-square central deviation*, or simply the root-mean-square deviation (if there is no danger of ambiguity). Since this term is rather inconvenient, we shall conform to current usage and call the square root of the variance the *standard deviation*[2], usually denoted by $\sigma(X)$ or simply σ. Thus

$$ \sigma^2(X) = \mathfrak{V}(X) = E\{[X - E(X)]^2\}. $$

It is clear that the variance of X is the same as that of X + b for any constant b, i.e. it does not depend on the choice of the origin on the x-axis.

EXAMPLE (4,2,5). Let us compute the standard deviation σ of the indicator X of an event E of probability Pr(E) = p, putting 1 - p = q; we know (Example (4,2,1)) that E(X) = p. By (4,2,22) with a = 0

$$ \sigma^2 = E(X^2) - [E(X)]^2 = E(X^2) - p^2 ; $$

but

$$ E(X^2) = 0^2 \times q + 1^2 \times p = p, $$

so that

$$ \sigma^2 = p - p^2 = pq. \qquad (4,2,23) $$

---

(1) Variance: German: Varianz; French: variance.

(2) Standard deviation: German: mittlere Abweichung; French, écart type.

Chebyshev's Theorem (2,7,2) is applicable to any point function. The foregoing reasoning may therefore be generalized in many ways; in particular, we can define various measures of dispersion other than the second moments. The most immediate generalization is the following.

Let X be a r.v. with d.f. F(x). The *absolute (algebraic) moments* of X are the absolute (algebraic) moments ${}_am^*_\alpha$ (${}_am_\alpha$) of F(x), as defined in III.9 by (3,9,1) ((3,9,2)). By (4,2,16), these formulas may be written in the form

$$ {}_am^*_\alpha = E(|X - a|^\alpha), \quad {}_am_\alpha = E[(X - a)^\alpha]. \qquad (4,2,24)$$

Putting a = E(X), we get the *central* moments, denoted in III.12 by μ^*_α, μ_α.

Note that the m.e. E(X) is simply the first algebraic moment, ${}_0m_1$, about the origin a = 0.

Let α be any positive number, and suppose that X has an absolute moment ${}_am^*_\alpha$ of order α. If this moment exists for one value of a, then it exists for any a. Apply Chebyshev's Theorem (2,7,2) to the probability law m(e) of X and the point function $|x - a|^\alpha$; then for any $\lambda > 0$ we obtain the following generalization of (4,2,21):

$$\Pr(|X - a| > \lambda) < \frac{{}_am^*_\alpha}{\lambda^\alpha}. \qquad (4,2,25)$$

As already indicated for $\alpha = 2$, this estimate of $\Pr(|X - a| > \lambda)$ is the best possible if the only available information on the probability law of X is the existence and the value of ${}_am^*_\alpha$. If additional information is at hand, for instance if F(x) is known to have a continuous derivative, or if additional moments are known, one can determine more precise properties of the probability law of X; however, this is of little practical use, and no more mention of it will be made.

It is clear from (4,2,25) that any absolute moment (if it exists) provides an estimate of dispersion. However, the following remarks are in order.

1) The assumption that a r.v. X has an absolute moment of order α is weaker, the smaller α.

2) The function $|x|$ is not algebraic; in particular, it is not differentiable at $x = 0$. Thus, as a rule, the absolute moment of order α is easy to handle only if it is at the same time an algebraic moment, i.e. if α is a positive even integer.

3) It will be seen in Volume II (and is also intuitively clear) that empirical estimation of a moment is less accurate and more difficult, the higher its order.

The fact that 2 is the smallest positive even integer is one of the reasons that probability-theoreticians have adopted second moments, in particular, variances, rather than absolute moments of other orders, as measures of dispersion. This is of course essentially a matter of convenience; however, there is a more fundamental reason. The importance of the notion of energy (or power) in physics is well known; energy or power is frequently expressed, except for a constant factor, as the square of some quantity. Examples: the power Ri^2 developed in a resistance R through which a current i passes is, up to the constant factor R, the square of i; the kinetic energy $mv^2/2$ of a mass m moving with velocity v, is, to the constant factor m/2, the square of v. To choose the moment of order two (or its square root, the root-mean-square deviation) as a measure of dispersion is the same as estimating the dispersion with reference to a "mean" energy or power. This procedure, both natural and legitimate, is often the only objectively significant one.

In the sequel we shall see that these considerations have had a strong guiding influence on certain parts of Probability Theory. Even now we can make the following remark.

Let X be a r.v. with a known d.f. F(X). Suppose that the (random) value of X is not known, either because the experiment or trial which determines it has not taken place, or, though it has, we have not recorded the outcome. This value is needed in order to adopt some decision. The difficulty may often be overcome by assuming that X will assume (or has assumed) a certain suitably chosen value $\hat{a}$. If the dispersion is estimated by moments of order 2, the best choice of $\hat{a}$ is that minimizing the function

$$E(|X - \hat{a}|^2),$$

i.e.,

$$\hat{a} = E(X)$$

as we have already seen. Thus E(X) is *the best estimate of* X *in the sense of minimum r.m.s.d.*, and this gives the m.e. a new and interesting meaning.

REMARK (4,2,3). We need only refer to II.12 and II.6 to see that a r.v. X reduces to a certain or almost certain number a if and only if there is some $\alpha > 0$ such that ${}_a m^*_\alpha = 0$.

Reduced random variables and laws. Given a r.v. with m.e. m, standard deviation σ, and d.f. F(X), the r.v. $Y = (X - m)/\sigma$ obviously has zero m.e. and standard deviation 1; it is the only r.v. of the same type as X which has these properties, and will be called the reduced r.v. of the type. Its probability law and d.f. $G(Y) = F(m + \sigma y)$ will be called *the reduced law and reduced d.f. of the type.*

Other central values and measures of dispersion. Not all probability laws of r.v. have the "bell" shape of Fig. (4,2,1). For example, consider the distribution of the students' arrival times at a school where lessons begin officially at eight. For various reasons, such as the threat of punishment, there are generally few latecomers and these will only be a few minutes late. Thus almost all the students arrive before 8 o'clock. On the other hand, they do not usually want to arrive too early. Thus one expects the proportion of students arriving at τ minutes to eight to increase as τ decreases. This is indeed usually so, and

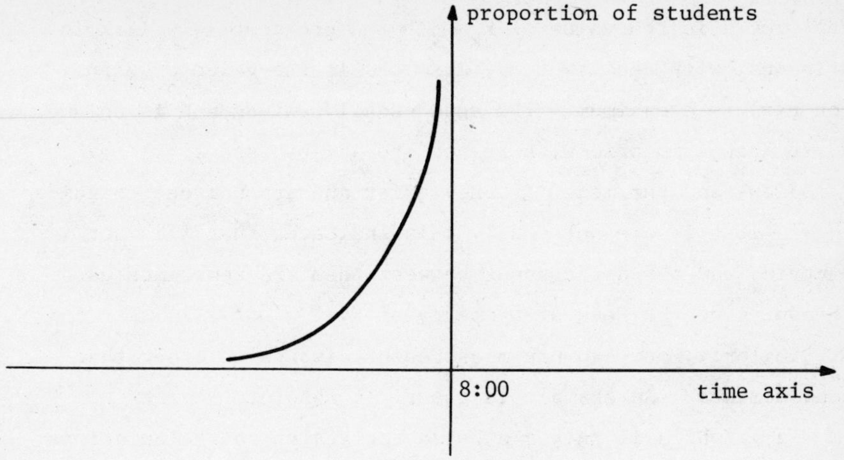

Fig. (4,2,2)

the result is a "J-shaped" distribution (see Fig. (4,2,2)). It is clear that for such a distribution, the m.e. (i.e. the abscissa of the center of gravity) and the standard deviation, even if they exist, are almost devoid of significance as central value and measure of dispersion, though they may retain other probabilistic meanings.

If we also remember that the m.e. does not always exist, it is obvious that sometimes other central values which arise in a natural way are used. We mention the median and the mode (or most probable value).

The median(1) of a r.v. X with d.f. F(x) is a number $\bar{\bar{X}}$ such that $F(\bar{\bar{X}}) = \frac{1}{2}$. It does not always exist; for instance, let X be the indicator of an event E with $Pr(E) = \frac{1}{2}$. There may also be more than one such number, if F(x) has the value ½ over a whole interval. Usually, however, $\bar{\bar{X}}$ exists and is unique; its meaning is obvious.

---

(1) Median; German: Zentralwert, Medianwert; French: valeur equiprobable.

Let X be discrete, and recall the notation of IV.1. The *mode*[1] of X is the value of x_k with the greatest p_k If X is continuous, with density f(x), its *mode* is the value of x for which f(x) is a maximum. The mode usually exists and is unique.

If X is symmetric with respect to an abscissa a, the m.e., the median and the mode, if they exist and are unique, coincide with a. If they are not equal, this indicates that X is not symmetric, and the differences between them are sometimes used as measures of the lack of symmetry of X.

Similarly, one can use measures of dispersion other than second moments. An example is the first absolute moment $E(|X - a|)$, which is rather awkward for analytical calculations, but for an empirically given distribution its numerical evaluation may be very simple. Note that $E(|X - a|)$ is a minimum when $a = \bar{\bar{X}}$, as is easily proved. The minimum $E(|X - \bar{\bar{X}}|)$ is known as the *mean deviation*[2] of X. By analogy with the median, the first and third *quartiles** $\underline{Q}$ and $\bar{Q}$ are defined by $F(\underline{Q}) = 1/4$, $F(\bar{Q}) = 3/4$; the inter-quartile range $\bar{Q} - \underline{Q}$ may be used as a measure of dispersion.

3. *Complex random variables*

A *complex* r.v. is a complex number Z whose value depends on chance, i.e. a r.e. with values in the set C of complex numbers, equivalent (cf. II.11) to a mapping $z = z(u)$ of the class $\mathfrak{U}$ of trials u into C; $z = z(u)$ denotes the value z taken by Z if u is the trial realized. We shall continue to use the term "random variable" for a real r.v., unless otherwise stated. As in

---

(1) Mode, most probable value: German, dichteste Wert, dominierende Wert; French: valeur la plus probable, dominante.

(2) Mean deviation: German: durchschnittliche Abweichung; French écart moyen.

* [Translator's note: The *second quartile* coincides with the median.]

Physics, where the introduction of complex quantities is convenient, if not necessary, we shall see that complex r.v. are very useful.

Let X and Y be the real and imaginary parts of a complex r.v. Z:

$$Z = X + i\,Y.$$

X and Y are two real r.v., and the study of Z is thus equivalent to that of the (real) 2-dimensional r.v. $\{X, Y\}$. This is related to the well-known fact that as a real vector space C is of dimension 2, while it is one-dimensional as a complex vector space. Many-dimensional r.v. will be studied in Chapter V. Here we shall disregard the interpretation of Z as a 2-dimensional r.v. $\{X, Y\}$, introducing only the notion of the m.e. E(Z) of a complex r.v. Z.

We shall say that E(Z) exists if and only if E(X) and E(Y) exist, and define

$$E(Z) = E(X) + i\,E(Y). \tag{4,3,1}$$

By a natural extension of (4,2,6), we can also define

$$E(Z) = \int_{\mathfrak{U}} z(u)\; p(du). \tag{4,3,2}$$

These two definitions are clearly equivalent. All the properties of integrals, as well as the results established in IV.2 for m.e. of real r.v., then apply to E(Z). Note, in particular, that

1) If $|Z|$ is almost certainly bounded, then both $|X|$ and $|Y|$ are almost certainly bounded, and E(Z) exists.

2) $|Z|$ is a real and nonnegative r.v. By Theorem (2,6,3),

$$|E(Z)| \leqslant E(|Z|).$$

3) Formulas (4,2,12), (4,2,15) remain true if X is a complex r.v. and μ the corresponding mapping of S (or $\mathfrak{Y}$) into C.

4) The statement of Theorem (4,2,1) remains true if X and Y are not real, but complex r.v. Similar remarks apply to many other propositions.

Let X be a r.v. with d.f. F(x). The characteristic function[1] of X is the function $\varphi(u)$ of the real variable u ($-\infty < u < +\infty$) defined as the Fourier transform of F(x):

$$\varphi(u) = \int_{-\infty}^{+\infty} e^{iux}\, dF(x) \qquad (4,3,3)$$

Fourier transforms of bounded distribution functions were studied in detail in III.7 and we shall confine ourselves to a few remarks.

1) The Fourier transform $\varphi(u)$ of a bounded d.f. (i.e., positive-definite and continuous function $\varphi(u)$) is the c.f. of a r.v. if and only if

$$\varphi(0) = 1. \qquad (4,3,4)$$

Indeed, by (4,3,3) we have $\varphi(0) = F(+\infty) - F(-\infty) = 1$. It follows that $|\varphi(u)| \leqq 1$ for any u [cf. (3,7,4)].

2) As we know, the Fourier transform $\varphi(u)$ of F(x) determines F(x) up to equivalence and therefore determines the probability law of X. Furthermore, F(x) is the only d.f. equivalent to the d.f. F(x) of the r.v. X which is normalized. Thus the c.f. $\varphi(u)$ of a r.v. determines the d.f. F(x) of X uniquely, not only up to equivalence.

3) For any real u, $Z = e^{iuX}$ is a complex r.v. which is a function of the (real) r.v. X. By (4,2,16) and (4,3,1) we can express (4,3,3) as

$$\varphi(u) = E(e^{iuX}) : \qquad (4,3,5)$$

the c.f. of a r.v. X is the m.e. of the (complex) r.v. e^{iuX}.

EXAMPLE (4,3,1). Let us determine the c.f. $\varphi(u)$ of the indicator X of an event E with probability Pr(E) = p; as usual, we put

---

(1) Characteristic function: abbreviation c.f.; German charakteristische Funktion; French: fonction caractéristique.

$1 - p = q$. By direct computation:

$$\varphi(u) = E(e^{iuX}) = e^{iu\times 0}\cdot q + e^{iu\times 1}\cdot p$$

$$= q + p\,e^{iu} = 1 + p\,(e^{iu} - 1). \qquad (4,3,7)$$

EXAMPLE (4,3,2). Given a r.v. X with c.f. $\varphi(u)$, let us compute the c.f. $\psi(u)$ of the r.v. $Y = aX + b$, where a and b are two certain numbers; Y is of the same type as X, provided $a > 0$. We have:

$$\psi(u) = E[e^{iu(aX+b)}] = E[e^{iuaX}\cdot e^{iub}]$$

$$= e^{iub}\cdot E[e^{iuaX}] = e^{iub}\,\varphi(au). \qquad (4,3,8)$$

Second characteristic function. In Chapter VI we shall see that the principal interest in introducing c.f. in Probability Theory is related to Theorem (3,8,2), which states that the Fourier transform of the convolution of two d.f. is the product of their Fourier transforms. It is then often convenient to consider the natural logarithm of the characteristic function $\varphi(u)$ of a r.v. X:

$$\log\varphi(u) = \log E(e^{iuX}) \qquad (4,3,9)$$

Since $\varphi(u)$ is generally complex, $\log\varphi(u)$ is an infinitely multiple-valued function. We define the *second characteristic function** $\psi(u)$ of X as that branch of $\log\varphi(u)$ that tends to 0 as u tends to 0, extended by continuity from $u = 0$.

The relations between the derivatives of $\varphi(u)$ and the moments of X, the expansions of $\varphi(u)$, either finite like (3,9,11), or in power series like (3,9,17), obviously imply corresponding properties for $\psi(u)$. To simplify notation, as in III.9, we use m_k, m_k^* for the moments ${}_0m_k$, ${}_0m_k^*$, and put $m_2 - m_1^2 = \sigma^2$, so that σ is the standard deviation of X. For instance, if X has a 2nd moment and we put

$$\psi(u) = i\,m_1\,u - \frac{\sigma^2}{2}\,u^2 + \rho(u)\,u^2, \qquad (4,3,10)$$

---

* [Translator's note: French *seconde caractéristique*; there is no generally accepted equivalent in English.]

then the function $\rho(u)$ has the property

$$\lim_{u\to 0} \rho(u) = 0 ; \tag{4,3,11}$$

this may be seen by using (3,9,10) with n = 2, (3,9,12), and the Taylor expansion of $\log(1+\alpha)$ for $|\alpha| < 1$.

Formula (4,3,11) can be made more precise if stronger assumptions are made concerning X. For instance, if X also has a 3rd absolute moment m_3^*, interesting estimates for $|\rho(u)|$ may be obtained by using (3,9,16) with n = 2. To simplify the notation, assume $m_1 = 0$, so that by (3,9,11) and (3,9,16)

$$\varphi(u) = 1 - \frac{m_2}{2} u^2 + \omega(u)\, u^2,$$

where

$$|\omega(u)| \leqq m_3^* |u|. \tag{4,3,12}$$

Put

$$v = (m_3^*)^{\frac{1}{3}} u , \quad \alpha = -\frac{m_2}{2} u^2 + \omega(u)\, u^2,$$

$$\psi(u) = \log(1+\alpha) = \alpha + \Omega(\alpha)\, \alpha^2. \tag{4,3,13}$$

For any number a > 1 there exists a number b > 0 such that

$$|\Omega(\alpha)| < \frac{a}{2}, \tag{4,3,14}$$

provided

$$|\alpha| < b. \tag{4,3,15}$$

By Theorem (3,9,1),

$$m_2 \leqq (m_3^*)^{\frac{2}{3}} ; \tag{4,3,16}$$

in view of (4,3,12) formula (4,3,15) is surely satisfied if

$$|v| < \min\left(b, \frac{1}{2}\right) = C.$$

Comparison of (4,3,10) and (4,3,13) shows that

$$\rho(u) = \omega(u) + \Omega(\alpha)\left[-\frac{m_2}{2} + \omega(u)\right]^2 u^2.$$

Therefore, by (4,3,12), (4,3,14), and (4,3,16) we get

$$|\rho(u)| \leqq m_3^* \, |u| + \frac{a}{8} \, |(m_3^*)^{\frac{2}{3}} + m_3^* \, |u||^2 \, |u|^2$$

$$\leqq m_3^* \, |u| + \frac{a}{8} \, |1 + |v||^2 \, |v| \, m_3^* \, |u|$$

$$\leqq \left[1 + \frac{a}{8} \, (1 + |v|)^2 \, |v| \right] \, m_3^* \, |u| \, ,$$

whence:

For any number $\varepsilon > 0$ there exists a number $\eta > 0$, *independent of the r.v.* X, such that

$$|\rho(u)| \leqq (1 + \varepsilon) \, m_3^* \, |u| \tag{4,3,17}$$

for any u such that

$$(m_3^*)^{\frac{1}{3}} \, |u| \leqq \eta . \blacksquare \tag{4,3,18}$$

In general, if $\varphi(u)$ has n derivatives, the same will be true for $\psi(u)$, provided $\varphi(u) \neq 0$; this condition is always satisfied if $|u|$ is sufficiently small, since $\varphi(u)$ is continuous and $\varphi(0) = 1$. The k-th derivative ψ_k of $\psi(u)$ at $u = 0$ is easily expressed in terms of the first k algebraic moments $m_1, m_2, \ldots, m_k$ of X; in particular, as is also clear from (4,3,10),

$$\psi_1 = i \, m_1, \quad \psi_2 = -\sigma^2 = m_1^2 - m_2 .$$

It follows from (4,3,8) that if $k > 1$ then ψ_k is unchanged when X is replaced by X + b, where b is a constant. In other words, for $k > 1$ the ψ_k are independent of the choice of the origin on the x-axis. We have already verified this for $\psi_2 = -\sigma^2$, since, as shown in IV.2, the variance of X does not depend on the origin on the x-axis.

If the series (3,9,17) has a positive radius of convergence, then obviously, for sufficiently small $|u|$,

$$\psi(u) = i \, m_1 \, u + \sum_{k=2}^{+\infty} \frac{\psi_k}{k\,!} \, u^k . \tag{4,3,19}$$

Moment-generating function, generating function

The d.f. F(x) of a r.v. X is a *bounded* d.f. Therefore its Fourier transform or characteristic function $\varphi(u)$ always exists, and for this reason it is the most frequently used of all the exponential transforms of F(x). Nevertheless, referring to Sections 6-10 of Chapter III, we can see that other exponential transforms may be used to the same effect, provided they exist.

Thus, for instance, we can introduce the Laplace transform

$$\mu(u) = \int_{-\infty}^{+\infty} e^{-ux}\, dF(x) = E[e^{-uX}] \tag{4,3,20}$$

for the values of u for which it exists (if there are any). $\mu(u)$ is called the *moment-generating function**; the *second moment-generating function* can be defined by

$$\lambda(u) = \log \mu(u). \tag{4,3,21}$$

If $\lambda(u)$ can be expanded in a power series in u,

$$\lambda(u) = \log \mu(u) = -\lambda_1 u + \frac{\lambda_2}{2} u^2 + \ldots + (-1)^k \frac{\lambda_k}{k!} u^k + \ldots, \tag{4,3,22}$$

then the coefficients λ_k, which can be expressed in terms of the algebraic moments of X and conversely, are independent of the choice of the origin on the x-axis, as were the coefficients ψ_k; the sole exception is λ_1, which is equal to m_1. The $\lambda_k (k > 2)$ are called the *semi-invariants* or *cumulants* of X; λ_2 is just the variance of X.

Introduction of the moment-generating function is particularly apt if X is almost certainly nonnegative, so that $F(x) = 0$ for $x \leqq 0$ and F(x) is a d.f. over $[0, +\infty)$; then $\mu(u)$ is defined by

$$\mu(u) = \int_0^{+\infty} e^{-ux}\, dF(x), \tag{4,3,23}$$

where the interval of integration $[0, +\infty)$ is closed at 0 if $F(+0) > 0$. By III.10, $\mu(u)$ exists and is absolutely monotone for all positive real u.

In the particular case when X is a discrete r.v. and can take only nonnegative integral values, it may be convenient to

---

* [Translator's note: French *caractéristique de Laplace*.]

introduce (cf. end of III.10) the *generating function* $g(z)$ defined by

$$g(z) = E(z^k) ; \qquad (4,3,24)$$

or, putting $p_k = \Pr(X = k)$ for nonnegative integers k,

$$g(z) = \sum_{k=0}^{+\infty} p_k z^k ; \qquad (4,3,25)$$

$g(z)$ is a power series in z, or a polynomial, if all p_k vanish for sufficiently large k.

We refer the reader to III.10 for the properties of the functions $\mu(u)$ and $g(z)$ defined by (4,3,23) and (4,3,25).

4. *Poisson laws*

A r.v. X is said to obey a *Poisson law with parameter* m if it is a discrete r.v. whose possible values are the nonnegative integers $x_k = k$ and

$$p_k = \Pr(X = k) = e^{-m} \frac{m^k}{k!} \qquad (k = 0, 1, 3, \ldots). \qquad (4,4,1)$$

There are thus *countably many* possible values of X. The parameter m is an arbitrary positive number. $m < 0$ is impossible, for otherwise $p_k < 0$ for odd k. If $m = 0$, then $\Pr(X = 0) = 1$, i.e. X reduces to the almost certain number 0, and one speaks of the *degenerate* Poisson law.

We shall confine ourselves to the nondegenerate case $m > 0$. Note first that (4,4,1) does indeed define the probability law of a r.v., since $p_k > 0$ and

$$\sum_{k=0}^{+\infty} p_k = \sum_{k=0}^{+\infty} e^{-m} \frac{m^k}{k!} = e^{-m} \sum_{k=0}^{+\infty} \frac{m^k}{k!}$$

$$= e^{-m} \cdot e^{m} = 1.$$

On the other hand, by (4,2,2) and (4,4,1),

$$E(X) = \sum_{k=0}^{+\infty} k \cdot e^{-m} \frac{m^k}{k!}.$$

The first term of this series ($k = 0$) is zero; therefore

$$E(X) = \sum_{k=1}^{+\infty} k\, e^{-m} \frac{m^k}{k\,!} = e^{-m} \sum_{k=1}^{+\infty} \frac{m^k}{(k-1)\,!}$$

$$= m \cdot e^{-m} \cdot \sum_{k=1}^{+\infty} \frac{m^{k-1}}{(k-1)\,!} \;;$$

but

$$\sum_{k=1}^{+\infty} \frac{m^{k-1}}{(k-1)\,!} = e^m ,$$

whence:

$$E(X) = m . \qquad (4,4,2)$$

Thus the parameter m of the Poisson law is the m.e. Computing $E(X^2)$, we have

$$E(X^2) = \sum_{k=0}^{+\infty} k^2\, e^{-m} \frac{m^k}{k\,!}$$

$$= m\, e^{-m} \cdot \sum_{k=1}^{+\infty} k \frac{m^{k-1}}{(k-1)\,!} .$$

Putting h = k - 1, we get

$$E(X^2) = m\, e^{-m} \sum_{h=0}^{+\infty} (h+1) \frac{m^h}{h\,!}$$

$$= m\, e^{-m} \left(\sum_{h=0}^{+\infty} h \frac{m^h}{h\,!} + \sum_{h=0}^{+\infty} \frac{m^h}{h\,!} \right).$$

Now

$$\sum_{h=0}^{+\infty} \frac{m^h}{h\,!} = e^m$$

and

$$\sum_{h=0}^{+\infty} h \frac{m^h}{h\,!} = \sum_{h=1}^{+\infty} \frac{m^h}{(h-1)\,!} = m \cdot \sum_{h=1}^{+\infty} \frac{m^{h-1}}{(h-1)\,!}$$

$$= m\, e^m ,$$

so that, finally,

$$E(X^2) = m^2 + m. \qquad (4,4,3)$$

Combined with (4,2,22), this yields the variance $\mathfrak{v}(X)$:

$$\mathfrak{v}(X) = E(X^2) - [E(X)]^2 = m^2 + m - m^2 = m . \qquad (4,4,4)$$

Thus the parameter m of the Poisson law is also the variance (or the square of the standard deviation).

It is clear that this computation of E(X) and $E(X^2)$ can be generalized to the evaluation of all the algebraic moments of X in terms of m; these moments clearly exist.

It is easy to determine the mode; indeed, the quotient p_{k+1}/p_k is equal to $m/(k + 1)$, and the mode is obviously the smallest value k_0 of k such that this quotient, which decreases with increasing k, is smaller than 1. k_0 lies between m - 1 and m; note, however, that in the particular case where $(k + 1)/m$ may equal 1, there are two modes. The median may be determined by using a table of the function $e^{-m} m^k/k!$ (see Tables [7], [8]).

The c.f. $\phi(u) = E(e^{iuX})$ may be computed on the basis of (4,2,16) and (3,3,3):

$$\varphi(u) = \sum_{k=0}^{+\infty} e^{iuk} \cdot e^{-m} \frac{m^k}{k!}$$

$$= e^{-m} \cdot \sum_{k=0}^{+\infty} \frac{(m\, e^{iu})^k}{k!} = e^{-m} \cdot e^{m \cdot e^{iu}} \qquad (4,4,5)$$

$$= e^{m(e^{iu}-1)};$$

whence the second characteristic function:

$$\psi(u) = \log\, \varphi(u) = m(e^{iu} - 1). \qquad (4,4,6)$$

$\psi(u)$ and $\phi(u)$ are entire functions of u, and may therefore be expanded in power series in u with infinite radius of convergence. The expansion of $\psi(u)$ is immediate. Using it one can find the coefficients of the expansion of $\phi(u)$, and then, by comparison with (3,9,17), the algebraic moments of X. This is easy for moments of lower order; in particular, formulas (4,4,2) and (4,4,3) are easily derived.

The generating function g(z), derived from $\phi(u)$ by putting $e^{iu} = z$, is very simple:

$$g(z) = e^{m(z-1)}. \qquad (4,4,7)$$

REMARK (4,4,1). If a r.v. X obeys a Poisson law, it may not continue to do so if subjected to a linear transformation $Y = Y_0 + \rho X$ (with $\rho > 0$, leaving the direction unchanged), for the possible values $y_0, y_1, \ldots, y_k, \ldots$ of Y, being an arithmetic progression with difference ρ and first term y_0,

$$y_k = y_0 + \rho k,$$

are no longer the nonnegative integers 0, 1, 2,..., k,... Thus the family of Poisson laws does not constitute a type. However,

$$\Pr(Y = y_k = y_0 + \rho k) = e^{-m} \frac{m^k}{k!} . \qquad (4,4,8)$$

It is thus natural to say that Y *obeys a generalized Poisson law* (possibly degenerate). The m.e. of Y is not m, but

$$E(Y) = y_0 + \rho m .$$

Every Poisson law with parameter m generates a type consisting of all generalized Poisson laws (4,4,8) for this m and arbitrary y_0, ρ

Normal distribution

A r.v. X is *normally distributed*[1] or normal if it is continuous and has the following probability density $l(x)$:

$$l(x) = \frac{1}{\sqrt{2\pi}} e^{-x^2/2} \qquad (-\infty < x < \infty) . \qquad (4,4,9)$$

We know (cf. III.6, in particular formula (3,6,22)), that (4,4,9) does indeed define the probability law of a continuous r.v. The function $l(x)$ is always positive and its graph has the typical "bell" shape of Fig. (4,4,1).

---

(1) Normal distribution; also: Gaussian law; German: Gaussche Verteilung; French: loi de Laplace, also: loi de Gauss, loi de Laplace-Gauss, loi normale. The term "Laplace law" is used almost exclusively in France; it is indeed more correct to associate this law with Laplace than with Gauss.

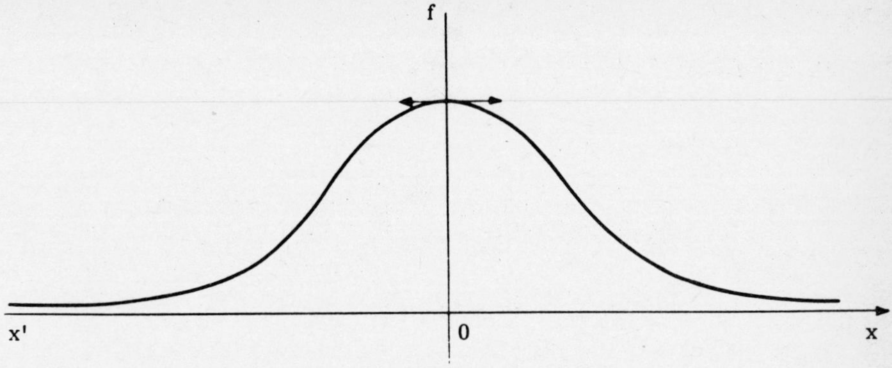

Fig. (4,4,1)

$l(x)$ is the density or derivative of the d.f. $L(x)$ defined by

$$L(x) = \frac{1}{\sqrt{2\pi}} \int_{-\infty}^{x} e^{-y^2/2}\, dy \,. \qquad (4,4,10)$$

A r.v. X with density $l(x)$ is obviously symmetric with respect to the origin. Its m.e., median and mode are all zero. Let us compute its standard deviation σ; integrating by parts, we get

$$\sigma^2 = E(X^2) = \frac{1}{\sqrt{2\pi}} \int_{-\infty}^{+\infty} x^2\, e^{-x^2/2}\, dx = \frac{1}{\sqrt{2\pi}} \int_{-\infty}^{+\infty} x.\, xe^{-x^2/2}\, dx$$

$$= \frac{1}{\sqrt{2\pi}} \cdot - xe^{-x^2/2} \Big]_{-\infty}^{+\infty} + \frac{1}{\sqrt{2\pi}} \int_{-\infty}^{+\infty} e^{-x^2/2}\, dx = 1 \,. \qquad (4,4,11)$$

Thus the normal distribution has standard deviation 1; it is the reduced distribution of its type.

The mean deviation $E(|X|)$ is

$$E(|X|) = \frac{1}{\sqrt{2\pi}} \int_{-\infty}^{+\infty} x\, e^{-x^2/2}\, dx$$

$$= \frac{2}{\sqrt{2\pi}} \int_{-0}^{+\infty} x\, e^{-x^2/2}\, dx$$

$$= \frac{2}{\sqrt{2\pi}} \cdot - e^{-x^2/2} \Big]_{0}^{+\infty} = \sqrt{\frac{2}{\pi}} = 0.798\ldots \,. \qquad (4,4,12)$$

The absolute moments about the origin may be computed by induction; we have

$$ {}_om^*_{\alpha+2} = E(|X|^{\alpha+2}) = \frac{2}{\sqrt{2\pi}} \int_o^{+\infty} x^{\alpha+2}\, e^{-x^2/2}\, dx $$

$$ = \frac{2}{\sqrt{2\pi}} \int_o^{+\infty} x^{\alpha+1} \cdot x\, e^{-x^2/2}\, dx $$

$$ = \frac{2}{\sqrt{2\pi}} \left| x^{\alpha+1} \times - e^{-x^2/2} \right|_o^{+\infty} + \frac{2(\alpha + 1)}{\sqrt{2\pi}} \int_o^{+\infty} x\, e^{-x^2/2}\, dx $$

$$ = (\alpha + 1)\, E(|X|^{\alpha}) = (\alpha + 1)\, {}_om^*_{\alpha}. \qquad (4,4,13) $$

Since $E(|X|^0) = 1$ and $E(|X|) = \sqrt{2/\pi}$, it follows that for $k = 0, 1, 2, 3, \ldots$

$$ {}_om^*_{2k} = E(|X|^{2k}) = 1 \times 3 \times 5 \times \ldots \times (2k - 1) $$

$$ = \frac{(2k)!}{2^k\, k!}\ ; \qquad (4,4,14) $$

$$ {}_om^*_{2k+1} = E(|X|^{2k+1}) = \sqrt{\frac{2}{\pi}} \times 2 \times 4 \times \ldots \times 2k $$

$$ = \sqrt{\frac{2}{\pi}}\, 2^k\, k!\ . \qquad (4,4,15) $$

As for algebraic moments, clearly

$$ {}_om_{2k+1} = E(X^{2k+1}) = 0, \qquad (4,4,16) $$

$$ {}_om_{2k} = E(X^{2k}) = E(|X|^{2k}) = {}_om^*_{2k} $$

$$ = \frac{(2k)!}{2^k\, k!}. \qquad (4,4,17) $$

We have already seen (cf. (3,6,23)) that the characteristic function $\phi(u)$ of X is

$$ \varphi(u) = e^{-u^2/2}, \qquad (4,4,18) $$

and its expansion in power series in u is

$$ \varphi(u) = 1 - u^2/2 + \cdots + (-1)^k \frac{u^{2k}}{2^k\, k!} + \cdots. \qquad (4,4,19) $$

Comparison of this expansion with (3,9,17) easily yields (4,4,15) and (4,4,16). The second characteristic function $\psi(u)$ is very simple:

$$\psi(u) = \log \varphi(u) = - u^2/2. \qquad (4,4,20)$$

The moment-generating function $\mu(u)$ is derived from $\phi(u)$ by replacing iu by -u:

$$\mu(u) = e^{u^2/2},$$

and the second moment-generating function $\lambda(u)$ is

$$\lambda(u) = \log \mu(u) = u^2/2 .$$

Comparison with (4,3,16) shows that the semi-invariants (cumulants) λ_k all vanish, except for the variance λ_2 which is 1.

In the sequel, we shall see that the normal distribution is of immense theoretical and practical importance. Naturally, a great number of related functions have been tabulated; in particular, tables are available for the following functions:

$$l(x) = \frac{1}{\sqrt{2\pi}} e^{-x^2/2} \qquad \text{(Tables [9], [14]);}$$

$$L(x) = \frac{1}{\sqrt{2\pi}} \int_{-\infty}^{x} e^{-y^2/2} dy \qquad \text{(Table [12]);}$$

$$\Phi(x) = \frac{1}{\sqrt{2\pi}} \int_{0}^{x} e^{-y^2/2} dy = L(x) - \frac{1}{2} \qquad \text{(Tables [11], [14]);}$$

$$\Theta(x) = \frac{2}{\sqrt{\pi}} \int_{0}^{x} e^{-y^2} dy = 2\Phi(\sqrt{2}\,x) = 2L(\sqrt{2}\,x)-1 \qquad \text{(Tables [10], [13]).}$$

For normally distributed r.v. X, asymptotic expressions for $P(x) = \Pr(|X| > x)$ for large x are often needed. We have

$$P(x) = 2\Pr(X > x) = \frac{2}{\sqrt{2\pi}} \int_{x}^{+\infty} e^{-y^2/2} dy.$$

By classical methods (cf. Exercise 4.5) it can be proved that

$$\Pr(X > x) = \frac{1}{\sqrt{2\pi}} \frac{e^{-x^2/2}}{x} \left[1 - \frac{2\eta(x)}{x^2} \right], \qquad (4,4,21)$$

where $0 < \eta(x) < 1$ for all x, and moreover

$$\lim_{x \to +\infty} \eta(x) = 1.$$

In particular,

$$P(1) = 0.3174\ldots, \quad P(2) = 0.0454,\ldots,$$
$$P(3) = 0.0026\ldots \qquad (4,4,22)$$

In other words, in more than 95 cases out of 100 the difference between X and its m.e. (zero) is at most 2 standard deviations; in more than 997 cases out of 1000 the difference is at most 3 standard deviations.

Chebyshev's inequality (4,2,22) gives only

$$P(x) \leqslant \frac{1}{x^2}. \qquad (4,4,23)$$

Thus the actual order of magnitude of P(x) given by (4,4,21) is far smaller than the estimate (4,4,23).

Tables are available for P(x) as a function of x (Table [14]) and, on the other hand, x as a function of P(x) (Table [15]); the latter function is often used.

Laws of Normal Type. The family of probability laws of the type defined by the normal law (4,4,9) will be called the *normal type*. We shall use the term *normal* for any normally distributed r.v., any d.f. or density of the normal type, the probability law of any r.v. of the normal type. It is easy to find the general form of distributions, more precisely, of densities of normal type, under an arbitrary change of origin and scale. Let the r.v. X be normally distributed with density (4,4,9); consider the r.v. $Y = m + \sigma X$, so that the m.e. is m instead of 0, and the standard deviation σ instead of 1. Then

$$F(y) = \Pr(Y < y) = \Pr(m + \sigma X < y) = \Pr\left(X < \frac{y - m}{\sigma}\right)$$

$$= L\left(\frac{y - m}{\sigma}\right) = \frac{1}{\sqrt{2\pi}} \int_{-\infty}^{\frac{y-m}{\sigma}} e^{-x^2/2}\, dx = \frac{1}{\sqrt{2\pi}\,\sigma} \int_{-\infty}^{y} e^{-\frac{(z-m)^2}{2\sigma^2}}\, dz.$$

(4,4,24)

Consequently, the probability density f(y) of Y is

$$f(y) = \frac{1}{\sqrt{2\pi}\sigma} e^{\frac{-(y-m)^2}{2\sigma^2}} = \frac{1}{\sigma} \, l\left(\frac{y - m}{\sigma}\right) . \qquad (4,4,25)$$

Formula (4,4,24) gives the general form of a normal density. The m.e. m and standard deviation σ are clearly indicated.

The *central* absolute and algebraic moments of Y are obtained by multiplying (4,4,14), (4,4,15), (4,4,16), (4,4,17) by σ^{2k} or σ^{2k+1}.

By (4,3,8) the characteristic function of the law (4,4,24) is obviously

$$\varphi(u) = e^{imu - \frac{\sigma^2}{2} u^2} . \qquad (4,4,26)$$

Henry's line. Consider two rectangular axes Oy, Oz. On Oy the usual (uniform) scale is used, while on Oz, the ruling $z = L(\alpha)$ is assigned to the point whose real coordinate is α. Putting $z = F(y)$ in (4,4,24), this equality becomes $z = L\left(\frac{y - m}{\sigma}\right)$. The graph of the function $z = F(y)$ is then a straight line $\alpha = (y-m)/\sigma$, which is known as *Henry's line*. The linearity of the graph is independent of the values of m and σ, depending only on the normal character of F(y). Hence an approximate but simple test for normality, even when the values of m and σ are not known; moreover, the slope of Henry's line and its intercept with Oy yield estimates for σ and m.

Degenerate normal distributions

Let m be fixed, and let σ tend to + 0; it is clear that the above d.f. F(y), with density (4,4,24), tends to the following d.f.:

$$F_o(y) = \begin{cases} 0 \text{ if } y \leqslant m, \\ 1 \text{ if } y > m; \end{cases}$$

i.e.,

$$F_o(y) = \Delta(y - m),$$

where $\Delta(x)$ denotes the Dirac d.f. (cf. (3,2,5)). $F_0(y) = \Delta(y - m)$ is the d.f. of a r.v. Y_0 that reduces to the almost certain number m; in terms of mass distributions, $G_0(y) = \Delta(y - m)$ corresponds to a unique unit mass at the abscissa m. This distribution, whose characteristic function is

$$\varphi(u) = e^{-0 \cdot u^2/2 + imu} = e^{imu},$$

may therefore be regarded as a distribution of normal type, though degenerate. We shall often adopt this viewpoint below.

II. AXIOM OF CONDITIONAL PROBABILITY

5. *Conditional probability of one event, given another*

In II.4 we observed that the probability of an event E depends on the class of trials to which it is referred, and also (see "Second Problem" in the same section) that this raises the question of the relations between the probabilities of the same event E with respect to different classes of trials. A new axiom will provide the answer; in this connection the basic notion will be that of the conditional probability of one event, given another.

As in II.2, and retaining the same notation, let us draw a card at random from a deck of n = 32 cards. Let A be the event "the card drawn is red (i.e. hearts or diamonds)" and B the event "the card drawn is either the king of hearts, the king of diamonds, or the king of spades." There are r(B) = 3 cards realizing B, so that

$$\Pr(B) = \frac{r(B)}{n} = \frac{3}{32}.$$

Suppose that we have not seen the card, but somebody informs us that it was red, i.e., the event A has occured. Then our argument will be as follows: there are n' = r(A) = 16 red cards, and the card drawn may be any one of them; since r'(B) = 2 of them (the kings of hearts and diamonds) realize B, the probability of B is not 3/32, but, by (2,2,15),

$$\frac{r(B)}{n'} = \frac{2}{16}.$$

This second probability of B, evaluated on the assumption that A has occured, is called the *conditional probability of* B, *given* A, and denoted by Pr(B/A).

Now recall II.3 (cf. also the beginning of II.4) concerning the statistical survey of the French population $\mathfrak{P}$, of cardinality n (=45,000,000), and the event B that the individual I is a metal worker. In the notation of II.3, the probability Pr(B) of B is

$$\Pr(B) = \frac{r(B)}{n},$$

where r(B) is about 500,000. Let A be the event "I belongs to the Paris region," whose population $\mathfrak{P}'$ has cardinality n' (=8,000,000); the probability of B, given that I belongs to the Paris region, i.e. given A, is

$$\frac{r'(B)}{n'},$$

where r'(B) (=200,000) is the number of metal workers in the Paris region. The second probability of B is its conditional probability Pr(B/A), given A.

In general, let $\mathfrak{U}$ be a class of trials u, A and B two events, i.e. subsets of $\mathfrak{U}$. Let Pr(B) be the probability of B relative to the class $\mathfrak{U}$. Consider another class of trials, consisting of all u that realize A. This class is precisely A, regarded as a subset of $\mathfrak{U}$, and we therefore call it A.

The conditional probability of B, given A, denoted by Pr(B/A), is the probability of B relative to the class A. Pr(B/A) is sometimes called the *a posteriori* probability of B, to indicate that it is the probability of B when it is known that A has occurred (or if A occurs, or on the assumption that A has occurred). To emphasize the difference, Pr(B) may be called the *a priori* probability of B.

Now return to the deck of cards, with the same notation and events A and B; r'(B) denotes the number of red cards, i.e. cards realizing A, that realize B. r'(B) is therefore the number of cards of the deck realizing $A \cap B$. Therefore, by the "principle of symmetry" (cf. II.2), it follows from n' = r(A) and Pr(A) = r(A)/n that

$$\Pr(A \cap B) = \frac{r'(B)}{n} \left(= \frac{2}{32}\right) = \frac{n'}{n} \cdot \frac{r'(B)}{n'}$$

$$= \frac{r(A)}{n} \cdot \frac{r'(B)}{n'} \qquad (4,5,1)$$

$$= \Pr(A) \Pr(B/A).$$

Recalling the statistical example concerning the metal workers, with the same notation, we note that $r'(B)$ is the number of individuals who are both metal workers and inhabitants of the Paris region, i.e. who realize $A \cap B$; note also that $n' = r(A)$, and therefore

$$\Pr(A) = \frac{r(A)}{n} = \frac{n'}{n}.$$

Consequently, as in the previous example,

$$\Pr(A \cap B) = \Pr(A) \Pr(B/A). \qquad (4,5,2)$$

Motivated by formulas (4,5,1) and (4,5,2), which were obtained for simple examples, we shall adopt the following additional axiom, which we call the Axiom of Conditional Probability.

Axiom of Conditional Probability (Provisional formulation)

Let $\mathfrak{U}$ be a class of trials, A and B two events, i.e. subsets of $\mathfrak{U}$. The probability $\Pr(A \cap B)$ (relative to $\mathfrak{U}$) of the intersection $A \cap B$ is equal to the product of the probability $\Pr(A)$ relative to $\mathfrak{U}$) of A by the conditional probability $\Pr(B/A)$ of B, given A:

$$\Pr(A \cap B) = \Pr(A) \Pr(B/A). \qquad (4,5,3)$$

Of course, in this formulation A and B play a symmetric role. We have called this formulation "provisional"; indeed, we shall see that it does not meet all requirements, and a more complete formulation is necessary. However, even this weak version is sufficient for a fairly large number of problems.

EXAMPLE (4,5,1). Consider the following problem from the chromosomic theory of heredity (cf. I.5 and the end of II.3): if two parents Q_1 and Q_2 are heterozygotes Aa, the probabilities that their offspring R will be either AA, Aa, or aa are 1/4, 1/2, 1/4, respectively. Assume that, when R was born, it was established that it is not an aa, but no more could be learned. What is the probability that R is a heterozygote Aa?

Let A denote the event that R is either AA or Aa and B the event that it is Aa. The required probability is precisely the conditional probability Pr(B/A) of B, given A; therefore, by (4,5,3), since obviously $A \cap B = B$ and $\Pr(A) = 1/2 + 1/4$, by the Axiom of Total Probability,

$$\Pr(B/A) = \frac{\Pr(A \cap B)}{\Pr(A)} = \frac{1/2}{1/2 + 1/4} = \frac{2}{3} .$$

EXAMPLE (4,5,2). *Lifetime of a radioactive atom and the exponential law*

Let us say that a radioactive atom is "alive" as long as fission has not occurred, fission constituting its "death," so to speak. Similarly, we use the term "birth" for its generation by another atom. Let $t_0 = 0$ and $X \geqq 0$ be the instants of birth and death, respectively. X is known as the *lifetime* of the atom. Occurrence of fission depends on certain phenomena inside the atom which cannot be predicted accurately, so that the lifetime X must be regarded as a random variable. Let F(x) be its d.f.; by definition, X is essentially nonnegative, $F(x) = 0$ for $x \leqq 0$. F(x) is therefore a d.f. over $[0, +\infty)$ (see III.10).

It is usually assumed that X obeys an *exponential law*, i.e.

$$F(x) = 1 - e^{-x/a} , \qquad (4,5,4)$$

where a is a positive constant depending on the element (or isotope) under consideration (cf. Example (3,10,1)). The hypothesis that F(x) has the form (4,5,4) has been confirmed by experimental physics; we now wish to ascertain the meaning and justification of this hypothesis.

In fact, the family of exponential laws, which depends on the parameter a, possesses a remarkable property. Suppose that an observer has verified that the atom is still alive at the instant $x > 0$. In other words, the atom has reached the age x, so that $X \geqq x$, and we can call $S = X - x$ its "survival time" beyond x. Let us find the *conditional d.f.* G(s), *given* $X \geqq x$ (for $s \geqq 0$), i.e., the conditional probability $G(s) = \Pr(S < s/X \geqq x)$.

The event $S < s$ is equivalent to the event $X < x + s$. By the Axiom of Conditional Probability (4,5,3), we have

$$\Pr(x \leqslant X < x + s) = \Pr(X \geqslant x)\Pr(X < x + s/X \geqslant x), \qquad (4,5,5)$$

or

$$F(x + s) - F(x) = [1 - F(x)] \cdot G(s) ;$$

whence by (4,5,4)

$$G(s) = \frac{e^{-x/a} - e^{\frac{-x+s}{a}}}{e^{-x/a}} = 1 - e^{-s/a}. \qquad (4,5,6)$$

G(s) is thus identical with F(s): the survival time S beyond any age x *actually reached* has the same probability law as the a priori life; the chances of the atom dying are no greater at an advanced age (already reached) than immediately after its birth. In other words, the atom *does not "age."*

We claim that the exponential distribution is the only one with this property. Indeed, let F(x) be the d.f. over $[0, +\infty)$ of a r.v. $X \geqq 0$ that does not age, i.e. for any $x \geqq 0$ and $s \geqq 0$,

$$\Pr(X < x + s/X \geqslant x) = F(s). \qquad (4,5,7)$$

By (4,5,5), formula (4,5,7) is equivalent to

$$F(x + s) - F(x) = [1 - F(x)]\, F(s)$$

or

$$\frac{F(x + s) - F(x)}{s} = [1 - F(x)]\, \frac{F(s)}{s} \quad (\text{for } s > 0). \qquad (4,5,8)$$

Now let s tend to + 0 in (4,5,8). Since F(x), as a d.f., is monotone nondecreasing, it is differentiable outside some exceptional set. Therefore,

$$\lim_{s \to +0} \frac{F(x + s) - F(x)}{s} = \frac{d}{dx} F(x)$$

exists for almost all x. Let x have any fixed (non-exceptional) value. Then

$$\lim_{s \to +0} \frac{F(s)}{s}$$

exists, irrespective of the way s tends to + 0. Put

$$\frac{1}{a} = \lim_{s \to +0} \frac{F(s)}{s}.$$

Now return to (4,5,8); we see that

$$\lim_{s \to +0} \frac{F(x + s) - F(x)}{s}$$

exists for any $x > 0$; therefore F(x) is continuous *from the right* for $x \geqq 0$, and has a *right derivative*:

$$\frac{d}{dx} F(x) = F'_+(x).$$

Moreover,

$$F'_+(x) = [1 - F(x)] \cdot \frac{1}{a}. \qquad (4,5,9)$$

But F(x), as a d.f., is continuous *from the left*; thus F(x) is continuous.

Now rewrite (4,5,8) putting $x + s = y$ for any $y > 0$; then

$$\frac{F(y) - F(y - s)}{s} = [1 - F(y - s)] \frac{F(s)}{s}.$$

For fixed y, let s tend to + 0; then

$$\lim_{s \to +0} \frac{F(y) - F(y - s)}{s} = [1 - F(y)] \cdot \frac{1}{a} ;$$

i.e., for any $y > 0$, $F(y)$ has both right and left derivatives satisfying (4,5,9), and they are equal. Thus it turns out that $F(x)$ is differentiable for any $x > 0$, and has the derivative

$$\frac{d}{dx} F(x) = F'(x) = [1 - F(x)] \cdot \frac{1}{a}. \qquad (4,5,10)$$

A priori, the case $1/a = 0$ or $a = +\infty$ is not excluded; but then, since $F(x)$ is a d.f., formula (4,5,10) would imply that $F(+0) = 1$, whereas $F(0) = 0$, and X would be almost-certainly zero. This case is clearly of no physical interest, and we shall disregard it. It follows directly from (4,5,9) that $F(x)$ has a right derivative at $x = 0$:

$$F'_+(0) = [1 - F(0)] \cdot \frac{1}{a} = \frac{1}{a}. \qquad (4,5,11)$$

It also follows from (4,5,10) that

$$[1 - F(x)]\, dx + a \cdot d[1 - F(x)] = 0,$$

$$1 - F(x) = C\, e^{-x/a},$$

where C is an arbitrary integration constant. But $F(x)$ is a d.f., and this is true if and only if $C = 1$, whence

$$F(x) = 1 - e^{-x/a},$$

as required.

The whole of the preceding argument could have been carried out for $a = 1$ without any loss of generality; the general case can always be reduced to this by a suitable choice of time unit. Note that if a r.v. X has the distribution (4,5,4), then

$$E(X) = \int_0^{+\infty} x \, d[1 - e^{-x/a}] = \int_0^{+\infty} \frac{a}{x} e^{-x/a} dx$$

$$= a \, ,$$

$$E(X^2) = \int_0^{+\infty} x^2 \, d[1 - e^{-x/a}] = \int_0^{+\infty} \frac{x^2}{a} e^{-x/a} dx$$

$$= 2a^2, \qquad (4,5,12)$$

and thus the variance of X is

$$\mathcal{V}(X) = \sigma^2(X) = a^2. \qquad (4,5,13)$$

These results can also be obtained from (3,10,6) by expanding $1/(u + 1)$ in a power series in u. Thus the parameter a of the exponential law (4,5,4) is both the m.e. and the standard deviation.

Case of more than two events

Repeated application of (4,5,3) yields the appropriate formulas for any finite number of events. For instance, consider three events A, B, C; by (4,5,3),

$$\Pr(A \cap B \cap C) = \Pr[(A \cap B) \cap C]$$

$$= \Pr(A \cap B) \Pr(C/A \cap B) \, ,$$

$$\Pr(A \cap B) = \Pr(A) \Pr(B/A) \, ,$$

whence

$$\Pr(A \cap B \cap C) = \Pr(A) \Pr(B/A) \Pr(C/A \cap B). \qquad (4,5,14)$$

Independent events. We have seen that, given two events A and B, the a priori probability P(B) of B and its conditional probability given A, Pr(B/A), are generally different. However, it may occur that the occurrence or non-occurrence of A has no effect on the probability of B, so that the conditional probability Pr(B/A) is equal to the a priori probability Pr(B). This is necessarily true when A and B are associated with experiments $\mathcal{E}_1$ and $\mathcal{E}_2$ which are absolutely independent from the physical viewpoint.

For instance, let $\mathcal{E}_1$ be a game of "écarté" played in Paris by Pierre and Simon, A being the event "Pierre draws a king"; let $\mathcal{E}_2$ be a game of poker played at the same time in Marseilles by Paul, André, Joseph, and Albert (with a different deck, of course) B being the event "Paul draws three aces." The probability of B is obviously not modified by the occurrence or non-occurrence of A.

REMARK (4,5,1). Note that in this example the trial u is the *pair* $(\mathcal{E}_1, \mathcal{E}_2)$ of "experiments" (or partial trials) $\mathcal{E}_1$ and $\mathcal{E}_2$. ▮

Now the equality Pr(B) = Pr(B/A) may hold even when A and B are associated with the same experiment, or with different but physically related experiments. For instance, in drawing a card from a deck of 32 cards, let B be the event "the card is a king," so that Pr(B) = 4/32. The 16 red cards include two kings (hearts and diamonds), so that if A is again the event "the card is red," then Pr(B/A) = 2/16, and since 2/16 = 4/32, we see that Pr(B) = Pr(B/A).

We thus arrive at the notion of *stochastic independence*[1], which is more general than physical independence, though it contains it as a particular case. A first attempt at a precise definition, suggested by the previous reasoning, might be: an event B is independent of an event A if

$$\Pr(B) = \Pr(B/A). \tag{4,5,15}$$

However, for reasons that will become clear presently, we shall adopt the following, almost equivalent, though different, definition:

(1) Independence: German: Unabhängigkeit; French: indépendance.

Two events A *and* B *are mutually independent if*

$$\Pr(A \cap B) = \Pr(A)\,\Pr(B). \qquad (4,5,16)$$

The independence is called *mutual* because A and B play entirely symmetric roles in the definition (4,5,16).

By this definition, if A is an almost impossible or almost certain event and B any event, then A and B are independent. Indeed, if for instance A is almost impossible, Pr(A) = 0, then $\Pr(A \cap B) = 0$ by Theorem (2,12,1), and

$$\Pr(A \cap B) = \Pr(A)\,\Pr(B) \; (= 0).$$

Compare (4,5,16) and (4,5,15); assume first that A and B satisfy (4,5,15). It follows immediately from (4,5,13) that (4,5,16) holds:

Condition (4,5,15) *always implies* (4,5,16), *i.e. the mutual independence of* A *and* B.

Now consider the converse: if (4,5,16) holds and Pr(A) > 0, it follows immediately from (4,5,3) that (4,5,15) holds. *However, formula* (4,5,16) *does not imply* (4,5,15) *if* Pr(A) = 0.

Assume then that A and B are such that Pr(A) = 0 and Pr(B) ≠ Pr(B/A). We have seen that A and B are independent, but this would be false had (4,5,15) been taken as the definition of independence, and this would be a disadvantage. The complementary event $\check{A}$ of A is almost certain; we expect it to occur (cf. II.12). By Theorem (2,12,1), $\Pr(\check{A} \cap B) = \Pr(B)$; therefore $\Pr(\check{A} \cap B) = \Pr(\check{A})\Pr(B)$ since $\Pr(\check{A}) = 1$, and moreover, by (4,5,3), Pr(B) = Pr(B/A). Under these conditions our intuitive notion of independence requires that A and B be independent; the fact that Pr(B) ≠ Pr(B/A) is no objection. Indeed, since A is almost impossible we expect it not to occur. Then, insofar as the only events considered apart from B itself, are A and $\check{A}$, the probability Pr(B/A) of B on the assumption that A occurs is of no interest, and it is immaterial whether or not it is equal to Pr(B).

Let A and B be two mutually independent events, i.e. events satisfying (4,5,16). Consider the complementary events $\check{A}$ and $\check{B}$; by (2,5,9), $\Pr(\check{A}) = 1 - \Pr(A)$, and by the Axion of Total Probability, $\Pr(A \cap B) + \Pr(\check{A} \cap B) = \Pr[(A \cup \check{A}) \cap B] = \Pr(B)$, or

$$\Pr(\check{A} \cap B) = \Pr(B) - \Pr(A \cap B);$$

or, using (4,5,15),

$$\begin{aligned}\Pr(\check{A} \cap B) &= \Pr(B) - \Pr(A)\Pr(B) = [1 - \Pr(A)]\Pr(B)\\ &= \Pr(\check{A})\Pr(B).\end{aligned}$$

This proves:

THEOREM (4,5,1). Let A and B be two events, $\check{A}$ and $\check{B}$ the corresponding complementary events. Then if A and B are mutually independent, so are $\check{A}$ and B, A and $\check{B}$, $\check{A}$ and $\check{B}$. $\blacksquare$

From the practical point of view, the essential property of independence is that, by (4,5,15), we can replace (4,5,3) by the formula

$$\Pr(A \cap B) = \Pr(A)\Pr(B),$$

avoiding explicit mention of the conditional probability $\Pr(B/A)$, since in this case it is the same as the a priori probability $\Pr(B)$.

Independent σ-algebras

Theorem (4,5,1) suggests an interpretation which proves very convenient for generalizations. Let $\mathfrak{U}$ denote a class of trials; as a subset of itself, $\mathfrak{U}$ is also the certain event; $\emptyset$ denotes both the empty subset of $\mathfrak{U}$ and the impossible event, the complementary event of $\mathfrak{U}$. Recall that all events can be regarded (cf. II.4) as subsets of $\mathfrak{U}$, and are assumed probabilized.

Note that if E is any event and $\check{E}$ its complementary event, the smallest σ-algebra (cf. II.5) of subsets of $\mathfrak{U}$ containing E obviously consists of the four events $\mathfrak{U}$, $\emptyset$, E, $\check{E}$. We denote this σ-algebra by $\Sigma(E)$.

Now consider the following definitions:

a) Let $\Sigma_1, \Sigma_2, \ldots, \Sigma_n$ be σ-algebras of subsets of $\mathfrak{U}$, where $n > 1$ is any finite number. These σ-algebras are said to be mutually independent if, for any n event $A_1, A_2, \ldots, A_n$ such that

$$A_j \in \Sigma_j \text{ for } j = 1, 2, \ldots, n,$$

we have

$$\Pr(A_1 \cap A_2 \cap \ldots \cap A_n) = \Pr(A_1) \times \Pr(A_2) \times \ldots \times \Pr(A_n). \tag{4,5,17}$$

b) Let $\mathfrak{F}$ be any family of (at least two) σ-algebras Σ of subsets of $\mathfrak{U}$. The σ-algebras of the family $\mathfrak{F}$ are said to be mutually independent if for any finite integer $n > 1$ and any n σ-algebras $\Sigma_1, \ldots, \Sigma_n$ of the family $\mathfrak{F}$, the n σ-algebras $\Sigma_1, \ldots, \Sigma_n$ are mutually independent in the sense of a).

Using these definitions, one immediately verifies that *two events* A *and* B *are independent if and only if the σ-algebras* $\Sigma(A)$ *and* $\Sigma(B)$ *are independent.*

The case of any finite number of events

Let n be any finite integer ($n > 2$) and consider n events A, B, C,..., L. The following definition is quite natural:

The n events A, B, C,..., L are mutually independent if the n σ-algebras $\Sigma(A), \Sigma(B), \Sigma(C), \ldots, \Sigma(L)$ are mutually independent.

For instance, consider the case of three events A, B, C, with complementary events $\check{A}$, $\check{B}$, $\check{C}$, respectively. Application of (4,5,17) leads to the following $2^3 = 8$ conditions:

$$\Pr(A \cap B \cap C) = \Pr(A)\Pr(B)\Pr(C)$$
$$\Pr(A \cap B \cap \check{C}) = \Pr(A)\Pr(B)\Pr(\check{C})$$
$$\Pr(A \cap \check{B} \cap C) = \Pr(A)\Pr(\check{B})\Pr(C)$$
$$\Pr(A \cap \check{B} \cap \check{C}) = \Pr(A)\Pr(\check{B})\Pr(\check{C})$$
$$\Pr(\check{A} \cap B \cap C) = \Pr(\check{A})\Pr(B)\Pr(C)$$
$$\Pr(\check{A} \cap B \cap \check{C}) = \Pr(\check{A})\Pr(B)\Pr(\check{C})$$
$$\Pr(\check{A} \cap \check{B} \cap C) = \Pr(\check{A})\Pr(\check{B})\Pr(C)$$
$$\Pr(\check{A} \cap \check{B} \cap \check{C}) = \Pr(\check{A})\Pr(\check{B})\Pr(\check{C})$$

A, B, C are mutually independent if and only if all these conditions are satisfied.

The reader may verify that this set of eight conditions is in fact equivalent to the simpler system consisting of the following four conditions:

$$\Pr(A \cap B) = \Pr(A)\Pr(B)$$
$$\Pr(B \cap C) = \Pr(B)\Pr(C)$$
$$\Pr(C \cap A) = \Pr(C)\Pr(A)$$
$$\Pr(A \cap B \cap C) + \Pr(A)\Pr(B)\Pr(C).$$

REMARK (4,5,2). The first three conditions of (4,5,18) express the obvious fact that mutually independent events are necessarily pairwise independent. The converse is false: three or more events may be pairwise independent without being mutually independent.

For instance, consider two random variables X and Y assuming only the values 1 or -1. Let

$$\Pr(X = 1) = \Pr(X = -1) = \Pr(Y = 1) = \Pr(Y = -1) = 1/2,$$

and suppose the events X = 1 and Y = 1 are independent; put

$$Z = X\,Y$$

It is easy to see that the events Z = 1 and X = 1 are independent, as are the events Z = 1 and Y = 1. Thus the three events X = 1, Y = 1, Z = 1 are pairwise independent. But they are not mutually independent, since, for instance,

$\Pr(Z = 1)$ $(=\frac{1}{2})$ is not equal to

$$\Pr[Z = 1/\ (X = 1) \cap (Y = 1)]\ (= 1).$$

It follows that when only two events are under consideration one can speak of their independence without the qualification "mutual." For more than two events, the term "mutual independence" is preferable in order to avoid ambiguity; however, when no confusion is possible the brief term "independence" is usually used.

The same remark obviously holds for independence of random elements, to be considered later.

The general case

Now consider an arbitrary family $\mathfrak{F}$ of events; we shall say that the events A of this family are mutually independent if the σ-algebras $\Sigma(A)$ $(A \in \mathfrak{F})$ are mutually independent.

Mutual independence of random elements

The notions of a random element and an event *associated* with a given random element were defined in II.11; the following definition is natural:

Two arbitrary random elements X *and* Y *are said to be mutually independent if any event* A *associated with* X *and any event* B *associated with* Y *are independent.*

Recall (cf. II.11) that the family of events associated with a r.e. X is a σ-algebra Σ_X of subsets of $\mathfrak{U}$, which we call *the σ-algebra* Σ_X *associated with* X. The above definition may thus be rephrased as follows:

The r.e. X and Y are mutually independent if the σ-algebras Σ_X and Σ_Y associated with X and Y, respectively, are mutually independent.

REMARK (4,5,3). Let E be an arbitrary event, and the r.v. X the indicator of E (cf. IV.1). The σ-algebra $\Sigma(E)$ defined above is simply the σ-algebra Σ_X associated with X, regarded as a r.e. ∎

More generally, let Λ be any family of r.e. X. We shall say that the r.e. X of the family Λ are mutually independent if the σ-algebras Σ associated with them are mutually independent.

It follows immediately from the definition that if two random elements X and Y are independent, the same holds for any two random elements which are functions of X and Y (cf. II.11).

These definitions will be applied, in particular, to random

variables in the following chapters.

EXAMPLE (4,5,3). Let X and Y be two independent r.v. with d.f. F(x) and G(y), respectively. For any x and y the two events $X < x$ and $Y < y$ are independent, hence

$$Pr[(X < x) \cap (Y < y)] = Pr(X < x)Pr(Y < y) = F(x)\,G(y)$$

Let Z denote the greatest of the numbers X and Y; the event $Z < z$ is equivalent to the event

$$(X < z) \cap (Y < z)$$

Hence the d.f. H(z) of Z is

$$H(z) = F(z)\,G(z). \tag{4,5,19}$$

6. *Application of the Axiom of Conditional Probability*

Formula (4,5,3) can be used in two ways:

1) Given Pr(A) and Pr(B/A), to compute $Pr(A \cap B)$.

2) Given Pr(A) and $Pr(A \cap B)$, i.e. the a priori probabilities, to compute the conditional probability Pr(B/A). By (4,5,3),

$$Pr(B/A) = \frac{Pr(A \cap B)}{Pr(A)} \tag{4,6,1}$$

and this solves problem 2) if $Pr(A) > 0$. Unfortunately, the case where A is almost impossible, i.e. $Pr(A) = 0$, occurs frequently; since then $Pr(A \cap B)$ also vanishes, the quotient (4,6,1) is not defined and problem 2) is not solved.

It is easily seen that the only possible solution is to introduce an axiom more complete than (4,5,3). The following remark will suggest one possible formulation of the new axiom.

Let S be a system of (pairwise mutually exclusive) events C (cf. II.1). Recall that an event A defined by S is a subset of the set S of elements C, i.e. a union of certain C's; conversely, any union of C's is an event defined by S; in particular, the union of all the C's is S regarded as a subset of itself. If we put $Pr(A) = n(A)$ for any event A defined by $S(A \subset S)$, then n(A) is a measure over S.

Let B be an arbitrary event, and consider the conditional probability Pr(B/C) for each $C \in S$.

Suppose first that S contains finitely or countably many events C_k (k = 1, 2, 3,...). For any $A \subset S$,

$$A = \bigcup_{\{k | C_k \in A\}} C_k ,$$

$$A \cap B = \bigcup_{\{k | C_k \in A\}} C_k \cap B .$$

Since the events C are pairwise mutually exclusive, so are the events $C_k \cap B$, and the Axiom of Total Probability gives

$$Pr(A \cap B) = \sum_{\{k | C_k \in A\}} Pr(C_k \cap B) ;$$

but by (4,5,3)

$$P(C_k \cap B) = Pr(B/C_k) \, Pr(C_k) ,$$

and therefore

$$Pr(A \cap B) = \sum_{\{k | C_k \in A\}} Pr(B/C_k) \, Pr(C_k). \tag{4,6,2}$$

Now in this specific case n(A) is a totally discontinuous measure, which assigns each C_k the mass $Pr(C_k)$; by (2,9,2), formula (4,6,2) can be written in the form

$$Pr(A \cap B) = \int_A Pr(B/C) \, n(dC). \tag{4,6,3}$$

Our final formulation of the Axiom of Conditional Probability states that (4,6,3) is valid for any system S:

Axiom of Conditional Probability (*Final Formulation*)

Let $\mathfrak{U}$ be a class of trials, which determines all the events under consideration, S a system of events C($S \subset \mathfrak{U}$). B an arbitrary event, A an event defined by S, and n(A) the measure over S defined by n(A) = Pr(A). Then

$$\Pr(A \cap B) = \int_A \Pr(B/C)\ n(dC). \qquad (4,6,4)$$

If in addition S is exhaustive, then, interpreted as an event, $S = \mathfrak{U}$ is certain, and the family of all subsets A of S is a σ-algebra of subsets of $\mathfrak{U} = S$ (cf. Example (2,5,1)). $S \cap B = B$ and $\Pr(S \cap B) = \Pr(B)$; thus, applying (4,6,4) with $A = S$ we deduce:

$$\Pr(B) = \int_S \Pr(B/C)\ n(dC). \qquad (4,6,5)$$

Formula (4,6,5) is obviously valid if S is only almost certain. For instance, if S consists of an event A and the complementary event $\check{A}$, then (4,6,5) gives

$$\Pr(B) = \Pr(A)\ \Pr(B/A) + \Pr(\check{A})\ \Pr(B/\check{A}) ; \qquad (4,6,6)$$

of course, this also follows from (4,5,3) alone.

Referring back to Remark (2,11,4) and Theorem (2,11,2), we can interpret (4,6,4) as follows. As often remarked, $\Pr(A \cap B) = 0$ if $n(A) = \Pr(A) = 0$. Consequently, given B, $\Pr(A \cap B)$, as a function of the set $A \subset S$, is a measure over S which is absolutely continuous with respect to the measure $n(A)$. Formula (4,6,4) then states that $\Pr(B/C)$ is the density of $\Pr(A \cap B)$ with respect to $n(A)$.

Thus, given the a priori probabilities, i.e. the probabilities of all the events relative to $\mathfrak{U}$, so that $n(A)$ and $\Pr(A \cap B)$ are known for any A, the Radon-Nikodym Theorem (2,11,2) yields the conditional probabilities $\Pr(B/C)$. In this way however, $\Pr(B/C)$ is determined (as a function of C) only up to n-equivalence. But in many applications this is no disadvantage, for the same reasons that one can often identify almost impossible and impossible events. Moreover, additional information about $\Pr(B/C)$ sometimes eliminates this indeterminacy.

EXAMPLE (4,6,1). Let X be a r.v. with d.f. F(x), and suppose that F(x) is absolutely continuous with density f(x). Let C_x denote the event X = x, and let S be the system of all events $C_x (-\infty < x < +\infty)$. Let B be an arbitrary event and let Pr(B/x) denote the (unknown) probability $Pr(B/C_x)$. To compute Pr(B/x), let A_x denote the event X < x, and apply (4,6,4) with $A = A_x$; thus

$$Pr(A_x \cap B) = \int_{-\infty}^{x} Pr(B/y)\, f(y)\, dy. \tag{4,6,7}$$

If $Pr(A_x \cap B)$ is known for all x, then by (4,6,7) it is a differentiable function of x and

$$Pr(B/x) = \frac{d}{dx} Pr(A_x \cap B)/f(x), \tag{4,6,8}$$

at least, for x such that $f(x) \neq 0$. Strictly speaking, the relation (4,6,8) may be false for some exceptional values of x; this may well be of no consequence for us. However, if f(x) is a positive continuous function, and $d/dx\, Pr(A_x \cap B)$ and Pr(B/x) are continuous functions of x, then (4,6,8) holds for any x and uniquely determines Pr(B/x).

REMARK (4,6,1). Consider a measure m(E) over $\mathfrak{U}$ ($E \subset \mathfrak{U}$) such that $m(\mathfrak{U}) = 1$. We derive a measure n(A) over S by defining n(A) = m(A) for any $A \subset S$. Given any subsets B of $\mathfrak{U}$, $m(A \cap B)$ (as a function of $A \subset S$) is an absolutely continuous measure with respect to n(A); it therefore has a density with respect to n(A), which is a function of $C \in S$. Now this density depends - like $m(A \cap B)$ itself - on B, and can therefore be denoted by $\lambda(C;B)$. For given C, $\lambda(C;B)$ (as a function of $B \subset U$) is a set function. We shall say that the triple ($\mathfrak{U}$, m, S) has property ($\mathfrak{R}$) if, except for events C whose union L is such that n(L) = 0 (i.e. the union is an almost impossible event), $\lambda(C;B)$ is a nonnegative completely additive function of B for any fixed C.

The reason for this definition is as follows. Regard m(E) as a probability on $\mathfrak{U}$, putting Pr(E) = m(E). Then, by the preceding arguments, $Pr(B/C) = \lambda(C;B)$ up to n-equivalence in C.

Now, as a function of B for fixed C, Pr(B/C) must be completely additive, because of the Axiom of Total Probability. Thus, if the triple ($\mathcal{U}$, m, S) does not have property ($\mathcal{R}$), there is no conditional probability compatible with this triple.

The problem of determining whether a triple ($\mathcal{U}$, m, S) does or does not have property ($\mathcal{R}$), in other words, the problem of the existence of a conditional probability, is now almost completely solved, but this has been done quite recently. It is indeed a delicate problem, and we shall confine ourselves to the following remarks.

1) There exist triples ($\mathcal{U}$, m, S) which do not have property ($\mathcal{R}$).

2) Sufficient conditions for a triple ($\mathcal{U}$, m, S) to have property ($\mathcal{R}$) are known; these conditions are so weak that they are always satisfied in practice.

Thus, we shall pay no more attention to the problem of the existence of conditional probability, assuming for any triple ($\mathcal{U}$, m, S) figuring in the discussion that it possesses property ($\mathcal{R}$) and therefore the conditional probability exists and satisfies (4,6,4).

REMARK (4,6,2). Return to (4,6,4) and suppose that B and C are independent for any $C \in S$, so that Pr(B/C) = Pr(B). Then (4,6,4) becomes

$$\Pr(A \cap B) = \int_A \Pr(B)\, n(dC) = \Pr(B) \cdot \int_A n(dC) = \Pr(B)\Pr(A),$$

and it follows that B and A are independent for any $A \subset S$. Conversely, suppose that B and A are independent for any $A \subset S$, so that

$$\Pr(A \cap B) = \Pr(B) \cdot \Pr(A) = \Pr(B) \cdot \int_A n(dC) \qquad (4,6,9)$$

$$= \int_A \Pr(B)\, n(dC).$$

Formula (4,6,9) shows that Pr(B) is the density of $Pr(A \cap B)$, regarded as a measure in A over S, with respect to the measure n(A). By the Radon-Nikodym Theorem (2,11,2) this density is unique up to n-equivalence. But by (4,6,4) Pr(B/C) is also the density of $Pr(A \cap B)$ with respect to n(A); therefore

$$Pr(B/C) = Pr(B), \tag{4,6,10}$$

except for certain events C whose union L is such that n(L) = 0 (i.e. L is an almost impossible event). Therefore, apart from the "exceptional" events C of L, B and C are independent.

EXAMPLE (4,6,2). Consider a sphere of radius R whose center O is the origin of a rectangular reference system Oxyz, Oz coinciding with the polar axis PQ. A point M on the sphere, defined by its longitude Φ and lattitude Θ, is chosen at random (cf. Fig. (4,6,1)). Let Γ denote the equator (the intersection of the sphere with the (xy)-plane), γ the great semicircle QMP, K the intersection of Γ and γ.

We have

$$\Phi = (Ox, \overrightarrow{OK}) \quad (0 \leq \Phi < 2\pi),$$
$$\Theta = (\overrightarrow{OK}, \overrightarrow{OM}) \quad (-\pi/2 \leq \Theta \leq \pi/2).$$

M is a random point on the sphere; its probability law is a measure or a mass distribution over the sphere, consisting of the probabilities $Pr(M \in \Sigma)$, where Σ is any subset of the surface of the sphere. Let us call this distribution uniform if $Pr(M \in \Sigma)$ is proportional to the area σ of Σ; since $Pr(M \in \Sigma) = 1$ when Σ is the entire surface of the sphere, whose area is $4\pi R^2$, it follows that in general

$$Pr(M \in \Sigma) = \frac{\sigma}{4\pi R^2}. \tag{4,6,11}$$

Suppose that the trial, i.e. the random selection of M, is performed in the following consecutive two steps:

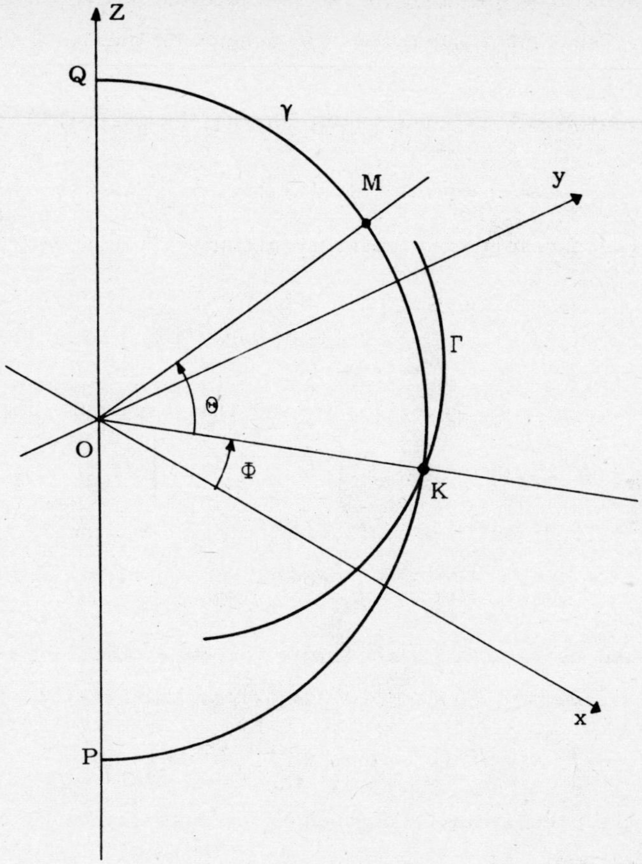

Fig. (4,6,1).

STEP I: Random selection of the value of Φ in $[0,2\pi[$ according to a uniform distribution, with density $1/2\pi$.

STEP II: Given the resulting value ϕ of Φ, the second step is the random selection of Θ in $[-\pi/2, \pi/2]$, according to a law $\mathcal{L}_\varphi$ depending on ϕ.

What must $\mathcal{L}_\varphi$ be in order that the resulting distribution of M on the sphere be the uniform law (4,6,11)?

Let $P(e/\phi)$ denote the probability, according to the law $\mathcal{L}_\varphi$, of the event $\Theta \in e$, where e is an arbitrary subset of $[-\pi/2, \pi/2]$. $P(e/\phi)$ is precisely the conditional probability $\Pr(\Theta \in e/\Phi = \phi)$.

For any subset ω of $[0, 2\pi[$, let A be the event $\Phi \in \omega$, B the event $\Theta \in e$, and Σ the subset of the sphere defined by $\Phi \in \omega$, $\theta \in e$.

The event $M \in \Sigma$ is then $A \cap B$, so that by (4,6,4)

$$\Pr(M \in \Sigma) = \int_\omega P(e/\varphi) \frac{d\varphi}{2\pi} . \qquad (4,6,12)$$

By a well-known result of elementary geometry, the area σ of Σ is given by

$$\sigma = \int_\omega \int_e R^2 \cos\theta \, d\varphi \, d\theta .$$

Comparing (4,6,11) and (4,6,12), we see that for any ω and e

$$\int_\omega P(e/\varphi) \, d\varphi = \int_\omega \left[\int_e \frac{\cos\theta}{2} \, d\theta \right] d\varphi. \qquad (4,6,13)$$

By Theorem (2,6,9) or Theorem (2,8,2), it follows that

$$P(e/\varphi) = \int_e \frac{\cos\theta}{2} \, d\theta . \qquad (4,6,14)$$

This can also be verified directly as follows. Let ω be the interval $[0, \alpha < 2\pi]$; then by (4,6,13), identically in α,

$$\int_0^\alpha P(e/\varphi) \, d\varphi \equiv \int_0^\alpha \left[\int_e \frac{\cos\theta}{2} \, d\theta \right] d\varphi$$

and since the α-derivatives of both sides must coincide, formula (4,6,14) follows.

Now (4,6,14) has the following implication:

1) First, the law $\mathcal{L}_\varphi = \mathcal{L}$ must be independent of ϕ, as is intuitively clear, for B is independent of the event $\Phi = \phi$, and B and A are independent for any ω. This is the situation described in Remark (4,6,2) above.

2) Second, the law $\mathcal{L}$ governing the choice of Θ has density $\cos\theta/2$ over $[-\pi/2, \pi/2]$; it is not the uniform distribution, as might be expected at first sight.

EXAMPLE (4,6,3). *The Axiom of Conditional Probability in Wave Mechanics*

When, as in the previous example, a trial consists of

several partial trials (or steps) (especially when these are performed in succession), one must take care not to forget that the results of certain partial trials may influence others. A frequent and serious source of error is to treat events A and B as independent when they are not, and to write $Pr(A \cap B) = Pr(A)Pr(B)$ when in actual fact $Pr(A \cap B) = Pr(A)Pr(B/A)$ and conditional probabilities should be used.

In fact, it is easy to think of examples in which the result of one step influences the next. In some card games, the dealer is at an advantage, and the winner of each "hand" deals out the cards in the next. In two successive hands of such a game the second is influenced by the first or connected with it.

The following example is somewhat subtler. Consider a gambler tossing an apparently symmetric coin several times. At first, when no further information is available, he will assign the same probability (½) to "heads" and "tails." Now suppose that in the first, say, hundred tosses "heads" come up far more often than "tails." Our gambler will then have to revise his opinion, to assume that the coin is loaded and assign a significantly greater probability to "heads." The converse situation obtains if "tails" come up more frequently than "heads." Thus, the 101st toss is bound up with the first 100 in that the probabilities involved in the former depend on the results of the latter (cf. II.4). Note that despite the influence of the previous trials on the 101st, the physical conditions of the trial are unchanged: it is always the same coin, tossed in the same way.

In our final example the question merits some thought; it is taken from Wave Mechanics, but it can be understood without a knowledge of the field. Let Σ be a physical system whose state is determined by chance. Let $\mathcal{A}$ be an operator over Σ, the "measure of A," which yields the measure of a real or complex magnitude A associated with Σ; similarly, $\mathcal{B}$ is another operator over Σ, the "measure of B," which yields the measure of a real or complex magnitude B associated with Σ. The values obtained for A or B by applying $\mathcal{A}$ or $\mathcal{B}$ to Σ cannot be predicted, and so A and B are r.v. In Wave Mechanics it is assumed that $\mathcal{A}$ and $\mathcal{B}$ are linear hermitian operators. Assuming that their spectra are totally discontinuous, let a_k, b_k be their eigenvalues and f_k, g_k their eingenfunctions, respectively. The two systems $\{f_k\}$ and $\{g_k\}$ are orthogonal, and there is no loss of generality in assuming that they are orthonormal. Finally, let Φ be the wave function of Σ. Φ can be expanded in series with respect to the f_k or the g_k. Let

$$\Phi = \sum_k p_k f_k = \sum_k q_k g_k$$

be these expansions. On the other hand, if we expand the functions f_k in series of g_k, we obtain

$$f_k = \sum_h \rho_{kh} g_h,$$

and then

$$q_h = \sum_k p_k \rho_{kh}.$$

Wave Mechanics assumes that if $\mathcal{A}$ is applied, under any circumstances, the r.v. A is discrete and its possible value are precisely the a_k. Similarly, the possible values of B are the b_k. Finally, it is assumed that if an observer O performs a measurement of A, and of A alone, the probability that he finds $A = a_k$ is $|p_k|^2$; similarly, if he performs a measurement of B alone, the probability of $B = b_k$ is $|q_k|^2$. If he performs a measurement of A, followed by a measurement of B, the conditional probability of $B = b_h$ given $A = a_k$, is $|\rho_{kh}|^2$. Under the same conditions, the *a priori* probability of $B = b_h$ is, by the Axiom of Conditional Probability (more precisely, by (4,6,2)),

$$\sum_k |p_k|^2 |\rho_{kh}|^2. \qquad (4,6,15)$$

Obviously, this value is generally different from $|q_h|^2 = |\sum_k p_k \rho_{kh}|^2$, since (4,6,15) involves only the moduli of the p_h and ρ_{kh}, whereas $|q_h|^2$ also involves their arguments.

The explanation is simple (cf. L. de Broglie [1]): $|q_h|^2$ is the probability of the event "the result of the measurement is $B = b_h$" on the assumption that O is measuring B alone, whereas $\sum_k |p_k|^2 |\rho_{kn}|^2$ is the probability of the same event on the assumption that O has first measured A. It is thus not surprising that these two probabilities are not equal. The source of our original surprise is that in Classical Physics it is assumed that the measurement, say, of A does not disturb the system Σ in any way, and consequently the probability law of B does not depend on whether A has or has not been measured first. The situation is different in Wave Mechanics, where it is assumed that any measurement operator applied to Σ, such as $\mathfrak{A}$ or $\mathfrak{B}$, modifies Σ. There is no well-defined category of trials unless the measurement operators are actually applied and the order in which they are applied is specified.

True, even in Wave Mechanics it may happen that $|q_h|^2 = \sum_k |p_k|^2 |\rho_{kh}|^2$ (though this a particular case). The magnitudes A and B are then said to be "permutable"; this implies that it is possible to measure them simultaneously, and that the result is the same whether they are measured simultaneously or successively (in any order).

Conditional mathematical expectation

Let $\mathfrak{U}$ be a class of trials u and X a r.v. referred to $\mathfrak{U}$; we shall assume that X is a real r.v., but the discussion extends immediately to complex r.v. Let $F(x) = \Pr(X < x)$ be the d.f. of X; we have defined the m.e. E(X) of X as

$$E(X) = \int_{-\infty}^{+\infty} x \, dF(x).$$

It will be convenient to call $F(x)$ and $E(X)$ the a priori d.f. and m.e. of X.

Let A be any event or subset of U; put $F(x/A) = Pr(X < x/A)$. As a function of x, $F(x/A)$ is the conditional d.f. of X, given A, or the d.f. of X relative to the class of trials realizing A. If we call $E(X/A)$ the m.e. of X relative to this class of trials, i.e. the conditional m.e. of X, given A, the definition gives

$$E(X/A) = \int_{-\infty}^{+\infty} x \; F(dx/A).$$

Let S be a system of events $C \subset \mathfrak{U}$. In the notation of the beginning of this section, A denotes any subset of S, i.e. any union of events C. By (4,6,4),

$$Pr[A \cap (X < x)] = \int_A F(x/C) \; n(dC) \; ,$$

or, by (4,5,3),

$$Pr(A) \; F(x/A) = \int_A F(x/C) \; n(dC). \qquad (4,6,16)$$

Multiply both sides of (4,6,16) by x, and integrate with respect to x from $-\infty$ to $+\infty$; the left-hand side gives

$$\int_{-\infty}^{+\infty} x \; [Pr(A) \; F(dx/A)] = Pr(A) \int_{-\infty}^{+\infty} x \; F(dx/A)$$

$$= Pr(A) \; E(X/A) \; ,$$

and the right-hand side

$$\int_{-\infty}^{+\infty} x \left[\int_A F(dx/C) \; n(dC) \right] . \qquad (4,6,17)$$

Now if we assume that the m.e. $E(X/C)$ exists for all C, i.e. the integrals

$$\int_{-\infty}^{+\infty} x \; F(dx/C)$$

are absolutely convergent, then it is easily verified that the order of integration in (4,6,17) may be changed. Thus, (4,6,17) becomes

$$\int_A \left[\int_{-\infty}^{+\infty} x \; F(dx/C) \right] n(dC) = \int_A \; E(X/C) \; n(dC).$$

Finally,

$$\Pr(A) \; E(X/A) = \int_A \; E(X/C) \; n(dC).$$

This result is not very interesting if $\Pr(A) = 0$; on the other hand, it is very important when $\Pr(A) > 0$. Its most important application is when $S = \mathfrak{U}$ is an exhaustive system and we set $A = S$. Then $\Pr(A) = 1$, $E(X/A) = E(X)$, and we get

$$E(X) = \int_S \; E(X/C) \; n(dC).$$

Define a mapping μ of S into the space R of real numbers by setting $y = \mu(C) = E(X/C)$ (whose existence is assumed). As seen in IV.2, this defines a r.v. Y, to which (4,2,12) can be applied, giving

$$E(Y) = \int_S \; \mu(C) \; n(dC) = \int_S \; E(X/C) \; n(dC).$$

This remark yields the following interpretation of (4,6,18): *the (a priori) m.e. of a r.v.* X *is equal to the m.e. of its conditional m.e.* E(X/C). This is of course subject to the assumption that the events C form an exhaustive system.

Bayes' Theorem

Let S be a system of pairwise mutually exclusive events C; assume that the union S of the events C is certain (i.e. S is exhaustive), or at least almost certain. Let B be an arbitrary event.

Assume first that S contains only finitely or countably many elements C, denoted by $C_1, C_2, \ldots, C_k, \ldots$ For any h ($h = 1, 2, \ldots$), we have, by (4,5,3),

$$\Pr(C_h \cap B) = \Pr(C_h)\Pr(B/C_h), \tag{4,6,19}$$

and

$$\Pr(C_h \cap B) = \Pr(B)\Pr(C_h/B). \tag{4,6,20}$$

Now by (4,6,2) with A = S,

$$\Pr(B) = \sum_k \Pr(C_k)\Pr(B/C_k). \tag{4,6,21}$$

Substitute (4,6,21) in the right-hand side of (4,6,20) and equate the resulting expression for $\Pr(C_h \cap B)$ to (4,6,19); then

$$\Pr(C_h/B) = \frac{\Pr(C_h)\ \Pr(B/C_h)}{\sum_k \Pr(C_k)\ \Pr(B/C_k)}, \tag{4,6,22}$$

provided $\Pr(B) \neq 0$.

This is the famous *Bayes' formula*; more precisely, formula (4,6,22) is the elementary form of Bayes' theorem, which has been proved using the incomplete formulation (4,5,3) of the Axiom of Conditional Probability. Later we shall present a complete formulation of Bayes' theorem, whose proof will require the complete formulation of the Axiom of Condition Probability.

The significance and applications of Bayes' theorem will be discussed in detail in Vol. II; the following example will suffice for the present.

EXAMPLE (4,6,4). Some mice have no teeth. This peculiarity is caused by a gene having two alleles a and A, where a is recessive and implies absence of teeth. On the basis of the Chromosomic Theory of Heredity, some elements of which were set forth in I.5 and II.3, we shall try to solve the following problem.

A female mouse F whose parents were not toothless is not toothless, but belonged to a litter including a toothless mouse. Mated with a male M which is not toothless but has previously begotten toothless offspring, F produces a litter of nine offspring none of which is toothless. What is the probability p that F is a heterozygote Aa?

Formula (4,6,22) provides the answer. Indeed, note first that since F is not toothless it is either a homozygote AA or a heterozygote Aa; let C_1 and C_2 denote these two possibilities: they are mutually exclusive and their union C logically certain.

Let us find the probabilities $Pr(C_1)$ and $Pr(C_2)$ of C_1 and C_2 when the only information at hand is that the parents of F were not toothless and that F has a toothless sibling. The parents of F may be only homozygotes AA or heterozygotes Aa; but if one or both were AA, they could not have produced a homozygote aa, i.e. a toothless offspring. The problem reduces to that discussed in Example (4,5,1) and it follows that $Pr(C_1) = 1/3$, $Pr(C_2) = 2/3$.

Since the male M is not toothless, but has begotten toothless offspring, it is necessarily a heterozygote Aa. Let B be the event, which actually occurs, that a litter of nine nice produced by the mating (F, M) does not include any toothless individual. The probability of B, given C_1, i.e. $Pr(B/C_1)$, is unity; the probability $Pr(B/C_2)$ of B, given C_2, is $(3/4)^9$; this follows immediately from the principles of the Chromosomic Theory described in II.3.

The required probability P is precisely $Pr(C_2/B)$; Bayes' formula (4,6,22), which is valid here, immediately gives

$$P = Pr(C_2/B) = \frac{Pr(C_2)Pr(B/C_2)}{Pr(C_1)Pr(B/C_1) + Pr(C_2)Pr(B/C_2)}$$

$$= \frac{2/3 \times (3/4)^9}{1/3 + 2/3 \times (3/4)^9} \approx 0.129. \blacksquare$$

Now consider the general case where S is any set, not necessarily finite or countable. Retain the previous notation; in particular, A will denote an arbitrary subset of S. Then (4,5,3), (4,6,4), and (4,6,5) immediately give:

THEOREM (4,6,1) (*Bayes*)

If S is a system of events C such that $Pr(S) = 1$, then for any event A determined by S and any event B such that $Pr(B) > 0$,

$$\Pr(A/B) = \frac{\int_A \Pr(B/C)\, n(dC)}{\int_S \Pr(B/C)\, n(dC)} . \qquad (4,6,23)$$

This formula includes (4,6,22) as a particular case; it adds no interesting information if Pr(A) = 0 (for then (4,6,23) becomes 0 = 0) or if Pr(B) = 0 (then the right-hand side of (4,6,23) has the indeterminate form 0/0), all the more so if both Pr(A) = 0 and Pr(B) = 0.

However, any required results may be derived directly from (4,6,23), or, what amounts to the same, from (4,6,4), of which (4,6,23) is merely an application. To this end it suffices to use the method whose main idea is embodied in Example (4,6,1). We shall verify this in V.2 p. 331, in a particular case; the reader will easily adapt the procedure to any other case.

BIBLIOGRAPHIC NOTES FOR CHAPTER IV

For random variables and the Axiom of Conditional Probability the reader may refer to the literature cited in the Bibliographic Notes for Chapter II, on the foundations and general elements of Probability Theory. See in particular P. Halmos [1], Chap. VII, and J. Neveu [1] for measures over product spaces, Fubini's Theorem, etc.

A very important reference for the problem of the existence of conditional probability (cf. Remark (4,6,1)), is M. Jirina [1]; an application may be found in R. Fortet [5].

EXERCISES

4.1. Let $0 < p < 1$, $q = 1 - p$. Consider a r.v. X assuming the positive integral values k = 1, 2, 3,..., with respective probabilities

$$p_k = \Pr(X = k) = p\, q^{k-1} \qquad (k = 1,2,3,\ldots) .$$

a) Prove that

$$\sum_k p_k = 1.$$

b) Compute the mathematical expectation and the standard deviation of X. (Answer: $E(X) = 1/p$; $\sigma^2(X) = q/p^2$).

c) Compute the characteristic function $\phi(u)$ of X and the first three terms of its expansion in Taylor series about $u = 0$. (Answer: $\phi(u) = pe^{iu}/(1 - qe^{iu})$).

4.2. a) A point M is chosen at random on a segment AB of unit length, according to the uniform probability law (i.e. the probability that M is in an arbitrary segment CD of AB is equal to the length of CD). Let L_1 and L_2 be the lengths of segments AM and MB, and X, Y respectively the greater and the smaller of the numbers L_1 and L_2. Find the d.f. $F(x)$, $G(y)$, $H(z)$ of the r.v. X, Y, $Z = X - Y$ respectively. A right triangle is constructed, with L_1 and L_2 as legs; let S be its area and Δ its hypotenuse. Find $E(S)$ and $E(\Delta^2)$. (Answers: $F(x) = 2x - 1$, $\frac{1}{2} \leq x \leq 1$; $G(y) = 2y$, $0 \leq y \leq \frac{1}{2}$; $H(z) = z$; $E(S) = 1/12$, $E(\Delta^2) = 2/3$).

b) Let $M_1, M_2, \ldots, M_n$ be n positions of M on AB, which are the outcomes of n *independent* trials, $Z_1, Z_2, \ldots, Z_n$ the lengths of the segments $AM_1, AM_2, \ldots, AM_n$, and T the smallest of the numbers Z_j. Find: the probability law of T; the conditional probability law of T, given $Z_1 = z$; the probability of the event $Z_1 = T$; the conditional probability law of Z_1, given $T = t$.

c) Let $\mathcal{I}_1, \mathcal{I}_2, \ldots, \mathcal{I}_n, \mathcal{I}_{n+1}$ denote the $n + 1$ segments determined on AB by $M_1, M_2, \ldots, M_n$, numbered in order from A to B. Let $L_1, L_2, \ldots, L_n, L_{n+1}$ be their respective lengths. What is the probability law of L_1? What is the probability law of L_2 when $n = 2$? Extend this result to the general case, showing that the probability law of L_k does not depend on k.

Let U be the greatest Z_j; what is the probability law of U? of $U - T$?

Exercise 4.2. can be applied to Example (5,1,1).

4.3. Complement to the Theorem of Borel-Cantelli (cf. Exercise 2.4). Let $E_1, E_2, \ldots, E_k, \ldots$ be an infinite countable sequence of mutually independent events; put $p_k = \Pr(E_k)$, and let A be the event: infinitely many E occur. Show that if $\sum_k p_k = +\infty$, then Pr(A) = 1 (Hint: Set $B_h = \sum_{k>h} E_k$, $A = \bigcap_h B_h$; $\check{A} = \bigcup_h \check{B}_h$, $\check{B}_h = \bigcup_{k>h} \check{E}_k$; because of the mutual independence of the E_k,

$$\Pr(\check{B}_h) = \prod_{k>h} (1 - p_k) \leqslant e^{-\sum_{k>h} p_k} = 0].$$

4.4. (After A. Renyi [3]). Let $X_1, X_2, \ldots, X_n, \ldots$ be an infinite countable sequence of mutually independent r.v., all having the same continuous d.f. F(x). Put

$$Y_n = \underset{1 \leqslant j \leqslant n}{\text{l.u.b.}} \ X_j .$$

Find the d.f. of Y_n (Answer: $[F(x)]^n$; this is an immediate generalization of (4,5,19)). X_n (n = 2, 3,...) is said to be *salient* if $X_n > X_j$ for $1 \leqq j < n$; denote the event "X_n is salient" by A_n. What is the conditional probability of A_{n+1}, given $Y_n = y$? What is the probability of A_{n+1}? Let C_k ($k = 1, 2, 3, \ldots, 2^{n-1}$) denote the 2^{n-1} events $\bigcap_{j=2}^{n} B_j$, where B_j is either A_j or $\check{A}_j$. Show that these events C_k, which form an exhaustive system, all have the same probability; hence deduce that $A_2, \ldots, A_n$ are mutually independent. (Hint: Since $\Pr(C_k \ A_{n+1})$ does not depend on k and is equal to $\Pr(C_k) \times \Pr(A_{n+1}/C_k)$, it follows that $\Pr(A_{n+1}/C_k)$ does not depend on k, therefore A_{n+1} is independent of each of the C_k).

4.5. Put

$$\varphi(x) = \int_x^{+\infty} e^{-u^2} \, du ;$$

$$H(x) = \int_0^{+\infty} \frac{e^{-v}\,dv}{\sqrt{1 + \frac{v}{x^2}}} ;$$

$$a_k = \frac{(2k)!}{2^{2k}\,k!} \qquad (k = 0, 1, 2, \ldots) ;$$

$$H_n(x) = \sum_{k=0}^{n} (-1)^k \frac{a_k}{x^{2k}} ;$$

$$H(x) = H_n(x) + (-1)^{n+1} \frac{a_{n+1}}{x^{2n+2}} \cdot \alpha_{n+1}(x),$$

where the last equality defines the functions $\alpha_{n+1}(x)$ for each n ($n = 0, 1, 2, \ldots$). Show that

a) $\varphi(x) = \dfrac{e^{-x^2}}{2x} \cdot H(x)$ (Set $u = \sqrt{x^2 + v}$);

b) $0 \leqq \alpha_{n+1}(x) \leqq 1$, and

$$\lim_{x \to +\infty} \alpha_{n+1}(x) = 1.$$

c) Putting $n = 0$, find a new proof of formula (4,4,21). (Hint: Use the Taylor expansion of the function $(1 + x)^{-\frac{1}{2}}$ about $x = 0$).

4.6. Each of n persons has an urn containing K identical balls numbered from 1 to K. Each draws a ball from his urn at random, independently of the others. Let α be a positive integer which divides n, and P_α the probability that at most α of the balls drawn from the urn are numbered identically. Compute P_1; compute P_α for $K = 4$, $\alpha = 1, 2, 3$, $n = 3\alpha$.

4.7. Every day, n copies of a certain newspaper are delivered to a newsagent, and each of X persons buys one copy. n is a fixed integer, X is random, and the probabilities $P_h = \Pr(X \leqq h)$ ($h = 0, 1, 2, 3, \ldots$) are given. Every copy sold results in a profit a, each copy not sold incurs a loss b ($a, b > 0$). Let Y be the net profit ($\gtrless 0$) for one day. Compute E(Y); for what value of n is E(Y) a maximum?

4.8. A positive-valued r.v. X is said to have a *log-normal* distribution if Y = log X is a normal r.v. If we set E(Y) = m, $\mathcal{V}(Y) = \sigma^2$, the d.f. F(x) and the density f(x) of X are given by

$$F(x) = L\left(\frac{\log x - m}{\sigma}\right), \quad f(x) = \frac{1}{\sqrt{2\pi}\,\sigma x}\, e^{-\frac{(\log x - m)^2}{2\sigma^2}}.$$

Show that $E(X^\alpha)$ may be deduced from (4,4,26) by putting $u = -i\alpha$.

4.9. Let X be a r.v. with a continuous d.f. F(x). Show that the r.v. Y = F(x) is uniformly distributed over (0, 1).

4.10. Let X be a nonnegative r.v. with probability density

$$f(x) = \begin{cases} \dfrac{1}{\Gamma(\alpha)}\, x^{\alpha-1} e^{-x} & \text{for } x > 0 \\ 0 & \text{for } x < 0 \end{cases}$$

where α is any positive number (cf. (1,1,2)). Find the moment-generating function $\phi(u)$ of X (Answer: $\phi(u) = (1 + u)^{-\alpha}$ for $u > -1$); hence deduce the m.e. and variance of X.

4.11. Let X be a normal r.v. with E(X) = 0, $E(X^2) = 1$; determine the density of the nonnegative r.v. $Y = X^2$ (cf. Exercise 4.10).

CHAPTER V

n-DIMENSIONAL

RANDOM VECTORS AND VARIABLES

1. *Distribution function of an n-dimensional random variable*

In II.11 we defined n-dimensional random vectors and random variables. We saw that an n-dimensional random variable may be interpreted either as the n-tuple of components of a random vector $\vec{X}$ in an n-dimensional vector space $\mathfrak{X}_n$, or as the n-tuple of coordinates of a random point N in a point space E_n. Since these interpretations are equivalent, we shall adopt the former in this chapter.

The study of a random vector $\vec{X}$ in $\mathfrak{X}_n$ may be envisaged in two ways. On the one hand, the probability law of a random element $\vec{X}$ with values in $\mathfrak{X}_n$ is determined by a measure m over $\mathfrak{X}_n$; we can therefore carry out the "*intrinsic*" study of this measure, i.e. without reference to a particular basis in $\mathfrak{X}_n$. On the other hand, we can choose a basis $\mathbf{b}$ in $\mathfrak{X}_n$ and relate the study of $\vec{X}$ to that of the n-tuple $\{X_1,X_2,\ldots,X_n\}$ of its components relative to $\mathbf{b}$, i.e. the n-dimensional random variable $\{X_1,X_2,\ldots X_n\}$. From the purely mathematical viewpoint, these two approaches are equivalent; however, from a practical viewpoint they are quite different, each suggesting a distinct direction of investigation.

Suppose that we wish to investigate the mutual influence of the cost P of raw metal and the turnover V in the electro-mechanical industry of France, regarding P and V as random variables. We can interpret these quantities as the components of a random vector in a 2-dimensional vector space $\mathfrak{X}_2$ relative to an orthonormal basis. In the course of the investigation, we are free, if the need arises, to utilize another basis $\mathbf{b}'$, or any other reference system. However, in the final analysis we shall have to return to the original basis $\mathbf{b}$, i.e. to the quantities P and V, which are the main objects of the investigation. In general, however, since P and V are quantities of quite different natures, the intrinsic geometric properties of $\mathfrak{X}_2$ should be employed with considerable caution, for they may well be devoid of any concrete meaning.∎

We begin by adopting the non-intrinsic viewpoint, i.e. studying n-dimensional random variables or n-tuples of real random variables $\{X_1, X_2, \ldots, X_n\}$. In principle, we shall confine ourselves to the case $n = 2$, since the reasoning is easily extended to the general case. Thus we shall consider an ordered pair $\{X, Y\}$ of real random variables, which we shall sometimes interpret as the components of a random vector $\vec{Z}$ in a two-dimensional vector space $\mathfrak{X}_2$ relative to a fixed *orthonormal* basis.

The probability law of this random vector consists of a measure or mass distribution $m(e)$ over $\mathfrak{X}_2$, $e \subset \mathfrak{X}_2$ denoting any subset of $\mathfrak{X}_2$. $m(e)$ is the probability that $\vec{Z} \in e$:

$$m(e) = \Pr(\vec{Z} \in e). \tag{5,1,1}$$

We shall only consider the case of a *nondegenerate* 2-dimensional random variable $\{X, Y\}$ (cf. II.12); in other words, we assume that the probability of the event E: "$\vec{Z}$ is a *finite* vector of $\mathfrak{X}_2$," is unity. Now if $\{e_k\}$ ($k = 1, 2, 3, \ldots$) is any sequence of bounded subsets of $\mathfrak{X}_2$ such that

$$e_k \subset e_{k+1} \ (k = 1, 2, \ldots), \quad \lim_{k \to +\infty} e_k = \bigcup_k e_k = \mathfrak{X}_2,$$

then E is the event

$$E = \bigcup_k (\vec{Z} \in e_k),$$

so that

$$\Pr(E) = m(\bigcup_k e_k) = m(\mathfrak{X}_2)$$

and thus

$$m(\mathfrak{X}_2) = 1. \tag{5,1,2}$$

In other words, $m(e)$ is a *bounded* measure.

We know from Chapter III that once a basis has been chosen in $\mathfrak{X}_2$ the measure m may be described by a 2-dimensional d.f. *We make the specific choice of the d.f.* $\rho(x, y)$ *associated with* m *by* (3,1,1), which we shall call the 2-dimensional d.f. of the 2-dimensional r.v. $\{X, Y\}$. By definition,

$$\rho(x, y) = \Pr[(X < x) \cap (Y < y)]. \tag{5,1,3}$$

The function $\rho(x, y)$ has all the general properties of a 2-dimensional d.f. In addition, it has the special properties implied by (5,1,2) and by the fact that it is associated with m by (3,1,1). We shall indicate the principal properties without proof, since they are immediate results of the arguments in III.1, in particular, of Theorem (3,1,1).

$$1) \quad \lim_{x \to +\infty, y \to +\infty} \rho(x, y) = \rho(+\infty, +\infty) = 1 ; \tag{5,1,4}$$

$$2) \quad \lim_{x \to -\infty} \rho(x, y) = \lim_{y \to -\infty} \rho(x, y) = 0 ; \tag{5,1,5}$$

$$3) \quad \lim_{\varepsilon \to +0, \eta \to +0} \rho(x - \varepsilon, y - \eta) = \rho(x, y). \tag{5,1,6}$$

Property 3) means that $\rho(x, y)$ is continuous from the left. Particularly interesting is the case in which m is absolutely continuous, with density $\rho(x, y)$. We know that then ρ itself is said to be absolutely continuous, with density $\rho(x, y)$ (cf. III.3), and by (3,1,13)

$$\rho(x, y) = \frac{\partial^2}{\partial x \, \partial y} \rho(x, y). \tag{5,1,7}$$

$\rho(x, y)$ is also known as the *probability density* of the 2-dimensional r.v. {X, Y}.

Conversely, again by III.1 and in particular by Theorem (3,1,3), $\rho(x, y)$ determines m. This is the content of formula (3,1,16), which becomes, in the present notation,

$$\Pr(\vec{Z} \in e) = m(e) = \int \int_e d\rho(x, y) \tag{5,1,8}$$

w..e the integral is an RS-integral (cf. III.10) if the set e is sufficiently simple. If ρ has a density ρ, formula (5,1,8) becomes

$$\Pr(\vec{Z} \in e) = m(e) = \int \int_e \rho(x, y) \, dx \, dy,$$

which shows, incidentally, that a function $\rho(x, y)$ defined on $\mathfrak{X}_2$ is the probability density of a 2-dimensional r.v. if and only

If it is nonnegative and

$$\int_{-\infty}^{+\infty} \int_{-\infty}^{+\infty} \rho(x , y) \, dx \, dy = 1. \qquad (5,1,9)$$

A priori or marginal probability laws of X *and* Y

We shall often have occasion to speak of the d.f. F(x) of the one-dimensional r.v. X; to distinguish it from the 2-dimensional d.f. ρ of the 2-dimensional r.v. {X, Y} we shall call it the a priori or *marginal* d.f. of X. It is easily derived from ρ. In fact, since the event $Y < +\infty$ is almost certain (the r.v. is nondegenerate), we have

$$F(x) = Pr(X < x) = Pr[(X < x) \cap (Y < +\infty)] = \rho(x , +\infty). \qquad (5,1,10)$$

In particular, if ρ has a density ρ, then (5,1,10) becomes

$$F(x) = \int_{-\infty}^{x} \left[\int_{-\infty}^{+\infty} \rho(\alpha , \beta) \, d\beta \right] d\alpha. \qquad (5,1,11)$$

Hence one can show that F(x) has density f(x) given by the formula

$$f(x) = \frac{d}{dx} F(x) = \int_{-\infty}^{+\infty} \rho(x , \beta) \, d\beta. \qquad (5,1,12)$$

Similarly, the a priori or marginal d.f. G(y) of Y is

$$G(y) = Pr(Y < y) = \rho(+\infty , y), \qquad (5,1,13)$$

and if ρ has density ρ the density g(y) of G(y) is

$$g(y) = \frac{d}{dy} G(y) = \int_{-\infty}^{+\infty} \rho(\alpha , y) \, d\alpha. \qquad (5,1,14)$$

Characteristic function and moments of a 2-dimensional random variable

The 2-dimensional characteristic function of the 2-dimensional r.v. {X, Y} is defined as the 2-dimensional Fourier transform $\phi(u, v)$ of the 2-dimensional d.f. ρ (cf. (3,12,2)):

$$\varphi(u , v) = \int_{-\infty}^{+\infty} \int_{-\infty}^{+\infty} e^{i(ux+vy)} \, d\rho(x , y), \qquad (5,1,15)$$

where u and v are real variables ($-\infty < u, v < +\infty$). The principal properties of the characteristic function are described in III.12. In particular, $\phi(u, v)$ is continuous and positive-definite. Moreover, by (5,1,4)

$$\varphi(0, 0) = 1. \tag{5,1,16}$$

A continuous positive-definite function of (u, v) is the characteristic function of a 2-dimensional r.v. if and only if it satisfies (5,1,16).

On the other hand, by (4,2,15) we can replace (5,1,15) by the formula

$$\varphi(u, v) = E[e^{i(uX+vY)}]. \tag{5,1,17}$$

Similarly, the moments of the 2-dimensional r.v. {X, Y} are by definition the moments of the 2-dimensional d.f. ρ as defined in III.12. For example, the absolute moment ${}_{a,b}m^*_{\alpha,\beta}$ of order (α, β) about a and b is (cf. (3,12,5))

$${}_{a,b}m^*_{\alpha,\beta} = \int_{-\infty}^{+\infty}\int_{-\infty}^{+\infty} |x - a|^{\alpha} \, |y - b|^{\beta} \, d\rho(x, y), \tag{5,1,18}$$

if the integral is convergent. The algebraic moment ${}_{a,b}m_{\alpha,\beta}$ of order (α, β) (where α and β are nonnegative integers) about a and b is (see (3,12,6))

$${}_{a,b}m_{\alpha,\beta} = \int_{-\infty}^{+\infty}\int_{-\infty}^{+\infty} (x - a)^{\alpha} \, (y - b)^{\beta} \, d\rho(x, y), \tag{5,1,19}$$

again, of course, provided the integral is convergent.

For the properties of these moments and their relation to the characteristic function $\phi(u, v)$ (formulas (3,12,7), (3,12,8)), we refer to III.12. Using (4,2,15), we can again express (5,1,18) and (5,1,19) as expectations:

$${}_{a,b}m^*_{\alpha,\beta} = E(|X - a|^{\alpha} \, |Y - b|^{\beta}), \tag{5,1,20}$$

$${}_{a,b}m_{\alpha,\beta} = E[(X - a)^{\alpha} \, (Y - b)^{\beta}]. \tag{5,1,21}$$

In particular,

$$ {}_{o,o}m_{1,o} = E(X)\ ,\ {}_{o,o}m_{o,1} = E(Y)\ ; \qquad (5,1,22)$$

$$ {}_{o,o}m_{2,o} = E(X^2),\ {}_{o,o}m_{1,1} = E(X\,Y),\ {}_{o,o}m_{o,2} = E(Y^2)\ . \qquad (5,1,23)$$

In III.12 we indicated the relation of the algebraic moments of order 1 to the notions of mean vector or center of gravity, and that of the moments of order 2 to the notion of moment of inertia. Thus E(X) and E(Y) are the components of the mean vector associated with the mass distribution or probability law m(e) of the random vector $\vec{Z} \in \mathfrak{X}_2$ relative to the basis chosen in $\mathfrak{X}_2$; similarly, the moments of order 2 are the moments of inertia of this probability law.

Suppose that E(X) and E(Y) exist, and put

$$X = E(X) + X',\ \ Y = E(Y) + Y'$$

Then clearly

$$E(X') = E(Y') = 0$$

In other words, there is no loss of generality in assuming that E(X) = E(Y) = 0, which will therefore be assumed from now on. Otherwise, one need only take a = E(X), b = E(Y) in (5,1,20) and (5,1,21).

Since E(X) and E(Y) vanish, it follows that the moments ${}_{o,o}m^{\bullet}_{\alpha,\beta}$, ${}_{o,o}m_{\alpha,\beta}$ equal the *central* moments $\mu^{\bullet}_{\alpha,\beta}$, $\mu_{\alpha,\beta}$. In particular, ${}_{o,o}m_{2,o}$ and ${}_{o,o}m_{o,2}$ are the respective variances of X and Y. If we denote the standard deviations of X and Y by σ_X and σ_Y, respectively, we have

$$E(X^2) = {}_{o,o}m_{2,o} = \sigma_X^2,\qquad E(Y^2) = {}_{o,o}m_{o,2} = \sigma_Y^2\ . \qquad (5,1,23)$$

Correlation coefficient

We now introduce the coefficient r defined by (3,12,15):

$$ {}_{o,o}m_{1,1} = E(X\,Y) = \mu_{1,1} = \sigma_X\,\sigma_Y\ r\ . \qquad (5,1,24)$$

Recall that r is dimensionless, and moreover

$$|r| \leqq 1. \tag{5,1,25}$$

It is called the *linear cross-correlation coefficient* of X and Y, or, more briefly, the correlation coefficient of X and Y.

We know that r = 0 if the orthonormal basis vectors chosen in $\mathfrak{X}_2$ are parallel to the axes of symmetry of the central ellipse of inertia of the mass distribution m(e). *Thus* r = 0 *is not an intrinsic property of* m(e), *but rather a property of the pair* {X, Y}.

The property (5,1,25) is derived by application of the Schwarz inequality to (3,12,14). It thus follows from Theorem (2,7,4) that $|r| = 1$ if and only if the functions $\lambda(x, y) = x$ and $\mu(x, y) = y$ are proportional up to ρ-equivalence (i.e. up to m-equivalence). In terms of probabilities, this means that there exist constants c and d such that cX = dY almost certainly. By (5,1,23), this means that $c^2\sigma_X^2 = d^2\sigma_Y^2$. In view of (5,1,24), returning to the general case in which E(X) and E(Y) are not necessarily zero, we have:

THEOREM (5,1,1). The correlation coefficient r of X and Y is ± 1 if and only if there is a linear relation between X and Y which they satisfy almost certainly; this relation is then necessarily

$$\pm \frac{X - E(X)}{\sigma_X} = \frac{Y - E(Y)}{\sigma_Y}. \tag{5,1,26}$$

EXAMPLE (5,1,1) (*a theory of vision*). Baumgardt [1] has proposed a theory of vision, from which we cite the following extract. "Excitation of a retinal ganglionic cell occurs, followed by the transmission of at least one influx propagated in the direc-

---

* [Translation editor's note: The reference seems to be incorrect - we have been unable to find this passage in Baumgardt [1].]

tion of the centers, when two photons are absorbed by two of the rods or one of the cones within the range of action of the ganglionic cell; the time interval between the absorption of these two photons should not exceed a fixed time τ." It is immaterial to us whether this theory is or is not in accord with the phenomenon of vision; it will merely provide us with an interesting and very simple illustration of the preceding arguments.

Among other things, Baumgardt assumes that a receiving cell in the retina is not activated unless it is hit by two consecutive photons separated by a time interval not exceeding a certain duration or threshold τ (which is evidently very small). In other words, any photon impinging upon the cell at an instant t, which is not preceded or followed by at least one more photon during the time interval $(t - \tau, t + \tau)$, is so to speak lost. Now the arrival times of the photons are random. On the average - but only on the average - a given luminous flux carries a constant number of photons per unit time; it can be assumed that if n photons hit the cell in unit time, say during the interval (0, 1), the arrival times of these photons are distributed at random, independently of each other, each obeying the uniform law over the interval (0, 1). We shall see in Volume II that this is equivalent to the assumption that the arrival times of the photons constitute a Poisson process.

It is thus natural to consider problems of the following type. Let $X_1, X_2, \ldots, X_n$ be the abscissas of n randomly selected points $M_1, M_2, \ldots, M_n$ in the interval (0, 1), independent of each other and governed by the uniform law. Let Y be the length of the smallest of the n + 1 subintervals defined in (0, 1) by these n points. What is the probability that $Y < \tau$, and, more generally, what is the probability law of Y? This problem has been solved by P. Lévy [3]. The solution may be obtained either directly (cf. Exercise 4.2) or by the following reasoning.

There are n! possible orderings of the points $M_1, M_2, \ldots, M_n$ in (0, 1); let these orderings be $\Omega_1, \Omega_2, \ldots, \Omega_{n!}$, where Ω_1 denotes the ordering corresponding to $X_1 \leqslant X_2 \leqslant \ldots \leqslant X_j \leqslant X_{j+1} \leqslant \ldots \leqslant X_n$, We shall use the symbol Ω_k to denote both the actual *ordering* Ω_k and the *event* that the M_j are situated in the order Ω_k. Since all orderings are equiprobable, the probability of each is 1/n!. Let us first compute the probability $\Pr(Y > \tau/\Omega_k)$ that $Y > \tau$ when Ω_k occurs. This probability is clearly independent of k, and it suffices to compute $\Pr(Y > \tau/\Omega_1)$. Now

$$\Pr(Y > \tau/\Omega_1) = \frac{\Pr[\Omega_1 \cap (Y > \tau)]}{\Pr(\Omega_1)}$$

by (4,5,3).

Thus we must evaluate $\Pr[\Omega_1 \quad (Y > \tau)]$; put

$$Z_1 = X_1 - \tau,\ Z_2 = X_2 - X_1 - \tau, \ldots,$$

$$Z_k = X_k - X_{k-1} - \tau, \ldots,\ Z_n = X_n - X_{n-1} - \tau.$$

Let us regard the X_k as the coordinates of a random point Q in an n-dimensional Euclidean space E_n, relative to an orthonormal reference system; then Q has an n-dimensional density which is constant and equal to unity. Regarding the Z_j as the coordinates relative to an orthonormal reference system of a random point R in an n-dimensional Euclidean space E'_n, we see that the same applies to the n-dimensional probability density of R, since the Jacobian of the X_k with respect to the Z_j is 1 (cf. (2,11,13)). Let D be the domain of E_n in which Q must lie if the events Ω_1 and $Y > \tau$ both occur, and D' the corresponding domain for R in E'_n. Then

$$\Pr[\Omega_1 \cap (Y > \tau)] = \underbrace{\int \cdots \int_{D'}}_{n\ fois} dz_1\, dz_2 \ldots dz_n$$

(n-tuple integral over D'). Thus

$$\Pr[\Omega_1 \cap (Y > \tau)] = \text{volume of } D'.$$

Now the conditions defining D are

$$\begin{cases} 0 \leqslant X_1 \leqslant X_2 \leqslant \ldots \leqslant X_k \leqslant X_{k+1} \leqslant \ldots \leqslant X_n \leqslant 1 ; \\ Y > \tau. \end{cases}$$

Thus the conditions defining D' are

$$\begin{cases} 0 \leqslant Z_j \quad (j = 1, 2, \ldots, n) ; \\ \sum_j Z_j < 1 - (n+1)\tau. \end{cases}$$

It follows that D' is a regular "tetrahedron" in E'_n, with vertex at the origin of the reference system in E'_n, such that the sides issuing from the vertex are mutually orthogonal (coinciding with the coordinate axes) and of length $1 - (n + 1)\tau$. The base of the tetrahedron is in the plane defined by the equation

$$\sum_j Z_j = 1 - (n+1)\tau.$$

It is well known that the volume of D' is

$$\frac{1}{n!}\,[1 - (n+1)\tau]^n.$$

Thus

$$\Pr(Y > \tau/\Omega_j) = [1 - (n+1)\tau]^n,$$

and

$$\Pr(Y > \tau) = \sum_j \Pr(\Omega_j) \cdot \Pr(Y > \tau/\Omega_j) = [1 - (n+1)\tau]^n,$$

and our problem is solved. Note that as $n \to +\infty$, $\Pr\,[(n + 1)^2 Y > \alpha]$ converges to $e^{-\alpha}$. Thus the probability law of the r.v. $(n + 1)^2 Y$ converges (cf. Example (4,5,2)) to the *exponential law* $1 - e^{-a}$.

The volume $(1/n!)\,[1 - (n + 1)\tau]^n$ of the tetrahedron D' may be derived by the following argument. We know that $\Pr(\Omega_1) = 1/n!$. The domain Δ in E_n in which Q must lie for Ω_1 to occur is precisely a regular tetrahedron whose edges issuing from the vertex are orthogonal and of length 1. Therefore

$$\text{Volume of } \Delta = \Pr(\Omega_1) = \frac{1}{n!}.$$

Since D' is similar to Δ, with similarity factor $[1 - (n + 1)\tau]$, we get the above result for the volume of D'.

In P. Lévy [3] one can find other, less elementary results concerning problems of this type.

2. *Conditional study of a random variable with respect to another*

Let G(y/x) denote the conditional d.f. of Y given X = x, where x is any given number:

$$G(y/x) = \Pr(Y < y/X = x). \tag{5,2,1}$$

Using (4,6,4) we see that

$$\wp(x, y) = \Pr[(X < x) \cap (Y < y)] = \int_{-\infty}^{x} G(y/\alpha)\, dF(\alpha). \tag{5,2,2}$$

This relation is valid for any x and y; in view of (5,1,10) it determines G(y/x) when $\wp$ is known. We can express the relation symbolically by $\wp(dx, y) = G(y/x)\, dF(x)$, or

$$G(y/x) = \frac{\wp(dx, y)}{dF(x)}; \tag{5,2,3}$$

it is particularly simple when $\wp$ has a density ρ, for then, by (5,1,12), we can write (5,2,2) in the form

$$\int_{-\infty}^{x}\left[\int_{-\infty}^{y} \rho(\alpha, \beta)\, d\beta\right] d\alpha = \int_{-\infty}^{x} G(y/\alpha)\, f(\alpha)\, d\alpha =$$

$$= \int_{-\infty}^{x} G(y/\alpha)\left[\int_{-\infty}^{+\infty} \rho(\alpha, \beta)\, d\beta\right] d\alpha,$$

which is a particular case of (4,6,7). Differentiating with respect to x, we get

$$G(y/x) = \frac{\int_{-\infty}^{y} \rho(x, \beta)\, d\beta}{f(x)} = \frac{\int_{-\infty}^{y} \rho(x, \beta)\, d\beta}{\int_{-\infty}^{+\infty} \rho(x, \beta)\, d\beta}, \tag{5,2,4}$$

which is a particular case of (4,6,8). It is clear from (5,2,4) that, under these conditions, the d.f. G(y/x) has a density g(y/x):

$$g(y/x) = \frac{d}{dy} G(y/x) = \frac{\rho(x\,,\,y)}{f(x)} = \frac{\rho(x\,,\,y)}{\int_{-\infty}^{+\infty} \rho(x\,,\,\beta)\,d\beta}\,, \tag{5,2,5}$$

which is the conditional probability density of Y given X = x. Needless to say, the roles of X and Y in this discussion may be reversed. For the d.f. F(x/y) and probability density f(x/y) (when it exists) we have the following analogues of (5,2,2), (5,2,3), (5,2,4), (5,2,5):

$$\rho(x\,,\,y) = \int_{-\infty}^{x} F(x/\beta)\,dG(\beta)\,, \qquad F(x/y) = \frac{\rho(x\,,\,dy)}{dG(y)}\,, \tag{5,2,6}$$

$$F(x/y) = \frac{\int_{-\infty}^{x} \rho(\alpha\,,\,y)\,d\alpha}{g(y)} = \frac{\int_{-\infty}^{x} \rho(\alpha\,,\,y)\,d\alpha}{\int_{-\infty}^{-\infty} \rho(\alpha\,,\,y)\,d\alpha}\,, \tag{5,2,7}$$

$$f(x/y) = \frac{d}{dx} F(x/y) = \frac{\rho(x\,,\,y)}{g(y)} = \frac{\rho(x\,,\,y)}{\int_{-\infty}^{+\infty} \rho(\alpha\,,\,y)\,d\alpha}\,. \tag{5,2,8}$$

Bayes' formula. Following (5,2,5) ρ(x, y) may be written as

$$\rho(x,\,y) = f(x)\,g(y/x).$$

Thus (5,2,8) becomes

$$f(x/y) = \frac{f(x)\,g(y/x)}{\int_{-\infty}^{+\infty} f(\alpha)\,g(y/\alpha)\,d\alpha}\,,$$

which is Bayes' formula (4,6,23) for the frequently arising particular case when the densities exist.

Conditional mathematical expectation. The conditional m.e. E(Y/X = x) of Y given X = x is defined by the formula

$$E(Y/X = x) = \int_{-\infty}^{+\infty} y\,G(dy/x); \tag{5,2,9}$$

if ρ has a density ρ, this formula becomes, by (5,2,5),

$$\begin{aligned} E(Y/X = x) &= \int_{-\infty}^{+\infty} y\,g(y/x)\,dy = \frac{1}{f(x)} \int_{-\infty}^{+\infty} y\rho(x\,,\,y)\,dy \\ &= \frac{\int_{-\infty}^{+\infty} y\rho(x\,,\,y)\,dy}{\int_{-\infty}^{+\infty} \rho(x\,,\,y)\,dy}\,. \end{aligned} \tag{5,2,10}$$

Since $E(Y/X = x)$ depends on the value of x, we can regard it as a function of x and write

$$\mu(x) = E(Y/X = x). \tag{5,2,11}$$

The function $y = \mu(x)$ is known as the *regression function* of Y. The curve obtained by plotting this function in a rectangular cordinate system Oxy is called the *regression curve* of Y on X.

Consider the r.v. $\mu(X)$, i.e. the random number which assumes the value $\mu(x)$ when X assumes the value x. A suggestive notation for this r.v. is $\mu(X) = E(Y/X)$.

By (4,6,18) we have

$$E(Y) = E[\mu(X)] = E[E(Y/X)], \tag{5,2,12}$$

thus proving:

THEOREM (5,2,1). The mathematical expectation of the conditional mathematical expectation of Y given X is equal to the a priori (marginal) mathematical expectation of Y.

More generally, let $\lambda(x, y)$ be a given function of x and y, and consider the r.v. $\Lambda = \lambda(X, Y)$, i.e. the r.v. that assumes the value $\lambda(x, y)$ when X assumes the value x *and* Y the value y. Conditionally, when $X = x$, $\Lambda = \lambda(x, Y)$ depends only on Y, and its conditional mathematical expectation given $X = x$ is given by

$$E(\Lambda/X = x) = \int_{-\infty}^{+\infty} \lambda(x, y)\, dG(y/x), \tag{5,2,13}$$

by virtue of (4,2,16). Let $E(\Lambda/X)$ denote the r.v. that assumes the value $E(\Lambda/X = x)$ when X assumes the value x; then by (4,6,18)

$$E(\Lambda) = E[E(\Lambda/X)], \tag{5,2,14}$$

and this formula is easily verified directly. For example, if $\lambda(x, y) = x + y$, $\Lambda = X + Y$, then Theorem (5,2,1) yields (4,2,18):

$$E(X + Y) = E(X) + E(Y).$$

Conditional variance. In general, one can consider conditional moments of Y given X = x, of any desired order, and the conditional characteristic function of Y given X = x. Here we confine the discussion to the conditional variance of Y given X = x. This number depends on the value of x, and we shall denote it by V(x):

$$V(x) = E[|Y - E(Y/X = x)|^2/X = x] = E[|Y - \mu(x)|^2/X = x],$$

or, by (4,2,22),

$$V(x) = E(Y^2/X = x) - |E(Y/X = x)|^2. \qquad (5,2,15)$$

Let V(X) and $E(Y^2/X)$ denote the r.v. that assume the values V(x) and $E(Y^2/X = x)$, respectively, when X assumes the value x. By (4,6,18), or (more precisely) by (5,2,12) applied to Y^2 instead of Y, we have

$$E[E(Y^2/X)] = E(Y^2),$$

and (5,2,13) immediately implies

$$E[V(X)] \leq E(Y^2). \qquad (5,2,16)$$

Now there is no loss of generality in assuming that E(Y) = 0, since otherwise we can always apply the preceding arguments to Y' = Y - E(Y); the m.e. of Y' is zero, while it has the same marginal and conditional variances as Y.

Now if E(Y) = 0, $E(Y^2)$ is the a priori variance of Y, and we have proved:

THEOREM (5,2,2). The mathematical expectation of the conditional variance of Y (given X) is at most equal to the a priori (or marginal) variance of Y. ▮

We shall comment on this result in the sequel.

REMARK (5,2,1). In order to facilitate the formulation of Theorems (5,2,1) and (5,2,2) we have employed the abbreviated terms "conditional expectation and variance of Y given X"; the precise meaning is clear from the context, but the reader should nonetheless remember that it is an abbreviation.

Case of independent X *and* Y

Recall the end of IV.5, formula (4,5,19). If X and Y are independent, then the events $X < x$ and $Y < y$ are independent, for any x and y. Therefore

$$\Pr[(X < x) \cap (Y < y)] = \rho(x, y) = \Pr(X < x) \cdot \Pr(Y < y) = F(x)G(y) \quad (5,2,17)$$

In other words the 2-dimensional d.f. ρ of the 2-dimensional r.v. {X, Y} is the product of two functions: one of x alone, and one of y alone; this is the case of separated variables, already encountered in (3,11,22) and (3,12,17). Conversely, suppose that $\rho(x, y)$ is the product of a function F(x) of x alone and a function G(y) of y alone. Then F(x) and G(y) are necessarily bounded d.f., continuous from the left, such that $F(-\infty) = G(-\infty) = 0$. Now since $\rho(x, y) = F(x)G(y)$, it follows that for every number $a > 0$: $\rho(x, y) = [a F(x)] \cdot \left[\frac{1}{a} G(y)\right]$; thus there is no loss of generality in assuming that $F(+\infty) = G(+\infty) = 1$.

Then, by (5,1,10), F(x) is precisely the marginal d.f. of X and G(y) the marginal d.f. of Y. Let μ and ν be any two subsets of the space R of real numbers. The event $X \in \mu$ is associated with X alone, and the event $Y \in \nu$ with Y alone. Then

$$\Pr[(X \in \mu) \cap (Y \in \nu)] = m(e = \mu \times \nu) = \int\int_{e=\mu\times\nu} d\rho(x, y),$$

and by (3,11,25),

$$\Pr[(X \in \mu) \cap (Y \in \nu)] = \left[\int_\mu dF(x)\right] \cdot \left[\int_\nu dG(y)\right] = \Pr(X \in \mu) \cdot \Pr(Y \in \nu). \quad (5,2,18)$$

Thus X and Y are independent. Note, incidentally, that a sufficient condition for X and Y to be independent is that the events $X < x$ and $Y < y$ be independent for any x and y.

If ρ has a density ρ, so that X and Y have marginal densities f(x) and g(y), then (5,2,17) implies that

$$\rho(x, y) = f(x)\, g(y). \quad (5,2,19)$$

On the other hand, by substituting (5,2,17) into (5,2,2) or (5,2,3) one can show that, *for every* x (except perhaps some "exceptional" values of x, cf. Remark (4,6,2)), the equality

$$G(y/x) = G(y) \qquad (5,2,20)$$

holds for every y. Conversely, if (5,2,20) holds for every x, then (5,2,2) implies (5,2,17): X *and* Y *are independent, if and only if the conditional probability law of* Y, *given* X = x, *is independent of* x, *and therefore identical to the a priori (marginal) probability law of* Y; of course one can invert the roles of X and Y in this statement.

Obviously, if ρ has the density ρ, (5,2,20) is equivalent to the statement: For every x (except perhaps some exceptional values of x), the equality

$$g(y/x) = g(y) \qquad (5,2,21)$$

holds for every y. It is very important to note that, if (5,2,17) holds, then, by virtue of (3,11,26),

$$E(XY) = \int_{-\infty}^{+\infty}\int_{-\infty}^{+\infty} xy \, d\rho(x, y) = \int_{-\infty}^{+\infty}\int_{-\infty}^{+\infty} xy \, dF(x)\, dG(y)$$

$$= \left[\int_{-\infty}^{+\infty} x \, dF(x)\right] \cdot \left[\int_{-\infty}^{+\infty} y \, dG(y)\right] = E(X) \cdot E(Y).$$

THEOREM (5,2,3): If the random variables X and Y are independent, the mathematical expectation E(XY) of their product is the product E(X) × E(Y) of their mathematical expectations. ▌

An immediate consequence is the following: Let X and Y be independent; then if E(X) = E(Y) = 0, E(XY) = $\mu_{1,1}$ = 0. Thus, by the definition (5,1,24) of the correlation coefficient r:

THEOREM (5,2,4). If two random variables are independent, their correlation coefficient, if it exists, is zero. ▌

Another result is the following: The 2-dimensional characteristic function $\phi(u, v)$ of the 2-dimensional r.v. {X, Y} is

$$E\left[e^{i(uX+vY)}\right].$$

If (5,2,17) holds,

$$\begin{aligned}\varphi(u, v) &= \int_{-\infty}^{+\infty}\int_{-\infty}^{+\infty} e^{i(ux+vy)}\, d\rho(x, y) \\ &= \int_{-\infty}^{+\infty}\int_{-\infty}^{+\infty} e^{iux} \cdot e^{ivy}\, dF(x)\, dG(y) \\ &= \left[\int_{-\infty}^{+\infty} e^{iux}\, dF(x)\right] \cdot \left[\int_{-\infty}^{+\infty} e^{ivy}\, dG(y)\right] \\ &= E(e^{iuX}) \cdot E(e^{ivY}).\end{aligned}$$

Then, from III.12 we obtain the following result:

THEOREM (5,2,5). The random variables X and Y are independent if and only if the 2-dimensional characteristic function $E[e^{i(uX+vY)}]$ of the 2-dimensional r.v. {X, Y} is the product of a function a(u) of u alone and a function b(v) of v alone; a(u) and b(v) are proportional to the characteristic functions $E(e^{iuX})$ and $E(e^{ivY})$ of X and Y, respectively. ▮

Recall (cf. IV.5) that, for any functions λ(x) and μ(y), if the r.v. X and Y are independent, then the r.v. λ(X) and μ(Y) are independent.

Comment. The intrinsic study of the 2-dimensional r.v. {X, Y} is in general not uniquely determined by the intrinsic study of the random variables X and Y separately. In other words, it is not sufficient to know the d.f. F(x), G(y) of X and Y, respectively, in order to determine the 2-dimensional d.f. ρ(x, y) of the pair {X, Y}. This is clear, for example, from (5,2,2), which shows that if we know, say, F(x), we need not the a priori d.f. G(Y) of Y but its conditional d.f. G(y/x), given X = x, for every x.

Only when X and Y are independent do F(x) and G(y) determine ρ(x, y). We see that two random variables X and Y can stand in either of the following relations to each other:

1) They may be *independent*, i.e. without relations, at least from the viewpoint of Probability Theory.

2) They may be *functions* of each other, or stand in a *functional relation*. For example, Y might be a function $\lambda(X)$ of X, which means that Y assumes the value $\lambda(x)$ almost certainly if X assumes the value x. Note that this is clearly incompatible with the assumption that ρ has a (2-dimensional) density ρ. Since the standard deviation of a certain or almost certain number is zero, a necessary and sufficient condition for Y to be a function of X is that the conditional variance of Y, given $X = x$, be identically zero as a function of x.

3) In the general (intermediate) case, midway between the above two extremes, X and Y are related, but not by a functional relation. We say that they are *stochastically related*, or *correlated*. It is clear that this correlation is stronger when closer to a functional relation, weaker when closer to independence.

Consideration of regression functions may be of use in this respect. When X and Y are independent, the regression function $y = \mu(x)$ of Y on X, defined by (5,2,11), clearly reduces to a constant; by symmetry, this is also the case for the regression function of X on Y. It may sometimes happen that $y = \mu(x)$ is a constant when X and Y are not independent. However, as a general rule, we can say that when X and Y are dependent, the regression function of Y on X describes this dependence qualitatively. Note that, in general, the regression curves of Y on X and of X on Y, plotted in the same coordinate system, are completely different.

But if X and Y stand in a functional relation, i.e. if $Y = \lambda(X)$ almost certainly, where $\lambda(x)$ is a certain function, it is clear that both regression curves coincide with the curve

$$y = \mu(x) = \lambda(x).$$

Hence, if the regression curves of Y on X and of X on Y are very close to each other, but not identical, one can assume that the relation between X and Y is very strict, almost functional.

Now consider the correlation coefficient r of X and Y, or rather its absolute value $|r|$. Let us recall its properties.

1) $|r|$ is always between 0 and 1.

2) If X and Y are independent, $r = 0$. On the other hand, $r = 0$ does not necessarily imply that X and Y are independent; it only implies that the axes of symmetry of the central inertia ellipse are parallel to the basis vectors.

3) If $|r| = 1$, there is a linear functional relation between X and Y (5,1,21); but, for every η such that $|\eta| < 1$, there exist pairs {X, Y} of r.v. X and Y, whose correlation coefficient r is η, but which are functionally related; this relation is clearly nonlinear.

These are the *rigorous* properties of the correlation coefficient r. Accordingly, one observes, in many practical applications, that $|r|$ is greater, the stronger the correlation; very small values of $|r|$ correspond to a situation close to independence, while values of $|r|$ very close to 1 correspond to a situation close to functional dependence. Later, in V.5, we shall see that this is due, in particular, to the fact that one often deals with 2-dimensional r.v. {X, Y} which are normally distributed or almost normally distributed. The correlation coefficient is therefore often used as an estimate of the degree of correlation.

Nevertheless, this cannot be done with absolute confidence, since $|r|$ is less than 1, possibly even very small, for a nonlinear functional relation, and $|r|$ may vanish in the dependent case. In any event, the idea of correlation is in itself purely qualitative, as is that of dispersion; one meets the same difficulties as for dispersion in trying to characterize it by a unique numerical parameter.

Estimation in the sense of minimum mean-square deviation

An interesting way to study the conditional behavior of a r.v. Y with respect to a r.v. X is the following. Suppose we measure dispersion, systematically, by 2nd moments, or, equivalently, by root-mean-square deviation. We saw in IV.2 that, provided the 2nd moments exist, this convention is both legitimate and convenient. Of course, it is still somewhat arbitrary; nevertheless, we mentioned in IV.2 (and we shall come back to this later) that at least for certain types of physical applications, it becomes quite natural and completely justified. For the moment, we adopt the convention without further discussion.

It naturally implies that we consider a r.v. X to be less "different" from a certain number a, the smaller $E(|X - a|^2)$, and, in general, that we consider two r.v. X and Y to be less "different," the smaller $E(|X - Y|^2)$.

In Chapter IV we saw that, according to this point of view, the best estimate of the actual value of the r.v. X, i.e. the value which minimizes $E(|X - a|^2)$, is precisely the m.e. E(X).

Let us consider a 2-dimensional r.v. {X, Y}, assuming that we have an experimental method for measuring the value assumed by X, but not that assumed by Y. Suppose we wish to determine the best estimate $\hat{y}$ of the value assumed by Y, *given the value assumed by* X. There are two ways to formulate this problem:

PROBLEM 1. Assume that the experiment has been performed and the value x assumed by X measured; we must then regard Y as a r.v. whose probability law is defined by the conditional d.f. G(y/x) of Y given X = x. It follows from the preceding remark that the best estimate $\hat{y}$ of Y is the number

$$\mu(x) = E(Y/X = x),$$

already encountered in (5,2,11). The probable order of magnitude of the error involved in this procedure is the conditional standard deviation of Y given X = x. This means that the quality of the estimate is measured by the conditional variance V(x) of Y given X = x, as defined by (5,2,15).

PROBLEM 2. Assume the experiment has not yet been performed, so that the value x assumed by X is not yet known. We can choose a function $\lambda(x)$ of x and agree to take $\hat{y} = \lambda(x)$ as an estimate of Y if the value assumed by X is x; let us call $\lambda(x)$ a conditional estimator with respect to X. Consider the r.v.

$$T = Y - \lambda(X).$$

It is clear that the best choice for the function $\lambda(x)$ is that which minimizes the "a priori" 2nd moment $E(T^2)$. This implies, of course, that the optimal function $\lambda$is being sought in the class of functions λ such that $E[\lambda(X)^2] < +\infty$. Assuming throughout that $E(X^2)$ and $E(Y^2)$ exist, i.e. $E(X^2) < +\infty$ and $E(Y^2) < +\infty$, note that since necessarily $V(X) \geqq 0$, formula (5,2,15) implies that

$$\mu(x)^2 \leqq E(Y^2/X = x),$$

so that

$$E[\mu(X)^2] \leqq E[E(Y^2/X)] = E(Y^2) < +\infty. \tag{5,2,22}$$

Now, to compute the moment $E(T^2)$ let us write T in the form

$$T = [Y - \mu\ X)] + [\mu(X) - \lambda(X)].$$

We first compute the conditional moment $E(T^2/X = x)$, using Theorem (4,2,1) and noting that

1) $E[Y - \mu(X)/X = x] = 0$;

2) $E[|Y - \mu(x)|^2/X = x] = V(x)$;

3) given X = x (i.e., conditionally), $\mu(X) - \lambda(X) = \mu(x) - \lambda(x$

is a certain number.

Thus

$$E[T^2/X = x] = V(x) + |\mu(x) - \lambda(x)|^2.$$

From Theorem (5,2,1) it is clear that $E(T^2)$ is a minimum when

$$\lambda(x) = \mu(x). \qquad (5,2,23)$$

By (5,2,22), this solution is admissible; therefore,

$$\text{minimum } E(T^2) = E[V(X)]. \qquad (5,2,24)$$

Equality (5,2,23) means that the best conditional estimator is precisely the regression function $y = \mu(x)$: the regression function thus acquires an augmented and more profound significance. On the other hand, formula (5,2,24) yields the following remark. following remark.

The best estimate of Y, a priori, i.e. without considering the value of X, is E(Y), and its quality or precision is measured by the a priori variance σ_Y^2 of Y. Intuitively, it seems clear that the best conditional estimate of Y given the value x assumed by X, whose quality is measured by V(x), should be more precise, at least not less precise, than the a priori estimate, since it takes more information into consideration; in other words,

$$V(x) \leqq \sigma_Y^2. \qquad (5,2,25)$$

However, it is easy to verify by means of examples that (5,2,25) may hold for some values of x, *but not for others*. Our intuition is therefore wrong.

Now Theorem (5,2,2) states that

$$E[V(X)] \leqslant \sigma_Y^2.$$

By using (5,2,24), this can be interpreted as follows: The conditional estimate of Y is not more precise than the a priori estimate for all values of x assumed by X. However, assuming that the value of X is not yet known and is therefore random, we can say that the conditional estimate of Y will *probably* be more precise than its a priori estimate; i.e., in non-mathematical, but nevertheless apt terms:

THEOREM 5.2.6. "On the average," the conditional estimate of Y is more precise than its a priori estimate.

The best conditional linear estimate

Let us return to Problem 2, whose solution is given by (5,2,23). Unfortunately, the function $\mu(x)$ is often complicated, unfamiliar, or even completely unknown. It may therefore be worthwhile to adopt a conditional estimator $\lambda(x)$, different from $\mu(x)$ but easier to handle. The estimate provided by this function will certainly be less precise than that provided by $\mu(x)$, but if the difference in precision is not too great it may well be compensated for by the resulting convenience.

The simplest conceivable function is, of course, the linear function

$$\lambda(x) = ax + b, \qquad (5,2,26)$$

where a and b are two constants. We thus have the following problem: what values $\hat{a}$ and $\hat{b}$ must be assigned to the coefficients a and b in order that the function

$$\hat{\lambda}(x) = \hat{a}x + \hat{b}$$

be the best conditional estimator of Y? Let

$$T = Y - (aX + b);$$

then $\hat{a}$ and $\hat{b}$ are the values of a and b that minimize $E(T^2)$. Now

$$E(T^2) = E[Y^2 - 2(aXY + bY) + (a^2X^2 + 2abX + b^2)].$$

By Theorem (4,2,1),

$$E(T^2) = E(Y^2) - 2aE(XY) - 2bE(Y) + a^2E(X^2) + 2abE(X) + b^2.$$

Without loss of generality, we assume that $E(X) = E(Y) = 0$. Then $E(X^2)$ is the variance σ_X^2 of X, and $E(Y^2)$ is the variance σ_Y^2 of Y; introducing the correlation coefficient r of X and Y, we can write $E(XY)$ as $\sigma_X\sigma_Y r$; therefore,

$$E(T^2) = \sigma_Y^2 - 2a\,\sigma_X\sigma_Y\, r + a^2\sigma_X^2 + b^2 ,$$

which is a minimum when

$$a = \hat{a} = \frac{\sigma_Y}{\sigma_X} r, \quad b = \hat{b} = 0.$$

Thus, the best linear conditional estimator of Y is

$$\hat{\lambda}(x) = \frac{\sigma_Y}{\sigma_X} r\, x .$$

If E(X) and E(Y) are not necessarily zero, then

$$\hat{\lambda}(x) = E(Y) + \frac{\sigma_Y}{\sigma_X} r[x - E(X)]. \qquad (5,2,27)$$

This result, together with Theorem (5,1,1), throws light on the precise meaning of the linear correlation coefficient, clearly displaying its possible merits and demerits as a measure of degree of correlation.

We shall return to the ideas and results of this section, in greater generality, in V.4, but we proceed first to other matters.

3. *Intrinsic study of an* n*-dimensional random vector*

To simplify the notation, we take $n = 2$ and consider the intrinsic study of a random vector $\vec{Z}$ (or Z) in $\mathfrak{X}_2$, i.e. independently of any definite reference system in $\mathfrak{X}_2$. Let m(e) be the measure on $\mathfrak{X}_2$ which defines the probability law of Z. We refer the reader to III.12 for the following concepts.

We shall call the mean vector of the distribution m, when it exists, the *mathematical expectation* of Z, denoted by E(Z). We know that if this vector exists, it is unique; in the notation of III.12, it is defined by (3,12,20), with M = 1. With every linear functional z^* on $\mathfrak{X}_2$, i.e. every element z^* of the dual space $\mathfrak{X}_2^*$ of $\mathfrak{X}_2$, we associate the r.v. $< z^*, Z >$. Thus, the integral in (3,12,20) is

$$\int_{\mathfrak{X}_2} < z^*, z > m(dz) = E(< z^*, Z >) , \qquad (5,3,1)$$

and E(Z) is defined by the equality

$$< z^*, E(Z) > \equiv E(< z^*, Z >) \quad \text{for all} \quad z^* \in \mathfrak{X}_2^* . \qquad (5,3,2)$$

The function $\xi(z^*)$ of $z^* \in \mathfrak{X}_2^*$ defined below will be called the *characteristic functional* of Z; $\xi(z^*)$ is the functional Fourier transform of m, i.e. the function $\xi(z^*)$ defined by (3,12,21):

$$\xi(z^*) = \int_{\mathfrak{X}_2} e^{i<z^*, z>} m(dz) , \qquad (5,3,3)$$

or

$$\xi(z^*) = E(e^{i<z^*, Z>}) . \qquad (5,3,4)$$

In other words, for every z^*, $\xi(z^*)$ is precisely the characteristic function of the r.v. $< z^*, Z >$.

Now consider some basis **b** in $\mathfrak{X}_2$ and its dual **b*** in $\mathfrak{X}_2^*$; call {X, Y} the components of Z relative to **b** and {u, v} those of z^* relative to **b***. Then $< z^*, Z > = uX + vY$, and therefore:

1) The components of E(Z) relative to **b** are the mathematical expectations E(X), E(Y) of the r.v. X, Y respectively. For E(Z) to exist it is necessary and sufficient that there exist two linearly independent functionals z_1^* and z_2^* such that the m.e. $E(< z_1^*, Z >)$ and $E(< z_2^*, Z>)$ exist.

2) From the analytic standpoint, $\xi(z^*)$ is the 2-dimensional Fourier transform

$$\varphi(u , v) = \int_{-\infty}^{+\infty} \int_{-\infty}^{+\infty} e^{i(ux+vy)} dF(x , y) \qquad (5,3,5)$$

of the 2-dimensional distribution function F(x, y) of m relative to the basis **b** (cf. (3,12,22)).

Consider a random vector Z in a real n-dimensional vector space $\mathfrak{X}_n$ and a random vector T in a real q-dimensional vector space $\mathfrak{Y}_q$. The pair (Z, T) can be regarded as a random vector in the (n + q)-dimensional vector space $\mathfrak{X}_n \times \mathfrak{Y}_q$, the cartesian product of $\mathfrak{X}_n$ and $\mathfrak{Y}_q$. Let $\mathfrak{X}_n^*$, $\mathfrak{Y}_q^*$ denote the dual spaces of $\mathfrak{X}_n$, $\mathfrak{Y}_q$ respectively, z^* an arbitrary linear functional on $\mathfrak{X}_n (z^* \in \mathfrak{X}_n^*)$, and t^* an arbitrary linear functional on $\mathfrak{Y}_q (t^* \in \mathfrak{Y}^*)$.

$\mathfrak{X}_n \times \mathfrak{Y}_q$ is the set of all pairs (z, t) consisting of an element z of $\mathfrak{X}_n$ and an element t of $\mathfrak{Y}_q$. We know that the most general linear functional on $\mathfrak{X}_n \times \mathfrak{Y}_q$ has the form

$$< z^*, z > + < t^*, t >.$$

The characteristic functional of the pair (Z, T), as a random (n + q)-dimensional vector in $\mathfrak{X}_n \times \mathfrak{Y}_q$, can therefore be regarded as a function $\eta(z^*, t^*)$ of the two variables z^* and t^* defined by

$$\eta(z^*, t^*) = E[\exp. \{i < z^*, Z > + i < t^*, T >\}]. \qquad (5,3,6)$$

An immediate extension of Theorem (5,2,5) is:

THEOREM (5,3,1). Two random vectors Z and T are independent if and only if the characteristic functional $\eta(z^*, t^*)$ of the pair (Z, T) is the product of a function $\xi(z^*)$ of z^* alone and a function $\zeta(t^*)$ of t^* alone; $\xi(z^*)$ and $\zeta(t^*)$ are then proportional to the characteristic functionals of Z and T, respectively. ▌

Now let λ be a linear mapping of $\mathfrak{X}_n$ into $\mathfrak{Y}_q$, and set $T = \lambda(Z)$. Let $\xi(z^*)$ and $\zeta(t^*)$ denote the characteristic functionals of Z and T, respectively. Then, letting λ^* denote the adjoint linear mapping of λ, we have

$$< t^*, \lambda(z) > \equiv < \lambda^*(t^*), z > \qquad (5,3,7)$$

for all $z \in \mathfrak{X}_n$ and $t^* \in \mathfrak{Y}_q^*$. Thus

$$\zeta(t^*) = \xi(\lambda^*(t^*)). \qquad (5,3,8)$$

An interesting application of (5,3,8) is the following. Assume that the probability law of Z, i.e. the measure m(e) (or the d.f.

F, if a basis $\mathfrak{b}$ has been chosen in $\mathfrak{X}_n$), tends to a limit m_0, which is the probability law of a nondegenerate random vector Z_0 in $\mathfrak{X}_n$. It then follows from Theorem (3,12,1) that $\xi(z^*)$ tends to the characteristic functional $\xi_0(z^*)$ of Z_0 (or of m_0), and the convergence is uniform in z^*. It then follows from (5,3,8) that $\zeta(t^*)$ tends uniformly in t^* to $\xi_0(\lambda^*(t^*))$, which is the characteristic functional of $\lambda(Z_0)$. By Theorem (3,12,1), we then have:

THEOREM (5,3,2). Let λ be a linear mapping of $\mathfrak{X}_n$ into $\mathfrak{Y}_q$; if the probability law of the random vector $Z \in \mathfrak{X}_n$ tends to the probability law of a random vector $Z_0 \in \mathfrak{X}_n$, then the probability law of $T = \lambda(Z) \in \mathfrak{Y}_q$ tends to that of $T_0 = \lambda(Z_0) \in \mathfrak{Y}_q$.

4. *Complex random variables*

We have already considered (IV.3) the case of a random element with values in the space C of complex numbers, i.e. a complex random variable Z. Denote the real part of Z by X and its imaginary part by Y, so that

$$Z = X + iY.$$

It is immediately clear that the study of Z is equivalent to that of the *real* 2-dimensional r.v. {X, Y}. We have already defined ((4,3,1) or (4,3,2)) the m.e. E(Z) of Z. Is it possible to introduce the concept of a moment of arbitrary order of Z?

We first consider the concept of the αth algebraic moment (α a positive integer) about the origin. The set of αth moments of the 2-dimensional r.v. {X, Y} is given by the $\alpha + 1$ moments

$$E(X^{\alpha}),\ E(X^{\alpha-1}Y),\ E(X^{\alpha-2}Y^2), \ldots,\ E(X^{\alpha-\beta}Y^{\beta}), \ldots,\ E(Y^{\alpha}). \quad (5,4,1)$$

The αth moments of Z should comprise a set of numbers equivalent to the set (5,4,1) (i.e. a set of numbers that uniquely determines (5,4,1) and vice versa), but expressed, as far as possible, in purely complex notation. Now, the set of $\alpha + 1$ complex numbers

$$E(Z^{\alpha}),\ E(Z^{\alpha-1}\bar{Z}),\ E(Z^{\alpha-2}\bar{Z}^2), \ldots,\ E(Z^{\alpha-\beta}\bar{Z}^{\beta}), \ldots,\ E(\bar{Z}^{\alpha}) \quad (5,4,2)$$

clearly satisfies these conditions. We define the numbers (5,4,2) to be the αth algebraic moments of Z, more precisely, the αth moments about the origin. The αth algebraic moments of Z about an arbitrary complex number z are clearly defined by

$$E[(Z - z)^{\alpha-\beta}\,(\overline{Z - z})^{\beta}] \quad (\beta = 0, 1, 2, \ldots, \alpha). \tag{5,4,3}$$

In particular, consider the 2nd moments about the origin. One of these is

$$E(Z\,\overline{Z}) = E(|Z|^2)\,, \tag{5,4,4}$$

which is precisely the 2nd moment of the real (nonnegative) r.v. $|Z|$. By the inequality (4,2,21),

$$\Pr(|Z| > \lambda) \leqslant \frac{E(|Z|^2)}{\lambda^2} \quad (\lambda > 0)\,. \tag{5,4,5}$$

More generally, if α is any positive number, and the αth moment $E(|Z|^{\alpha})$ of the (real and nonegative) r.v. $|Z|$ exists, then by (4,2,25) we have

$$\Pr(|Z| > \lambda) \leqslant \frac{E(|Z|^{\alpha})}{\lambda^{\alpha}} \quad (\lambda > 0). \tag{5,4,6}$$

This justifies the following remark. For any real number $\alpha > 0$, any complex number z, and complex r.v. Z, we define only one α^{th} *absolute moment* of Z about z, as the m.e.

$$E(|Z - z|^{\alpha})$$

of $|Z - z|^{\alpha}$.

Of course, the algebraic or absolute moments do not necessarily exist, but it is clear that the problem of their existence reduces to that of the moments of the real 1-dimensional r.v. $|Z|$, or those of the real 2-dimensional r.v. $\{X, Y\}$. Using, among other things, Theorem (2,7,5), we see that an αth algebraic moment of Z exists if and only if the αth absolute moment of Z exists. If Z has an αth absolute moment ($\alpha > 0$), all α'^{th} algebraic or absolute moments of Z exist, for $\alpha' \leqq \alpha$.

Independence of complex random variables

Let Z_1 and Z_2 denote two complex r.v. with real parts X_1, X_2 and imaginary parts Y_1, Y_2:

$$Z_1 = X_1 + i\,Y_1\,,\; Z_2 = X_2 + i\,Y_2\,.$$

It is clear that Z_1 and Z_2 are independent if and only if the real 2-dimensional r.v. $\{X_1, Y_1\}$ and $\{X_2, Y_2\}$ are independent (see IV.5). Now, since

$$Z_1 Z_2 = (X_1 X_2 - Y_1 Y_2) + i(X_1 Y_2 + Y_1 X_2),$$

it follows that

$$E(Z_1 Z_2) = [E(X_1 X_2) - E(Y_1 Y_2)] + i[E(X_1 Y_2) + E(Y_1 X_2)]\,.$$

If Z_1 and Z_2 are independent, then so are X_1 and X_2, X_1 and Y_2, Y_1 and X_2, and Y_1 and Y_2. It then follows from Theorem (5,2,3) that

$$\begin{aligned} E(Z_1 Z_2) &= [E(X_1)\,E(X_2) - E(Y_1)\,E(Y_2)] + i[E(X_1)\,E(Y_2) + E(Y_1)\,E(X_2)] \\ &= [E(X_1) + i\,E(Y_1)] \cdot [E(X_2) + i\,E(Y_2)] \\ &= E(Z_1) \cdot E(Z_2)\,. \end{aligned}$$

Theorem (5,2,3) may therefore be generalized to complex r.v., and it states that if two complex r.v. Z_1 and Z_2 are independent, the mathematical expectation $E(Z_1 Z_2)$ of their product is the product of their mathematical expectations $E(Z_1)$ and $E(Z_2)$.

Hilbert space of second-order complex random variables

A complex random variable X is of *second order* if it possesses a 2nd moment, i.e.,

$$E(|X|^2) < +\infty\,. \qquad (5,4,7)$$

This definition is applicable, in particular, when X is a real r.v., i.e. a complex r.v. with an imaginary part which vanishes certainly (or almost certainly). From the preceding discussion it follows that every second-order r.v. X has a m.e. E(X). Using Theorem (2,7,5), we also have

$$|E(X)|^2 \leqq [E(|X|)]^2 \leqq E(|X|^2). \tag{5,4,8}$$

X-E(X) is also a second-order r.v.; we shall call

$$E(|X - E(X)|^2) = \mathfrak{V}(X) = \sigma_x^2 \tag{5,4,9}$$

the *variance* of X, and $\sigma_x = \sqrt{\mathfrak{V}(X)}$ the *standard deviation* of X.

Let a be any number; from the equalities $E(\bar{X}) = \overline{E(X)}$ and

$$|X - a|^2 = |[X - E(X)] + [E(X) - a]|^2$$
$$= \{[X - E(X)] + [E(X) - a]\} \cdot \{[\bar{X} - E(\bar{X})] + [E(\bar{X}) - \bar{a}]\}$$

it follows that

$$E(|X - a|^2) = \sigma_x^2 + |E(X) - a|^2, \tag{5,4,10}$$

which extends (4,2,22) to complex r.v.

REMARK (5,4,1). Every real normally distributed r.v. X is a second-order r.v.; if X is a degenerate normally distributed r.v., its variance is zero. ▮

Let $\mathcal{H}$ denote the set (space) of all second-order r.v. associated with a given class of trials $\mathfrak{U}$ with elements u. Let p(E) denote the probability law (measure or mass distribution) over $\mathfrak{U}$, where $E \subset \mathfrak{U}$ is an event (arbitrary subset) of $\mathfrak{U}$. Then a complex r.v. X is a mapping $x = x(u)$ of $\mathfrak{U}$ into the space C of complex numbers.

It thus follows from (4,2,6) and its generalization (4,3,2) that

$$E(|X|^\alpha) = \int_{\mathfrak{U}} |x(u)|^\alpha \, p(du),$$

$$E(X^{\alpha-\beta} \bar{X}^\beta) = \int_{\mathfrak{U}} x(u)^{\alpha-\beta} \, \overline{x(u)}^\beta \, p(du),$$

and so on. We see, in particular, that X is a second-order r.v. if and only if

$$\int_{\mathfrak{U}} |x(u)|^2 \, p(du) < +\infty \,. \qquad (5,4,11)$$

Therefore, the space $\mathcal{H}$ of second-order r.v. is precisely the Hilbert space of square-integrable functions over $\mathfrak{U}$ (with the measure p(E); c.f. Remark (2,10,3)). In this space $\mathcal{H}$, which is in general not separable, the Hermitian product of an element x (x(u)) and an element y (y(u)) is defined by

$$\int_{\mathfrak{U}} x(u) \, \overline{y(u)} \, p(du) \,. \qquad (5,4,12)$$

If we regard the functions or mappings x = x(u), y = y(u) of $\mathfrak{U}$ into C as second-order r.v. X, Y and interpret (5,4,12) as the m.e. $E(X\overline{Y})$ of the r.v. $X\overline{Y}$, it follows that the space $\mathcal{H}$ of second-order r.v. is a Hilbert space, in which the Hermitian product of elements $X \in \mathcal{H}$ and $Y \in \mathcal{H}$ is defined by $E(X\overline{Y})$.

If X is a second-order r.v., then every r.v. Y equivalent to X, i.e. such that X = Y a.e. (cf. II.12), is also of second order; as an element of $\mathcal{H}$ it is identical with X.

We now refer the reader to II.10, which is devoted to Hilbert spaces. The concepts and results of II.10 are therefore applicable to $\mathcal{H}$. We call the mathematical expectation of the product $X\overline{Y}$, i.e. the hermitian product of X and Y as elements of the Hilbert space $\mathcal{H}$, the *covariance* of the ordered pair (X, Y) of the two second-order r.v. X and Y.

For real r.v. we explained in IV.2 how, at least for a certain type of physical applications, it was not only natural and convenient, but also legitimate to measure the difference between two random variables X and Y by the 2nd moment $E[(X - Y)^2]$; these physical applications often involve complex quantities and the result is a complex random variable. In the case of a complex quantity, our interpretation of a squared real quantity as energy carries over to the square $|Z|^2$ of its absolute value $|Z|$. It is then convenient to measure the difference between two complex r.v. X and Y by the 2nd absolute moment $E(|X - Y|^2)$. Therefore, in the case of two second-order r.v., $E(|X - Y|^2)$ is the square of the norm of X - Y in the Hilbert space $\mathcal{H}$. Thus, in the space $\mathcal{H}$ of second-order r.v., the topology induced by our physical arguments coincides with the Hilbert-space topology, which was defined on $\mathcal{H}$ by purely mathematical means. This harmonious encounter of physical necessity and the appropriate mathematical tool is noteworthy.

In V.2, in considering the case of two real r.v. X and Y with E(X) = 0 and E(Y) = 0, we associated the correlation coefficient r of X and Y

$$r = \frac{E(X\,Y)}{\sqrt{E(|X|^2) \cdot E(|Y|^2)}}$$

with the covariance E(XY). We saw that $|r| \leqq 1$, and r has properties which make it, to some extent, a measure of the degree of mutual dependence of X and Y. We then encountered r again in a more natural way, and with a better defined meaning, in solving the problem of the best linear conditional estimate (cf. (5,2,27)). How do these considerations carry over to the case of two second-order *complex* r.v. X and Y?

First, we must distinguish between the covariance $E(X\bar{Y})$ of the ordered pair (X, Y) and the covariance $E(Y\bar{X})$ of the ordered pair (Y, X); these two covariances are not equal, but complex conjugates of each other. Assuming that E(X) = E(Y) = 0, we might define the "correlation coefficient of Y with respect to X" as the complex number

$$r = \frac{E(Y\,\bar{X})}{\sqrt{E(|X|^2) \cdot E(|Y|^2)}} . \qquad (5,4,13)$$

The correlation coefficient of X with respect to Y is then the complex conjugate $\bar{r}$ of r.

It follows immediately from Schwarz's inequality in $\mathcal{H}$ that the coefficient r defined by (5,4,22) satisfies the inequality

$$|r| \leqslant 1,$$

and $|r| = 1$ if and only if X and Y are "parallel" as vectors of $\mathcal{H}$, i.e. in probability-theoretic terms, if there exist two certain numbers a and b such that, almost certainly,

$$a X + b Y = 0.$$

If X and Y are independent it is clear that $r = 0$. The converse is false; $r = 0$ implies only that, as vectors of $\mathcal{H}$, X and Y are *orthogonal*.

The covariance matrix

For any natural number n, let

$$X = \{X_1, X_2, \ldots, X_n\}$$

be an n-tuple of arbitrary second-order r.v. X_k; in particular, we are not assuming that $E(X_k) = 0$. We might call X a second-order n-dimensional r.v. One can also interpret X as an $n \times 1$ random *matrix* with elements X_k.

Let γ_{jk} denote the covariance of the ordered pair (X_j, X_k) $(j, k = 1, \ldots, n)$:

$$\gamma_{jk} = E(X_j \overline{X}_k) ;$$

Let Γ be the square $n \times n$ matrix with elements γ_{jk}. We call Γ the *covariance matrix* of the second-order n-dimensional r.v. $X = \{X_1, \ldots, X_n\}$.

It is clear that $\gamma_{jk} = \overline{\gamma}_{kj}$, and so Γ is a hermitian matrix (cf. III.5). Now let $u_1, \ldots, u_n$ be n arbitrary complex numbers, and Z the second-order r.v. defined by

$$Z = \sum_j u_j X_j .$$

Then

$$|Z|^2 = Z\bar{Z} = \left(\sum_j u_j X_j\right)\left(\sum_k \bar{u}_k \bar{X}_k\right)$$

$$= \sum_{jk} X_j \bar{X}_k u_j \bar{u}_k \geqslant 0 .$$

Thus

$$E(|Z|^2) = E\left(\sum_{jk} X_j \bar{X}_k u_j \bar{u}_k\right)$$

$$= \sum_{jk} E(X_j \bar{X}_k) u_j \bar{u}_k \qquad (5,4,14)$$

$$= \sum_{jk} \gamma_{jk} u_j \bar{u}_k \geqslant 0 .$$

Now (5,4,14) is the hermitian form associated with Γ, and if we let u denote the $n \times 1$ matrix with elements u_j, it can be expressed in matrix notation (cf. III.5) as

$$\sum_{jk} \gamma_{jk} u_j \bar{u}_k = {}^{\circ}u \, \Gamma \, u , \qquad (5,4,15)$$

while Z can be expressed in matrix notation as

$$Z = {}^{\circ}u X .$$

Thus Γ is defined by the following identity in u:

$$E(|{}^{\circ}u X|^2) = {}^{\circ}u \, \Gamma \, u .$$

From this we immediately deduce the following theorem:

THEOREM (5,4,1). The covariance matrix of a second-order n-dimensional random variable is hermitian and positive-definite. ▮

Let T be a square $n \times n$ nonsingular matrix, and consider the second-order n-dimensional r.v. (or $n \times 1$ random matrix) Y =

$\{Y_1, \ldots, Y_n\}$ defined by

$$X = T\,Y \,. \tag{5,4,16}$$

If v is any $n \times 1$ matrix, the covariance matrix Γ' of Y is defined by the following identity in v:

$$E(|{}^\circ v\, Y|^2) = {}^\circ v\, \Gamma' v \,.$$

Now if $v = {}^\circ T\, u$, i.e. ${}^\circ v = {}^\circ u\, T$, we have ${}^\circ v\, Y = {}^\circ u\, X$, and therefore

$${}^\circ v\, \Gamma'\, v = {}^\circ v\, T^{-1}\, \Gamma\, {}^\circ T^{-1}\, v,$$

so that

$$\Gamma' = T^{-1}\Gamma\, {}^\circ T^{-1}, \quad \Gamma = T\,\Gamma'\, {}^\circ T. \tag{5,4,17}$$

Now let ${}^\circ T^{-1}$ be a matrix which diagonalizes Γ (cf. III.5); then Γ' is diagonal (cf. III.5), and its elements γ'_{jk} have the following properties:

$$\gamma'_{jk} = E(Y_j\, \overline{Y}_k) = 0 \text{ if } j \neq k \; (j,\, k = 1\,,2\,,\ldots,\, n)\;; \tag{5,4,18}$$

$$\gamma'_{jk} = E(|Y_h|^2) \text{ real } \& \geqslant 0 \quad (h = 1,\, 2,\ldots,\, n). \tag{5,4,19}$$

Formula (5,4,18) means that, as elements of $\mathcal{H}$, the Y_h are pairwise orthogonal.

Suppose, in particular, that ${}^\circ T^{-1}$ is a unitary diagonalizing matrix for Γ (we know that at least one such matrix exists). Then the numbers $\gamma'_{hh} = E(Y_h^2)$ are the characteristic roots of Γ.

Given a positive-definite $n \times n$ hermitian matrix Γ with characteristic roots $s_1, \ldots, s_n$ (not necessarily distinct), we can always find (even in many different ways) a second-order n-dimensional r.v. $Y = \{Y_1, \ldots, Y_n\}$ such that

$$E(Y_j\, \overline{Y}_k) = 0 \text{ for } j \neq k\,,\; E(|Y_h|^2) = s_h\,.$$

For example, consider real and mutually independent Y_h, with $E(Y_h) = 0$ and $E(Y_h^2) = s_h$. If T is a unitary diagonalizing matrix for Γ, then Γ is the covariance matrix of the n-dimensional r.v. $X = \{X_1, \ldots, X_n\}$ defined by

$$X = T\,Y.$$

Thus:

THEOREM(5,4,2). An $n \times n$ matrix is the covariance matrix of at least one second-order n-dimensional r.v. if and only if it is Hermitian and positive-definite. ∎

If ρ is the rank of Γ, we shall also call ρ the rank of the second-order n-dimensional r.v. X; $\rho < n$ if and only if there exists a nonzero $n \times 1$ matrix u such that $E(|{}^{\circ}uX|^2) = 0$. Using this and the preceding arguments, we easily deduce the following statements.

1) ρ is also the rank of $\{X_1, \ldots, X_n\}$ as an n-tuple of vectors in $\mathcal{H}$; if $\mathcal{H}'$ is the Hilbert subspace of $\mathcal{H}$ generated by the (finite) family of the X_h (cf. III.13), i.e., the set of all linear combinations (with certain coefficients) of the X_h, $\mathcal{H}'$ is of dimension ρ.

2) There exist $n - \rho$ homogeneous and independent linear relations among the X_h, which hold almost certainly.

3) There are at most $n - \rho$ homogeneous and independent linear relations among the X_h which hold almost certainly.

Best linear estimate. Let us return to the problem of the best linear estimate, formulated and solved in V.2, but under the following, more general conditions.

Note first that every complex r.v. X which reduces to a certain or almost certain number x is a second-order r.v. Let $\mathcal{H}_0$ be the subset of $\mathcal{H}$ consisting of all second-order r.v. which reduce to almost certain numbers. It is clear that $\mathcal{H}_0$ is a linear subspace of $\mathcal{H}$, more precisely, it is a 1-dimensional linear subspace; for if X is a nonzero element of $\mathcal{H}_0$, i.e. X is almost certainly equal to a nonzero number x, and Y any element of $\mathcal{H}_0$, i.e. Y is almost certainly equal to a number y, there exists exactly one scalar (number) ρ such that

$$Y = \rho X \text{ almost certainly;}$$

this scalar is $\rho = y/x$.

$\mathcal{H}_0$ is thus *closed* in the Hilbert-space topology of $\mathcal{H}$, and is therefore a *Hilbert subspace* of $\mathcal{H}$:

THEOREM (5,4,3). The almost certain numbers constitute a 1-dimensional Hilbert subspace of the space $\mathcal{H}$ of second-order random variables. ∎

Let X be a second-order r.v. and b a certain number. We require a value $\hat{b}$ of b which minimizes $E(|X - b|^2)$. In $\mathcal{H}$, this reduces to the following problem: among the elements b of $\mathcal{H}_0$, to find an element $\hat{b}$ which is the orthogonal projection of X on $\mathcal{H}_0$; $\hat{b}$ is thus determined by the conditions:

1) $\hat{b} \in \mathcal{H}_0$, 2) $(X - \hat{b})$ is orthogonal in $\mathcal{H}$ to any element b of $\mathcal{H}_0$.
This clearly implies

$$\hat{b} = E(X), \tag{5,4,20}$$

thus extending the results of IV.2 for real r.v. to complex r.v.

Now let X and Y be two second-order r.v. What values $\hat{a}$ and $\hat{b}$ must be assigned to the complex numbers a and b in order to minimize $E[|Y - (aX + b)|^2]$?

The previous reasoning easily implies that the set of r.v. $aX + b$, where a and b are arbitrary given numbers, is a Hilbert subspace $\mathcal{H}'$ of $\mathcal{H}$, of dimension at most 2. $\hat{a}X + \hat{b}$ is the orthogonal projection of Y on $\mathcal{H}'$; i.e. $Y - (\hat{a}X + \hat{b})$ is orthogonal in $\mathcal{H}$ to all elements of $\mathcal{H}'$; $\hat{a}$ and $\hat{b}$ are thus determined by the condition:

$$E\{[Y - (\hat{a}X + \hat{b})] \cdot (\bar{a}\bar{X} + \bar{b})\} = 0$$

for any numbers a and b. This is equivalent to

$$\hat{a}E(|X|^2) + \hat{b}E(\bar{X}) = E(Y\bar{X}), \tag{5,4,21}$$

$$\hat{a}E(X) + \hat{b} = E(Y). \tag{5,4,22}$$

If we set

$$X' = X - E(X), \quad Y' = Y - E(Y) ,$$

$$\sigma_X^2 = E(|X'|^2), \quad \sigma_Y^2 = E(|Y'|^2),$$

$$\sigma_X \sigma_Y r = E(Y' \bar{X}') ,$$

it follows from (5,4,21) and (5,4,22) that

$$\hat{a} = \frac{\sigma_Y}{\sigma_X} r \; ; \; \hat{b} = E(Y) - \frac{\sigma_Y}{\sigma_X} r \, E(X),$$

and the best conditional linear estimate $\hat{\lambda}(X)$ of Y with respect to X is, therefore,

$$\hat{\lambda}(X) = E(Y) + \frac{\sigma_Y}{\sigma_X} r \, [X - E(X)]. \qquad (5,4,23)$$

This generalizes (5,2,27).

Now consider the following, more general problem. Given a positive natural number n, let $X = \{X_1, \ldots, X_n\}$ be a given second-order n-dimensional r.v., i.e. an ordered n-tuple of given second-order r.v. X_k $(k = 1, \ldots, n)$; let Y be a given second-order r.v. Then the random variables

$$Z = a_1 X_1 + a_2 X_2 + \ldots + a_n X_n + b ,$$

where $a_1, a_2, \ldots, a_n$, b are arbitrary (nonrandom) complex numbers, constitute a subset $\mathcal{H}'$ of $\mathcal{H}$. Again, it is easily shown that is a Hilbert subspace of $\mathcal{H}$, of dimension at most n + 1. What values b and a_j must be assigned to $\hat{b}$ and $\hat{a}_j$ $(j = 1, \ldots, n)$ in order that the quantity

$$e = E\left(\left|Y - \left(\sum_{j=1}^{n} a_j X_j + b\right)\right|^2\right)$$

be a minimum? The r.v.

$$\hat{Z} = Y - \left(\sum_{j=1}^{n} \hat{a}_j X_j + \hat{b}\right)$$

must be the orthogonal projection, in $\mathcal{H}$, of Y on $\mathcal{H}'$; it is therefore determined by the condition

$$E\left[\hat{Z}\left(\sum_{k=1}^{n} \bar{a}_k \bar{X}_k + \bar{b}\right)\right] = 0$$

for any numbers a_k and b. This is equivalent to the following system of equations:

$$(\Sigma) \quad \begin{cases} \sum_{j=1}^{n} E(X_j \bar{X}_k)\, \hat{a}_j + E(\bar{X}_k)\, \hat{b} = E(Y \bar{X}_k), \\ \qquad (k = 1, 2, \ldots, n) & (5,4,24) \\ \sum_{j=1}^{n} E(X_j)\, \hat{a}_j + \hat{b} = E(Y). & (5,4,25) \end{cases}$$

We know that the orthogonal projection $\hat{Z}$ of Y on $\mathcal{H}'$ exists and is unique. The system (Σ) has, therefore, at least one solution, and every solution of (Σ) determines the same $\hat{Z}$; but the representation

$$Y - \hat{Z} = \sum_{j=1}^{n} \hat{a}_j X_j + \hat{b} \qquad (5,4,26)$$

of $Y - \hat{Z}$ as a linear (non-homogeneous) form in the X_j is not necessarily unique, and (Σ) may have several solutions. Let us study the system (Σ). First note that once the values $\hat{a}_j$ are determined, $\hat{b}$ is uniquely determined by (5,4,25). Multiplying (5,4,25) by $E(\bar{X}_k)$ and subtracting from (5,4,24), we obtain the following system for the $\hat{a}_j$:

$$(\Sigma') \quad \sum_{j=1}^{n} [E(X_j \bar{X}_k) - E(X_j)\, E(\bar{X}_k)]\, \hat{a}_j = E(Y \bar{X}_k) - E(Y)\, E(\bar{X}_k)$$

$$(k = 1, 2, \ldots, n). \qquad (5,4,27)$$

Set

$$X'_j = X_j - E(X_j) \quad (j = 1, 2, \ldots, n),$$

$$E(X'_j \overline{X'_k}) = \gamma_{jk} \;;\; E(Y \overline{X'_k}) = \rho_k \quad (j, k = 1, 2, \ldots, n),$$

and introduce the following notation:

X': the $n \times 1$ matrix with elements X'_j $(j = 1, 2, \ldots, n)$;

ρ: the $n \times 1$ matrix with elements ρ_k $(k = 1, 2, \ldots, n)$;

Γ: the square $n \times n$ matrix with elements γ_{jk} $(j, k = 1, 2, \ldots, n)$;

Γ is thus the covariance matrix of $X' = \{X'_1, \ldots, X'_n\}$;

A: the $n \times 1$ matrix with elements $a_1, \ldots, a_n$, where the a_j are arbitrary certain numbers;

$\hat{A}$: the n × 1 matrix with elements $\hat{a}_j$.

The system (Σ') can be written as

$$\sum_{j=1}^{n} \gamma_{jk}\, \hat{a}_j = \rho_k \quad (k = 1, 2, \ldots, n),$$

or, in matrix notation,

$$\Gamma \hat{A} = \rho. \tag{5,4,28}$$

If we now ask for what value $\hat{A}$ of A the quantity

$$e' = E(|Y - {}^{\circ}X'\, A|^2) \tag{5,4,29}$$

is a minimum, the answer is precisely equation (5,4,28). Moreover, ρ and e' are unchanged if we replace Y by Y + c, where c is an arbitrary certain number; in other words, there would have been no loss in generality had the original problem been formulated on the assumption that $E(Y) = E(X_j) = 0$ $(j = 1, 2, \ldots, n)$ (as could have easily been shown directly).

Now return to (5,4,29). If Γ is a *nonsingular* matrix, i.e. X' is of rank n, there is a unique solution

$$\hat{A} = \Gamma^{-1}\, \rho. \tag{5,4,30}$$

Since $Y - {}^{\circ}X'\hat{A}$ and ${}^{\circ}X'\hat{A}$ are orthogonal in $\mathcal{H}$, it follows from Theorem (3,13,3) that the minimum e'_m of e_m is

$$\begin{aligned} e'_m &= E(|Y|^2) - E(|{}^{\circ}X'\, \hat{A}|^2) \\ &= E(|Y|^2) - \sum_{jk} \gamma^{-1}_{jk} \rho_j\, \bar{\rho}_k, \end{aligned}$$

where γ^{-1}_{jk} is the (j, k) element of Γ^{-1}, and we have used the fact that Γ^{-1} is hermitian (since Γ is). In matrix notation,

$$e'_m = E(|Y|^2) - {}^{\circ}\bar{\rho}\,\Gamma^{-1}\, \rho. \tag{5,4,31}$$

We leave it to the reader to discuss (Σ') in the case where Γ is a singular matrix, i.e. X' is of rank < n.

5. n-*dimensional normal random vectors and variables*

We now return to *real* r.v. or random vectors. Let Z be a random vector in the real n-dimensional vector space $\mathfrak{X}_n$ with zero element θ and dual $\mathfrak{X}_n^*$. We call the random vector Z normal if, for every linear functional $z^* \in \mathfrak{X}_n^*$, the r.v. $\langle z^*, Z \rangle$ is a (possibly degenerate) normally distributed r.v.

Let $\mathfrak{X}_n$ and $\mathfrak{Y}_q$ be two vector spaces of finite dimensions n and q, respectively. Let $\lambda(x)$ be a *linear* mapping of $\mathfrak{X}_n$ into $\mathfrak{Y}_q$, Z a random vector with values in $\mathfrak{X}_n$, and T the random vector with values in $\mathfrak{Y}_q$ defined by

$$T = \lambda(Z) .$$

If t^* is any element of the dual $\mathfrak{Y}_q^*$ of $\mathfrak{Y}_q$, we know (cf. V.3) that, for any $z \in \mathfrak{X}_n$,

$$\langle t^*, \lambda(z) \rangle \equiv \langle \lambda^*(t^*), z \rangle , \qquad (5,5,1)$$

where λ^* denotes the *adjoint* of λ. In effect, the adjoint may be considered to be defined by (5,5,1); it is a linear mapping of $\mathfrak{Y}_q^*$ into the dual $\mathfrak{X}_n^*$ of $\mathfrak{X}_n$.

It follows that *if* Z *is a normal random vector in* $\mathfrak{X}_n$, *then* $T = \lambda(Z)$ *is a normal random vector in* $\mathfrak{Y}_q$.

This remark is true, in particular, when n = q and λ is an isomorphism (linear one-to-one mapping of $\mathfrak{X}_n$ onto $\mathfrak{Y}_{q=n}$); moreover, since in this case the inverse λ^{-1} of λ is also a linear mapping of $\mathfrak{Y}_{q=n}$ onto $\mathfrak{X}_n$, it follows that if one of the two vectors Z or T is normal then so is the other.

These results will be made more precise later, but we can already supplement them as follows. If Z is a normal random vector in $\mathfrak{X}_n$ and z is any nonrandom fixed vector in $\mathfrak{X}_n$ it is clear that the vector z + Z is a normal random vector in $\mathfrak{X}_n$. Then, in the above notation, we have the following proposition.

Let λ be a mapping of $\mathfrak{X}_n$ into $\mathfrak{Y}_q$, of the form

$$t = \lambda(z) = t_o + \lambda_o(z),$$

where λ_0 is a linear mapping of $\mathfrak{X}_n$ into $\mathfrak{Y}_q$ and t_0 is an arbitrary fixed nonrandom vector in $\mathfrak{Y}_q$; if Z is a normal random vector in $\mathfrak{X}_n$, then

$$T = \lambda(Z) = t_o + \lambda_o(Z)$$

is a normal random vector in $\mathfrak{Y}_q$. ▌

The set of all ordered n-tuples $\{X_1, X_2, \ldots, X_n\}$ of real numbers is an n-dimensional vector space R^n. We shall call an n-dimensional r.v. $\{X_1, X_2, \ldots, X_n\}$ normal if, as a vector of R^n, it is a normal random vector. Every n-dimensional vector space $\mathfrak{X}_n$ is isomorphic to R^n. In particular, if $\mathfrak{b}$ is any basis in $\mathfrak{X}_n$, then the mapping λ which maps any $z \in \mathfrak{X}_n$ onto the n-tuple $\{X_1, X_2, \ldots, X_n\}$ of its components relative to $\mathfrak{b}$ is an isomorphism of $\mathfrak{X}_n$ onto R^n. Thus:

If $\{X_1, X_2, \ldots, X_n\}$ is a normal n-dimensional r.v., and we interpret the X_j $(j = 1, 2, \ldots, n)$ as the components of a random vector Z of some vector space $\mathfrak{X}_n$, relative to some basis $\mathfrak{b}$, then Z is a normal random vector in $\mathfrak{X}_n$. Conversely, if Z is a normal random vector in some given n-dimensional vector space $\mathfrak{X}_n$, the n-tuple $\{X_1, X_2, \ldots, X_n\}$ of its components relative to any fixed basis $\mathfrak{b}$ is a normal r.v.

Retaining the notation of the beginning of this section, we remark that the existence of normal random vectors in $\mathfrak{X}_n$ is not self-evident. Assuming, however, that they exist, let Z be one of them, and let us study the consequences of the assumption that Z is normal.

Since every normal r.v., even degenerate, has a m.e., we see that the mathematical expectation E(Z) exists, and

$$\langle z^*, E(Z) \rangle = E(\langle z^*, Z \rangle). \qquad (5,5,2)$$

Set

$$Z' = Z - E(Z).$$

The m.e. $E(Z')$ of Z' is θ, i.e.

$$E(\langle z^*, Z' \rangle) = 0 \qquad\qquad z^* \in \mathfrak{X}_n^* .$$

Thus $\langle z^*, Z \rangle$ is a normal r.v. with m.e. $\langle z^*, E(Z) \rangle$ and variance

$$E(|\langle z^*, Z' \rangle|^2) .$$

This variance exists (and is, of course, nonnegative), by virtue of the assumption that $\langle z^*, Z \rangle$ is normal. Moreover, by (4,4,21),

$$E(e^{i\langle z^*, Z\rangle}) = e^{-\frac{1}{2} E(|\langle z^*, Z'\rangle|^2) + i\langle z^*, E(Z)\rangle},$$

and therefore the characteristic functional $\xi(z^*)$ of Z is

$$\xi(z^*) = \exp. \left\{ -\frac{1}{2} E(|\langle z^*, Z' \rangle|^2) + i \langle z^*, E(Z) \rangle \right\} . \quad (5,5,3)$$

We now proceed to an analytical computation of this functional. In so doing, and indeed throughout the rest of V.5, we shall frequently use matrix notation. Let $\mathbf{b}$ be a basis in $\mathfrak{X}_n$; let $\{x_1, x_2, \ldots, x_n\}$ be the components of some vector z in $\mathfrak{X}_n$ relative to $\mathbf{b}$. We regard the x_j as the elements of an $n \times 1$ matrix x. We know that, for every j ($j = 1, 2, \ldots, n$), there exists a well-defined linear functional z_j^* such that, for all $z \in \mathfrak{X}_n$,

$$\langle z_j^*, z \rangle = x_j = j^{\text{th}} \text{ component of } z \text{ relative to } \mathbf{b}.$$

The z_j^* ($j = 1, 2, \ldots, n$) constitute a basis $\mathbf{b}^*$ for $\mathfrak{X}_n^*$. This basis is the *dual* basis of $\mathbf{b}$ in $\mathfrak{X}_n$.

Let $\{X_1, X_2, \ldots, X_n\}$, $\{X_1', X_2', \ldots, X_n'\}$, $\{\mu_1, \mu_2, \ldots, \mu_n\}$ be the components relative to $\mathbf{b}$ of Z, Z' and $E(Z)$, respectively. Let us regard the X_j, X_j', and μ_j as the elements of $n \times 1$ matrices X, X', and μ respectively. X, X', μ are vectors in R^n; μ is a nonrandom vector. Clearly, i.e., $X_j = \mu_j + X'_j$ for every j (j = 1, 2,..., n)

It follows from previous arguments that X and X' are normal random vectors in R^n, or normal n-dimensional r.v., and μ is the m.e. of X.

The X_j are normal r.v. with m.e. μ_j. The X'_j are normal r.v. with m.e. zero. The variance of both X_j and X'_j is $E(|X'_j|^2)$. Hence, incidentally, every normal n-dimensional r.v. $X = \{X_1, X_2,..., X_n\}$ is a second-order n-dimensional r.v.

Set

$$\gamma_{jk} = E(X'_j X'_k) = E[(X_j - \mu_j)(X_k - \mu_k)]. \tag{5,5,4}$$

Recall that if σ_j is the standard deviation of X_j (or X'_j), and r_{jk} the correlation coefficient of X_j and X_k (or X'_j and X'_k), then

$$r_{jk} = r_{kj}, \; r_{jj} = 1 \tag{5,5,5}$$

and

$$\gamma_{jk} = \sigma_j \sigma_k r_{jk}. \tag{5,5,6}$$

Let Γ be the $n \times n$ matrix (γ_{jk}) and R the $n \times n$ matrix (r_{jk}). From (5,5,5) it follows that Γ and R are *symmetric* matrices. In fact, since Γ is the covariance matrix of the ordered n-tuple of the X'_j, it follows from Theorem (5,4,1) that Γ is not only symmetric but also *positive-definite*. It follows immediately that R is also positive-definite.

Now let u denote the real $n \times 1$ matrix whose elements u_h (h = 1, 2,..., n) are the components of z* relative to **b***; then

$$<z^*, E(Z)> = \sum_h u_h \mu_h = {}^\circ u \, \mu ,$$

$$<z^*, Z'> = \sum_h u_h X'_h = {}^\circ u \, X' ,$$

$$|<z^*, Z'>|^2 = \sum_{jk} X'_j X'_k u_j u_k ,$$

$$E(|<z^*, Z'>|^2) = \sum_{jk} \gamma_{jk} u_j u_k = {}^\circ u \, \Gamma \, u . \tag{5,5,7}$$

It is thus clear that (5,5,7) is the quadratic form associated with Γ. In view of (5,5,2), the characteristic functional $\xi(z^*)$ becomes a function $\phi(u_1, u_2, \ldots, u_n)$ of the n real variables u_h, or, symbolically, a function $\phi(u)$ of u. To be explicit,

$$\xi(z^*) = \varphi(u_1, u_2, \ldots, u_n) = \varphi(u)$$

$$= \exp. \left\{- \frac{1}{2} \,{}^\circ u \,\Gamma u + i \,{}^\circ u \,\mu\right\}. \tag{5,5,8}$$

Note first that, as a function of the variables u_h, the expression (5,5,8) for $\xi(z^*)$ is precisely the (n-dimensional) Fourier transform of the n-dimensional d.f. $F(x_1, \ldots, x_n)$, relative to the basis **b** of the n-dimensional r.v. $\{X_1, X_2, \ldots, X_n\}$. In other words, the function $\phi(u)$ of the u_n defined by (5,5,8) is the characteristic function of the n-dimensional r.v. $\{X_1, X_2, \ldots, X_n\}$. Thus:

THEOREM (5,5,1). If Z is a normal random vector in $\mathfrak{X}_n$, with components $X = \{X_1, X_2, \ldots, X_n\}$ relative to some basis **b** of $\mathfrak{X}_n$, then the n-dimensional characteristic function $\phi(u)$ ($u = \{u_1, u_2, \ldots, u_n\}$) of the n-dimensional r.v. X has the form

$$\varphi(u) = \exp. \left\{- \frac{1}{2} \,{}^\circ u \,\Gamma u + i \,{}^\circ u \,\mu\right\},$$

where ${}^\circ u \,\mu = \sum_h u_h \mu_h$ is a homogeneous linear form in the u_n, and ${}^\circ u \,\Gamma u = \sum_{jk} \gamma_{jk} u_j u_k$ is a homogeneous positive-definite quadratic form in the u_h; furthermore,

$$\mu_h = E(X_h) \text{ et } \gamma_{jk} = E[(X_j - \mu_j)(X_k - \mu_k)]. \blacksquare$$

Thus, the probability law of a normal n-dimensional random variable $X = \{X_1, X_2, \ldots, X_n\}$ is completely defined by the mathematical expectations $E(X_h)$ and the covariance matrix of the r.v.

$$X'_h = X_h - \mu \quad (h = 1, 2, \ldots, n).$$

Conversely:

THEOREM (5,5,2). Let $X = \{X_1, X_2, \ldots, X_n\}$ be an n-dimensional r.v. whose n-dimensional characteristic function $\phi(u_1, u_2, \ldots, u_n) = \phi(u)$ has the form

$$\varphi(u) = \exp. \left\{ -\frac{1}{2} {}^\circ u \, \Gamma \, u + i \, {}^\circ u \, \mu \right\},$$

where ${}^\circ u\mu$ is a linear homogeneous form in the u_h and ${}^\circ u\Gamma u$ is a positive-definite homogeneous quadratic form in the u_h. If the X_j are regarded as components of a random vector Z in $\mathfrak{X}_n$, relative to a basis **b**, then Z is a normal random vector (or: X is a normal n-dimensional r.v.).

In fact, if z^* is a linear functional on $\mathfrak{X}_n$ and $\alpha = \{\alpha_1, \alpha_2, \ldots, \alpha_n\}$ is the $n \times 1$ matrix of the components α_h of z^* relative to the dual basis **b*** of **b**, then

$$< z^*, Z > = {}^\circ\alpha X$$

Thus the characteristic function $a(\rho)$ of the r.v. $< z^*, Z >$ is

$$\begin{aligned} a(\rho) &= E(e^{i\rho< z^*, Z >}) = E(e^{i< \rho z^*, Z >}) \\ &= E[e^{i(\rho {}^\circ\alpha)X}] = \varphi(\rho \, {}^\circ\alpha) \\ &= \exp. \left\{ -\frac{1}{2} ({}^\circ\alpha \, \Gamma \, \alpha)\rho^2 + i({}^\circ\alpha \, \mu)\rho \right\} \end{aligned} \qquad (5,5,9)$$

Now ${}^\circ\alpha\Gamma\alpha$ is a nonnegative number, since Γ is positive-definite. Thus formula (5,5,9) clearly has the form (4,4,25) of the characteristic function of a normal r.v. (degenerate if ${}^\circ\alpha \, \Gamma \, \alpha = 0$). Thus $< z^*, Z >$ is normal and Z is a normal random vector. ∎

Retaining the previous notation, let Z be a normal random vector in $\mathfrak{X}_n$, and again choose a basis **b**. Let T be any non-singular $n \times n$ matrix, T_k ($k = 1, 2, \ldots, n$) the kth column of T (as an $n \times 1$ matrix). Let us regard the elements of T_k as the components (relative to the basis **b**) of a vector z'_k in $\mathfrak{X}_n$. Then the n-tuple of vectors $\{z'_1, z'_2, \ldots, z'_n\}$ constitutes a basis $\mathbf{b}'$ in $\mathfrak{X}_n$, and we call T the *change-of-basis matrix* which makes it possible to go from the basis $\mathbf{b}$ to the basis $\mathbf{b}'$. Conversely, let $\mathbf{b}'$ be any given basis of $\mathfrak{X}_n$ consisting of vectors $\{z'_1, z'_2, \ldots, z'_n\}$, T_k the n × 1 matrix whose elements are the components of z'_k (k = 1, 2,..., n) relative to $\mathbf{b}$, and T the n × n matrix whose k^{th} column (k = 1, 2,..., n) is T. Then T is nonsingular and it is the change-of-basis matrix from $\mathbf{b}$ to $\mathbf{b}'$.

Let $\mathbf{b}'^*$ denote the basis of $\mathfrak{X}_n^*$ dual to $\mathbf{b}'$, and let $\{Y_1, Y_2, \ldots, Y_n\}$, $\{Y'_1, Y'_2, \ldots, Y'_n\}$, $\{\nu_1, \nu_2, \ldots, \nu_n\}$ be the respective components relative to $\mathbf{b}'$ of Z, Z', and E(Z). Let $\{v_1, \ldots, v_n\}$ be the components of z* relative to $\mathbf{b}'^*$, and Y, Y', ν, v the n × 1 matrices with elements Y_j, Y'_j, ν_j, v_j (j = 1, 2,..., n), respectively.

The formulas for change of basis are well known:

$$X = TY, \; X' = TY', \; \mu = T\nu, \; {}^{\circ}v = {}^{\circ}uT. \qquad (5,5,10)$$

Note that

$$Y_j = Y'_j + \nu_j \quad (j = 1, 2, \ldots, n)$$

and

$$\langle z^*, E(Z)\rangle = {}^{\circ}v\,\nu, \; \langle z^*, Z'\rangle = {}^{\circ}v\,Y'.$$

Set

$$\gamma' = E(Y'_j\; Y'_k) = E[(Y_j - \nu_j)(Y_k - \nu_k)],$$

i.e., γ'_{jk} is the covariance of Y'_j and Y'_k, and let Γ' be the n × n matrix (γ'_{jk}), i.e. the covariance matrix of the n-dimensional r.v. $Y' = \{Y'_1, Y'_2, \ldots, Y'_n\}$; thus Γ' is symmetric and positive-definite. Using (5,4,17), we obtain

$$T\Gamma'\,{}^{\circ}T = \Gamma$$

or

$$\Gamma' = T^{-1}\Gamma\,{}^{\circ}T^{-1}. \tag{5,5,11}$$

Now suppose $^{\circ}T^{-1}$ belongs to the family $\mathcal{D}_\Gamma$ of diagonalizing matrices for Γ (cf. III.5). Then Γ' is diagonal; if $s_1, s_2, \ldots, s_n$ are the n elements, not necessarily distinct, on its principal diagonal, these are also its characteristic roots, and since Γ' is symmetric and positive-definite, the s_h are real and non-negative. We thus obtain the following simple expression for the characteristic functional $\xi(z^*)$:

$$\xi(z^*) = \psi(v_1, \ldots, v_n) = \psi(v)$$

$$= \exp\left\{-\frac{1}{2}\sum_h s_h v_h^2 + i\sum_h \nu_h v_h\right\}, \tag{5,5,12}$$

which we call a *reduced* analytic form of $\xi(z^*)$.

For this reduced form we know, either directly (cf. V.4) or via Theorem (5,5,1), that

$$\gamma_{jk} = E(Y'_j Y'_k) = 0 \text{ if } j \neq k\,;$$

$$\gamma_{hh} = E(|Y'_h|^2) = s_h. \tag{5,5,13}$$

Setting

$$\psi_h(v_h) = e^{-\frac{1}{2}s_h v_h^2 + i\nu_h v_h}, \tag{5,5,14}$$

we see that

$$\psi(v) = \psi(v_1, v_2, \ldots, v_n) = \prod_{h=1}^{n} \psi_h(v_h). \tag{5,5,15}$$

Theorem (5,2,5) then implies:

1) The Y_h are mutually independent;

2) $\psi_h(v_h)$ is the c.f. of Y_h; therefore Y_h is normal (which we already know) with m.e. ν_h and variance s_h; Y_h is a degenerate normal r.v. if $s_h = 0$.

As there exist n-tuples of mutually independent r.v. $Y_1, Y_2, \ldots, Y_n$, each of which is normal with given m.e. ν_h and variance $s_h \geqq 0$, it follows that *there exist normal random vectors*. By Theorem (5,5,2), it also follows that every function $\phi(u) = \phi(u_1, u_2, \ldots, u_n)$ of the form (5,5,8), where $^\circ u\mu$ is any linear form in the u_h and $^\circ u\Gamma u$ any positive-definite homogeneous quadratic form in the u_h, is a positive-definite function of the u_h. To be precise, it is the n-dimensional c.f. of a normal n-dimensional r.v. $\{X_1, X_2, \ldots, X_n\}$ with the properties indicated in Theorem (5,5,1).

Considering, in particular, the case in which the $n \times 1$ matrix μ is zero, we obtain:

THEOREM (5,5,3). If Γ is any real, symmetric and positive-definite $n \times n$ matrix, there exists one normal n-dimensional r.v. $X = \{X_1, \ldots, X_n\}$, and only one, such that $E(X_j) = 0$ $(j = 1, 2, \ldots, n)$ and Γ is its covariance matrix. ■

This theorem is an improvement of Theorem (5,4,2), but only for the case of a real matrix Γ. Later we shall see whether Theorem (5,5,3) may be extended to the complex case.

Returning to the diagonalization of Γ, we recall that the family $\mathfrak{W}_\Gamma$ includes at least one orthogonal matrix. Let U be one of these. In the preceding argument, we can set $^\circ T^{-1} = U$, i.e. $T = U$ since U (being orthogonal) satisfies $^\circ U^{-1} = U$. The numbers s_h are then the characteristic roots of Γ. If **b** is an orthonormal basis of $\mathfrak{X}_n$ (this can always be assumed, without loss of generality), then **b'** and **b'*** are orthonormal bases of $\mathfrak{X}_n$ and $\mathfrak{X}_n^*$, respectively. Starting with an orthonormal basis **b**, it is then always possible to obtain the reduced form (5,5,12) by means of an orthonormal basis **b'**, i.e. using an orthogonal matrix T.

THEOREM (5,5,4). If Z is a normal random vector in $\mathfrak{X}_n$, there exist bases in $\mathfrak{X}_n$, in particular at least one orthonormal basis, such that the components $\{Y_1, Y_2, \ldots, Y_n\}$ of Z relative to this basis are n mutually independent normal r.v. The n-dimensional characteristic function $\psi(v_1, v_2, \ldots, v_n)$ of the n-dimensional r.v. $\{Y_1, Y_2, \ldots, Y_n\}$ then has the form (5,5,12), which we call a *reduced* form of the characteristic functional $\xi(z^*)$ of Z.

Rank of a normal random variable

We define the *rank* of the normal random vector Z to be the rank r of the matrix Γ. This definition is legitimate, since r is independent of the specific basis **b** used to obtain Γ, as follows from the properties reviewed in III.5. r is also the number of positive characteristic roots (not necessarily distinct) of Γ; $r \leqq n$. r is also the rank, whatever basis **b** is used, of the second-order n-dimensional r.v. $X = \{X_1, X_2, \ldots, X_n\}$ formed by the components of Z relative to **b**. We call Z degenerate if $r < n$, nondegenerate if $r = n$.

Let Z be a normal random vector of rank r. Consider the reduced form (5,5,12); there are exactly r numbers $s_h > 0$, and the other $n - r$ numbers s_h are zero. To fix ideas, suppose that $s_h > 0$ for $h \leqq r$ and $s_h = 0$ for $h > r$. Then it is clear that $Y_{r+1}, Y_{r+2}, \ldots, Y_n$ are degenerate normal random variables.

Distribution function and probability density of a normal n-dimensional random variable

We start with the nondegenerate case $r = n$, using the reduced form (5,5,12) relative to an orthonormal basis; let the corresponding normal n-dimensional r.v. be $\{Y_1, Y_2, \ldots, Y_n\}$.

Y_h has the distribution function

$$G_h(y_h) = \frac{1}{\sqrt{2\pi s_h}} \int_{-\infty}^{y_h} e^{-\frac{(\alpha-\nu_h)^2}{2s_h}} \, d\alpha,$$

and density

$$g_h(y_h) = \frac{1}{\sqrt{2\pi s_h}} e^{-\frac{(y_h-\nu_h)^2}{2s_h}}$$

Since the Y_h are independent, the n-dimensional d.f. of the n-dimensional r.v. $\{Y_1, \ldots, Y_n\}$ is

$$G(y_1, \ldots, y_n) = \prod_{h=1}^{n} G_h(y_h), \tag{5,5,16}$$

and the density is

$$g(y_1, \ldots, y_n) = e^{-\frac{1}{2}\sum_{h=1}^{n}\frac{(y_h-\nu_h)^2}{s_h}} \Big/ (2\pi)^{\frac{n}{2}}\left(\prod_{h=1}^{n} s_h\right)^{1/2}. \tag{5,5,17}$$

We shall call (5,5,17) a *reduced* density for the normal random vector Z.

Let y and y' be the n × 1 matrices with elements y_h, $y'_h = y_h - \nu_h$, respectively, and A' the n × n diagonal matrix whose diagonal elements are $1/s_h$; then

$$\Phi'_2(y'_1, \ldots, y'_n) = \sum_h \frac{y'^2_h}{s_h}$$

is the homogeneous positive-definite quadratic form in the y'_h associated with A', and, in matrix notation:

$$\Phi'_2(y'_1, \ldots, y'_n) = {}^\circ y'\, A'\, y' .$$

But since Γ' is the n × n diagonal matrix whose diagonal elements are the s_h, it is clear that $A' = \Gamma'^{-1}$. We therefore get the following expressions for $g(y_1, \ldots, y_n)$:

$$g(y_1, \ldots, y_n) = \frac{1}{(2\pi)^{\frac{n}{2}}\left(\prod_{h=1}^{n} s_h\right)^{\frac{1}{2}}} \exp.\left\{-\frac{1}{2}\, {}^\circ y'\, \Gamma'^{-1}\, y'\right\}, \tag{5,5,18}$$

$$g(y_1, \ldots, y_n) = \frac{1}{(2\pi)^{\frac{n}{2}}\left(\prod_{h=1}^{n} s_h\right)^{\frac{1}{2}}} \exp.\left\{-\frac{1}{2}\, {}^\circ(y-\nu)\, \Gamma'^{-1}\, (y-\nu)\right\}. \tag{5,5,19}$$

$g(y_1, \ldots, y_n)$ is the density of the random vector Z relative to the specific basis **b'**. We now try to determine its density $F(x_1, \ldots, x_n)$ relative to a basis **b**, which is also orthonormal but otherwise arbitrary. $f(x_1, \ldots, x_n)$ is, as we know, the n-dimensional density of the n-dimensional r.v. $\{X_1, \ldots, X_n\}$ formed by the components of Z relative to **b**. Setting $x'_h = x_h - \mu_h$ and letting x and x' denote the n × 1 matrices with elements x_h, x'_h, respectively, we pass from $g(y_1, \ldots, y_n)$ to $f(x_1, \ldots, x_n)$ by a change of variables, as in (5,5,10); thus

$$\mu = T\nu\,,\ x = Ty\,,\ x' = Ty'\,, \tag{5,5,20}$$

where T is an orthogonal matrix. The Jacobian of the y_h with respect to the x_h is precisely the determinant of T^{-1}, and is therefore equal to unity in absolute value. Using (2,11,13) we obtain

$$f(x_1, \ldots, x_n) = \frac{1}{(2\pi)^{\frac{n}{2}} \left(\prod_{h=1}^{n} s_h\right)^{\frac{1}{2}}} \ \exp. \left\{ -\frac{1}{2}\, {}^{\circ}x'\, T\, \Gamma'^{-1}\, T^{-1} x' \right\}.$$

Since ${}^{\circ}T^{-1} = T$ (T and T^{-1} being orthogonal), we have, by (5,5,20): $y' = T^{-1}x'$, ${}^{\circ}y' = {}^{\circ}x'\,T$. But, by (5,5,11), $\Gamma'^{-1} = {}^{\circ}T\,\Gamma^{-1}\,T = T^{-1}\,\Gamma^{-1}\,T$. Moreover, since the s_h are the characteristic roots of Γ, $1/s_h$ are those of Γ^{-1}, and the product $\prod_{h=1}^{n}\left(\frac{1}{s_h}\right)$ is thus the determinant of Γ^{-1}. Denoting the elements of Γ^{-1} by γ^{-1}_{jk} and its determinant by Δ, we finally obtain

$$f(x_1, \ldots, x_n) = \frac{\sqrt{\Delta}}{(2\pi)^{n/2}} \ \exp. \left\{ -\frac{1}{2}\, {}^{\circ}x'\, \Gamma^{-1} x \right\}, \tag{5,5,21}$$

$$= \frac{\sqrt{\Delta}}{(2\pi)^{n/2}} \ \exp. \left\{ -\frac{1}{2}\, {}^{\circ}(x-\mu)\, \Gamma^{-1}\, (x-\mu) \right\}, \tag{5,5,22}$$

$$= \frac{\sqrt{\Delta}}{(2\pi)^{n/2}} \ \exp. \left\{ -\frac{1}{2} \sum_{jk} \gamma^{-1}_{jk}\, (x_j - \mu_j)(x_k - \mu_k) \right\}. \tag{5,5,23}$$

The reader should have no difficulties in deriving the corresponding results for the case of arbitrary (not necessarily orthonormal) bases **b'** and **b**.

Retaining the assumption that **b'** and **b** are orthonormal, we remark that it is also interesting to express $f(x_1, \ldots, x_n)$ in terms of the matrix R with the elements r_{jk} defined below, rather than in terms of Γ:

$$\sigma_j\, \sigma_k\, r_{jk} = \gamma_{jk}.$$

Let δ denote the determinant of R; the assumption that Z is of rank n implies that δ is not zero, neither are any of the σ_j. Let r^{-1}_{jk} denote the elements of R^{-1}; then

$$\gamma^{-1}_{jk} = \frac{r^{-1}_{jk}}{\sigma_j \, \sigma_k}, \quad \frac{1}{\Delta} = \left(\prod_{h=1}^{n} \sigma_h\right)^2 \delta, \tag{5,5,24}$$

and therefore

$$f(x_1, \ldots, x_n) = \frac{1}{(2\pi)^{\frac{n}{2}} \sqrt{\delta} \prod_{h=1}^{n} \sigma_h} \exp. \left\{ -\frac{1}{2} \sum_{jk} r^{-1}_{jk} \left(\frac{x_j - \mu_j}{\sigma_j}\right) \left(\frac{x_k - \mu_k}{\sigma_k}\right) \right\}. \tag{5,5,25}$$

We shall call (5,5,25) the *canonical form* of the probability density of the normal random vector Z relative to the (orthonormal) basis **b**. Alternatively, we call (5,5,25) the canonical form of the normal n-dimensional r.v. $\{X_1, \ldots, X_n\}$ defined by the matrices μ and Γ (or R).

We now turn to a problem which is, in a certain sense, the converse of the above. Let $\{X_1, \ldots, X_n\}$ be an n-dimensional r.v. whose density $f(x_1, \ldots, x_n)$ has the form

$$f(x_1, \ldots, x_n) = H \exp. \left\{ -\frac{1}{2} \Omega(x_1, \ldots, x_n) \right\}, \tag{5,5,26}$$

where H is a constant and $\Omega(x_1, \ldots, x_n)$ is a quadratic form, not necessarily homogeneous, in the x_h. Without loss of generality we shall assume that

$$\Omega(x_1, \ldots, x_n) = \Phi_2(x_1, \ldots, x_n) - 2\,\Phi_1(x_1, \ldots, x_n), \tag{5,5,27}$$

where

$$\Phi_2(x_1, \ldots, x_n) = \sum_{jk} a_{jk} x_j x_k \quad (a_{jk} = a_{kj})$$

is a *homogeneous* quadratic form in the x_h, and

$$\Phi_1(x_1, \ldots, x_n) = \sum_h \tau_h x_h$$

is a *homogeneous* linear form in the x_h. Let A be the $n \times n$ symmetric matrix of the form Φ_2, i.e. the matrix (a_{jk}), and τ the $n \times 1$ matrix with elements τ_h; retaining the previous notation, in particular, matrix notation, we can write

$$f(x_1, \ldots, x_n) = H \exp. \left\{ -\frac{1}{2}\, {}^\circ x A x + {}^\circ \tau x \right\}.$$

The assumption that $F(\mathbf{x}_1,\ldots, x_n)$, as given by (5,5,26), is a probability density imposes on the forms Φ_1, Φ_2 and the constant H certain conditions, which we now describe in detail.

Let us regard $\{X_1,\ldots, X_n\}$ as the components of a random vector Z in an n-dimensional space, relative to an orthonormal basis **b**. Let $\lambda_1, \lambda_2,\ldots, \lambda_n$ be the characteristic roots, not necessarily distinct, but real, of the matrix A. Let **b'** be an orthonormal basis of $\mathfrak{X}_n$, T the (orthogonal) change-of-basis matrix from the basis **b** to the basis **b'**, and $\{Y_1,\ldots, Y_n\}$ the components of Z relative to **b'**. By (2,11,13), the formulas for change of basis (5,5,10), (5,5,20), and the fact that T is orthogonal, the density of $\{Y_1, Y_2,\ldots, Y_n\}$ is

$$g(y_1,\ldots,y_n) = H \ \exp. \left\{-\frac{1}{2}\,{}^{\circ}y\,T^{-1}A\,T\,y + {}^{\circ}\tau\,T\,y\right\}.$$

Let us choose **b'**, i.e. T, in such a way that the matrix $A' = T^{-1}AT$ is diagonal, with the λ_n as the elements on its principal diagonal (we know that this is possible).

Then, letting τ'_h denote the elements of the $1 \times n$ matrix ${}^{\circ}\tau T$, we write $g(y_1,\ldots, y_n)$ in the form

$$g(y_1,\ldots,y_n) = H \prod_{h=1}^{n} e^{-\frac{1}{2}\lambda_h y_h^2 + \tau'_h y_h}. \qquad (5,5,28)$$

For the function $f(x_1,\ldots, x_n)$ defined by (5,5,26) to be a probability density, the integral

$$\underbrace{\int_{-\infty}^{+\infty}\cdots\int_{-\infty}^{+\infty}}_{(n\ \text{fois})} f(x_1,\ldots,x_n)\,dx_1\ldots dx_n \qquad (5,5,29)$$

must be finite; equivalently, the integral

$$\int_{-\infty}^{+\infty}\cdots\int_{-\infty}^{+\infty} g(y_1,\ldots,y_n)\,dy_1\ldots dy_n \qquad (5,5,30)$$

must be finite. Since the integral (5,5,30) can be expressed in the form

$$H \cdot \prod_{h=1}^{n}\int_{-\infty}^{+\infty} e^{-\frac{1}{2}\lambda_h y_h^2 + \tau'_h y_h}\,dy_h\ ;$$

the integral (5,5,29) is finite if and only if each of the integrals

$$\int_{-\infty}^{+\infty} e^{-\frac{1}{2}\lambda_h y_h^2 + \tau_h^\circ y_h} dy_h \qquad (h = 1, 2, \ldots, n)$$

is finite, and this is true if and only if $\lambda_h > 0$. Therefore, it is necessary and sufficient that the quadratic form Φ_2 or its symmetric matrix A be *nondegenerate* and positive-definite. Assume that this is the case; this implies, in particular, that A is nonsingular, has an inverse A^{-1}, and its determinant Δ is positive. Denoting by μ the $n \times 1$ matrix whose elements μ_h are defined by

$$\mu = A^{-1}\tau, \qquad (5,5,31)$$

one uses elementary matrix calculus to show that $f(x_1, \ldots, x_n)$ can be put in the form

$$f(x_1, \ldots, x_n) = H\, e^{\frac{1}{2}{}^\circ\tau A^{-1}\tau} \exp.\left\{-\frac{1}{2}\,{}^\circ(x-\mu)\,A(x-\mu)\right\}$$

$$= H\, e^{\frac{1}{2}{}^\circ\tau A^{-1}\tau} \exp.\left\{-\frac{1}{2}\sum_{jk} a_{jk}(x_j-\mu_j)(x_k-\mu_k)\right\}.$$

(5,5,32)

Comparison of (5,5,22) and (5,5,23) proves:

THEOREM (5,5,5). The function $f(x_1, \ldots, x_n)$ of n variables $\{x_1, \ldots, x_n\}$ defined by

$$f(x_1, \ldots, x_n) = H \exp.\left\{-\frac{1}{2}\sum_{jk} a_{jk}\, x_j\, x_k + \sum_h \tau_h\, x_h\right\},$$

where H, τ_h and a_{jk} are constants such that $a_{jk} = a_{kj}$ (j, k, h = 1, 2,..., n), is an n-dimensional probability density if and only if

1) the symmetric matrix A with elements a_{jk} and determinant Δ is nonsingular and positive-definite;

2) the constant H satisfies the relation

$$\frac{\sqrt{\Delta}}{(2\pi)^{n/2}} = H\, e^{\frac{1}{2}\,{}^{\circ}\tau A^{-1}\tau}, \tag{5,5,33}$$

where τ is the $n \times 1$ matrix with elements τ_h.

If these conditions are satisfied, $f(x_1, \ldots, x_n)$ is the probability density of a nondegenerate normal n-dimensional random variable, and may be expressed in the form (5,5,32), where the μ_h are the elements of the matrix $\mu = A^{-1}\tau$, or in the canonical form (5,5,25).

REMARK (5,5,1). Part of the above results can also be expressed as follows:

Let A be a given $n \times n$ symmetric, nondegenerate and positive-definite real matrix, μ any given real $n \times 1$ matrix and x the $n \times 1$ matrix with variable elements x_h. Then the integral

$$\underbrace{\int_{-\infty}^{+\infty} \cdots \int_{-\infty}^{+\infty}} \exp.\left\{-\frac{1}{2}\,{}^{\circ}(x - \mu)\, A\, (x - \mu)\right\} dx_1 \ldots dx_n$$

exists, and its value is

$$\frac{(2\pi)^{n/2}}{\sqrt{\Delta}},$$

where Δ is the determinant of A.

REMARK (5,5,2). Consider again the expression (5,5,26), where Ω is given by (5,5,27). Let us regard the x_h as coordinates of a point N in an n-dimensional point space E_n, relative to a reference system [0, **b**]. Then the equation

$$\Omega(x_1, \ldots, x_n) = \rho, \tag{5,5,34}$$

where ρ is an arbitrary constant, is the equation of a surface Q_ρ in E_n. It is a quadratic surface, since Ω is a second-degree polynomial. Now the symmetric matrix A (or its quadratic form $\Phi_2(x_1, \ldots, x_n)$) is nondegenerate positive-definite if and only if Q_ρ is an *ellipsoid* (not necessarily real). The elements μ_h of the matrix μ are the coordinates of the center G of the ellipsoid Q_ρ. G is independent of ρ, and the Q_ρ $(-\infty < \rho < +\infty)$ form a family of similar concentric ellipsoids (with G as the common center). From (5,5,34) and (5,5,26) it follows that the Q_ρ are the surfaces on which $f(x_1, \ldots, x_n)$ is constant. Thus, any normal n-dimensional nondegenerate probability law or mass distribution has a family of similar and concentric ellipsoids with center at the mean point G as equidensity surfaces. The axes of these ellipsoids are clearly the central axes of inertia of the mass distribution.

Case of a degenerate normal random variable

We return to the previous notation and conventions, in particular, those mentioned under the heading "Rank of a normal random variable." Let us consider a normal random variable Z with rank $r < n$; $\{Y_1, \ldots, Y_r\}$ is a nondegenerate normal r-dimensional r.v., i.e. of rank r; its r-dimensional d.f. is obviously

$$\prod_{h=1}^{r} G_h(y_h) ;$$

but $Y_{r+1}, Y_{r+2}, \ldots, Y_n$ are $n - r$ degenerate normal r.v.; in other words, they reduce to almost certain numbers, to be precise,

$$Y_h = \nu_h \text{ almost certainly for } h > r.$$

The n-dimensional d.f. of $\{Y_1, \ldots, Y_n\}$ is

$$G(y_1, \ldots, y_n) = \left[\prod_{h=1}^{r} G_h(y_h)\right] \cdot \left[\prod_{h>r} \Delta(y_h - \nu_h)\right] \qquad (5,5,35)$$

(cf. (3,2,5) and IV.4); of course, this d.f. has no density.

Since $\{Y_1, \ldots, Y_r\}$ is a nondegenerate normal r-dimensional r.v., we can say that Z reduces to a nondegenerate normal random vector, but of dimension r.

If we now pass to the basis **b**, i.e. to the n-dimensional r.v. $\{X_1, \ldots, X_n\}$ in accordance with the formulas for change of variables (5,5,10), the Y_h, as functions of the X_h, are independent homogeneous linear forms. Hence, we easily deduce:

THEOREM (5,5,6). If a normal n-dimensional random variable $\{X_1, \ldots, X_n\}$ has the rank $r < n$, almost certainly, there exist among the X_h exactly $n - r$ independent linear functional relations. The r.v. $\{X_1, \ldots, X_n\}$ reduces to a nondegenerate normal r-dimensional random variable.

The case $n = 2$. As an example, let us consider the case $n = 2$, and study a normal 2-dimensional r.v. $\{X_1, X_2\}$. To simplify the notation, we assume that

$$E(X_1) = \mu_1 = E(X_2) = \mu_2 = 0,$$

since we can always achieve this by a change of variables:

$$X_1 = \mu_1 + X_1', \quad X_2 = \mu_2 + X_2'.$$

Set

$$\gamma_{12} = \gamma_{21} = \sigma_1\sigma_2 r,$$

where r is the correlation coefficient of X_1 and X_2. The c.f. $\phi(u_1, u_2)$ of $\{X_1, X_2\}$ is

$$\varphi(u_1, u_2) = \exp. \left\{ -\frac{1}{2} \ \sigma_1^2 u_1^2 + 2r\, u_1\, u_2 + \sigma_2^2 u_2^2) \right\} \qquad (5,5,36)$$

by Theorem (5,5,1). It then follows from Theorem (5,2,5) that:

THEOREM (5,5,7). Let X_1 and X_2 be two random variables such that $\{X_1, X_2\}$ is a normal 2-dimensional random variable. Then X_1 and X_2 are independent if and only if their correlation coefficient is zero (i.e., the condition is not only necessary but also sufficient). ∎

The 2×2 matrix Γ is

$$\Gamma = \begin{pmatrix} \sigma_1^2 & \sigma_1 \sigma_2 r \\ \sigma_1 \sigma_2 r & \sigma_2^2 \end{pmatrix},$$

and the 2×2 matrix R

$$R = \begin{pmatrix} 1 & r \\ r & 1 \end{pmatrix},$$

whose determinant is

$$\delta = 1 - r^2 ;$$

R^{-1} is thus

$$R^{-1} = \begin{pmatrix} \frac{1}{1 - r^2} & -\frac{r}{1 - r^2} \\ -\frac{r}{1 - r^2} & \frac{1}{1 - r^2} \end{pmatrix} = \frac{1}{1 - r^2} \begin{pmatrix} 1 & -r \\ -r & 1 \end{pmatrix}.$$

$\{X_1, X_2\}$ is degenerate only in one of the following cases:

$$\sigma_1 = 0 \; ; \; \sigma_2 = 0 \; ; \; r = \pm 1.$$

If $\{X_1, X_2\}$ has rank 2, the canonical form (5,5,25) of its density $f(x_1, x_2)$ is

$$f(x_1, x_2) = \frac{1}{2 \pi \sqrt{1 - r^2}\, \sigma_1 \sigma_2} e^{-\frac{1}{2(1-r^2)} \left(\frac{x_1^2}{\sigma_1^2} + 2r \frac{x_1}{\sigma_1} \frac{x_2}{\sigma_2} + \frac{x_2^2}{\sigma_2^2} \right)}. \quad (5,5,37)$$

Conservation of normality

Let $\mathfrak{X}_n$ be a real n-dimensional vector space with elements z, and $\mathcal{Y}_q$ a real q-dimensional vector space with elements t, where n and q are finite. Let λ_0: $t = \lambda_0(z)$ be a linear mapping of $\mathfrak{X}_n$ into $\mathcal{Y}_q$, t_0 some fixed element of $\mathcal{Y}_q$, and λ: $t = \lambda(z)$ the mapping of $\mathfrak{X}_n$ into $\mathcal{Y}_q$ defined by

$$t = \lambda(z) = t_o + \lambda_o(z).$$

Let Z be a normal random vector of rank r in $\mathfrak{X}_n$. We have already seen (at the beginning of this section) that the random vector T in $\mathcal{Y}_q$ defined by

$$T = \lambda(Z) = t_0 + \lambda_0(Z) \tag{5,5,38}$$

is normal; by using the preceding arguments, this result can easily be improved as follows.

THEOREM (5,5,8). If Z is a normal random vector of rank r in $\mathfrak{X}_n$, then the random vector T in $\mathcal{Y}_q$ defined by (5,5,38), where λ_0 is a linear mapping of rank ρ of $\mathfrak{X}_n$ into $\mathcal{Y}_q$, is normal, and its rank is equal to the smaller of the two numbers r and ρ. Moreover, the (n + q)-dimensional random variable {Z, T} is clearly normal (though degenerate, of rank r).

EXAMPLE (5,5,1). Let **b** be an orthonormal basis in $\mathfrak{X}_n$, and a an element in whose vector of components relative to **b** is $°\alpha = \{\alpha_1, \ldots, \alpha_n\}$. Let us assume that a is a unit vector:

$$\sum_{k=1}^{n} \alpha_k^2 = 1 \,. \tag{5,5,39}$$

In the dual $\mathfrak{X}_n^*$ of $\mathfrak{X}_n$, consider the basis **b*** dual to **b**. Let a* be the element in $\mathfrak{X}_n^*$ whose vector of components relative to **b*** is $°\alpha = (\alpha_1, \ldots, \alpha_n)$. Then, by (5,5,39)

$$°\alpha\, \alpha = \langle a^*, a \rangle = 1. \tag{5,5,40}$$

Consider the random vector $Z \in \mathfrak{X}_n$ whose characteristic functional $\xi(z^*)$ has the analytical expression

$$\xi(z^*) = \exp. \left\{ -\frac{1}{2} H(u) \right\} \tag{5,5,41}$$

relative to the basis **b**, where H(u) is the quadratic form

$$H(u) = °u\, u - (°u\, \alpha)^2 = \sum_{k=1}^{n} u_k^2 - \Big(\sum_{k=1}^{n} \alpha_k u_k\Big)^2 .$$

Since

$$\Big|\sum_{k=1}^{n} \alpha_k u_k\Big|^2 = |°u\, \alpha|^2 \leqq \Big(\sum_{j=1}^{n} \alpha_j^2\Big)\Big(\sum_{k=1}^{n} u_k^2\Big),$$

it follows from (5,5,39) that H(u) is positive-definite. Therefore Z is normal, but (5,5,40) implies that H(u) vanishes for $u = \alpha$, which is a nonzero matrix; thus H(u) is degenerate, and Z is degenerate of rank $n - 1$. To be precise, since by definition (cf. (5,5,7)), $H(u) = E(< z^*, Z >^2)$, we see that

$$E(< a^*, Z >^2) = 0,$$

i.e.,

$$< a^*, Z > = 0 \qquad (5,5,42)$$

Let $\mathfrak{X}_{n-1}$ be the $(n - 1)$-dimensional subspace of $\mathfrak{X}_n$ consisting of those elements t of $\mathfrak{X}_n$ such that $< a^*, t > = 0$. We see that the normal mass distribution defined by (5,5,41) is not distributed over the entire space $\mathfrak{X}_n$, but only over $\mathfrak{X}_{n-1}$, and there it defines an $(n - 1)$-dimensional nondegenerate normal law $\mathfrak{L}$. It is easy to determine $\mathfrak{L}$. Let $\mathbf{b}'$ be an orthonormal basis in $\mathfrak{X}_n$ *whose* n^{th} *element* is a, its other elements arbitrary. Let T be the $n \times n$ matrix for change of basis $\mathbf{b}$ to $\mathbf{b}'$ in $\mathfrak{X}_n$. T is orthogonal, ${}^\circ T = T^{-1}$. Using formulas (5,5,10), (5,5,11) etc., and the corresponding notation, we see that the analytical expression of $\xi(z^*)$ relative to $\mathbf{b}'$ is

$$\xi(z^*) = \exp. \left\{ -\frac{1}{2} K(v) \right\}, \qquad (5,5,43)$$

where

$$K(v) = {}^\circ v\, T^{-1}\, {}^\circ T^{-1}\, v - ({}^\circ v\, T^{-1}\, \alpha)^2 = {}^\circ v\, v - ({}^\circ v\, T^{-1}\, \alpha)^2.$$

But $T^{-1}\alpha$ is the matrix ${}^\circ\{0, 0, \ldots, 0, 1\}$, and so

$$K(v) = {}^\circ v\, v - v_n^2 = \sum_{k=1}^{n} v_k^2 - v_n^2 = \sum_{k=1}^{n-1} v_k^2. \qquad (5,5,44)$$

If we denote the components of Z relative to $\mathbf{b}'$ by $Y = \{Y_1, Y_2, \ldots, Y_n\}$, formula (5,5,42) becomes

$$Y_n = 0 \text{ almost certainly.}$$

The required law $\mathfrak{L}$ is that of the $(n - 1)$-dimensional r.v. $\{Y_1, \ldots, Y_{n-1}\}$. From (5,5,43), (5,5,44) and (5,5,17) it follows that the probability density of $\mathcal{L}$ is

$$g(y_1, \ldots, y_{n-1}) = \frac{1}{(2\pi)^{\frac{n-1}{2}}} \exp. \left\{ -\frac{1}{2} \sum_{k=1}^{n-1} y_k^2 \right\}. \qquad (5,5,45)$$

Let $°x = \{x_1, \ldots, x_n\}$ be the components, relative to **b**, of a vector $z \in \mathfrak{X}_n$ which has components $°y = \{y_1, \ldots, y_{n-1}, y_n\}$ relative to **b'**. Since a is a unit vector and T is orthogonal,

$$°\alpha x = \langle a^*, x \rangle = y_n = \sum_k \alpha_k x_k ; \quad \sum_{k=1}^{n} y_k^2 = \sum_{k=1}^{n} x_k^2 . \qquad (5,5,46)$$

Now if $z \in \mathfrak{X}_{n-1}$, i.e. $y_n = \sum_{k=1}^{n} \alpha_k x_k = 0$, formula (5,5,46) becomes

$$\sum_{k=1}^{n-1} y_k^2 = \sum_{k=1}^{n} x_k^2 .$$

We thus see that, relative to the initial basis **b** and in terms of the initial variables $\{x_1, \ldots, x_n\}$, the density $f(x_1, \ldots, x_n)$ of $\mathcal{L}$ is

$$f(x_1, \ldots, x_n) = \frac{1}{(2\pi)^{\frac{n-1}{2}}} \exp. \left\{ -\frac{1}{2} \sum_{k=1}^{n} x_k^2 \right\}, \qquad (5,5,47)$$

where the x_k satisfy the relation

$$\sum_{k=1}^{n} \alpha_k x_k = 0. \qquad (5,5,48)$$

6. *Conditional probability laws of normal* n*-dimensional random variables*

Let Z be a normal random vector in $\mathfrak{X}_n$, and let $x_1^*, x_2^*, \ldots, x_s^*$ be linear functionals on $\mathfrak{X}_n$, i.e. s elements of the dual $\mathfrak{X}_n^*$ of $\mathfrak{X}_n$. We wish to study the probability law of Z, under the condition that the s numbers $\langle x_h^*, Z \rangle$ have arbitrary given values x_h:

$$\langle x_h^*, Z \rangle = x_h \quad (h = 1, 2, \ldots, s).$$

There is no essential loss of generality in assuming that the functionals x_h^* are linearly independent. This implies $s \leqq n$, and we shall assume $s < n$, since for $s = n$ the problem formulated above is trivial.

Let us choose a basis $\mathbf{b}^*$ in $\mathfrak{X}_n^*$ whose first s elements are $x_1, \ldots, x_s^*$. Let **b** be the basis in $\mathfrak{X}_n$ whose dual basis is $\mathbf{b}^*$, and $\{X_1, X_2, \ldots, X_n\}$ the components of Z relative to **b**; then

$$< x^*_h, Z > = X_h \quad (h = 1, 2, \ldots, s).$$

If Y denotes the s-dimensional r.v. $\{X_1, \ldots, X_s\}$ and T the (n - s)-dimensional r.v. $\{X_{s+1}, X_{s+2}, \ldots, X_n\}$ the problem reduces to the following:

a) to determine the probability law of Y;

b) to determine the conditional probability of T, given an arbitrary fixed value of Y.

Of the previous notation we shall use the following, with the same meaning:

We introduce the following additional notation:

ν = $s \times 1$ matrix with elements μ_h $(h = 1, 2, \ldots, s)$;

ω = $(n - s) \times 1$ matrix with elements μ_h $(h = s + 1, \ldots, n)$;

X' = $X - E(X)$ = $n \times 1$ matrix with elements $X_h - \mu_h$ $(h = 1, 2, \ldots,$ r

Y' = $Y - E(Y)$ = $s \times 1$ matrix with elements $Y_h - \mu_h$ $(h = 1, 2, \ldots,$ s

T' = $T - E(T)$ = $(n - s) \times 1$ matrix with elements $X_h - \mu_h$ $(h = s + 1, s + 2, \ldots, n)$;

B = $s \times s$ matrix with elements γ_{jk} $(j, k = 1, 2, \ldots, s)$;

C = $(n - s) \times (n - s)$ matrix with elements γ_{jk} $(j, k = s + 1, \ldots,$

L = $s \times (n - s)$ matrix with elements γ_{jk} $(j = 1, 2, \ldots, s$); $k = s + 1, s + 2, \ldots, n)$;

y = $s \times 1$ matrix with elements x_h $(h = 1, 2, \ldots, s)$;

t = $(n - s) \times 1$ matrix with elements x_h $(h = s + 1, \ldots, n)$;

v = $s \times 1$ matrix with elements u_h $(h = 1, 2, \ldots, s)$;

w = $(n - s) \times 1$ matrix with elements u_h $(h = s + 1, \ldots, n)$;

Δ_s = determinant of the matrix B^{-1} (if it exists);

$\mathcal{B}$ = $s \times s$ matrix with elements a_{jk} $(j, k \leqq s)$;

$\mathcal{C}$ = $(n - s) \times (n - s)$ matrix with elements a_{jk} $(j, k > s)$;

$\mathcal{L}$ = $s \times (n - s)$ matrix with elements a_{jk} $(j \leqq s, k > s)$;

θ = $s \times 1$ matrix with elements τ_h $(h \leqq s)$;

ε = $(n - s) \times 1$ matrix with elements τ_h $(h > s)$;

Δ_{n-s} = determinant of $\mathcal{C}$.

Y is clearly obtained from Z by a linear mapping of rank s; by Theorem (5,5,8), Y is therefore a normal s-dimensional r.v. The covariance matrix of Y' is B. By Theorem (5,5,1) the characteristic function $\psi(v)$ of Y is

$$\psi(v) = \exp. \left\{ -\frac{1}{2} \,{}^\circ v \, B \, v + i \,{}^\circ v \, \nu \right\}. \tag{5,6,1}$$

If Y is nondegenerate, its probability density $g(x_1, \ldots, x_n)$ is, by (5,5,22),

$$g(x_1, \ldots, x_n) = \frac{\sqrt{\Delta_s}}{(2\pi)^{s/2}} \exp. \left\{ -\frac{1}{2} \,{}^\circ(y - \nu) \, B^{-1} \, (y - \nu) \right\}. \tag{5,6,2}$$

There is no difficulty in examining the case in which Y is degenerate. By the same arguments we obtain the characteristic function and (if it exists) the a priori probability density of T.

A necessary condition for Y and T to be independent is that each X_j with $j \leqq s$ be independent of each X_k with $k > s$, i.e., if $\gamma_{jk} = 0$ for $j \leqq s$ and $k > s$, so that L is the zero matrix. Conversely, if L is the zero matrix, we use the equality $\gamma_{jk} = \gamma_{kj}$ to show that

$$-\frac{1}{2} \,{}^\circ u \, \Gamma u + i \,{}^\circ u \, \mu = \left(-\frac{1}{2} \,{}^\circ v \, B \, v + i \,{}^\circ v \, \nu \right) + \left(-\frac{1}{2} \,{}^\circ w \, C \, w + i \,{}^\circ w \, \omega \right).$$

From (5,5,8) and Theorem (5,3,1), using Theorem (5,5,7), we obtain:

THEOREM (5,6,1). Let $Y = \{X_1, \ldots, X_s\}$ be an s-dimensional random variable and $T = \{X_{s+1}, \ldots, X_n\}$ an (n - s)-dimensional random variable.

1) If the n-dimensional random variable (Y, T) is normal (this implies that Y is normal s-dimensional and T is normal (n - s)-dimensional), then for Y and T to be independent it is not only necessary, but also sufficient, that each X_j with $j \leqq s$ be independent of each X_k with $k > s$.

2) If both Y and T are normal, and Y and T are independent, then the pair (Y, T) is a normal n-dimensional random variable. ∎

We resume the general case, where L is not necessarily zero.

Let $\mathcal{H}'$ be the Hilbert subspace of the space $\mathcal{H}$ of all second-order random variables, consisting of all homogeneous linear combinations of the X'_h (h = 1,..., s). Consider the orthogonal projection of $X'_{s+\alpha}$ (α = 1, 2,..., n - s) on $\mathcal{H}'$. It can be represented by

$$°Y' \Lambda_\alpha = °\Lambda_\alpha Y'$$

where Λ_α is a suitable s × 1 matrix with elements $\lambda_{h\alpha}$ (h = 1, 2,..., s). Let Λ be the s × (n - s) matrix with elements $\lambda_{h\alpha}$ (h = 1, 2,..., s; α = 1,..., n - s), so that Λ_α is the αth column of Λ. Set

$$X_{s+\alpha} = \mu_{s+\alpha} + °\Lambda_\alpha Y' + Q_\alpha \qquad (\alpha = 1, 2, \ldots, n-s), \qquad (5,6,3)$$

and denote by Q the random matrix $°\{Q_1, \ldots, Q_{n-s}\}$, so that, in matrix notation, (5,6,3) becomes

$$T = \omega + °\Lambda(Y - \nu) + Q . \qquad (5,6,4)$$

Note that

$$E(Q_\alpha) = 0 \qquad (\alpha = 1, 2, \ldots, n-s),$$

and each Q_α is orthogonal in $\mathcal{H}$ to each X'_j ($j \leqq s$). Q is obtained from Z by a linear transformation. It follows from Theorems (5,5,8) and (5,6,1) that Q is a normal (n - s)-dimensional r.v., independent of Y and Y'. Therefore,

$$E(T/Y = y) = \omega + °\Lambda(y - \nu), \qquad (5,6,5)$$

and, on the other hand,

$$E(Q_\alpha Q_\beta) = \gamma_{s+\alpha, s+\beta} - \sum_{h,l=1}^{s} \lambda_{h\alpha}\lambda_{l\beta}\gamma_{hl} = \gamma_{s+\alpha, s+\beta} - °\Lambda_\alpha B \Lambda_\beta .$$

Hence the covariance matrix of Q is

$$C - °\Lambda B \Lambda , \qquad (5,6,6)$$

and the characteristic function of Q is

$$\exp.\left\{ - \frac{1}{2} °w(C - °\Lambda B \Lambda) w \right\}. \qquad (5,6,7)$$

Therefore, given fixed values x_h of the r.v. Y_h $(h = 1, 2, \ldots, s)$, the conditional probability law of T is normal, and its characteristic function $\eta(w/y)$ is easily obtained from (5,6,5) and (5.6.6):

$$\eta(w/y) = \exp.\left\{-\frac{1}{2}\,{}^\circ w(C - {}^\circ\Lambda B \Lambda)\,w + i{}^\circ w\,[\omega + {}^\circ\Lambda(y - \nu)]\right\}. \quad (5,6,8)$$

We note two important points:

1) $E(T/Y = y)$ is a linear function of y; more precisely, formulas (5,6,4) and (5,6,5) show that for each $X_{s+\alpha}$ $(\alpha = 1, 2, \ldots, n - s)$ the best conditional estimator $X_{s+\alpha}$ with respect to Y (of type (5,2,12)), in the sense of minimum mean-square deviation, coincides with the best *linear* estimator (of type (5,2,27) or (5,4,32)).

2) The conditional covariance matrix (5,6,6) of $T - E(T/Y = y)$, given $Y = y$, is *independent of* y.

Recall that in V.4 we saw how to determine the Λ_α and thus Λ. Now consider, in particular, the case where Y is nondegenerate. By (5,4,30),

$$\Lambda_\alpha = B^{-1} L_\alpha ,$$

where L_α is the αth row of L. Thus

$$\Lambda = B^{-1} L .$$

Since ${}^\circ B = B$, the conditional m.e. $E(T/Y = y)$ (5,6,5) can be expressed in the form

$$E(T/Y = y) = \omega + {}^\circ L\, B^{-1}(y - \nu), \quad (5,6,9)$$

while the covariance matrix (5.6,6) is

$$C - {}^\circ L\, B^{-1} L , \quad (5,6,10)$$

and the conditional characteristic function $\eta(w/y)$ becomes

$$\eta(w/y) = \exp.\left\{-\frac{1}{2}\,{}^\circ w(C - {}^\circ L\, B^{-1} L)\,w + i\,{}^\circ w\,[\omega + {}^\circ L\, B^{-1}(y - \nu)]\right\}. \quad (5,6,11)$$

If the covariance matrix (5,6,10) has rank $n - s$, an expression for the conditional density $h(x_{s+1}, \ldots, x_n/y)$ of T, given $Y = y$, is easily deduced from (5,6.11).

These results solve the problems formulated above, for the probability law of Z is defined by Γ and μ, which were assumed given. Now let us assume that Z has rank n, and that its probability law is defined by a given probability density $f(x_1, \ldots, x_n)$ of the form (5,5,26), i.e. (see (5,5,32)) by given matrices A and τ. One can then reduce the problem to the previous formulas (5,6,1), (5,6,5), (5,6,8), (5,6,11), using

$$\Gamma = A^{-1} \text{ et } \mu = A^{-1}\tau.$$

However, a direct approach is also possible. Assume that the s functionals x_j^* (j = 1, 2,..., s) are linearly independent, so that Y has rank s and T rank n - s; then

$$f(x_1, \ldots, x_n) = \frac{\sqrt{\Delta}}{(2\pi)^{n/2}} \exp. \left\{ -\frac{1}{2} K \right\},$$

where

$$K = {}^\circ x\, A\, x - 2\, {}^\circ\tau\, x + {}^\circ\tau\, A^{-1}\, \tau .$$

We immediately see that

$$K = {}^\circ y\, \mathcal{B}\, y - 2\, {}^\circ\theta\, y + {}^\circ\tau\, A^{-1}\, \tau + {}^\circ t\, \mathcal{C}\, t - 2\, {}^\circ\varepsilon\, t + 2\, {}^\circ y\, \mathcal{L}\, t .$$

Now set

$$t = \mathcal{C}^{-1} (\varepsilon - {}^\circ\mathcal{L}\, y) + \bar{t},$$

$$y = (\mathcal{B} - \mathcal{L}\, \mathcal{C}^{-1}\, {}^\circ\mathcal{L})^{-1} (\theta - \mathcal{L}\, \mathcal{C}^{-1}\, \varepsilon) + y',$$

$$K_o = {}^\circ\tau\, A^{-1}\, \tau - {}^\circ\varepsilon\, \mathcal{C}^{-1} \varepsilon - {}^\circ(\theta - \mathcal{L}\mathcal{C}^{-1}\, \varepsilon)\, (\mathcal{B} - \mathcal{L}\, \mathcal{C}^{-1}\, {}^\circ\mathcal{L})\, (\theta - \mathcal{L}\mathcal{C}^{-1}\, \varepsilon) .$$

Then

$$K = {}^\circ\bar{t}\, \mathcal{C}\, \bar{t} + {}^\circ y' (\mathcal{B} - \mathcal{L}\, \mathcal{C}^{-1}\, {}^\circ\mathcal{L})\, y' + K_o,$$

$$f(x_1, \ldots, x_n) = \frac{\sqrt{\Delta}}{(2\pi)^{n/2}}\, e^{-\frac{1}{2} K_o} \times e^{-\frac{1}{2} {}^\circ\bar{t}\, \mathcal{C}\, \bar{t}} \times e^{-\frac{1}{2}\, {}^\circ y' (\mathcal{B} - \mathcal{L}\, \mathcal{C}^{-1}\, {}^\circ\mathcal{L}) y'} .$$

Using Remark (5,5,1), formulas (5,1,12) and (5,2,5), and comparing the results with (5,6,2) and (5,6,8), we obtain:

RULE (5,6,1). Let Z have rank n and Y rank s. If the probability law of the pair (Y, T) is given by the matrices A and τ, then the matrix μ, and therefore also the matrices ν and ω, are determined by $\mu = A^{-1}\tau$. The a priori probability density $g(x_1, \ldots, x_s)$ of Y is then

$$g(x_1, \ldots, x_s) = \frac{\sqrt{\Delta_s}}{(2\pi)^{s/2}} \exp.\left\{-\frac{1}{2}\,{}^\circ(y-\nu)\,(\mathcal{B} - \mathcal{L}\,\mathcal{C}^{-1}\,{}^\circ\mathcal{L})\,(y-\nu)\right\}, \quad (5,6,12)$$

where Δ_s is the determinant of the matrix $(\mathcal{B} - \mathcal{L}\,\mathcal{C}^{-1}\,{}^\circ\mathcal{L})$; the conditional mathematical expectation $E(T/Y = y)$ of T with respect to Y is given by

$$E(T/Y = y) = \omega - \mathcal{C}^{-1}\,{}^\circ\mathcal{L}\,(y - \nu). \quad (5,6,13)$$

and the conditional probability density $h(x_{s+1}, \ldots, x_n/y)$ of T, given Y, is

$$h(x_{s+1}, \ldots, x_n/y) = \frac{\sqrt{\Delta_{n-s}}}{(2\pi)^{(n-s)/2}} \exp.\left\{-\frac{1}{2}\,{}^\circ[t-E(T/Y=y)]\,\mathcal{C}\,[t-E(T/Y=y)]\right\}. \quad (5,6,14)$$

The case n = 2. We now consider the case n = 2, using the above notation with n = 2 and assuming $\mu_1 = \mu_2 = 0$, σ_1 and $\sigma_2 \neq 0$. The matrices B, C, L, ν, ω are 1×1 matrices, with single elements

$$\sigma_1^2,\ \sigma_2^2,\ \sigma_1\sigma_2 r,\ 0,\ 0;$$

respectively. B^{-1} is the 1×1 matrix with element $\frac{1}{\sigma_1^2}$ and $\Delta_s = \frac{1}{\sigma_1^2}$; the characteristic function $\psi(v)$ of $Y = X_1$ and its density $g(x_1)$ are

$$\psi(v) = e^{-\frac{1}{2}\sigma_1^2 v^2}, \quad g(x_1) = \frac{1}{\sqrt{2\pi}\,\sigma_1} e^{-\frac{x_1^2}{2\sigma_1^2}}$$

by (5,6,1) and (5,6,2). By (5,6,9), the conditional m.e. of $T = X_2$, given $X_1 = x_1$, is

$$E(X_2/X_1 = x_1) = \frac{\sigma_2}{\sigma_1} r x_1; \quad (5,6,15)$$

the characteristic function $\eta(w/x_1)$ and the conditional probability density $h(x_2/x_1)$ of X_2, given $X_1 = x_1$ are, by (5,6,11),

$$\eta(w/y) = \exp.\left\{-\frac{1}{2}\sigma_2^2(1-r^2)\,w^2\right\}, \tag{5,6,16}$$

$$h(x_2/x_1) = \frac{1}{\sqrt{2\pi}\,\sigma_2\sqrt{1-r^2}}\exp.\left\{-\frac{\left(x_2-\frac{\sigma_2}{\sigma_1}r\,x_1\right)^2}{2\,\sigma_2^2\,(1-r^2)}\right\}. \tag{5,6,17}$$

Complex normal random variables

A *complex* r.v. Z with real part X and imaginary part Y,

$$Z = X + iY,$$

is said to be normal if the *real* r.v. $\{X, Y\}$ is normal 2-dimensional.

Let $Z = \{Z_1,\ldots, Z_n\}$ be a *complex* n-dimensional r.v., where Z_j has real part X_j and imaginary part Y_j; we call the *complex* r.v. $\{Z_1,\ldots, Z_n\}$ normal n-dimensional, if the *real* r.v. $T = \{X_1,\ldots, X_n; Y_1,\ldots, Y_n\}$ is normal 2n-dimensional.

We may assume, without loss of generality, that $E(X_j) = E(Y_j) = 0$ $(j = 1, 2,\ldots, n)$. Let Γ be the (complex) covariance matrix of Z. Even in the case n = 1, and a fortiori for arbitrary n, Γ is not sufficient to determine the probability law of Z, i.e. the probability law of T, in contrast to the real case.

Let Γ_{XX} and Γ_{YY} denote the covariance matrices of the real n-dimensional r.v. $X = \{X_1,\ldots, X_n\}$, $Y = \{Y_1,\ldots, Y_n\}$; matrices with elements $E(X_jY_k)$, $E(Y_jX_k)$, respectively $(j, k = 1, 2,\ldots, n)$; Ω the covariance matrix of T. The probability law of T is determined by Ω, which is equivalent to the quadruple $(\Gamma_{XX}, \Gamma_{YY}, \Gamma_{XY}, \Gamma_{YX})$ But for a given quadruple $(\Gamma_{XX}, \Gamma_{YY}, \Gamma_{XY}, \Gamma_{YX})$ to define a matrix Ω which is a covariance matrix, i.e. symmetric and positive-definite, the following conditions are necessary and sufficient:

a) Γ_{XX}, Γ_{YY} are symmetric;

b) $\Gamma_{YX} = {}^{\circ}\Gamma_{XY}$;

c) for any real $n \times 1$ matrices u and v,

$$°u\,\Gamma_{XX}\,u + °v\,\Gamma_{YY}\,v + °v\,\Gamma_{XY}\,u + °u\,\Gamma_{YX}\,v \geqslant 0. \qquad (5,6,18)$$

On the other hand,

$$E(Z_j\,\overline{Z}_k) = E[(X_j + iY_j)\,(X_k - iY_k)],$$

and by separating the real part $\frac{1}{2}(\overline{\Gamma} + \Gamma)$ and the imaginary part $\frac{1}{2}(\overline{\Gamma} - \Gamma)$ of Γ we see that

$$\Gamma_{XX} + \Gamma_{YY} = \frac{1}{2}(\overline{\Gamma} + \Gamma),\; \Gamma_{YX} - \Gamma_{XY} = \frac{i}{2}(\overline{\Gamma} - \Gamma). \qquad (5,6,19)$$

Any quadruple $(\Gamma_{XX}, \Gamma_{YY}, \Gamma_{XY}, \Gamma_{YX})$ satisfying conditions a), b), c) and (5,6,19) determines a normal probability law of Z, which is compatible with the covariance matrix Γ. For given Γ, except for special cases, there are obviously many quadruples $(\Gamma_{XX}, \Gamma_{YY}, \Gamma_{XY}, \Gamma_{YX})$ satisfying conditions a), b), c) and (5,6,19); one of these is defined by

$$\Gamma_{XX} = \Gamma_{YY} = \frac{1}{4}(\Gamma + \overline{\Gamma}),\; \Gamma_{YX} = -\Gamma_{XY} = \frac{i}{4}(\overline{\Gamma} - \Gamma). \qquad (5,6,20)$$

In fact, since Γ is symmetric and positive-definite, the quadruple (5,6,20) clearly satisfies (5,6,19), a) and b); moreover, the first term of (5,6,18) is, up to a factor 2, $°(u + iv)\Gamma(u - iv)$, which is clearly real and nonnegative for any u and v.

REMARK (5,6,1). Consider the case n = 2, and assume that $Z_2 = \overline{Z}_1$; then the r.v. Z_1 and Z_2 are clearly not independent. We have

$$E(Z_1\,\overline{Z}_2) = E(Z_1^2) = E(X_1^2 - Y_1^2 + 2i\,X_1\,Y_1).$$

Assume that X_1 and Y_1 are independent normal real r.v. such that $E(X_1) = E(Y_1) = 0$, $E(X_1^2) = E(Y_1^2)$; $Z = \{Z_1, Z_2\}$ is clearly a normal 2-dimensional complex r.v., and the covariance $E(Z_1\overline{Z}_2)$ of Z_1 with Z_2 is *zero*; thus:

If $\{Z_1, Z_2\}$ is a normal 2-dimensional complex r.v., then a necessary condition for Z_1 and Z_2 to be independent is that the correlation coefficient of Z_1 with respect to Z_2 vanish; however, in contrast to the real case, *this condition is not sufficient.*

BIBLIOGRAPHIC NOTES FOR CHAPTER V

More or less detailed accounts of n-dimensional r.v., particularly normal r.v., may be found in all works on the elements of Probability Theory; cf., e.g., those mentioned in the Bibliographical Note for Chapter II, in particular H. Cramer [4].

The concept of the best estimate, in particular, the best linear estimate in the sense of minimum mean-square deviation, and its relation to the geometry of the Hilbert space of second-order random variables, arises more or less explicitly in many problems of mathematical statistics; on this subject, cf. H. Cramer [4], M.G. Kendall and A. Stuart [1], C. R. Rao [1], E.L. Lehman [1], and also H. Wold [1], A. Blanc-Lapierre and R. Fortet [1], U. Grenander and M. Rosenblatt [1].

The concepts of vector spaces, in particular finite-dimensional and of matrices, which are useful in the study of n-dimensional r.v. can be found, for example, in A Lichnerowicz [1], R. Fortet [4].

EXERCISES

5.1. Let R and Φ be two independent r.v.; R is nonnegative, with probability density f(r) and finite variance. Φ is a random angle in $[0, 2\pi[$, uniformly distributed on this interval. Define coordinate axes Ox, Oy in the plane, and let M be a random point with cartesian coordinates (X, Y) and polar coordinates (R, Φ). Express the probability density, m.e. and variance of X in terms of the density f(r). Also, determine the m.e. of the product XY, and the conditional probability density of Y, given X = x. Examine the special case $f(r) = 2re^{-r^2}$.

What conditions must f(r) satisfy for X and Y to be independent? (*Hint*: Go from the 2-dimensional r.v. {X, Y} to the 2-dimensional r.v. {R, Φ} by the change of variables $x = r \cos \phi$; $y = r \sin \phi$; use (2,11,13) and then the results of V.1 and V.2).

5.2. Let X_1, $X_2, \ldots, X_n$ be n mutually independent r.v., each uniform in the interval [0, 1]. Let Y_1, $Y_2, \ldots, Y_n$ be the values of the X_j, but in order of increasing magnitude (cf. Exercise 4.2). Find the probability law of the n-dimensional r.v. $\{Y_1, \ldots, Y_n\}$, that of Y_1, that of the 2-dimensional r.v. $\{Y_1, Y_n\}$, the conditional probability law of Y_n given $Y_1 = y$, the regression curve of Y_n on Y_1.

5.3. Let $X = \{X_1, \ldots, X_n\}$ and $Y = \{Y_1, \ldots, Y_n\}$ be complex second-order n-dimensional r.v. Set $Z_k = X_k + Y_k$ $(k = 1, 2, \ldots, n)$ and $Z = \{Z_1, \ldots, Z_n\}$. Let R, S, T be the $n \times n$ matrices with elements $E(X_j\bar{X}_k)$, $E(\ _j\bar{Y}_k)$, $E(X_j\bar{Y}_k)$, respectively. Assuming R, S, T to be given, find the covariance matrix Γ of Z. Let w_j $(j = 1, \ldots, n)$ be n given positive real numbers and M an $n \times n$ complex matrix (m_{jk}). Set $e = E\left(\sum_{j=1}^{n} w_j \left|X_j - \sum_{k=1}^{n} m_{jk} Z_k\right|^2\right)$, and let $\hat{M}$ be the matrix M which minimizes e. Show that $\hat{M}$ is independent of the w_j; find $\hat{M}$ and the corresponding minimum of e (cf. V.4).

5.4. Let $\{X, Y, Z\}$ be a normal 3-dimensional real r.v. with $E(X) = E(Y) = E(Z) = 0$. Set $E(X^2) = \sigma_X^2$, $E(Y^2) = \sigma_Y^2$, $E(Z^2) = \sigma_Z^2$, $E(XY) = \sigma_X \sigma_Y r_{XY}$, $E(XZ) = \sigma_X \sigma_Z r_{XZ}$, $E(YZ) = \sigma_Y \sigma_Z r_{YZ}$. What condition must these moments satisfy so that, for every $x \in (-\infty, +\infty)$, Y and Z are conditionally independent given $X = x$? (Answer: $r_{YZ} = r_{XY}r_{XZ}$).

5.5. Let X and Y be two independent r.v. with distribution functions F(x) and G(y), respectively, and Z the r.v. $Z = Y/X$. Assume that $\Pr(X = 0) = F(+0) - F(0) = 0$, so that Z is nondegenerate.

1) Assuming that F(x) and G(y) are given, find the d.f. H(z) and the characteristic function $\phi(u) = E(e^{iuZ})$ of Z.

2) Show, in particular, that if X and Y are normal, $E(X) = E(Y) = 0$, $E(X^2) = E(Y^2) = 1$, then H(z) is absolutely continuous,

with density

$$h(z) = \frac{1}{\pi} \frac{1}{1 + z^2} \quad \text{(Cauchy's law, cf. VI.2)}$$

(Answer: 1) $H(z) = \int_{-\infty}^{0} [1 - G(zx + 0)]\, dF(x) + \int_{0}^{-\infty} G(zx)\, dF(x);$

$$\phi(u) = \int_{-\infty}^{+\infty}\int_{-\infty}^{+\infty} e^{iuy/x}\, dF(x)\, dG(y);$$

2) $\phi(u) = \frac{1}{\pi}\int_{-\infty}^{+\infty} e^{iuz} \frac{dz}{1 + z^2}$ (cf. Exercise 3.5)).

CHAPTER VI

ADDITION OF INDEPENDENT RANDOM VARIABLES;

STOCHASTIC CONVERGENCE,

LAWS OF LARGE NUMBERS, ERGODIC THEOREMS;

CONVERGENCE TO A NORMAL LAW,

CONVERGENCE TO A POISSON LAW;

GENERALIZATIONS

I. ADDITION OF INDEPENDENT RANDOM VARIABLES

1. *Addition of random variables*

Let $X_1, X_2, \ldots, X_k$ be random variables; then their sum S,

$$S = X_1 + X_2 + \cdots + X_k,$$

i.e., the quantity which by definition assumes the value $x_1 + x_2 + \ldots + x_k$ when $X_1, \ldots, X_k$ assume the values $x_1, \ldots, x_k$, (cf. Example (2,11,6)) is also a r.v. Given the probability law of the k-dimensional r.v. $\{X_1, \ldots, X_k\}$ that of S is uniquely determined. The main interest here lies in determining what may be said of the probability law of S on the basis of partial information concerning the probability law of $\{X_1, \ldots, X_k\}$.

This problem is one of the principal subjects of probability theory, first because addition is the simplest operation one can perform on numbers, whether random or not. Second, the problem arises unavoidably in connection with the common case in which the differences $Y_j = X_j - E(X_j)$ remain relatively small. Indeed, let $\lambda(x_1, \ldots, x_k)$ be any function of the k variables $x_1, \ldots, x_k$ and consider the r.v. Z defined by

$$Z = \lambda(X_1, \ldots, X_k). \tag{6,1,1}$$

Under quite broad conditions, we have the approximation

$$\lambda(x_1, \ldots, x_k) \;\#\; \lambda(m_1, \ldots, m_k) + \sum_{j=1}^{k} \frac{\partial\lambda}{\partial x_j\,[x_h = m_h]} \cdot y_j\,, \tag{6,1,2}$$

where we have set $m_j = E(X_j)$, $y_j = x_j - m_j$. Thus, if the approximation (6,1,2) is admissible, the study of Z, apart from the certain term $\lambda(m_1, \ldots, m_k)$, reduces to the study of the addition of the r.v.

$$Z_j = \frac{\partial\lambda}{\partial x_j\,[x_h = m_h]} \cdot Y_j\,,$$

where $\dfrac{\partial\lambda}{\partial x_j\,[x_h = m_h]}$ is a certain number and $Y_j = X_j - E(X_j)$.

Obviously, study of the sum of k r.v. for any k follows easily from the results for k = 2: in the sequel we shall confine ourselves to studying the sum

$$S = X + Y$$

of two r.v. X and Y.

We have already touched upon this subject in Theorem (4,2,1); because of its importance we restate the theorem:

THEOREM (4,2,1). If the mathematical expectations E(X) and E(Y) of two random variables exist, the expectation E(S) of their sum S = X + Y also exists, and is the sum of E(X) and E(Y):

$$E(S) = E(X + Y) = E(X) + E(Y). \blacksquare \qquad (6,1,3)$$

This theorem is both simple and general, in the sense that it imposes no restrictions on the 2-dimensional r.v. {X, Y} other than the existence of E(X) and E(Y). Unfortunately, as we shall soon see, it is the only simple result having this degree of generality. Since the theorem deals with the first (algebraic) moment of S, let us consider the second moment. Since $S^2 = X^2 + Y^2 + 2XY$, Theorem (4,2,1) gives

$$E(S^2) = E(X^2) + E(Y^2) + 2\,E(XY). \qquad (6,1,4)$$

Here, the term E(XY), the covariance of X and Y (which may be calculated from the correlation coefficient of X and Y), takes account of the mutual dependence of X and Y and, depending on the nature of this dependence, may produce quite different results This remark is valid a fortiori for higher-order moments, and for the probability law of S as a whole and the various related quantities.

There is, nevertheless, a simple case: that in which X and Y are mutually independent. Here the law of the pair {X, Y} (hence that of S) is indeed determined by the probability laws of X and Y; i.e., data relevant only to X, together with data relevant only to Y, enable one to draw conclusions about S. This will often arise in applications of Theorem (5,2,3), which is worth restating here:

THEOREM (5,2,3). If X and Y are independent random variables with mathematical expectations E(X) and E(Y), then the expectation E(XY) of their product XY exists and is equal to the product of E(X) and E(Y):

$$E(XY) = E(X)\,E(Y). \blacksquare \tag{6,1,5}$$

In particular, application of Theorem (5,2,3) to (6,1,3) yields

$$E(S^2) = E(X^2) + E(Y^2) + 2\,E(X)\cdot E(Y). \tag{6,1,6}$$

There is no loss of generality in assuming E(X) = E(Y) = 0 here, which implies that E(S) = 0; $E(X^2)$, $E(Y^2)$, and $E(S^2)$ are then the variances of X, Y, and S, respectively, so that

$$\mathcal{V}(S) = \mathcal{V}(X) + \mathcal{V}(Y), \tag{6,1,7}$$

and hence the simple but very important result:

THEOREM (6,1,1). The sum S = X + Y of two real, second-order, mutually independent random variables X and Y is a second-order r.v., and its variance is the sum of the variances of X and Y. $\blacksquare$

Henceforth, therefore, we shall limit ourselves to the case of independent X *and* Y.

We first generalize (6,1,6) by considering the αth algebraic moment of S about the origin 0 (α is any positive integer). We have

$$S^\alpha = (X + Y)^\alpha = \sum_{j=0}^{\alpha} C_\alpha^j\, X^j\, Y^{\alpha-j}.$$

Applying Theorems (4,2,1) and (5,2,3) we get

$$E(S^\alpha) = \sum_{j=0}^{\alpha} C_\alpha^j\, E(X^j)\, E(Y^{\alpha-j}) . \tag{6,1,8}$$

Thus, if X and Y have algebraic moments of order up to and including α, their sum S has algebraic moments of order up to α, and the latter may be evaluated by formula (6,1,8), using the moments of X and Y of order up to α.

Two special cases of (6,1,7) are worthy of note: that already seen in (6,1,6), and the other, $\alpha = 3$ and $E(X) = E(Y) = 0$, gives

$$E(S^3) = E[(X + Y)^3] = E(X^3) + E(Y^3). \qquad (6,1,9)$$

REMARK (6,1,1). Let us return to the r.v. Z defined by (6,1,1) assuming $X_1, \ldots, X_k$ to be mutually independent. If the approximation (6,1,2) is admissible, Theorem (6,1,1) gives

$$\mathcal{V}(Z) \# \sum_{j=1}^{k} \left| \frac{\partial \lambda}{\partial x_j \; [x_h = m_h]} \right|^2 \mathcal{V}(X_j), \qquad (6,1,10)$$

an *approximate* formula often utilized in Physics.

The heads-or-tails scheme

Consider the well-known game of heads or tails. Let $\mathcal{E}_1, \mathcal{E}_2, \ldots, \mathcal{E}_j, \ldots$ be a sequence of tosses, and let E_j denote the event "in $\mathcal{E}_j$ the coin turns up heads." Under normal conditions, the E_j are mutually independent, and $Pr(E_j)$ has a fixed value p (equal to 1/2). This example suggests the following more general model:

A *heads-or-tails scheme* is a countable sequence $E_1, \ldots, E_j, \ldots$ of mutually independent events, where E_j is the event "heads on the jth toss." Let X_j be the indicator of E_j, i.e., the r.v. which is 1 if E_j occurs and 0 otherwise; study of the E_j's then amounts to study of the mutually independent random variables X_j. Set

$$p_j = Pr(E_j) = Pr(X_j = 1)$$

$$q_j = Pr(\check{E}_j) = Pr(X_j = 0) = 1 - p_j .$$

We know (cf. Example (4,2,5)) that

$$E(X_j) = p_j, \quad \mathcal{V}(X_j) = p_j\, q_j \,. \tag{6,1,11}$$

Let

$$R_k = X_1 + X_2 + \cdots + X_k, \quad F_k = R_k/k \,. \tag{6,1,12}$$

We shall call R_k the *recurrence of heads* in the first k tosses - it is in fact equal to the number of events $E_1, \ldots, E_k$ which occur. F_k is the (relative) *frequency of heads* for the first k throws.

We shall use the term *Bernoulli case* for the particular case in which $p_j = p$ is independent of j, so that $q_j = 1 - p_j = q$ is also independent of j; the "true" game of heads-or-tails is the Bernoulli case with $p = 1/2$. The interpretation of R_k and F_k as recurrence and frequency, respectively, is particularly apt in the Bernoulli case. For the general case, in which p_j may depend on j, we shall use the term *Poisson case*.

We emphasize the great importance of the heads-or-tails scheme, even in the very special Bernoulli case: it is the basis for every discussion of the frequency of an arbitrary event E connected with an arbitrary, *independently repeated* trial, where the number of repetitions is arbitrary, even infinite.

Formula (6,1,12) shows that study of the heads-or-tails scheme amounts to studying a sum of mutually independent r.v. In particular, via Theorems (4,2,1) and (6,1,1) it follows from (6,1,11) that

$$E(R_k) = \sum_{j=1}^{k} p_j, \quad E(F_k) = \frac{1}{k}\sum_{j=1}^{k} p_j, \tag{6,1,13}$$

$$\mathcal{V}(R_k) = \sum_{j=1}^{k} p_j\, q_j\,, \quad \mathcal{V}(F_k) = \frac{1}{k^2}\sum_{j=1}^{k} p_j\, q_j. \tag{6,1,14}$$

In the Bernoulli case ($p_j = p$, $q_j = q$) these formulas become

$$E(R_k) = k\,p, \qquad E(F_k) = p, \tag{6,1,15}$$

$$\mathcal{V}(R_k) = k\,p\,q, \qquad \mathcal{V}(F_k) = pq/k. \tag{6,1,16}$$

Bernoulli Law. Remaining in the Bernoulli case, we can easily obtain the complete probability law of R_k: R_k is a discrete r.v. with possible values 0, 1, 2,..., k; $R_k = h$ ($0 \leqq h \leqq k$) if and only if heads turn up in exactly h of the first k tosses (say on tosses $j_1, j_2, \ldots, j_h$, where $0 \leqq j_\alpha \leqq k$ and $j_\alpha \neq j_\beta$ if $\beta \neq \alpha$). Now, since the tosses are independent, the probability of the event $A(j_1, \ldots, j_h)$, "heads turn up in tosses $j_1, j_2, \ldots, j_h$ and in no others," is

$$\Pr[A(j_1, j_2, \ldots, j_h)] = p^h\, q^{k-h}.$$

This probability does not depend on k or h, and is independent of the values of $j_1, j_2, \ldots, j_h$. There are C_k^h events $A(j_1, j_2, \ldots, j_h)$, i.e., combinations of k throws h at a time; they are pairwise mutually exclusive, and their union is the event $R_k = h$. The Axiom of Total Probability thus gives

$$\Pr(R_k = h) = C_k^h\, p^h\, q^{k-h} \qquad (h = 0, 1, \ldots, k), \tag{6,1,17}$$

where the right-hand side is just the $(h + 1)^{th}$ term in the binomial expansion of $(p + q)^k$ (cf. (1, 2, 6); recall that in the applications $p + q = 1$). The discrete probability law defined by (6,1,17) is often called the *Bernoulli law* or the *binomial law*, and it has been tabulated (Tables [16] and [18]). It is of great theoretical and practical importance. In this connection, note that

$$\Pr(R_k \geqq h) = \sum_{j=h}^{k} C_k^j\, p^j\, q^{k-j}$$

$$= B_p(h, k - h + 1)/B(h, k - h + 1), \tag{6,1,18}$$

where B_p denotes the incomplete B-function (cf. (1, 1,18) and Exercise 1.4). The function (6,1,18) has also been tabulated; see Tables [17], [18], [19].

1) Derive (6,1,15) and (6,1,16) directly from (6,1,17).

2) In the Poisson case, let a and b be arbitrary constants, and set $Y_j = (a - b) X_j + b$, so that Y_j equals a if $X_j = 1$, b if $X_j = 0$. Show that

$$E(Y_j) = a\,p_j + b\,q_j\,, \qquad \mathcal{V}(Y_j) = |b - a|^2\, p_j\, q_j\,. \qquad (6,1,19)$$

3) Rayleigh describes an idealized Brownian motion in which a particle M moves along an axis from the origin 0 by successive, independent steps, all of the same size l, in either the positive or negative direction, with probability 1/2 for either direction. Letting Y_j denote the increment of the abscissa of M in the jth step and Z_k its abscissa after the kth step, calculate the m.e. and variance of Z_k. (*Answer*: $E(Z_k) = 0$; $\mathcal{V}(Z_k) = kl^2$; note that $Z_k = \sum_{j=1}^{k} Y_j$, and use (6,1,19) with $a = l$, $b = l$, $p_j = q_j = 1/2$).

2. *Convolutions of distribution functions and probability densities of independent random variables*

Let $F(x) = \Pr(X < x)$ and $G(y)$ denote the distribution functions of independent r.v. X and Y, and

$$H(s) = \Pr(S < s) = \Pr(X + Y < s)$$

that of their sum S.

We first consider the conditional probability

$$\Pr(S < s/X = x) = \Pr(X + Y < s/X = x)\,.$$

Given $X = x$, the event $X + Y < s$ coincides with the event $Y < s - x$, so that

$$\Pr(S < s/X = x) = \Pr(Y < s - x/X = x)\,.$$

But, since the r.v. X and Y are independent, it follows from V.2 that

$$\Pr(Y < s - x/X = x) = \Pr(Y < s - x) = G(s - x)\,,$$

and by (4,6,5)

$$\Pr(S < s) = \int_{-\infty}^{+\infty} \Pr(S < x/X = x)\, dF(x)$$

$$= \int_{-\infty}^{+\infty} G(s - x)\, dF(x)\ .$$

Since X and Y are clearly interchangeable, we get the formulas

$$H(s) = \int_{-\infty}^{+\infty} G(s - x)\, dF(x) = \int_{-\infty}^{+\infty} F(s - y)\, dG(y)\ . \qquad (6,2,1)$$

Referring back to III.8, in particular to (3,8,4), we see that:

THEOREM (6,2,1). The distribution function H(s) of the sum $S = X + Y$ of two independent random variables X and Y with distribution functions F(x) and G(y), respectively, is the convolution F*G of F and G. $\blacksquare$

By a direct application of the concepts and results of III.8 (and also III.6), we obtain the following results:

1) If one of the r.v. X and Y, say Y, possesses a probability density g(y), then S is continuous, with density h(s) given by

$$h(s) = \int_{-\infty}^{+\infty} g(s - x)\, dF(x)\ . \qquad (6,2,2)$$

2) If the r.v. X and Y have probability densities f(x) and g(y), respectively, then the probability density h(s) of S is given by

$$h(s) = \int_{-\infty}^{+\infty} g(s - x)\, f(x)dx = \int_{-\infty}^{+\infty} f(s - y)\, g(y)dy. \qquad (6,2,3)$$

In other words, h is the convolution of f and g (cf. III.6).

3) If X and Y are both discrete, then so is S. Let $x_1, x_2, \ldots, x_k, \ldots$ be the possible values of X, $y_1, y_2, \ldots, y_k, \ldots$ the possible values of Y, and $s_1, s_2, \ldots, s_k, \ldots$ the possible values of S, and let

$$a_k = Pr(X = x_k),\ b_k = Pr(Y = y_k),\ c_k = Pr(S = s_k)$$

Now s_h is a possible value of S if and only if there is at least one pair (j, k) of indices such that $x_j + y_k = s_h$. Let (j_1, k_1), $(j_2, k_2), \ldots, (j_\alpha, k_\alpha)$ be the pairs (j, k) satisfying this condition.

Then

$$c_h = Pr(X + Y = s_h) = \sum_\alpha Pr[(X = x_{j_\alpha}) \cap (Y = h_{k_\alpha})]$$

$$= \sum_\alpha a_{j_\alpha} b_{k_\alpha}. \qquad (6,2,4)$$

4) Let

$$\lambda(u) = E[e^{iuX}] = \int_{-\infty}^{+\infty} e^{iux}\, dF(x),$$

$$\mu(u) = E[e^{iuY}] = \int_{-\infty}^{+\infty} e^{iuy}\, dG(y),$$

$$\nu(u) = E[e^{iuS}] = E[e^{iu(X+Y)}] = \int_{-\infty}^{+\infty} e^{ius}\, dH(s)$$

(u real, $-\infty < u < +\infty$) denote the characteristic functions of X, Y, and S, respectively; then Theorem (3,8,2) (or, since e^{iuX} and e^{iuY} are independent, Theorem (5,2,3)) shows that

$$\nu(u) = \lambda(u)\,\mu(u). \qquad (6,2,5)$$

Hence:

THEOREM (6,2,2). The characteristic function of a sum of independent random variables is the product of their characteristic functions.

COROLLARIES (6,2,1)

1) The second characteristic function (cf. IV.3) of a sum of independent random variables is the sum of their second characteristic functions.

2) The moment-generating function (generating function) of a sum of independent random variables is the product of their moment-generating functions (generating functions) (see IV.3).

3) The semi-invariants (or cumulants, cf. IV.3) of a sum of independent random variables are the sums of their semi-invariants of the same order.

REMARK (6,2,1). It may happen that the characteristic function of the sum $S = X + Y$ of two r.v. X and Y is the product of the characteristic functions of X and Y *without* X and Y being independent; for an example, see H. Cramer [4], p. 317.

EXERCISE (6,2,1). Using (4,3,7), show that the characteristic function $\phi_k(u)$ of the recurrence R_k defined in (6,1,12) for the heads-or-tails scheme is

$$\varphi_k(u) = \prod_{j=1}^{k} [1 + p_j(e^{iu} - 1)], \qquad (6,2,6)$$

while in the Bernoulli case

$$\varphi_k(u) = [1 + p(e^{iu} - 1)]^k$$

$$= (q + pe^{iu})^k = \sum_{h=0}^{k} C_k^h \, p^h \, q^{k-h} \cdot e^{iuh},$$

which is again (6,1,17).

THEOREM (6,2,3). If two independent random variables X and Y obey Poisson laws, then so does their sum S.

Indeed, if X and Y obey Poisson laws, then they - therefore also S - are discrete random variables whose possible values are the nonnegative integers 0, 1, 2,...; setting $a = E(X)$ and $b = E(Y)$, we get

$$\Pr(X = h) = e^{-a} \frac{a^h}{h\,!}, \quad \Pr(Y = h) = e^{-b} \frac{b^h}{h\,!},$$

and Theorem (6,2,3) is then easily proved from (6,2,4). But we can also employ (6,2,5), since by (4,4,5) the characteristic functions $\lambda(u)$ and $\mu(u)$ of X and Y are, respectively,

$$\lambda(u) = \exp.\{a(e^{iu} - 1)\}, \quad \mu(u) = \exp.\{b(e^{iu} - 1)\},$$

so that the characteristic function $\nu(u)$ of S is

$$\nu(u) = \exp.\{(a + b)(e^{iu} - 1)\},$$

which is the characteristic function of a Poisson law; as expected, $E(S) = E(X) + E(Y) = a + b$.

There is a converse to Theorem (6,2,3), which we state here without proof (see D.A. Raįkov [1]):

THEOREM (6,2,4) (Raikov). If a random variable S which obeys a Poisson law is the sum of two independent random variables X and Y, then X and Y obey (possible degenerate) Poisson laws.

THEOREM (6,2,5). If two random variables X and Y obey normal laws, then their sum S also obeys a normal law.

Indeed, since X and Y are normal, they are continuous; if we set

$$E(X) = a, \; E(Y) = b, \; \mathcal{V}(X) = \sigma^2, \; \mathcal{V}(Y) = \rho^2,$$

their respective densities f(x) and g(y) are

$$f(x) = \frac{1}{\sqrt{2\pi}\,\sigma} e^{-\frac{(x-a)^2}{2\sigma^2}}, \quad g(y) = \frac{1}{\sqrt{2\pi}\,\rho} e^{-\frac{(y-b)^2}{2\rho^2}}.$$

Formula (6,2,3) then yields the density of their sum S; but there is no need to carry out the calculation, since by Theorems (4,2,1) and (6,1,1) we know in advance the $E(S) = a + b$ and $\mathcal{V}(S) = \sigma^2 + \rho^2$. On the other hand, since X and Y are normal and independent, the pair {X, Y} is a normal 2-dimensional r.v. (see V.1). As S is a linear combination of X and Y, S is itself normal (by Theorem (5,5,8) and we have

$$h(s) = \frac{1}{\sqrt{2\pi(\sigma^2 + \rho^2)}} e^{-\frac{(s-a-b)^2}{2(\sigma^2+\rho^2)}}. \qquad (6,2,8)$$

It is also true that by (4,4,25) the characteristic functions $\lambda(u)$ and $\mu(u)$ of X and Y are, respectively,

$$\lambda(u) = \exp . \left\{ - \frac{\sigma^2 u^2}{2} + iau \right\} ,$$

$$\mu(u) = \exp . \left\{ - \frac{\rho^2 u^2}{2} + ibu \right\} .$$

Hence the characteristic function $\nu(u)$ of S is

$$\nu(u) = \exp . \left\{ - \frac{(\sigma^2 + \rho^2) u^2}{2} + i(a + b) u \right\} , \qquad (6,2,9)$$

which is indeed the characteristic function corresponding to the probability density (6,2,8).

Theorem (6.2,5) has a converse, which we state without proof (see H. Cramer [1]):

THEOREM (6,2,6) (H. Cramer). If a normal random variable S is the sum of two independent random variables X and Y, then X and Y each obey (possibly degenerate) normal laws.

Closed types of laws

In IV.1 we defined the concept of a type of distribution functions (or type of random variables). Now Theorem (6,2,5) indicates a very interesting property of the type of normal laws: if two distribution functions belong to the normal type, so does their convolution. In general, we call a type $\mathcal{C}$ of d.f. *closed* if, for any d.f. F and G in $\mathcal{C}$ the convolution F*G also belongs to $\mathcal{C}$; i.e., if F and G are any two independent r.v. of type $\mathcal{C}$, then their sum is also of type $\mathcal{C}$.

The type of normal laws is thus closed. Note that the random variables which reduce to an almost certain number (or, equivalently, their distribution functions) clearly constitute a closed type.

Consider a d.f. F(x) with characteristic function $\phi(u)$, belonging to type $\mathcal{C}$. Now $\mathcal{C}$ is closed if and only if, for arbitrary constants $a_1 > 0$, $a_2 > 0$, b_1, b_2, there exist two constants $a > 0$ and b such that

$$F(a_1x + b_1) * F(a_2x + b_2) = F(ax + b) . \qquad (6,2,10)$$

The characteristic functions of $F(a_1x + b_1)$, $F(a_2x + b_2)$, $F(ax + b)$ are $e^{-i\frac{b_1u}{a_1}}\varphi\left(\frac{u}{a_1}\right)$, $e^{-i\frac{b_2u}{a_2}}\varphi\left(\frac{u}{a_2}\right)$, $e^{-i\frac{b}{a}u}\varphi\left(\frac{u}{a}\right)$, respectively. By Theorem (6,2,2), equality (6,2,10) is equivalent to the statement: A type $\mathcal{C}$ is closed if and only if, for any positive constants a_1 and a_2, there exist two constants $a > 0$ and β such that

$$e^{i\beta u}\,\varphi\left(\frac{u}{a_1}\right)\varphi\left(\frac{u}{a_2}\right) = \varphi(au). \qquad (6,2,11)$$

It can be shown that

THEOREM (6,2,7). A characteristic function $\phi(u)$ is of a closed type, i.e. satisfies condition (6,2,11), if and only if there exist four real constants α, β, γ, c such that

a) $0 < \alpha \leqslant 2, \; -1 \leqslant \beta \leqslant 1, \; c \geqslant 0$,

b) $$\text{Log } \varphi(u) = i\gamma u - c\,|u|^{\alpha}\left[1 + i\beta\frac{u}{|u|}\,\omega_{\alpha}(u)\right], \qquad (6,2,12)$$

where $\omega_{\alpha}(u)$ denotes the constant $\tan(\pi\alpha/2)$ if $\alpha \neq 1$, and the function $\frac{2}{\pi}\log|u|$ if $\alpha = 1$. $\alpha = 2$ clearly corresponds to the normal type, $c = 0$ to the type of random variables reducible to almost certain numbers.

Cauchy law and type

If $\alpha = 1$ and $\beta = \gamma = 0$, formula (6,2,12) gives

$$\varphi(u) = e^{-c|u|} , \qquad (6,2,13)$$

which, by Exercise 3.5, is the characteristic function of an absolutely continuous distribution function with density

$$f(x) = \frac{1}{\pi}\frac{c}{c^2 + x^2} . \qquad (6,2,14)$$

The probability law of any absolutely continuous r.v. with density

$$f(x) = \frac{1}{\pi} \frac{1}{1 + x^2} \tag{6,2,15}$$

is known as the *Cauchy law*, and the corresponding type is called the Cauchy type. Since the d.f. with density (6,2,14) is clearly of the Cauchy type, the latter is *closed*.

II. STOCHASTIC CONVERGENCE, LAWS OF LARGE NUMBERS, ERGODIC THEOREMS

3. *Laws of large numbers and stochastic convergence*

Let us return to the heads-or-tails scheme as defined in VI.1, in particular, to the Bernoulli case. Now, our intuition leads us to believe that the *frequency* F_k of heads in a run of k tosses should be about the same as the *probability* p of heads, provided, at least, that k is sufficiently large. In considering the statistical standpoint, the same intuition (see II.3) enabled us to adopt the concept of probability based on observed frequencies. There seems to be no special reason for the difference $F_k - p$ to be particularly small for small values of k; it is only for "sufficiently large" values of k that we expect $F_k - p$ to be small - even to become smaller as k increases. To phrase this intuition more precisely, we must resort to a statement of the type: as $k \to +\infty$, F_k approaches p - in a sense yet to be defined.

We remark in passing that a statement of this kind is a typical example of what is known as a *law of large numbers*. In the sequel, we shall return to laws of large numbers in a more general form.

For the moment we note that in the above statements the word "approaches," irrespective of its ultimate meaning, alludes to the convergence of a sequence $\{F_k\}$ of *random* variables. This kind of convergence is not the same as that of a sequence of numbers in Analysis, but rather of a kind which we shall call *stochastic convergence*.

Moreover, regardless of what stochastic convergence turns out to be, the limit L of a sequence of random variables must also be a r.v. In the Bernoulli case, the limit of the F_k is apparently a nonrandom number p, but this only means that in this case the limit r.v. L happens to reduce to an almost certain number p, an interesting special case but hardly a general rule.

Stochastic convergence would thus seem to be convergence in a space of r.v., which will usually be associated with a topology in this space. Conversely, any concept of convergence in a space of r.v. - in particular, any topology in such a space - will be a type of stochastic convergence. This means that one can envisage and introduce an endless variety of types of stochastic convergence; it turns out, however (for a variety of reasons), that there are only a few important types, which we shall now consider.

4. *Convergence in quadratic mean*

Returning to the frequencies F_k in Bernoulli case, we see that (6,1,15) and (6,1,16) give

$$E(|F_k - p|^2) = \frac{pq}{k}, \qquad (6,4,1)$$

whence, by (4,2,1), we have for all $\varepsilon > 0$

$$\Pr(|F_k - p| > \varepsilon) \leqslant \frac{pq}{\varepsilon^2 k}. \qquad (6,4,2)$$

This is our first justification for the intuitive argument discussed in VI.3: for arbitrarily small $\varepsilon > 0$, the probability that F_k differs from p by more than ε may be made as small as we wish, provided k is chosen sufficiently large. This is simply a consequence of (6,4,1), i.e., of the fact that the second moment of $|F_k - p|$ becomes infinitesimal as $k \to +\infty$. We must nevertheless bear in mind that it is not *impossible* for F_k to differ appreciably from p for some very large value of k; it is just highly unlikely.

Now let us return to some general concepts with which we are already familiar - those of V.4, IV.2, II.10 and II.8. In the present section and the following two (VI.5 and VI.6), it will be convenient to deal with *complex* r.v.

Let $\mathcal{H}$ be the Hilbert space of second-order complex random variables; the Hilbert topology in $\mathcal{H}$ induces the following definition of convergence:

A countable sequence $\{X_k\}$ of second-order random variables converges to a r.v. X (possibly reducing to an almost certain number) as $k \to +\infty$ if

$$\lim_{k \to +\infty} E(|X_k - X|^2) = 0 . \qquad (6,4,3)$$

Now, since

$$X = (X - X_k) + X_k ,$$

formulas (6,4,3) and (2,10,4) imply that $X \in \mathcal{H}$. The properties of convergence as defined by (6,4,3) follow immediately from those of convergence in Hilbert spaces - we need only recall that the Cauchy property and Theorem (2,10,6) provide the necessary criteria for the convergence of a sequence $\{X_k\}$.

Now let us relax the restriction that the complex random variables in the sequence $\{X_k\}$ be second-order r.v., and let X be any complex r.v. We shall say that the sequence $\{X_k\}$ converges to the limit r.v. X in *quadratic mean*(1) if

$$\lim_{k \to +\infty} E(|X_k - X|^2) = 0. \qquad (6,4,4)$$

This is a broader definition than that of (6,4,3), since here the X_k and X need not be second-order r.v.. Note, however, that $Y_k = X_k - X$ must be second-order r.v. (at least, for sufficiently large k) in order for (6,4,4) to hold; formula (6,4,4) then says that, in $\mathcal{H}$, the Y_k converge to the r.v. which is almost-certainly 0, the zero element of $\mathcal{H}$.

Thus, convergence in quadratic mean, though sometimes meaningful for random variables which are not second-order r.v., reduces to Hilbert convergence in $\mathcal{H}$.

As usual, let $\mathfrak{U}$ denote the class of trials, with elements u, and $p(E) = Pr(E \subset \mathfrak{U})$ the measure or probability law on $\mathfrak{U}$. Recall

---

(1)Quadratic mean: abbreviation, q.m.; German: im quadratischen Mittel; in French: en moyenne quadratique.

that a complex r.v. X is simply a mapping $x = x(u)$ of $\mathfrak{U}$ into the space C of complex numbers. Let $x_k(u)$ and $x(u)$ be the mappings of $\mathfrak{U}$ into C corresponding to X_k and X; then (6,4,4) is equivalent to

$$\lim_{k\to+\infty} \int_{\mathfrak{U}} |x_k(u) - x(u)|^2 \; p(du) = 0 . \qquad (6,4,5)$$

Thus, stochastic convergence in quadratic mean coincides with convergence in quadratic mean (or in 2nd mean) as defined in II.8 for "point functions."

The fundamental properties of convergence in q.m. follow immediately; in particular, we mention the following:

1) If X_k converges to X in q.m. and each X_k reduces to an almost certain number x_k, then X reduces to an almost certain number x, and $\lim\limits_{k\to+\infty} x_k = x$.

2) If X_k converges to X in q.m. and X_k also converges to X' in q.m., then X and X' are equivalent (i.e., almost-certainly equal; see II.12). If X and X' are equivalent and X_k converges to X in q.m., then X_k also converges to X' in q.m. (recall that if X is a second-order r.v. and X' is equivalent to X, then X' is also a second-order r.v. and is identified with X as an element of $\mathcal{H}$ (see II.10)).

3) If X_k converges to X in q.m. and Y_k converges to Y in q.m., then $X_k + Y_k$ converges to X + Y in q.m. However, X_kY_k need not converge to XY, but if X_k converges to X in q.m. and $\{a_k\}$ is a sequence of nonrandom numbers converging to a limit a, then a_kX_k converges to aX in q.m.

4) Assume that X_k converges to X in q.m.; if $E(|X|^2) < +\infty$, then

$$\lim_{k\to+\infty} E(|X_k|^2) = E(|X|^2) , \qquad (6,4,6)$$

and

$$\lim_{k\to+\infty} E(X_k) = E(X) . \qquad (6,4,7)$$

Indeed, (6,4,6) follows from Theorem (2,10,6), while (6,4,7) may be derived from Theorem (2,7,5) (or from (3,9,5)). On the other hand, if $\alpha > 2$ and the moments $E(|X_k|^\alpha)$ exist, then $\lim_{k\to+\infty} E(|X_k|^\alpha) = E(|X|^\alpha)$, need not hold - $E(|X|^\alpha)$ may not even exist.

5) Assume that X_k converges to X in q.m.; if $E(|X|^2) = +\infty$, then $\lim_{k\to+\infty} E(|X_k|^2) = +\infty$.

We mentioned in IV.2 (and again in V.4) the particular interest attaching specifically to convergence in quadratic mean from the standpoint of physical applications - it may be interpreted as a convergence concept deriving from the concept of energy.

Now, from the standpoint of Probability Theory, the justification for convergence in q.m. lies in its convenience. Moreover, since (4,2,21) implies

$$\Pr(|X_k - X| > \varepsilon) \leqslant \frac{E(|X_k - X|^2)}{\varepsilon^2}, \tag{6,4,8}$$

we see that if $X_k \to X$ in q.m., then, for arbitrary small ε, we can make the probability that $|X_k - X|$ exceeds ε as small as desired, provided k is chosen sufficiently large. Nevertheless, convergence in q.m. has no intrinsic probability-theoretic significance.

In the Bernoulli case, formula (6,4,1) implies that F_k converges to p in quadratic mean as $k \to +\infty$.

Convergence in α^{th} mean

For any positive real α convergence of "point functions" in α^{th} mean, defined in II.8, translates into probability-theoretic language as stochastic convergence in α^{th} mean: a complex r.v. X_k converges to a complex r.v. X as $k \to +\infty$ if

$$\lim_{k\to+\infty} E(|X_k - X|^\alpha) = 0. \tag{6,4,9}$$

5. *Convergence in probability*

To obtain stochastic convergence concepts which do have intrinsic significance for Probability Theory, we need only consider convergence in measure and convergence almost everywhere, as defined in II.8 for "point functions," and translate them into the language of probability.

In the notation of (6,4,5), relative to the measure p(E), the function $x_k(u)$ converges to $x(u)$ in measure if, for all ε and $\eta > 0$,

$$p\,[E_\varepsilon(k)] < \eta$$

for all sufficiently large k, where $E_\varepsilon(k)$ is the set of u such that $|x_k(u) - x(u)| > \varepsilon$.

In probability-theoretic language, convergence in measure is known as *convergence in probability*[1], and may be defined as follows: a sequence of complex random variables $\{X_k\}$ converges in probability to a complex r.v. X if, for any positive real ε and η, for all sufficiently large k,

$$\Pr(|X_k - X| > \varepsilon) \leqslant \eta. \qquad (6,5,1)$$

Either referring back to II.8 or reasoning directly, the reader should have no difficulty in proving the following assertions:

1) If X_k converges to X i.p. and each X_k reduces to an almost-certain number x_k, then X reduces to an almost-certain number x and $\lim\limits_{k\to+\infty} x_k = x$.

2) If X_k converges to X i.p. and also to X' i.p., then X and X' are equivalent; if X and X' are equivalent and X_k converges to X i.p., then X_k also converges to X' i.p.

3) If X_k converges to X i.p. and Y_k converges to Y i.p. then $X_k + Y_k$ converges i.p. to X + Y and X_kY_k converges i.p. to XY. In particular, if X_k converges to X i.p. and $\{a_k\}$ is a sequence of nonrandom numbers converging to a limit a, then a_kX_k converges i.p. to aX.

4) Assume that X_k has an algebraic (absolute) moment of arbitrary order for all (sufficiently large) k; then the fact that X_k converges to X i.p. neither implies that the moment converges to the corresponding moment of X, nor even that the latter exists.

THEOREM (6,5,1). If the real random variable X_k converges in probability to the real random variable X as $k \to +\infty$, the distribution

---

[1] In probability: abbreviation: i.p.; German: im Wahrschein= lichkeit; French: en probabilité.

function $F_k(x)$ of X_k converges to the distribution function $F(x)$ of X.

In fact, the event $(X_k < x)$ may be expressed in the form

$$(X_k < x) = [(X_k < x) \cap (|X_k - X| \leqslant \varepsilon)] \cup [(X_k < x) \cap (|X_k - X| > \varepsilon)].$$

Thus

$$F_k(x) = \Pr[(X_k < x) \cap (|X_k - X| \leqslant \varepsilon)] + \Pr[(X_k < x) \cap (|X_k - X| > \varepsilon)].$$

Now, for any events A and B, $\Pr(A \cap B) \leq \Pr(A)$; thus

$$F_k(x) \leqslant \eta + \Pr[(X_k < x) \cap (|X_k - X| \leqslant \varepsilon)].$$

But

$$[(X_k < x) \cap (|X_k - X| \leqslant \varepsilon)] \subset (X < x + \varepsilon),$$

so that

$$F_k(x) \leqslant \eta + \Pr(X < x + \varepsilon) = \eta + F(x + \varepsilon). \qquad (6,5,2)$$

By interchanging the roles of X_k and X and replacing x by $x - \varepsilon$, we get

$$F(x - \varepsilon) \leqslant \eta + F_k(x). \qquad (6,5,3)$$

If we assume $F(x)$ continuous for the value x of the variable, formulas (6,5,2) and (6,5,3) yield the theorem (see III.4).

REMARK (6,5,1). Given a sequence of real random variables X_k with d.f. F_k, and a real r.v. X with d.f. F, the fact that F_k converges to F depends only on F and on each of the F_k, not on the probability laws of X and the X_k. The possible correlations among the X_k or between the X_k and X are immaterial; in particular, X and the X_k may be mutually independent.

However, let us assume that X_k converges to X in some way, say in quadratic mean or in probability. Disregarding the special case in which X reduces to an almost-certain number, we may say, in qualitative terms, that for sufficiently large k the convergence of X_k to X must imply that X_k is correlated with X, moreover, that this correlation is on the whole stronger, the greater k. Another result is that, for any $h > 0$, X_k and X_{k+h} are strongly correlated for large k. ∎

Return to the general case of complex random variables. Formula (6,4,8) shows that if X_k converges to X in quadratic mean, then X_k also converges to X in probability; moreover, if X_k converges to X in quadratic mean and to Y in probability, the random variables X and Y must be equivalent.

On the other hand, the following example will show that if X_k converges to X in probability it need not converge to X in quadratic mean.

EXAMPLE (6,5,1). Let α be any fixed positive number; assume that the only possible values of X_k are 0, k, and -k:

$$\Pr(X_k = 0) = 1 - \frac{1}{k^{\alpha}}, \quad \Pr(X_k = k) = \Pr(X_k = -k) = \frac{1}{2k^{\alpha}} .$$

Clearly, X_k converges to 0 in probability as $k \to +\infty$. On the other hand,

$$E(|X_k - 0|^2) = E(|X_k|^2) = k^{2-\alpha} ,$$

which shows that, for $\alpha \leq 2$, X_k does not converge to 0 in quadratic mean; in fact, if $\alpha < 2$, not only does X_k not converge to 0 in quadratic mean, but its second moment is not even bounded.

EXAMPLE (6,5,2). As before, let us assume that X_k has a Cauchy-type distribution (see (6,2,14)) with density

$$f_k(x) = \frac{1}{\pi} \frac{k}{1 + k^2x^2} .$$

For all $\varepsilon > 0$,

$$\Pr(|X_k| > \varepsilon) = \frac{2}{\pi} \int_{\varepsilon}^{+\infty} \frac{kdx}{1 + k^2x^2} = 2\left(\frac{\pi}{2} - \operatorname{arctg} k\varepsilon\right),$$

so that X_k converges to 0 in probability. However, X_k clearly does not converge to 0 in quadratic mean; moreover, it has no finite second moment.

6. *Almost sure convergence*

Let us return to the notation of (6,4,5). A function $x_k(u)$ converges to a function x(u) almost everywhere with respect to the measure p(E) if the set $\breve{\Omega}$ of all u for which the sequence of *numbers* $\{x_k(u)\}$ does *not* converge to the *number* $x(u)$ has measure zero: $p(\check{\Omega}) = 0$.

In probability-theoretic terminology, convergence almost-everywhere is known as *almost sure*[1] (or almost certain) *convergence*, and may be defined as follows: Let Ω be the event: the sequence of numerical values attributed by chance to X_k converges (in the sense of the convergence of number sequences in Analysis) to the numerical value attributed by chance to X; we say that X_k *converges to X almost surely* (*almost certainly*) if $\Pr(\Omega) = 1$. In other words, the probability $\Pr(\check{\Omega})$ of the event complementary to Ω (i.e., the event $\check{\Omega}$ that the sequence of values assumed by the X_k does not converge to the value assumed by X) is zero.

It is now easy to verify the following assertions, either directly or on the basis of II.8:

1) If X_k converges a.s. to X and each X_k reduces to an almost-certain number x_k, then X reduces to an almost-certain number x and $\lim\limits_{k\to+\infty} x_k = x$.

2) If X_k converges a.s. to both X and X', then X and X' are equivalent. If X and X' are equivalent and X_k converges a.s. to X, then X_k also converges a.s. to X'.

3) If X_k converges a.s. to X and Y_k converges a.s. to Y, then $X_k + Y_k$ converges a.s. to $X + Y$ and X_kY_k converges a.s. to XY. In particular, if X_k converges a.s. to X and a_k is a sequence of non-random numbers converging to a limit a, then a_kX_k converges a.s. to aX.

REMARK (6,6,1). Consider a sequence $\{X_k\}$ of r.v. and a random variable X; in order to ascertain whether X_k converges to X in quadratic mean or in probability, we need only consider the probability laws of each of the r.v. $X_k - X$; in other words, the possible correlations among the r.v. $X_k - X$ are immaterial. However, it is

---

[1] Almost sure (almost certain) convergence: abbreviation: a.s. (or a.c.); German: fast sicher Konvergenz; French: convergence presque-sûre (or presque-certain).

obvious that X_k cannot converge a.s. to X (i.e., X_k - X to 0) without some fairly strong correlation among the r.v. X_k - X, since convergence to 0 of the numerical values attributed by chance to X_k - X is a property of this number sequence as a whole, not of each of its terms in isolation. This means that to establish that X_k converges - or does not - a.s. to X, we must study the mutual dependence of the X_k - X; this dependence may be quite varied in form. ■

Under these conditions there can be no simple necessary and sufficient condition for almost sure convergence. Nevertheless, there is a very useful sufficient condition:

THEOREM (6,6,1). A sequence $\{X_k\}$ of complex random variables converges a.s. to 0 if there exist an integer K and a number $\alpha > 0$ such that

$$\sum_{k>K}^{+\infty} E(|X_k|^{\alpha}) < +\infty.$$

This is simply Theorem (2,8,2), restated in probability terminology.

THEOREM (6,6,2). If a sequence $\{X_k\}$ of complex random variables converges almost surely to a complex random variable X, it also converges to X in probability.

Indeed, let ε be any given positive number and k an arbitrary fixed integer; define events

$E_k(\varepsilon)$: $|X_h - X| \leq \varepsilon$ for all $h > k$,

$E(\varepsilon)$: $|X_k - X| \leq \varepsilon$ for all sufficiently large k.

Clearly,

$$E_k(\varepsilon) \subset E_{k+1}(\varepsilon), \quad E(\varepsilon) = \bigcup_k E_k(\varepsilon), \quad E(\varepsilon) \supset \Omega,$$

where Ω has the same meaning as above. If X_k converges a.s. to X, then, by Theorem (2,6,1),

$$1 = \Pr(\Omega) = \Pr[E(\varepsilon)] = \lim_{k\to+\infty} \Pr[E_k(\varepsilon)],$$

proving the theorem. ■

However, as the following example shows, almost sure convergence does not necessarily imply convergence in quadratic mean.

EXAMPLE (6,6,1). Recall Example (6,5,1), with $\frac{3}{2} < \alpha \leq 2$. Since $E(|X_k|^{1/2}) = \frac{1}{k^{\alpha-1/2}}$, it follows from Theorem (6,6,1) that X_k converges a.s. to 0. But $E(|X_k|^2) = \frac{1}{k^{\alpha-2}}$, so that X_k does not converge to 0 in quadratic mean; in fact, $E(|X_k|^2)$ is not even bounded for $\alpha < 2$.

It is intuitively clear that neither convergence in probability, nor even convergence in quadratic mean, imply almost sure convergence. The following example should make this clear.

EXAMPLE (6,6,2). Let $\{Y_k\}$ be the sequence of mutually independent real random variables defined by

$$\Pr(Y_k = 0) = 1 - 1/k^{1/4}, \qquad \Pr(Y_k = 1) = \Pr(Y_k = -1) = \frac{1}{2k^{1/4}}$$

Let E_k be the event: all Y_h for $k - \sqrt{k} \leq h < k$ vanish, $\check{E}_k$ the event complementary to E_k. For each k, let X_k be the r.v. defined by

$$\Pr(X_k = Y_k/\check{E}_k) = 1, \ \Pr(X_k = 1/E_k) = \Pr(X_k = -1)/E_k) = \frac{1}{2}\cdot$$

Clearly, $E(X_k) = 0$. But X_k does not converge a.s. to 0. Indeed, since the only possible values of X_k are 0, 1, and -1, almost sure convergence of X_k to 0 would mean that, almost certainly, all X_k vanish for sufficiently large k. Now, $X_k = 0$ only if $\check{E}_k$ occurs; but then $X_k = Y_k$, whence a.c. all Y_k vanish for sufficiently large k. But then it is a.c. that all E_k for sufficiently large k occur, i.e. no $\check{E}_k$ occurs; this is a contradiction.

Nevertheless, X_k converges to 0 in probability, even in quadratic mean. Indeed, setting $\Pr(E_k) = p_k$, we get

$$p_k = \prod_{k-\sqrt{k}\leq h<k} \left(1 - \frac{1}{h^{1/4}}\right) = \exp.\ \{-o(k^{1/4})\},$$

so that $\lim_{k\to+\infty} p_k = 0$. Now, by (4,6,18),

$$E(|X_k|^2) = \Pr(E_k)\ E(X_k^2/E_k) + \Pr(\check{E}_k)\ E(X_k^2/\check{E}_k)$$

$$= p_k \times 1 + (1 - p_k) \times \left(\frac{1}{k^{1/4}}\right),$$

which indeed converges to 0 as $k \to +\infty$.

REMARK (6,6,2). The following point is worthy of attention. The assumption that X_k converges a.s. to X means that for all $\varepsilon > 0$ the probability is 1(i.e., it is certain, or at least almost certain) that there exists a *finite* K such that the inequality $|X_k - X| \leqq \varepsilon$ holds (simultaneously) for all $k > K$. However, K is in general not fixed in advance, but *random*, and the value it takes may vary from one trial to another. It is possible, though not very probable, for K to be very large in some particular trial. Should this happen, the values x_k that the X_k assume in this trial will not confirm convergence of the sequence $\{x_k\}$ unless observation is continued beyond $k > K$. Since in practice it is generally not possible to prolong observation beyond certain limits, and K is not known in advance, convergence of the sequence $\{x_k\}$ may well remain undetected by the observer.

Equivalently, we might say: given an observation of length $k_0 (1 \leqq k \leqq k_0)$, arbitrarily long but fixed, the fact that the sequence of observed values x_k of the X_k $(k = 1, 2, \ldots, k_0)$ displays no tendency to converge is no decisive argument against the hypothesis that the sequence $\{x_k\}$ converges almost certainly.

Application to the heads-or-tails scheme

Let us return to the heads-or-tails scheme as defined in VI.1, using the same notation, and consider the event E that heads turns up infinitely many times - i.e., that infinitely many E_k occur. E is also the event that, for arbitrarily large K, there occurs at least one E_k with $k > K$.

From Exercise 4.3 we know that if $\sum_{k=1}^{+\infty} p_k = +\infty$ the probability of E is not only positive, but even unity. Now, this may be the case for p_k that become infinitesimal as $k \to +\infty$. For example, in the Bernoulli case, $\Pr(E) = 1$ no matter how small the probability p of heads, provided, of course, that it is not zero.

Thus, even if the probability of each event in a given sequence E_k $(k = 1, 2, \ldots)$ is very small, the probability of the occurrence of events E_k with arbitrarily large k may be positive, even unity.

In particular, consider the Bernoulli case. We already know that F_k converges in probability to p, which means that if ε is a given positive number and A_k the event $|F_k - p| > \varepsilon$, then $Pr(A_k)$ is arbitrarily small for sufficiently large k. Now, by the foregoing argument, the probability that events A_k with arbitrarily large k will occur may nevertheless be positive, even unity. Were this to be true, at least for small ε, it would mean that F_k does not converge a.s. to p. This is not the case, and we shall see below that F_k converges to p not only in p. and in q.m., as we already know, but also in a.s. All this emphasizes how and to what extent convergence a.s. is stronger than convergence in p.

7. *Laws of large numbers*

Let $X_1, X_2, \ldots, X_j, \ldots$ be an infinite sequence of random variables (either real or complex), and set

$$S_k = X_1 + X_2 + \cdots + X_k , \qquad Y_k = S_k/k = \frac{1}{k}(X_1 + X_2 + \cdots + X_k) .$$

A *law of large numbers* is any statement which asserts that, under certain conditions, the arithmetic mean Y_k of the first k terms X_j converges stochastically to a limit r.v. Y as $k \to +\infty$.

Depending on the type of stochastic convergence in question, we must distinguish between a.s. laws of large numbers (also called strong laws of large numbers), laws of large numbers in p., in q.m., etc. Laws of large numbers also differ in the conditions for their validity, which may be more or less restrictive, etc. We shall indicate those of greatest practical interest, i.e., those whose conditions of validity, while fairly broad, are easily verified.

First assume that, for all j, X_j has a finite absolute second moment. This means that $E(X_j)$ exists, and without loss of generality we may assume that

$$E(X_j) = 0 ; \qquad (6,7,1)$$

Set

$$E(|X_j|^2) = \sigma_j^2 < +\infty \; ; \tag{6,7,2}$$

$$a_{h,k}^2 = E\left(\left|\sum_{j=h+1}^{h+k} X_j\right|^2\right) \quad (h = 0, 1, 2, \ldots, k = 1, 2, \ldots) \; ;$$

then

$$E(|Y_k|^2) = a_{0,k}^2/k^2. \tag{6,7,3}$$

Thus:

THEOREM (6,7,1) (*law of large numbers in quadratic mean*). Let X_j ($j = 1, 2, \ldots$) be real or complex random variables with zero expectation and finite variance. If $E\left(\left|\sum_{j=1}^{k} X_j\right|^2\right) = o(k^2)$ as $k \to +\infty$, the arithmetic mean $Y_k = \frac{1}{k}\sum_{j=1}^{k} X_j$ of the first k X_j converges to 0 in quadratic mean. ∎

The condition

$$E\left(\left|\sum_{j=1}^{k} X_j\right|^2\right) = o(k^2) \tag{6,7,4}$$

is easily verified when the X_k are *mutually independent*, since in this case:

$$a_{h,k}^2 = \sum_{j=h+1}^{h+k} \sigma_j^2 . \tag{6,7,5}$$

Then for (6,7,4) to hold it suffices that the σ_j have an upper bound independent of j. A common example of this in applications is the special case in which the X_j are identically distributed, so that $\sigma_j = \sigma$ is independent of j.

Assume there exist two real constants λ and α such that

$$\lambda > 0, \; 0 \leqslant \alpha < 2, \; a_{h,k}^2 \leqslant \lambda k^\alpha \text{ for all } h \geqslant 0 \text{ and all } k \geqslant 1 \tag{6,7,6}$$

Note that (6,7,6) implies (6,7,4). If β is any fixed positive number, let us associate with each $k \geqq 1$ the integer h_k defined by

$$h_k^\beta \leqslant k < (h_k + 1)^\beta \; ; \tag{6,7,7}$$

as $k \to +\infty$, $h_k \to +\infty$, $h_k^\beta/k \to 1$ and $[h_k^\beta]/k \to 1$.

From (6,7,3) and (6,7,6) it follows that

$$E(|Y_{[h_k^\beta]}|^2) \leqslant \lambda'/h_k^{\beta(2-\alpha)}, \tag{6,7,8}$$

where λ' is a suitable number greater than λ. Let β be such that

$$\beta(2 - \alpha) > 1 ; \tag{6,7,9}$$

Theorem (6,6,1) then states that $Y_{h_k^\beta}$ converges almost surely to 0 as $k \rightarrow +\infty$.

On the other hand, if we set

$$Z_k = \frac{1}{k} \sum_{j=[h_k^\beta]+1}^{k} X_j ,$$

we can write, for all k,

$$Y_k = \frac{1}{k} \sum_{j=1}^{k} X_j = \frac{1}{k} \sum_{j=1}^{[h_k^\beta]} X_j + Z_k$$

$$= \left(\frac{[h_k^\beta]}{k}\right) Y_{[h_k^\beta]} + Z_k .$$

By (6,7,6),

$$E(|Z_k|^2) \leqslant \lambda' \frac{(k - h_k^\beta)^\alpha}{k^2} < \lambda' \frac{[(h_k + 1)^\beta - h_k^\beta]^\alpha}{k^2}$$

$$< \lambda' \frac{h_k^{\alpha\beta}}{k^2} \left[\left(1 + \frac{1}{h_k}\right)^\beta - 1\right]^\alpha ,$$

where λ' is again a suitable number greater than λ. There thus exists a constant $\mu > 0$ such that, for all k,

$$E(|Z_k|^2) \leqslant \mu/k^{2-\alpha+\alpha/\beta} . \tag{6,7,10}$$

If

$$2 - \alpha + \alpha/\beta > 1, \tag{6,7,11}$$

it follows from Theorem (6,6,1) that Z_k converges almost surely to 0, and therefore Y_k converges almost surely to 0. Inequality (6,7,11) always holds for $\alpha \leqq 1$; for $\alpha > 1$ it is compatible with (6,7,9) only if $\alpha^2 - \alpha - 1 \leqq 0$, i.e., if

$$\alpha < \frac{1}{2} (1 + \sqrt{5}) ;$$

whence:

THEOREM (6,7,2) (*Almost sure law of large numbers*). Let X_j ($j = 1, 2, \ldots$) be real or complex random variables with zero expectation and finite variance. If there exist two real constants λ and α such that

$$\begin{cases} \lambda > 0, \quad 0 \leqslant \alpha < \frac{1}{2}\,(1 + \sqrt{5}) \\ E\left(\left|\sum_{j=h+1}^{h+k} X_j\right|^2\right) \leqslant \lambda k^{\alpha} \quad \text{for all } h \geqslant 0 \text{ and all } k \geqslant 1, \end{cases} \tag{6,7,12}$$

then the arithmetic mean $Y_k = \frac{1}{k}\sum_{j=1}^{k} X_j$ of the first k terms X_j converges almost surely to 0 as $k \to +\infty$. $\blacksquare$

If the X_j are mutually independent, verification of (6,7,12) is particularly easy, because of (6,7,5). In particular, condition (6,7,12) is satisfied for $\alpha = 1$, if the X_j are mutually independent and identically distributed.

Application to the heads-or-tails scheme

Consider the heads-or-tails scheme defined in VI.1, using the same notation. For all j,

$$0 \leqslant p_j \leqslant 1, \quad q_j = 1 - p_j,$$

and as a result

$$p_j\, q_j \leqslant 1/4\,. \tag{6,7,13}$$

Now the m.e. $E(X_j)$ of the indicator X_j of the event "heads on the jth toss" is p_j, not 0; but the r.v. $Z_j = X_j - p_j$ has zero expectation, so that the preceding discussion is applicable. The Z_j are mutually independent, the repetition R_k of heads and its relative frequency F_k are, respectively,

$$R_k = \sum_{j=1}^{k} Z_j + \sum_{j=1}^{k} p_j, \qquad F_k = \frac{1}{k}\sum_{j=1}^{k} Z_j + \frac{1}{k}\sum_{j=1}^{k} p_j\,;$$

moreover, by (6,7,13),

$$E\left(\left|\sum_{j=h+1}^{h+k} Z_j\right|^2\right) \leqslant k/4\,. \tag{6,7,14}$$

Theorems (6,7,1) and (6,7,2) thus imply the following result:

As $k \longrightarrow +\infty$, $F_k - \frac{1}{k}\sum_{j=1}^{k} p_j$ converges to 0 both in quadratic mean and almost surely, so that in the Bernoulli case, where $p_j = p$ is independent of j, F_k converges to p in quadratic mean and almost surely. It is not *logically* impossible for some sequence of tosses to turn up heads exclusively, but this possible event is almost impossible. Here is yet another justification for the use of the term "almost sure convergence" in place of "sure convergence." ∎

Consider a countable sequence of nonrandom numbers x_k (k = 1, 2, 3,...) and let A be a property which this sequence may or may not possess. We call A an *asymptotic property* if any change in the values of finitely many x_k has no effect on the sequence's possessing or not possessing the property A. Examples of asymptotic properties are the fact that $\lim_{k \to +\infty} x_k = 0$, the fact that x_k has some kind of limit as $k \longrightarrow +\infty$, the fact that the sequence of arithmetic means

$$y_k = \frac{1}{k}\sum_{j=1}^{k} x_j$$

is convergent, and the fact that the series $\sum_k x_k$ is convergent. The following theorem is true:

THEOREM (6,7,3) (Kolmogorov)

Let $\{X_j\}$ (j = 1, 2,...) be a countable sequence of mutually independent real or complex random variables, and let A be an asymptotic property. Then the probability that the sequence of values attributed by chance to the X_k possesses property A is either 0 or 1. ∎

A proof and several extensions of this at first sight surprising and widely applicable theorem may be found in P. Lévy [1]. For example, since the X_k are by assumption mutually independent, Theorem (6,7,3) implies that the probability that the arithmetic means $Y_k = \frac{1}{k}\sum_{j=1}^{k} X_j$ converge (in the analytical sense) is either 0 or 1. Laws of large numbers, such as those given above, then make the result precise by stating that, under certain conditions, this probability is 1 and not 0.

As an application, consider a countable sequence $\{X_j\}$ of mutually independent real r.v., all with the same *continuous* d.f. $F(x)$. Let $I_j(x)$ denote the indicator of the event $X_j < x$, and set

$$G_k(x) = \frac{1}{k} \sum_{j=1}^{k} I_j(x) . \tag{6,7,15}$$

As a function of x, $G_k(x)$ is a d.f. for all k; but it is random. If we place masses 1/k at the k points (distinct or not) $X_1, X_2, \ldots, X_k$ on the x-axis, the result is a totally discontinuous mass distribution; $G_k(x)$ is its associated d.f. in the sense of III.1, and $\sum_{j=1}^{k} I_j(x)$ is the number of those abscissas $X_1, X_2, \ldots, X_k$ which are smaller than x. This interpretation justifies the term *empirical* distribution function for $G_k(x)$. Of course, the mass distribution and its d.f. $G_k(x)$ are random; but, for a given trial, an observer successively recording the values assumed by $X_1, X_2, \ldots, X_k, \ldots$ can construct (for this trial) successive distribution functions $G_1(x), G_2(x), \ldots, G_k(x), \ldots$.

For the Bernoulli case of the heads-or-tails scheme, $G_k(x)$ converges a.s. to $E[I_j(x)] = F(x)$ as $k \to +\infty$, for any fixed x. Let $\{x_h\}$ be a countable set of values of x which is dense in the interval $(-\infty, +\infty)$ (e.g., the set of rational numbers). Now the union of countably many events all having probability zero has probability zero (see Remark (2,12,1)). It is therefore almost certain that $G_k(x_h) \to F(x_h)$ as $k \to +\infty$ for all x_h. Referring to III.4, we see that the continuity of $F(x)$ implies:

THEOREM (6,7,4) (Glivenko-Cantelli)

Let X_j $(j = 1, 2, 3, \ldots)$ be mutually independent real random variables all having the same continuous distribution function $F(x)$. It is almost certain that, as $k \to +\infty$, the empirical distribution function $G_k(x)$ defined by (6,7,15) converges uniformly in x to $F(x)$ over $(-\infty, +\infty)$. ∎

An equivalent formulation is: the random distance $\Delta(G_k, F)$ from the d.f. G_k to the d.f. F (see III.4) converges a.s. to 0.

The value of Theorem (6,7,4) for empirical determination of an unknown d.f. F(x) is evident.

EXERCISE (6,7,1)

Study the convergence of $G_k(x)$ to $F(x)$ when $F(x)$ is not continuous.

8. *Ergodic theory*

Let $\mathfrak{X}$ be a real or complex vector space, with elements x; let $\mathfrak{L}$ be the set of linear operators on $\mathfrak{X}$ (i.e., the set of linear mappings of $\mathfrak{X}$ into itself). Let A be an element of $\mathfrak{L}$ which maps each $x \in \mathfrak{X}$ onto an element $y = A(x) \in \mathfrak{X}$. Letting I denote the identity operator $I(x) = x$, let us consider the nonnegative integer powers $A^0 = I$, $A^1 = A$, $A^2, \ldots, A^j, \ldots$, of A.

The aim of ergodic theory[1] is to study the asymptotic behavior of the operator B_k defined by

$$B_k = \frac{1}{k}\,(A^1 + A^2 + \cdots + A^k)\,. \qquad (6,8,1)$$

as $k \to +\infty$. This problem is meaningful only with respect to some topology defined either in all of $\mathfrak{L}$ or (at least) in the subset containing the B_k; it is usually derived from a previously-defined topology in $\mathfrak{X}$.

As one might expect from the fact that both deal with the asymptotic behavior of arithmetic means, there is a connection between ergodic theory and laws of large numbers; the following remarks will make this clear.

Depending on the nature of $\mathfrak{X}$ and the topologies involved, there is naturally a great variety of ergodic theorems. We shall limit ourselves to two theorems, which are both typical and useful in Probability Theory.

---

[1] Ergodism, ergodic, ergodic theory; German: ergodische Eigenschaft, ergodisch, ergodische Theorie; French: ergodisme, ergodique, théorie ergodique.

Assume $\mathfrak{X}$ is a Hilbert space with zero element θ, and let the linear operator A be *bounded* (for terminology and notation, cf. II.10).

Then A, its powers A^j, and the B_k all belong to the subspace $\mathfrak{L}_f$ of bounded linear mappings in $\mathfrak{L}$. It is known from the general theory of normed vector spaces that the linear-operator norm in $\mathfrak{L}_f$ makes $\mathfrak{L}_f$ a complete normed vector space.

Consider the norm N(A) of A. If $N(A) < 1$, then, since $N(A^k) \leqq [N(A)]^k$, it is clear that the operators A^k, thus also B_k, converge (in the sense of the norm in $\mathfrak{L}_f$) to the zero operator.

Now assume $N(A) = 1$; this means that $N(B_k) \leqq 1$ for all k. Let $\mathfrak{Z}_0$ be the set of all $x \in \mathfrak{X}$ for which there exists at least one $y \in \mathfrak{X}$ such that

$$x = A(y) - y . \qquad (6,8,2)$$

Now, by (6,8,2), we have for all $x \in \mathfrak{Z}_0$

$$B_k(x) = \frac{1}{k} [A^{k+1}(y) - A(y)],$$

and therefore

$$\lim_{k \to +\infty} B_k(x) = \theta . \qquad (6,8,3)$$

Let $\mathfrak{Z}$ be the set of all elements x of $\mathfrak{X}$ which are limits of elements of $\mathfrak{Z}_0$, i.e., $\mathfrak{Z}_0$ is the "closure" of $\mathfrak{Z}$. Clearly, $\mathfrak{Z}$ is a Hilbert subspace of $\mathfrak{X}$. For any two elements x and x' of $\mathfrak{X}$ we have

$$N_2[B_k(x) - B_k(x')] = N_2[B_k(x - x')] \leqslant N(B_k) \cdot N_2(x - x')$$
$$\leqslant N_2(x - x') . \qquad (6,8,4)$$

Let $x \in \mathfrak{Z}$; apply the estimate (6,8,4), which is uniform in k, to an element x' of $\mathfrak{Z}_0$ which converges to x. Then, by (6,8,3),

$$\lim_{k \to +\infty} B_k(x) = \theta \qquad \text{for all} \quad x \in \mathfrak{Z} . \qquad (6,8,5)$$

Clearly, if $x \in \mathfrak{Z}_0$ then $A(x) \in \mathfrak{Z}_0$, whence it follows easily that $x \in \mathfrak{Z}$ implies $A(x) \in \mathfrak{Z}$.

Now let $\mathfrak{Y}$ be the set of all $x \in \mathfrak{X}$ such that

$$A(x) = x.$$

$\mathfrak{Y}$ is obviously a Hilbert subspace of $\mathfrak{X}$. If $x \in \mathfrak{Y}$, then, for all k,

$$B_k(x) = x . \tag{6,8,6}$$

Comparison of (6,8,5) with (6,8,6) shows that the only element common to $\mathfrak{Z}$ and $\mathfrak{Y}$ is θ.

Let x be an element of $\mathfrak{X}$ whose normal projection on $\mathfrak{Z}$ is θ (i.e., x is orthogonal to every element of $\mathfrak{Z}$). Set

$$y = A(x) - x$$

and note that $y \in \mathfrak{Z}_0 \subset \mathfrak{Z}$; since

$$A(x) = x + y,$$

it follows that y is the normal projection of A(x) on $\mathfrak{Z}$. Were $y \neq \theta$, the uniform convexity of $\mathfrak{X}$ would imply

$$N_2[A(x)] > N_2(x),$$

contradicting N(A) = 1. Hence $y = \theta$ and A(x) = x, i.e., $x \in \mathfrak{Y}$.

Conversely, if x is an element of $\mathfrak{Y}$ and x its normal projection on $\mathfrak{Z}$, then

$$A(x - z) = x - A(z),$$

with $A(z) \in \mathfrak{Z}$. Since x has only one normal projection on $\mathfrak{Z}$, it follows that, if $A(z) \neq z$,

$$N_2[A(x - z)] = N_2[x - A(z)] > N_2(x - z),$$

which is impossible since N(A) = 1. Thus A(z) = z, i.e., $z \in \mathfrak{Y}$. However, since by definition $z \in \mathfrak{Z}$, it follows that $z = \theta$, i.e., the normal projection of x on $\mathfrak{Z}$ is zero (x is orthogonal to every element in $\mathfrak{Z}$).

Let H be the mapping of $\mathfrak{X}$ into itself (actually, the mapping of $\mathfrak{X}$ onto $\mathfrak{Z}$) which maps every element x of $\mathfrak{X}$ onto its normal projection z on $\mathfrak{Z}$, and let L be the mapping of $\mathfrak{X}$ into itself which maps x onto x - z = x - H(x). The following assertions follow from the preceding argument:

1) L is a mapping of $\mathfrak{X}$ onto $\mathfrak{Y}$.

2) $\lim_{k \to +\infty} B_k(x) = L(x)$, so that L is a linear mapping; and since $H(x) = x - L(x)$, we see that H is also a linear mapping, and

$$I = L + H.$$

$\mathfrak{Z}$ is precisely the set of all $x \in \mathfrak{X}$ such that

$$\lim_{k \to +\infty} B_k(x) = \theta .$$

3) For any $x \in \mathfrak{X}$, $LH(x) = HL(x) = \theta$ - in this situation the linear mappings L and H are said to be orthogonal.

4) $AL = LA = L$; $L^2 = L$, $H^2 = H$.

5) $AH = HA$; if we set $AH = HA = M$, then L and M are orthogonal, $A = L + M$, and $M^k = A^kH = HA^k$, and the result is

$$A^k = L + M^k \text{ et } B_k = L + \frac{1}{k} \sum_{j=1}^{k} M^j.$$

In the light of properties of Hilbert spaces (cf. II.10), $\mathfrak{Y}$ and $\mathfrak{Z}$ are clearly *complementary* Hilbert subspaces of $\mathfrak{X}$, L and H are the respective *projections* on $\mathfrak{Y}$ and $\mathfrak{Z}$. We can summarize these results as follows:

THEOREM (6,8,1). Let $\mathfrak{X}$ be a Hilbert space, A a bounded linear operator on $\mathfrak{X}$ with norm at most 1, and $B_k = \frac{1}{k} \sum_{j=1}^{k} A^j$. There exist two bounded linear operators L and H such that

1) $L^2 = AL = LA = L$; $H^2 = H$; $AH = HA$; $L + H = I$;

2) if $M = AH$, then $A^k = L + M^k$ for all k;

3) $\lim_{k \to +\infty} B_k(x) = L(x)$ for all $x \in \mathfrak{X}$;

4) the set $\mathfrak{Y}$ of all $x \in \mathfrak{X}$ such that $A(x) = x$ and the set $\mathfrak{Z}$ of all $x \in \mathfrak{X}$ such that $\lim_{k \to +\infty} B_k(x) = \theta$ are complementary Hilbert subspaces of $\mathfrak{X}$, and L and H are the projections on $\mathfrak{Y}$ and $\mathfrak{Z}$, respectively.

REMARK (6,8,1). The only property of $\mathfrak{X}$ actually used in the preceding reasoning is its uniform convexity; this means that Theorem (6,8,1) may easily be extended to uniformly convex complete normed vector spaces in general.

Second-order stationary sequences

Let $S = \{X_j\}$ be a unilateral ($j = 0, 1, 2, 3,\ldots$) or bilateral ($j = 0, \pm 1, \pm 2,\ldots$) sequence of second-order complex r.v. with the following properties:

1) $E(X_j)$ for all j;

2) the covariance $E(X_j \overline{X}_k) = r_{k-j}$ depends only on the difference $h = k - j$.

We shall call such a sequence a *second-order stationary sequence*.

Note that any given sequence $S = \{X_j\}$ of second-order r.v. X_j may always be reduced to the case $E(X_j) = 0$ for all j, by considering the r.v. $Y_j = X_j - E(X_j)$ instead of X_j.

Let $\mathcal{H}[S]$ be the Hilbert subspace generated by the family $\{X_j\}$ in the space $\mathcal{H}$ of second-order r.v. (cf. II.10).

Let $\mathcal{H}'$ be the subspace of $\mathcal{H}[S]$ consisting of all linear combinations of finitely many X_j. $\mathcal{H}'$ is a vector subspace but not complete: $\mathcal{H}[S]$ is the closure of $\mathcal{H}'$. Let U' be the linear operator on $\mathcal{H}'$ defined by

1) If $X = X_j$, $U'(X) = U'(X_j) = X_{j+1}$;

2) if $X = \sum_{j=1}^{s} \alpha_j X_j$ is any finite linear combination of the X_j, then

$$U'(X) = U'\left(\sum_{j=1}^{s} \alpha_j X_j\right) = \sum_{j=1}^{s} \alpha_j U'(X_j) = \sum_{j=1}^{s} \alpha_j X_{j+1}.$$

The hermitian product in $\mathcal{H}$ of an element X and an element Y is the covariance $E(X\overline{Y})$. If X and Y are both in $\mathcal{H}'$, i.e., X and Y are both linear combinations

$$X = \sum_{j=1}^{s} \alpha_j X_j, \qquad Y = \sum_{k=1}^{t} \beta_k X_k ,$$

then

$$E[U'(X)\,\overline{U'(Y)}] = E\left[\sum_{j,k=1}^{s,t} \alpha_j \bar{\beta}_k X_{j+1} \overline{X}_{k+1}\right] = \sum_{j,k=1}^{s,t} \alpha_j \bar{\beta}_k r_{k-j}$$

$$= E\left[\sum_{j,k=1}^{s,t} \alpha_j \bar{\beta}_k X_j \bar{X}_k\right] = E(X\,\bar{Y}) . \qquad (6,8,7)$$

In other words, U' preserves hermitian products in $\mathcal{H}'$. Let U be the linear operator in $\mathcal{H}[S]$ which maps each element $X \in \mathcal{H}[S]$ onto the element $Y = U(X) \in \mathcal{H}[S]$ defined as follows:

a) if $X \in \mathcal{H}'$, $Y = U(X) = U'(X)$;

b) if $X \notin \mathcal{H}'$ there exists a sequence $\{Z_h\}$ of elements of $\mathcal{H}'$ which converges to X. If we then set

$$Y = U(X) = \lim_{h\to+\infty} U'(Z_h) , \qquad (6,8,8)$$

Remark (2,10,1) shows that U is well-defined and preserves hermitian products in $\mathcal{H}[S]$, so that if U is a homeomorphism, then U is a *unitary* operator or isometry.

Now let Y_k be the arithmetic mean:

$$Y_k = \frac{1}{k}\sum_{j=1}^{k} X_j .$$

Since $X_1 = U(X_0)$, we have for all j

$$X_j = U^j(X_o),$$

and therefore

$$Y_k = \frac{1}{k}[U^1 + U^2 + \ldots + U^k](X_o) .$$

Now since $\mathcal{H}[S]$ is a Hilbert space, and any operator U in a Hilbert space which preserves hermitian products has norm 1, Theorem (6,8,1) implies:

THEOREM (6,8,2). Assume that the second-order complex r.v. $\{X_j\}$ ($j = 0, 1, 2,\ldots$ or $j = 0, \pm1, \pm2,\ldots$), with m.e. zero, form a second-order stationary sequence. Then the arithmetic mean $Y_k = \frac{1}{k}\sum_{j=1}^{k} X_j$ converges in quadratic mean to a second-order random variable Y with m.e. zero. ∎

The variance $E(|Y|^2)$ of the limit Y may be evaluated from the covariances r_h as follows:

$$E(|Y|^2) = \lim_{k\to+\infty} E(|Y_k|^2) \ ;$$

$$E(|Y_k|^2) = \frac{1}{k^2} E\left[\sum_{\alpha,\beta=1}^{k} X_\alpha \overline{X}_\beta\right] = \frac{1}{k^2} \sum_{\alpha,\beta=1}^{k} r_{\alpha-\beta},$$

whence

$$E(|Y|^2) = \lim_{k\to+\infty} \frac{1}{k^2}\left[k\, r_0 + \sum_{j=1}^{k-1} (k-j)\,(r_j + r_{-j})\right]. \qquad (6,8,9)$$

Note that a sequence $\{X_j\}$ of mutually independent second-order r.v., all having m.e. zero and identical variances, is a second-order stationary sequence. Here $r_h = 0$ for $h \neq 0$, and (6,8,9) shows that $E(|Y|^2) = 0$: Y_k converges to 0 in q.m., a result which could also have been deduced from Theorem (6,7,1).

9. *Birkhoff's ergodic theorem*

Let $\mathcal{V}$ be an arbitrary set with elements v, p(E) ($E \subset \mathcal{V}$) a *bounded* measure on $\mathcal{V}$, and $\mathcal{L}_1$ the set of point functions, or mappings $x = x(v)$ of $\mathcal{V}$ into the space R of real numbers, such that

$$M_1(x) = N_1(x) = \int_{\mathcal{V}} |x(v)|\; p(dv) < +\infty. \qquad (6,9,1)$$

$\mathcal{L}_1$ is a vector space, with norm $N_1(x)$; by reasoning analogous to that of Theorem (2,10,7) we can show that $\mathcal{L}_1$ is complete in this norm.

Now let φ be a mapping of $\mathcal{V}$ into itself, which maps each element $v \in \mathcal{V}$ onto an element $v' = \varphi(v) \in \mathcal{V}$. For an element x of $\mathcal{L}_1$, let y be the point function derived from x as follows:

$$y(v) = x[\varphi(v)] .$$

The function y is obtained from x by a linear mapping, which we shall denote by A: $y = A(x)$. Since

$$N_1(y) = \int_{\mathcal{V}} |y(v)|\; p(dv) = \int_{\mathcal{V}} |x[\varphi(v)]|\; p(dv), \qquad (6,9,2)$$

$y = A(x)$ is not necessarily in $\mathcal{L}_1$.

Now let φ be a *one-to-one* mapping of $\mathcal{V}$ onto itself, with inverse φ^{-1}, and assume that it *preserves the measure* p; i.e., given any subset E of $\mathcal{V}$, let $F = \varphi(E)$ denote the set of all $v \in \mathcal{V}$ such that $\varphi^{-1}(v) \in E$; then $p(F) = p(E)$. In other words, the measure $q(F)$ which φ induces (see II.11) on $\mathcal{V}$, corresponding to the measure $p(E)$,

$$q(F) = p[\varphi^{-1}(F)], \quad q(dv) = p[\varphi^{-1}(dv)],$$

is such that $q(F) = p(F)$, and so

$$p(F) = p[\varphi^{-1}(F)], \quad p(dv) = p[\varphi^{-1}(dv)]. \qquad (6,9,3)$$

Substituting v' for $\varphi(v)$ in the second integral in (6,9,2), we get

$$N_1(y) = \int_{\mathcal{V}} |x(v')| \; p[\varphi^{-1}(dv')]$$

$$= \int_{\mathcal{V}} |x(v')| \; p(dv') = N_1(x) \; , \qquad (6,9,4)$$

which shows that A is a linear operator on $\mathcal{L}_1$ with norm 1. However, Theorem (6,8,1) does not apply, even in the generalized form indicated in Remark (6,8,1), since $\mathcal{L}_1$ is not uniformly convex.

Nevertheless, setting $B_k = \frac{1}{k} \sum_{j=1}^{k} A^j$, one can prove:

THEOREM (6,9,1) (Birkhoff). Under the above conditions, as $k \to + \infty$, the point function $y_k = B_k(x)$ for all $x \in \mathcal{L}_1$ converges a.e. on $\mathcal{V}$, relative to the measure p, to a function $y \in \mathcal{L}_1$ such that

$$\int_{\mathcal{V}} y(v) \; p(dv) = \int_{\mathcal{V}} x(v) \; p(dv). \; \blacksquare$$

Rather than prove the theorem directly, we shall first translate it into probability terminology.

Strictly stationary sequences. Let $\mathcal{W}$ be the set of bilateral sequences w of real numbers w_j ($j = 0, \pm 1, \pm 2, \ldots$):

$$\mathbf{w} = \{\ldots, w_{-1}, w_o, w_1, \ldots\} \; ,$$

and let $\mathbf{w}' = \psi(\mathbf{w})$ be the mapping of $\mathcal{W}$ into itself which maps the element $\mathbf{w} = \{\ldots, w_{-1}, w_o, w_1, \ldots\}$ of $\mathcal{W}$ onto the element $\mathbf{w}' = \{\ldots, w'_{-1}, w'_o, w'_1, \ldots\}$ of $\mathcal{W}$ defined by

$$w'_j = w_{j+1} \quad (j = 0, \pm 1, \pm 2, \ldots) \; ; \qquad (6,9,5)$$

note that ψ is one-to-one.

Any bilateral sequence (j = 0, ±1, ±2,...) of real r.v. may be regarded as a random element with values in $\mathcal{W}$. We call such a sequence *strictly stationary* if, for any subset $\Omega \in \mathcal{W}$,

$$\Pr[W \in \Omega] = \Pr[W \in \psi(\Omega)], \tag{6,9,6}$$

where $\psi(\Omega)$, as usual, denotes the set of all w such that $\psi^{-1}(w) \in \Omega$. In particular, this implies that, for any positive integer n, any n integers $j_1, j_2, \ldots, j_n$, and any integer h, the probability law of the n-dimensional r.v. $\{X_{j_1+h}, X_{j_2+h}, \ldots, X_{j_n+h}\}$ is independent of h.

A simple example of a strictly stationary sequence is one in which the X_j are mutually independent and identically distributed.

We leave it to the reader to extend the definition of strictly stationary sequences to sequences $W = \{X_j\}$ of complex r.v. X_j.

In sequences of numbers (real or complex), whether certain sequences $w = \{w_j\}$ or random sequences $W = \{X_j\}$, interpretation of the index j as an instant of time is suggestive, especially in view of the concept of stationarity. w_j and X_j are then (certain or random) numbers associated with the instant of time j, where j is in the past with respect to the instant j + 1 and the latter is in the future with respect to j. With this interpretation an arithmetic mean such as $Y_k = \frac{1}{k} \sum_{j=1}^{k} X_j$ assumes the meaning of a *time average*.

Referring to this interpretation in a general and qualitative manner, the term "stationarity" is used to express the fact that certain properties of a sequence remain invariant under an arbitrary translation of the time axis (cf. the beginning of III.6, especially p. 171). For a random sequence $W = \{X_j\}$, the properties involved will of course be statistical properties.

Strict stationarity means that *all* the statistical properties (probabilities, moments if any, etc.) of the sequence $W = \{X_j\}$ are invariant: strict stationarity is an absolute and intrinsic property.

Nevertheless, a great variety of restricted types of stationarity easily come to mind - involving the invariance of only certain specific statistical properties. Let $(\mathcal{R})$ be the set of those statistical properties whose invariance is implied by some type of restricted stationarity: we see that this stationarity is essentially relative, relative to $(\mathcal{R})$. Second-order stationarity, as defined above, is an example: it consists in the invariance of the first and second moments only (assuming they exist).

If $W = \{X_j\}$ is strictly stationary and $E(|X_j|^2) = +\infty$, then W cannot be second-order stationary. But if $E(|X_j|^2) < +\infty$, then if W is strictly stationary it is a fortiori second-order stationary (at least, after making $E(X_j)$ vanish by a suitable change of origin).

On the other hand, the fact that $W = \{X_j\}$ is second-order stationary does not necessarily imply that it is strictly stationary.

Case of normal sequences

We shall call a unilateral or bilateral sequence $W = \{X_j\}$ of random variables normal if, for any positive integer n and any n integers $j_1, j_2, \ldots, j_n$, the r.v. $\{X_{j_1}, X_{j_2}, \ldots, X_{j_n}\}$ is a normal n-dimensional r.v.(real or complex, depending on whether the X_j are real or complex). Theorem (5,5,1) immediately implies:

THEOREM (6,9,2). If $W = \{X_j\}$ is a bilateral normal sequence $(j = 0, \pm 1, \pm 2, \ldots)$ of real random variables X_j and W is second-order stationary, then it is strictly stationary.$\blacksquare$

We shall say that a subset Ω of $\mathcal{W}$ belongs to class $\mathfrak{G}$ if there exist a positive integer n, n integers $j_1, j_2, \ldots, j_n$, and a subset e of R^n such that $w \in \Omega$ if and only if $\{\mathbf{w}_{j_1}, \mathbf{w}_{j_2}, \ldots, \mathbf{w}_{j_n}\} \in \mathbf{e}$.

In fact, Theorem (5,5,1) merely states that (6,9,6) holds for every Ω in class $\mathfrak{G}$. To prove Theorem (6,9,2) rigorously, therefore, it remains to prove that (6,9,6) also holds for every subset $\Omega \subset \mathcal{W}$ (more precisely, for every Ω such that the event is probabilized); we shall not give the proof here (it may be based on a theorem of A. Kolmogorov [1], p. 29).

Now consider the following theorem:

THEOREM (6,9,3). If a bilateral ordered sequence $W = \{X_j\}$ of real random variables X_j ($j = 0, \pm1, \pm2, \ldots$) is strictly stationary and $E(X_j)$ exists, then, as $k \to +\infty$, the arithmetic mean $Y_k = \frac{1}{k}\sum_{j=1}^{k} X_j$ converges almost surely to a random variable Y such that $E(Y) = E(X_j)$. ∎

We shall now prove two lemmas:

LEMMA (6,9,1). Theorem (6,9,3) implies Theorem (6,9,1). To prove this, let us return to the previous notation. Since p is a bounded measure and the case $p(\mathcal{V}) = 0$ is trivial, there is no loss of generality in assuming that $p(\mathcal{V}) = 1$, regarding $\mathcal{V}$ as a class of trials and p as a probability measure on $\mathcal{V}$. Now, A has an inverse A^{-1} defined by

$$A^{-1}(x) = y \quad \text{or} \quad y(v) = x[\varphi^{-1}(v)] .$$

Thus the jth powers φ^j and A^j of the mappings φ and A are defined for all integers j, both negative and positive; for $j = 0$, we set

φ° = the identity mapping of $\mathcal{V}$ onto itself;

A° = the identity operator I in $\mathcal{L}_1$.

Now set

$$x_j = A^j(x), \quad \text{i.e.,} \quad x_j(v) = x[\varphi^j(v)] ;$$

$$y_k = \frac{1}{k}\sum_{j=1}^{k} x_j, \quad \text{i.e.,} \quad y_k(v) = \frac{1}{k}\sum_{j=1}^{k} x_j(v)$$

for all integers j and all $k > 0$.

Let λ be the mapping $w = \lambda(v)$ of $\mathcal{V}$ into $\mathcal{W}$ which maps each element $v \in \mathcal{V}$ onto the element $\mathbf{w} = \{\ldots, w_{-1}, w_o, w_1, \ldots\}$ of $\mathcal{W}$ defined by

$$w_j = x_j(v) = x[\varphi^j(v)] ; \qquad (6,9,7)$$

we see that

$$x_j[\varphi(v)] = x_{j+1}(v) = w_{j+1},$$

and so

$$\lambda[\varphi(v)] = \psi[\lambda(v)]. \qquad (6,9,8)$$

It follows that if E is the set of all $v \in \mathcal{V}$ such that $\mathbf{w} = \lambda(v) \in \Omega$, then $\varphi(E)$ is the set of all $v \in \mathfrak{V}$ such that $\mathbf{w} = \lambda(v) \in \psi(\Omega)$.

Each x_j is a mapping of $\mathfrak{V}$ into R, and may be regarded as a real r.v. X_j; the ordered sequence $W = \{X_j\}$ is a random element with values in $\mathfrak{W}$, defined by the mapping λ of $\mathfrak{V}$ into $\mathfrak{W}$. With the above notation we have

$$\Pr(W \in \Omega) = p(E), \ \Pr[W \in \psi(\Omega)] = p[\varphi(E)],$$

and (6,9,3) implies that (6,9,6) holds: the sequence $W = \{X_j\}$ is strictly stationary. Furthermore, by (4,2,12)

$$E(X_j) = \int_{\mathfrak{V}} x_j(v)\ p(dv)$$

exists.

Since the arithmetic mean $Y_k = \frac{1}{k}\sum_{j=1}^{k} X_j$ is precisely the r.v. defined by the mapping y_k of $\mathfrak{V}$ into R, it is easy to see that Theorem (6,9,3) implies Theorem (6,9,1).∎

Conversely:

LEMMA (6,9,2). Theorem (6,9,1) implies Theorem (6,9,3).

Indeed, relative to some class of trials $\mathfrak{U}$, consider a sequence $W = \{X_j\}$ $(j = 0, \pm 1, \pm 2, \ldots)$ of real r.v. which is strictly stationary - hence satisfies (6,9,6) - and such that $E(X_j)$ exists. Let $m(\Omega)$ be the probability law of W, so that $m(\Omega)$ is a bounded measure on $\mathfrak{W}$. By (6,9,6) ψ is a one-to-one mapping of $\mathfrak{W}$ onto itself which preserves the measure $m(\Omega)$.

Let $\mathfrak{L}_1$ be the set of point functions, or mappings $x = x(w)$ of $\mathfrak{W}$ into R, such that

$$\int_{\mathfrak{W}} |x(\mathbf{w})|\ m(d\mathbf{w}) < +\infty . \tag{6,9,9}$$

For each j we define a mapping x_j of $\mathfrak{W}$ into R by

$$\mathbf{x}_j = \mathbf{x}_j(\mathbf{w}) = \mathbf{w}_j . \tag{6,9,10}$$

The r.v. X_j coincide with the r.v. $x_j(W)$, and the assumption that $E(X_j)$ exists becomes

$$\int_{\mathfrak{W}} |\mathbf{x}_j(\mathbf{w})|\ m(d\mathbf{w}) < +\infty,$$

In other words, the x_j belong to $\mathcal{L}_1$, and, by the very definition of ψ, $x_{j+1}(w) = w_j[\psi(w)]$. Let A be the linear operator on $\mathcal{L}_1$ defined by

$$y = A(x), \quad \text{or} \quad y(w) = x[\psi(w)],$$

Then, for all j,

$$x_j = A^j(x_o) .$$

Lemma (6,9,2) then follows, since the r.v.

$$Y_k = \frac{1}{k}\sum_{j=1}^{k} X_j$$

coincides with the mapping

$$y_k = \frac{1}{k}\sum_{j=1}^{k} A^j(x_o)$$

of $\mathcal{W}$ into R. ∎

It thus follows that *Theorems* (6,9,1) *and* (6,9,3) *are equivalent.*

Following A. Kolmogorov [2], we shall now prove Theorem (6,9,3), retaining his notation:

$$Y_{h,k} = \frac{1}{k}(X_{h+1} + X_{h+2} + \ldots + X_{h+k}) ;$$

$$\overline{Y} = \lim_{k\to+\infty}\sup. Y_{o,k}, \qquad \underline{Y} = \lim_{k\to+\infty}\inf. Y_{o,k} .$$

Since

$$E(Y_{o,k}) = E(X_k),$$

it follows from Theorem (2,7,3) that

$$\Pr(\underline{Y} < +\infty) = 1 \tag{6,9,11}$$

and

$$E(\underline{Y}) \leqq E(X_k). \tag{6,9,12}$$

Assume that

$$\Pr(\lim_{k\to+\infty} Y_{o,k} \text{ exists}) < 1 ; \tag{6,9,13}$$

then there exists a positive number λ such that

$$\Pr(\overline{Y} > \underline{Y} + \lambda) > 0 .$$

By (6,9,11), there exists a number μ such that

$$\Pr[(\underline{Y} < \mu) \cap (\overline{Y} > \mu + \lambda)] > 0 .$$

Henceforth we set $\alpha = \mu$ and $\beta = \mu + \lambda > \alpha$.

Let us call a pair (h,k) (k > 0) *singular* if

$$Y_{h,j} \leqq \beta \quad \text{for all} \quad j < k \; ; \; Y_{h,k} > \beta . \tag{6,9,14}$$

If two pairs (h,k) and (h',k') are singular, they cannot separate each other, e.g., the inequality

$$h + 1 < h' + 1 < h + k < h' + k' \tag{6,9,15}$$

cannot hold. For (6,9,15) would then imply

$$Y_{h,k} = \frac{h' - h}{k} Y_{h,h'-h} + \frac{h + k - h'}{k} Y_{h',h+k-h'} ,$$

and $Y_{h,k} > \beta$ would then imply that either $Y_{h,h'-h} > \beta$ or $Y_{h',h+k-h'} > \beta$, neither of which is compatible, by (6,9,15), with the assumption that the pairs (h,k) and (h',k') are singular.

Let s be an arbitrary integer; we shall call a pair (h,k) s-singular if the pair is singular, $k \leqq s$ and there is no singular pair (h',k') such that

$$k' \leqslant s, \quad h' \leqslant h < h + k < h' + k'.$$

If the pair (0,t) is singular, then for all $s \geqq t$ there exists an s-singular pair (h,k) such that $h < 0 < h + k$.

Let G be the event $(\underline{Y} < \alpha) \cap (\overline{Y} > \beta)$; by (6,9,13) its probability is positive.

H_s the event "there exists an s-singular pair (h,k) such that $h < 0 < h + k$";

$H_{s,p,q}$ the event "the pair (-p, -p + q) is s-singular";

$G_s = G \cap H_s$;

$G_{s,p,q} = G \cap H_{s,p,q}$.

Note that

1) $\lim_{s \to +\infty} G_s = G$ so that $\Pr(G_s) > 0$ for all sufficiently large s;

2) $H_s = \bigcup_{\substack{q=1,2,\ldots,s \\ p=1,2,\ldots,q-1}} H_{s,p,q}$;

3) $G_s = \bigcup_{\substack{q=1,2,..,s \\ p=1,2,..,q-1}} G_{s,p,q}$.

By (4,6,17),

$$\Pr(G_s)\, E(X_o/G_s) = \sum_{\substack{q=1,2,..,s \\ p=1,2,..,q-1}} \Pr(G_{s,p,q})\, E(X_o/G_{s,p,q}) ,$$

and strict stationarity implies that

$$\Pr(G_{s,p,q}) = \Pr(G_{s,o,q}),\ E(X_o/G_{s,p,q}) = E(X_p/G_{s,o,q}) .$$

Therefore

$$\Pr(G_s)\, E(X_o/G_s) = \sum_{\substack{q=1,2,..,s \\ p=1,2,..,q-1}} \Pr(G_{s,o,q})\, E(X_p/G_{s,o,q})$$

$$\Pr(G_s)\, E(X_o/G_s) = \sum_{q=1,2,..,s} \Pr(G_{s,o,q}) \left[\sum_{p=1,2,..,q-1} E(X_p/G_{s,o,q}) \right]$$

$$= \sum_{q=1,2,..,s} \Pr(G_{s,o,q})\, E\left[\left(\sum_{p=1,2,..,q-1} X_p \right) / G_{s,o,q} \right]$$

$$\geqslant \sum_{q=1,2,..,s} \Pr(G_{s,o,q})\, (q-1)\beta . \qquad (6,9,16)$$

Now,

$$\Pr(G_s) = \sum_{\substack{q=1,2,..,s \\ q=1,2,..,q-1}} \Pr(G_{s,p,q}) = \sum_{q=1,2,..,s} \left[\sum_{p=1,2,..,q-1} \Pr(G_{s,p,q}) \right]$$

$$= \sum_{q=1,2,..,s} \sum_{p=1,2,..,q-1} \Pr(G_{s,o,q})$$

$$= \sum_{q=1,2,..,s} (q-1) \Pr(G_{s,o,q}) . \qquad (6,9,17)$$

It follows from (6,9,16) and (6,9,17) that

$$\Pr(G_s)\, E(X_o/G_s) \geqslant \beta \Pr(G_s)$$

$$E(X_o/G_s) \geqslant \beta .$$

Now, using the same method we can prove that $E(X_0/G_S) \leqq \alpha$. The assumption (6,9,13) thus implies a contradiction and the first part of Theorem (6,9,3) is proved.

Retaining the notation of the statement of the theorem, we must still prove that $E(Y) = E(X_j)$.

First assume that the X_j are bounded, i.e., there exists a finite number M such that almost-certainly, $|X_j| < M$ for all j. The equality $E(Y) = E(X_j)$ then follows immediately from Theorem (2,8,4).

Now note that

a) If the X_j are almost certainly nonnegative then Y is almost certainly nonnegative and, by Theorem (2,8,5),

$$E(Y) \leqslant E(X_j) . \qquad (6,9,18)$$

b) If the X_j form a strictly stationary sequence, so do their absolute values $|X_j|$. Let Z be the a.s. limit of the arithmetic means $\frac{1}{k}\sum_{j=1}^{k}|X_j|$; then, clearly $|Y| \leqq Z$ a.c., and so, by (6,9,18),

$$|E(Y)| \leqslant E(|Y|) \leqslant E(Z) \leqslant E(|X_j|) . \qquad (6,9,19)$$

Now let T be any r.v., M an arbitrary positive number, $\rho_M(T)$ the r.v. defined by

$$\rho_M(T) = \begin{cases} T, \text{ if } |T| \leqslant M , \\ M, \text{ if } T > M , \\ -M, \text{ if } T < -M , \end{cases}$$

and set

$$\chi_M(T) = T - \rho_M(T) .$$

If the X_j form a strictly stationary sequence, then so do the $\rho_M(X_j)$ on the one hand and the $\chi_M(X_j)$ on the other. Set

$$\rho_M = \lim_{k\to+\infty} \frac{1}{k}\sum_{j=1}^{k} \rho_M(X_j) \quad \text{a.s.} \ ;$$

$$\chi_M = \lim_{k\to+\infty} \frac{1}{k}\sum_{j=1}^{k} \chi_M(X_j) \quad \text{a.s.},$$

so that

$$Y = \rho_M + \chi_M .$$

Now, by the very definition of the integral (see II.6), for arbitrarily small ε and all sufficiently large M we have

$$E(|X_j - \rho_M(X_j)|) = E|\chi_M(X_j)| < \varepsilon \qquad (6,9,20)$$

for any j, because of stationarity; hence, by (6,9,19),

$$E(|\chi_M|) < \varepsilon . \qquad (6,9,21)$$

On the other hand, the $\rho_M(X_j)$ are bounded, and so

$$E(Y) = E(\rho_M) + E(\chi_M) = E[\rho_M(X_j)] + E(\chi_M)$$

$$= E(X_j) - E[\chi_M(X_j)] + E(\chi_M),$$

and by (6,9,19) and (6,9,20) this means that

$$|E(Y) - E(X_j)| < 2\varepsilon,$$

which proves the theorem. ∎

Now let us add the restriction that the X_j are mutually independent. The same then holds for the bounded r.v. $\rho_M(X_j)$, so that Theorem (6,7,1) implies

$$\rho_M = E[\rho_M(X_j)] \quad \text{a. s.}$$

By (6,9,21), there exists a sequence $\{M_h\}$ $(h = 1, 2, 3, \ldots)$ of values of M, converging to $+\infty$, such that

$$\sum_h E(|\chi_{M_h}|) < +\infty \quad \text{and} \quad \lim_{h\to+\infty} \rho_{M_h} = E(X_j) . \qquad (6,9,22)$$

Using Theorem (6,6,1), we thus conclude that

THEOREM (6,9,4). If $\{X_j\}$ $(j = 1, 2, 3, \ldots)$ is a sequence of mutually independent real random variables obeying the same law, with $E(X_j) < +\infty$, then as $k \to +\infty$ the arithmetic mean $Y_k = \frac{1}{k}\sum_{j=1}^{k} X_j$ converges to $E(X_j)$ almost surely.

EXERCISE (6,9,1). Extend Theorems (6,9,1), (6,9,3) and (6,9,4) to complex "point functions" or r.v.

III. CONVERGENCE TO A LAW OF NORMAL TYPE; CONVERGENCE TO A POISSON LAW

10. *Convergence to a normal law in the Bernoulli case*

In our discussion of the addition of a large number of random variables, our attention has been focussed mainly on the laws of large numbers and the question they attempt to answer; we have yet to consider the asymptotic behavior of the probability laws themselves. The remainder of this chapter will be devoted to this problem.

As a first example we consider the Bernoulli case of the heads-or-tails scheme, since it gives us occasion to throw light on some points which have a general bearing on the formulation of the problem.

Recall the notation of VI.1 for the Bernoulli case. Consider, say, the arithmetic mean F_k; its variance pq/k converges to 0 as $k \longrightarrow +\infty$; moreover, we know that F_k converges to p in q.m. and a.s. It follows immediately that its d.f. converges to the degenerate Dirac d.f. $\Delta(x - p)$. However, this adds nothing to what we already know. Considering the repetition R_k, we see that its expectation kp and variance kpq tend to infinity: the probability law of R_k must become degenerate, the degeneracy being in some way opposite to that of the Dirac d.f. It thus seems that we may obtain results of value for our present purposes only by concentrating on some quantity whose value remains bounded and whose dispersion remains finite (i.e., neither too large nor too small) - a quantity derived from R_k (or from F_k) by a suitable centering, followed by normalization.

One such quantity immediately comes to mind; set

$$Z_k = \frac{1}{\sqrt{kpq}}(R_k - kp) = \sqrt{\frac{k}{pq}}(F_k - p)$$

$$= \frac{1}{\sqrt{kpq}} \sum_{j=1}^{k} (X_j - p) .$$

Then the expectation of Z_k is 0 and its variance unity except, of course, in the degenerate cases p = 0 and p = 1.

In the sequel we shall show that the d.f. of Z_k converges to a limit d.f. H(z) as $k \longrightarrow +\infty$. This is by no means self-evident; by contrast, it is a priori evident that, if a limit d.f. H(z) exists, it must belong to a closed type (see VI.2). For example, we shall show that the convolution of $H(z\sqrt{2})$ (which belongs to the type of H(x)) with itself is a d.f. of the same type - namely H(z) itself. To do this, consider the even values of k, k = 2h, and set

$$Z_k^1 = \frac{1}{\sqrt{kpq}} \sum_{j=1}^{h} (X_{2j} - p), \quad Z_k^2 = \frac{1}{\sqrt{kpq}} \sum_{j=0}^{h-1} (X_{2j+1} - p).$$

If Z_k has a limit d.f. H(z), the same holds for

$$\sqrt{2}\, Z_k^1 = \frac{1}{\sqrt{hpq}} \sum_{j=1}^{h} (X_{2j} - p)$$

and

$$\sqrt{2}\, Z_k^2 = \frac{1}{\sqrt{hpq}} \sum_{j=0}^{h-1} (X_{2j+1} - p),$$

since the mutually independent X_j obey the same probability law; i.e., Z_k^1 and Z_k^2 have a limit d.f. $H(z\sqrt{2})$. Since $Z_k = Z_k^1 + Z_k^2$, Z_k^1 and Z_k^2 are mutually independent and Z_k has limit d.f. H(z), it is clear that the convolution of $H(z\sqrt{2})$ with itself is H(z).

By an obvious extension of this procedure we can show that, in the general case, the convolution of two distribution functions H(az + b) and H(a'z + b') belonging to the type of H(z) is a d.f. of the same type. This re-emphasizes the importance of the notion of closed types.

We shall now prove rigorously that, as $k \longrightarrow +\infty$, the d.f. of Z_k converges to the normal d.f., whose type we already know to be closed (see Theorem (6,2,5). This means that, in the notation of (4,4,10), we must prove that, for any constant z,

$$\lim_{k\to+\infty} \Pr(Z_k < z) = L(z) = \frac{1}{\sqrt{2\pi}} \int_{-\infty}^{z} e^{-y^2/2}\, dy . \qquad (6,10,1)$$

Now, if a is an arbitrary negative number, the Bienaymé inequality shows that, for any k,

$$\Pr(Z_k \leqslant a) < \frac{1}{a^2}.$$

Thus, for arbitrarily small ε, there is a fixed negative a with absolute value so large that, for any k,

$$\Pr(Z_k \leqslant a) < \varepsilon \quad \text{and} \quad L(a) < \varepsilon . \tag{6,10,2}$$

Thus, to prove (6,10,1) we need only show that, for fixed z_1 and z_2 ($z_1 < z_2$), if we set

$$P_k(z_1, z_2) = \Pr(z_1 < Z_k < z_2),$$

then

$$\lim_{k\to+\infty} P_k(z_1, z_2) = \frac{1}{\sqrt{2\pi}} \int_{z_1}^{z_2} e^{-y^2/2}\, dy = L(z_2) - L(z_1).$$

Now the event $z_1 < Z_k < z_2$ is the same as the event

$$kp + z_1\sqrt{kpq} < R_k < kp + z_2\sqrt{kpq} .$$

Let $h_1, h_2, \ldots, h_s$ be integers defined as follows:

$$h_1 = [kp + z_1\sqrt{kpq} + 1],\ h_2 = h_1 + 1,\ h_3 = h_2 + 1, \ldots,\ h_s = h_{s-1} + 1,$$

and h_s the greatest integer $< kp + z_2\sqrt{kpq}$.
For every integer $h (0 \leqq h \leqq k)$, set $p_h(k) = \Pr(R_k = h)$. We know (cf. (6,1,17)) that

$$p_h(k) = C_k^h p^h q^{k-h} = \frac{k!}{h!\,(k-h)!}\, p^h q^{k-h} ; \tag{6,10,3}$$

it is obvious that

$$P_k(z_1, z_2) = \sum_{j=1}^{s} p_{h_j}(k). \tag{6,10,4}$$

Setting

$$h = kp + y\sqrt{kpq}$$

and, in particular

$$h_j = kp + y_j\sqrt{kpq} \qquad (j = 1, 2, \ldots, s);$$

we note that

$$z_1 < y_j < z_2 \quad (j = 1, 2, \ldots, s), \quad y_{j+1} - y_j = \frac{1}{\sqrt{kpq}} \; (j = 1, 2, \ldots s-1). \quad (6{,}10{,}$$

We wish to find an asymptotic expression for $p_h(k)$ as $k \longrightarrow +\infty$, valid uniformly in y for $y \in]\, z_1, z_2 [$ (note that since the h_j are integers the y_j cannot remain fixed as $k \longrightarrow +\infty$). Under these conditions, since neither p nor q is zero, h and k - h are infinitely large, of the same order as k:

$$h = kp + y\sqrt{kpq} = kp\left(1 + y\sqrt{\frac{q}{kp}}\right),$$

$$k - h = kq + y\sqrt{kpq} = kq\left(1 + y\sqrt{\frac{p}{kq}}\right). \qquad (6{,}10{,}$$

In our expression (6,10,3) for $p_h(k)$ we may thus replace k!, h!, and (k - h)! by their Stirling approximations (1,1,9). This gives

$$p_h(k) = \frac{1}{\sqrt{2\pi}} \sqrt{\frac{k}{h(k-h)}} \frac{k^k}{h^h(k-h)^{k-h}} p^h q^{k-h} [1 + \omega_h(k)],$$

where $\lim\limits_{k\to+\infty} \omega_h(k) = 0$ uniformly in $y \in]\, z_1, z_2[$.
It remains to consider the quantity

$$A_h(k) = \sqrt{\frac{k}{h(k-h)}} \frac{k^k}{h^h(k-h)^{k-h}} p^h q^{k-h}.$$

Consider its (natural) logarithm; if we replace h and k - h by their expressions (6,10,6) (remembering that $y\sqrt{\frac{p}{kq}}$ and $y\sqrt{\frac{q}{kp}}$ are infinitesimal, so that $\log\left(1 - y\sqrt{\frac{p}{kq}}\right)$ and $\log\left(1 - y\sqrt{\frac{q}{kp}}\right)$ may be replaced by their second-order expansions), we get

$$\log A_h(k) = \log\frac{1}{\sqrt{kpq}} - \frac{y^2}{2} + \eta_h(k),$$

where

$$\lim_{k\to+\infty} \eta_h(k) = 0, \quad \text{uniformly in} \quad y \in]\, z_1, z_2[\,;$$

It follows that

$$p_h(k) = \frac{e^{-y^2/2}}{\sqrt{2\pi\, kpq}}\,[1 + \chi_h(k)], \qquad (6,10,7)$$

where

$$\lim_{k\to+\infty} \chi_h(k) = 0\,, \quad \text{uniformly in} \quad y \in\,]z_1, z_2[\,,$$

and, in particular,

$$p_{h_j}(k) = \frac{e^{-\frac{1}{2}y_j^2}}{\sqrt{2\pi\, kpq}}\,[1 + \chi_{h_j}(k)],$$

where

$$\lim_{k\to+\infty} \chi_{h_j}(k) = 0\,, \quad \text{uniformly in } j.$$

Using (6,10,4) and (6,10,5), we thus get

$$\begin{aligned} P_k(z_1, z_2) &= \sum_{j=1}^{s} \frac{1}{\sqrt{2\pi}}\, e^{-\frac{1}{2}y_j^2}\,(y_{j+1} - y_j) \\ &\quad + \sum_{j=1}^{s} \frac{1}{\sqrt{2\pi}}\, e^{-\frac{1}{2}y_j^2}\,\chi_{h_j}(k)\,(y_{j+1} - y_j)\,. \end{aligned} \qquad (6,10,8)$$

The second term on the right-hand side of the equality clearly converges to zero, while the first converges to

$$\frac{1}{\sqrt{2\pi}} \int_{z_1}^{z_2} e^{-y^2/2}\, dy,$$

and this proves (6,10,1). It is also easy to see that $P_k(z_1, z_2)$ converges to $L(z_2) - L(z_1)$ *uniformly in* z_1 *and* z_2; in other words, for all $\varepsilon > 0$ there exists a number K, independent of z_1 and z_2, such that for all $k > K$

$$|P_k(z_1, z_2) - [L(z_2) - L(z_1)]| < \varepsilon \qquad (6,10,9)$$

for any z_1 and z_2. This means that (6,10,9) will hold even if z_1 and z_2 are not fixed but vary arbitrarily with k.

This follows from (6,10,7), which is of independent interest, since it states that for large k the probability $p_h(k)$ is approximately

$$\frac{1}{\sqrt{2\pi kpq}}\, e^{-y^2/2}\,. \qquad (6,10,10)$$

An idea of the order of magnitude of the error caused by replacing $p_h(k)$ by the approximation (6,10,10) may be obtained from the following table, which gives the values of $p_h(k)$, and the corresponding values of $\frac{1}{\sqrt{2\pi\, kpq}} e^{-y^2/2}$, for $k = 10$, $p = q = 1/2$, and various values of y.

y	$p_h(k)$	$\frac{1}{\sqrt{2\pi\, kpq}} e^{-y^2/2}$
0.000	0.2461	0.2523
0.632	0.2051	0.2066
1.265	0.1172	0.1134
1.897	0.0440	0.0417
2.530	0.0098	0.0103
3.162	0.0010	0.0017

We see that the approximation is quite good, despite the fact that 10 is by no means a large number. It is true, as we shall see in VI.13, that $p = q = 1/2$ is the most favorable case. For the moment it is intuitively clear that, since $\frac{1}{\sqrt{2\pi kpq}} e^{-y^2/2}$ is an even function of y, the approximation should be especially good when $p_h(k)$ itself is an even function of y - and this is precisely what happens when $p = q = 1/2$. Were p noticeably different from q we should expect a considerably greater error (cf. VI.13).

Generalizations. Let $\{X_j\}$ $(j = 1, 2, \ldots)$ be a countable sequence of mutually independent second-order real random variables (there is no loss of generality in assuming that $E(X_j) = 0$), and set

$$\sigma_j^2 = \mathcal{V}(X_j) \qquad (\sigma_j \geqslant 0), \quad S_k = \sum_{j=1}^{k} X_j,$$

$$\mathcal{V}(S_k) = \sum_{j=1}^{k} \sigma_j^2 = a_k^2 \qquad (a_k \geqslant 0),$$

$$Z_k = S_k / a_k .$$

a_k is nondecreasing as k increases, and we may assume $a_k > 0$ for all k, as this excludes only trivial cases. Note that $E(Z_k) = 0$ and $\mathcal{V}(Z_k) = 1$. The preceding example (the Bernoulli case) motivates formulation of the following problem: to find (at least) sufficient conditions under which, as $k \longrightarrow +\infty$, the d.f. $H_k(z)$ of Z_k converges to the normal d.f. L(z) (defined by (4,4,10)).

However, it is clearly difficult to extend the direct-calculation method of the Bernoulli case, and we shall therefore adopt another method. Introduce the following notation:

1) $\phi_j(u) = E[e^{iuX_j}]$: the characteristic function of X_j;

2) $\psi_j(u)$: the second characteristic function $\psi_j(u) = \log \phi_j(u)$;

3) $\lambda_k(u)$: the characteristic function of Z_k;

4) $\mu_k(u) = \log \lambda_k(u)$: the second characteristic function of Z_k.

Direct calculation gives

$$\mu_k(u) = \sum_{j=1}^{k} \psi_j\left(\frac{u}{a_k}\right),$$

and by virtue of (4,3,10) we may write

$$\psi_j(u) = -\frac{\sigma_j^2}{2} u^2 + \rho_j(u)\, u^2,$$

where

$$\lim_{u \to 0} \rho_j(u) = 0 . \qquad (6,10,11)$$

Set

$$A_k(u) = \sum_{j=1}^{k} \rho_j\left(\frac{u}{a_k}\right). \qquad (6,10,12)$$

Then

$$\mu_k(u) = -\frac{u^2}{2} + \frac{A_k(u)}{a_k^2} u^2 . \qquad (6,10,13)$$

Since $-u^2/2$ is the second characteristic function of the d.f. L(z) (cf. (4,4,20)), it follows from Theorem (3,8,1) that, if

$\lim_{k\to+\infty} \frac{A_k(u)}{a_k^2} = 0$ for all u and there exists a number $\eta > 0$ such that the convergence is uniform in u over the interval $[-\eta, +\eta]$, then the d.f. $H_k(z)$ converges to $L(z)$.

Now $A_k(u)/a_k^2$ is easy to study in certain cases:

1) Assume that the X_j all obey the same law; the functions $\rho_j(u) = \rho(u)$ and $\sigma_j^2 = \sigma^2$ are independent of j, so that

$$\frac{A_k(u)}{a_k^2} = \frac{1}{\sigma^2}\,\rho\left(\frac{u}{\sigma\sqrt{k}}\right).$$

Thus:

THEOREM (6,10,1). If $\{X_j\}$ ($j = 1, 2, 3, \ldots$) is a sequence of mutually independent second-order random variables, all obeying the same law, such that $E(X_j) = 0$ and $E(X_j^2) = \sigma^2 > 0$, then the distribution function $H_k(z)$ of the random variable $Z_k = \frac{1}{\sigma\sqrt{k}}\sum_{j=1}^{k} X_j$ converges to the normal distribution function $L(z)$ as $k \longrightarrow +\infty$. ∎

2) Assume that the X_j have absolute 3rd moments, and set

$$\beta_j^3 = E(|X_j|^3), \qquad b_k^3 = \sum_{j=1}^{k} \beta_j^3 \ ;$$

then, by (4,3,17) for $\varepsilon = 1$,

$$\left|\rho_j\left(\frac{u}{a_k}\right)\right| \leqslant 2\,\beta_j^3 \left|\frac{u}{a_k}\right|, \tag{6,10,14}$$

provided that

$$\frac{\beta_j}{a_k}|u| < \eta, \tag{6,10,15}$$

where η is the fixed positive number of (4,3,18) which assures the validity of (4,3,17). Since $\beta_j < b_k$ for $j \leqq k$, inequality (6,10,15) holds uniformly in j if

$$|u| \leqslant \eta\,\frac{a_k}{b_k}\ . \tag{6,10,16}$$

Now assume that

$$\lim_{k\to+\infty} \frac{b_k}{a_k} = 0\ . \tag{6,10,17}$$

For any u, inequality (6,10,16) will hold for sufficiently large k. It follows that

$$\left|\frac{A_k(u)}{a_k^2}\right| \leqslant 2\,\frac{b_k^3}{a_k^3}\,|u|,$$

whence

$$\lim_{k\to+\infty} \frac{A_k(u)}{a_k^2} = 0,$$

uniformly in u over any bounded interval. Therefore:

THEOREM (6,10,2). Let $\{X_j\}$ $(j = 1, 2, 3,\ldots)$ be a sequence of mutually independent random variables such that $E(|X_j|^3) < +\infty$ and $E(X_j) = 0$; set

$$a_k^2 = \sum_{j=1}^{k} E|X_j|^2, \qquad b_k^3 = \sum_{j=1}^{k} E(|X_j|^3), \text{ and assume that } \lim_{k\to+\infty} \frac{b_k}{a_k} = 0.$$

Then the distribution function $H_k(z)$ of $Z_k = \frac{1}{a_k}\sum_{j=1}^{k} X_j$ converges to the normal distribution function $L(z)$.

11. *The general problem of convergence to a limit law*

Hitherto we have assumed that the $E(X_j)$ are zero. If the $E(X_j)$ have arbitrary values, consideration of the r.v.

$$Z_k = \frac{1}{a_k}\sum_{j=1}^{k} X_j - \frac{1}{a_k}\sum_{j=1}^{k} E(X_j) \qquad (6,11,1)$$

brings us back to the previous case. Now, setting $c_k = \frac{1}{a_k}\sum_{j=1}^{k} E(X_j)$ and $Y_{k,j} = X_j/a_k$, we can rewrite (6,11,1) as $Z_k = \sum_{j=1}^{k} Y_{k,j} - c_k$

In this expression the constants c_k ensure appropriate centering, while the normalization factors a_k are included in the $Y_{k,j}$. The most general formulation of our problem - convergence of the sum of an increasing number of mutually independent random variables to a limit law - may thus be phrased as follows:

Assume that with each k $(k = 1, 2, 3,\ldots)$ are associated

1) a positive integer h_k such that $\lim_{k\to+\infty} h_k = +\infty$;

2) h_k mutually independent random variables $Y_{k,j}$ $(j = 1, 2,\ldots, h_k)$.

Under what conditions (concerning the probability laws of the $Y_{k,j}$) do there exist constants c_k ($k = 1, 2, 3, \ldots$) such that, as $k \longrightarrow + \infty$, the d.f. of the r.v.

$$Z_k = \sum_{j=1}^{h_k} Y_{k,j} - c_k \tag{6,11,2}$$

converges to a (nondegenerate, i.e. not Dirac) d.f. $H(z)$?

Now, without suitable restrictions this question is meaningless. For instance, assume that for each k one of the $Y_{k,j}$ - say $Y_{k,1}$ - is much greater than the sum of all the others, so that if $c_k = 0$ the d.f. H_k coincides (approximately) with the d.f. $G_{k,1}$ of $Y_{k,1}$. There is no reason that the $G_{k,1}$ should not behave in any desired fashion as $k \longrightarrow + \infty$, and the same will then hold for the H_k.

To make this restriction more precise in the sequel, we introduce the following definition:

DEFINITION (6,11,1). The $Y_{k,j}$ are said to be *infinitesimal* if, for all $\varepsilon > 0$,

$$\lim_{k \to +\infty} \ \underset{1 \leq j \leq h_k}{\text{l.u.b.}} \ \Pr(|Y_{k,j}| \geqslant \varepsilon) = 0 . \tag{6,11,3}$$

Infinitely divisible laws and distribution functions

Let $G(x)$ be a bounded d.f. such that $G(-\infty) = 0$. For every positive integer k we define the kth convolution power of G, denoted by G^{*k}, as the d.f. obtained by convoluting G with itself (see III.8) k times:

$$G^{*k} = \underbrace{G * G * \ldots * G}_{k \text{ times}}.$$

The probability law of a r.v. X and its d.f. $F(x)$ are said to be *infinitely divisible* if, for all k ($k = 1, 2, 3, \ldots$), there exists a bounded normalized d.f. $G_k(x)$ such that

$$F = G_k^{*k}. \tag{6,11,4}$$

Now, (6,11,4) implies $G_k(+\infty) = 1$, since $F(+\infty) = 1$; therefore (6,11,4) is equivalent to the assertion that, for all k, there exist k mutually independent r.v. $Y_{k,j}$ ($j = 1, 2, 3, \ldots$), all having the same d.f. G_k, such that

$$X = \sum_{j=1}^{k} Y_{k,j} .$$

Let $\phi(u)$ be the characteristic function of $F(x)$; then (6,11,4) is equivalent to the statement: for every k there exists a positive-definite (and necessarily continuous) function $\phi_k(u)$ such that

$$\varphi(u) = [\varphi_k(u)]^k. \qquad (6,11,5)$$

We shall use the letter D for the class of infinitely divisible distribution functions (or probability laws of r.v.). Note that if a d.f. is infinitely divisible (belongs to class D), then it is normalized and bounded, and $F(+\infty) = 1$. The following theorems follow immediately from the definition.

THEOREM (6,11,1). If a sequence F_h $(h = 1, 2, 3,...)$ of infinitely divisible distribution functions converges as $h \longrightarrow +\infty$ to a limit distribution function F such that $F(+\infty) = 1$, then F is infinitely divisible (use Theorem (3,4,1)).

THEOREM (6,11,2). If F and G are two infinitely divisible distribution functions, then their convolution $F * G$ is infinitely divisible. ■

Theorem (6,11,2) thus means that the class D is closed with respect to convolution (or addition of independent r.v.).

THEOREM (6,11,3). Every distribution function F belonging to a closed type is infinitely divisible. ■

A normal law is thus infinitely divisible. Note that by Theorem (6,2,3), any Poisson law is also infinitely divisible. One can prove:

THEOREM (6,11,4). A distribution function F with characteristic function $\phi(u)$ is infinitely divisible if and only if there exist a bounded distribution function $G(\alpha)$ of the real variable α $(-\infty < \alpha < +\infty)$ and a constant μ such that

$$\log \varphi(u) = i\mu u + \int_{-\infty}^{+\infty} \Theta(u\,;\alpha)\, dG(\alpha), \qquad (6,11,6)$$

where

$$\Theta(u\,;\alpha) = \begin{cases} \left(e^{iu\alpha} - 1 - \dfrac{iu\alpha}{1+\alpha^2}\right)\dfrac{1+\alpha^2}{\alpha^2} & \text{if } \alpha \neq 0\,; \\ \lim\limits_{\alpha \to 0} \left(e^{iu\alpha} - 1 - \dfrac{iu\alpha}{1+\alpha^2}\right)\dfrac{1+\alpha^2}{\alpha^2} = -\dfrac{u^2}{2} & \text{if } \alpha = 0. \end{cases}$$

The representation (6,11,6) of $\log \phi(u)$ is then unique. ∎

We shall not prove this theorem; it clearly implies:

COROLLARY (6,11,1). If a distribution function F with characteristic function $\phi(u)$ is infinitely divisible, then $\phi(u)$ never vanishes $(-\infty < u < +\infty)$.

THEOREM (6,11,5). A distribution function F with characteristic function $\phi(u)$ is infinitely divisible if and only if there exist

1) constants μ and V, $V \geqq 0$,

2) a monotone nondecreasing function $M(\alpha)$ on $(-\infty, 0[$ and a monotone nondecreasing function $N(\alpha)$ on $]0, +\infty)$ such that

a) $\int_{-a}^{0} \alpha^2\, dM(\alpha) + \int_{0}^{a} \alpha^2\, dN(\alpha) < +\infty$ for all $a > 0$;

b) $$\log \varphi(u) = i\mu u - \frac{V}{2}u^2 + \int_{-\infty}^{0} \left(e^{iu\alpha} - 1 - \frac{iu\alpha}{1+\alpha^2}\right) dM(\alpha)$$

$$+ \int_{0}^{+\infty} \left(e^{iu\alpha} - 1 - \frac{iu\alpha}{1+\alpha^2}\right) dN(\alpha). \qquad (6,11,7)$$

Indeed, we get (6,11,7) from (6,11,6) by setting

$$V = G(+0) - G(-0)\,;$$

$$dM(\alpha) = \frac{1+\alpha^2}{\alpha^2}\, dG(\alpha) \qquad \text{for } \alpha < 0\,;$$

$$dN(\alpha) = \frac{1+\alpha^2}{\alpha^2}\, dG(\alpha) \qquad \text{for } \alpha > 0.$$

The representation (6,11,7) is unique (of course, up to equivalence as regards the monotone functions $M(\alpha)$ and $N(\alpha)$). ∎

Those distribution functions $F(x)$ of class D which have finite variance, i.e., such that

$$\int_{-\infty}^{+\infty} x^2\, dF(x) < +\infty,$$

form a subclass which we shall denote by D_2. One can prove:

THEOREM (6,11,6). A distribution function F (normalized, bounded, and such that $F(+\infty) = 1$) with characteristic function $\phi(u)$ belongs to the subclass D_2 if and only if there exist a constant ν and a bounded normalized distribution function $K(\alpha)$ $(-\infty < \alpha < +\infty)$ such that

$$\log \varphi(u) = i\nu u + \int_{-\infty}^{+\infty} (e^{iu\alpha} - 1 - iu\alpha)\frac{1}{\alpha^2}\, dK(\alpha) ; \qquad (6,11,8)$$

and then

$$\nu = \int_{-\infty}^{+\infty} x\,dF(x) , \qquad K(+\infty) = \int_{-\infty}^{+\infty} x^2 dF(x). \qquad (6,11,9)$$

In fact, (6,11,8) may be derived from (6,11,6) as a special case by setting

$$dK(\alpha) = (1 + \alpha^2)\, dG(\alpha). \qquad (6,11,10)$$

The representation (6,11,8) is unique.

EXAMPLE (6,11,1). Consider the normal-type d.f. with density

$$f(x) = \frac{1}{\sqrt{2\pi}\sigma} e^{-\frac{(x-m)^2}{2\sigma^2}}.$$

To derive the representations (6,11,6) and (6,11,8) we set

$$\mu = \nu = m , \; G(\alpha) = K(\alpha) = \begin{cases} 0 & \text{for } \alpha \leqslant 0 ; \\ \sigma^2 & \text{for } \alpha > 0. \end{cases}$$

As for (6,11,7), we set

$$\mu = m , \; V = \sigma^2, \; M(\alpha) \equiv 0, \; N(\alpha) \equiv 0.$$

EXAMPLE (6,11,2). Consider the most general Poisson law, for which, by (4,4,6)

$$\log \varphi(u) = m(e^{iu} - 1) . \qquad (6,11,11)$$

We can derive (6,11,11) in three ways:

1) from (6,11,6), by setting

$$\mu = \frac{m}{2} , \; G(\alpha) = \begin{cases} 0 & \text{for } \alpha \leqslant 1 ; \\ \frac{m}{2} & \text{for } \alpha > 1 ; \end{cases}$$

2) from (6,11,7), by setting

$$\mu = \frac{m}{2} , \; V = 0 , \; M(\alpha) \equiv 0 \; ; \; N(\alpha) = \begin{cases} -m & \text{for } \alpha \leqslant 1 ; \\ 0 & \text{for } \alpha > 1 ; \end{cases}$$

3) from (6,11,8), by setting

$$\nu = m , \; K(\alpha) = \begin{cases} 0 & \text{for } \alpha \leqslant 1 ; \\ m & \text{for } \alpha > 1. \end{cases}$$

Thus every Poisson law is infinitely divisible; in particular, it belongs to the subclass D_2.

REMARK (6,11,1). Formulas (6,11,6) and (6,11,7) and the more particular form (6,11,8) are often called "canonical forms." But it is clearly possible to transform them into equivalent but different forms in various ways, e.g. by a change of variable in the integrals. For certain applications these equivalent forms may be more interesting than the "canonical" forms. ∎

The importance of infinitely divisible laws rests on the following theorem:

THEOREM (6,11,7). Let $Y_{k,j}$ ($j = 1, 2, \ldots, h_k$) be mutually independent infinitesimal r.v. (cf. Definition (6,11,1)). Then, if there exist constants c_k ($k = 1, 2, 3, \ldots$) such that the distribution function H_k of the random variable Z_k defined by (6,11,2) converges to a limit distribution function H as $k \longrightarrow +\infty$, H is infinitely divisible. Conversely, every infinitely divisible distribution function H is the limit of distribution functions H_k of random variables Z_k of the form (6,11,2), where $Y_{k,j}$ $(j = 1, 2, \ldots, h_k)$ are mutually independent and infinitesimal. ∎

We shall not go into the proof of this theorem. To complete the theorem (still assuming that the $Y_{k,j}$ are mutually independent and infinitesimal) one can state necessary and sufficient conditions on the probability laws of the $Y_{k,j}$ for the existence of constants c_k such that

1) H_k converges to a limit;

2) H_k converges to the normal d.f. L;

3) H_k converges to the Poisson d.f. with parameter $m (m > 0)$.

If we assume in addition that the $Y_{k,j}$ are second-order r.v. and $E(Y_{k,j}) = 0$, we can prove more precise theorems, giving necessary and sufficient conditions on the $Y_{k,j}$ for the existence of constants c_k such that

1) H_k converges to a limit $H \in D_2$ and

$$\lim_{k \to +\infty} \mathcal{V}(Z_k) = \text{ the variance associated with } H;$$

2) H_k converges to the normal d.f. L, and

$$\lim_{k \to +\infty} \mathcal{V}(Z_k) = 1 \text{ (variance associated with } L);$$

3) H_k converges to a Poisson law with parameter m, and

$$\lim_{k \to +\infty} \mathcal{V}(Z_k) = m .$$

We shall only prove a special case of these results. Assume that the only possible values of the $Y_{k,j}$ are the nonnegative integers 0, 1, 2, 3,..., and set

$$\Pr(Y_{k,j} = 1) = q_j(k), \; \Pr(Y_{k,j} > 1) = r_j(k),$$

$$\Pr(Y_{k,j} = s) = p_{j,s}(k) \quad (s = 0, 1, 2, \ldots).$$

Note that

$$p_{j,o}(k) = 1 - q_j(k) - r_j(k), \quad q_j(k) = p_{j,1}(k), \quad r_j(k) = \sum_{s>1} p_{j,s}(k) .$$

Let $\phi_{k,j}(u)$ denote the characteristic function of $Y_{k,j}$, $\psi_{k,j}(u) = \log \phi_{k,j}(u)$ its second characteristic function, and $\phi_k(u)$ and $\psi_k(u)$ the characteristic function and second characteristic function of

$$S_k = \sum_{j=1}^{h_k} Y_{k,j} ;$$

then

$$\psi_k(u) = \sum_{j=1}^{h_k} \psi_{k,j}(u) .$$

Now,

$$\begin{aligned}\varphi_{k,j}(u) &= \sum_{s=0}^{+\infty} e^{isu}\, p_{j,s}(k) \\ &= [1 - q_j(k) - r_j(k)] + e^{iu} q_j(k) + \sum_{s>1} e^{isu}\, p_{j,s}(k) .\end{aligned} \tag{6,11,12}$$

If we set

$$\omega_{k,j}(u)\, r_j(k) = \sum_{s>1} e^{isu}\, p_{j,s}(k) ,$$

$$\rho_{k,j}(u) = \omega_{k,j}(u) - 1,$$

it is clear that

$$|\omega_{k,j}(u)| \leqslant 1, \quad |\rho_{k,j}(u)| \leqslant 2 , \tag{6,11,13}$$

and we may write (6,11,12) as

$$\varphi_{k,j}(u) = 1 + (e^{iu} - 1)\, q_j(k) + \rho_{k,j}(u)\, r_j(k) . \tag{6,11,14}$$

There exists a number $\alpha > 0$ such that

$$|\log(1 + x) - x| < x^2 \qquad \text{if } |x| < \alpha.$$

Therefore, if we assume that

$$q_j(k) < \frac{\alpha}{4}, \quad r_j(k) < \frac{\alpha}{4} , \tag{6,11,15}$$

we may write

$$\psi_{k,j}(u) = \log \varphi_{k,j}(u) = (e^{iu} - 1)\, q_j(k) + \chi_{k,j}(u) , \tag{6,11,16}$$

where, by (6,11,13)

$$|\chi_{k,j}(u)| \leqslant 2r_j(k) + 4\,|q_j(k) + r_j(k)|^2. \qquad (6,11,17)$$

Recall that, by assumption, $\lim h_k = +\infty$. Assume that

1) $$\lim_{k\to+\infty} \underset{1\leqslant j\leqslant h_k}{\text{l.u.b.}} [q_j(k) + r_j(k)] = 0\,, \qquad (6,11,18)$$

2) $$\lim_{k\to+\infty} \sum_{j=1}^{h_k} q_j(k) = m \geqslant 0\,; \qquad (6,11,19)$$

3) $$\lim_{k\to+\infty} \sum_{j=1}^{h_k} r_j(k) = 0\,; \qquad (6,11,20)$$

we shall disregard the trivial case m = 0.

By virtue of (6,11,18), inequality (6,11,15) is valid for all sufficiently large k and all j; thus,

$$\psi_k(u) = \left[\sum_{j=1}^{h_k} q_j(k)\right](e^{iu} - 1) + \chi_k(u),$$

where, by (6,11,17),

$$|\chi_k(u)| \leqslant 2\sum_{j=1}^{h_k} r_j(k) + 4\sum_{j=1}^{h_k} |q_j(k) + r_j(k)|^2.$$

An obvious result of (6,11,18), (6,11,19) and (6,11,20) is that

$$\lim_{k\to+\infty} \psi_k(u) = m(e^{iu} - 1)$$

uniformly in u. By (4,4,6) and Theorem (3,8,1) we get:

THEOREM (6,11,8). If $\lim_{k\to+\infty} h_k = +\infty$ and the $Y_{k,j}$ $(j = 1, 2, \ldots, h_k)$ are mutually independent infinitesimal random variables which assume nonnegative integer values, then a sufficient condition for the probability law of the sum $S_k = \sum_{j=1}^{h_k} Y_{h,j}$ to converge as $k \longrightarrow +\infty$ to the Poisson law with parameter $m > 0$ is

$$\lim_{k\to+\infty} \sum_{j=1}^{h_k} \Pr(Y_{k,j} = 1) = m\,, \quad \lim_{k\to+\infty} \sum_{j=1}^{h_k} \Pr(Y_{k,j} > 1) = 0\,.$$

EXAMPLE (6,11,3)

We know from Remark (6,8,1) that, given random variables X_k

whose d.f. $F_k(x) = Pr(X_k < x)$ converge to a limit d.f. F, it does not necessarily follow that the moments of X_k (i.e. of F_k), assuming they exist, converge to the corresponding moments of F.

The following example is a simple illustration. In the preceding notation, take $h_k = k$ and assume that the $Y_{k,j}$ have the probability law defined by

$$Pr(Y_{k,j} = 0) = 1 - \frac{1}{k} - \frac{1}{k^2},$$

$$Pr(Y_{k,j} = 1) = q_j(k) = \frac{1}{k},$$

$$Pr(Y_{k,j} = k^{\alpha}) = r_j(k) = \frac{1}{k^2} \qquad (\alpha > 0).$$

Now the conditions of Theorem (6,11,8) are satisfied for m = 1, and the d.f. of S_k converges to the Poisson d.f. with parameter m, i.e. with expectation 1 and variance 1. Clearly,

$$E(S_k) = k \cdot E(Y_{k,j}) = 1 + k^{\alpha-1}$$

$$\mathcal{V}(S_k) = k\,\mathcal{V}(Y_{k,j}) = 1 + k^{2\alpha-1} - \frac{1}{k}(1 + k^{\alpha-1})^2,$$

and so

if $\quad 0 < \alpha < \frac{1}{2}, \ \lim_{k\to+\infty} E(S_k) = 1, \ \lim_{k\to+\infty} \mathcal{V}(S_k) = 1;$

if $\quad \frac{1}{2} < \alpha < 1, \ \lim_{k\to+\infty} E(S_k) = 1, \ \lim_{k\to+\infty} \mathcal{V}(S_k) = +\infty;$

if $\quad 1 < \alpha, \ \lim_{k\to+\infty} E(S_k) = +\infty, \ \lim_{k\to+\infty} \mathcal{V}(S_k) = +\infty.$

EXERCISE (6,11,1)

Let the r.v. X_k have distribution functions F_k converging to a limit d.f. F, and assume that the X_k are almost-certainly uniformly bounded in k; show that, for any $\alpha > 0$, the (algebraic or absolute) α^{th} moment of X_k (i.e. of F_k) converges to the αth moment of F.

EXAMPLE (6,11,4). As an illustration, let us return to the heads-or-tails scheme in the Bernoulli case. Suppose that the number k of tosses is large, but the probability of heads is small, say of the order of 1/k. To fix ideas, assume $p = m/k$, where m is a positive constant. Now, although the number k of tosses is large, the fact that the probability p of heads is small means that we can expect the repetition R_k of heads to have a finite order of magnitude. This is confirmed by Theorem (6,11,8), which shows that, approximately,

$$\Pr(R_k = h) \# e^{-m} \frac{m^h}{h!} \quad (h \text{ an integer } > 0),$$

as may be verified directly from the exact expression (6,1,17) for $\Pr(R_k = h)$.

This example explains why Poisson laws were often called "laws of rare events."

REMARK (6,11,2). There is a fundamental difference between normal-type laws and Poisson laws. This is particularly clear from their canonical representations as infinitely divisible laws (see Examples (6,11,1) and (6,11,2)). At the beginning of VI.10 we saw that, under certain conditions which we shall call (C_L), the Bernoulli law (6,1,17) converges to a law of normal type, while we have just seen that under other conditions (C_p) it converges to a Poisson law. At the same time, the distinction between (C_L) and (C_p) tends to disappear if $m = kp$ in conditions (C_p) is large. We should therefore expect only a small numerical difference between the Poisson law defined by

$$\Pr(X = h) = e^{-m} \frac{m^h}{h!} \quad (h = 0, 1, 2, \ldots)$$

and a normal law, when m is large (at least, for values of h such that $\frac{h - m}{\sqrt{m}} = y$ remains bounded). Direct calculation easily shows that, as $m \longrightarrow +\infty$,

$$e^{-m} \frac{m^h}{h!} \sim \frac{1}{\sqrt{2\pi m}} e^{-y^2/2},$$

uniformly in y over any bounded interval.

12. *Convergence to a limit law for cumulative sums of a large number of independent random variables*

The problem just posed in its most general form in VI.11 includes a special case which we shall call "the problem of cumulative sums," which may be formulated as follows:

Given an infinite sequence X_j ($j = 1, 2, 3,\ldots$) of mutually independent r.v., we shall call the sums $S_k = \sum_{j=1}^{k} X_j$ ($k = 1, 2, 3,\ldots$) "cumulative sums." Under what conditions do there exist constants c_k and positive constants a_k ($k = 1, 2, 3,\ldots$) such that the d.f. $H_k(z)$ of the r.v.

$$Z_k = S_k/a_k - c_k = \frac{1}{a_k}\sum_{j=1}^{k} X_j - c_k \qquad (6,12,1)$$

converges to a limit d.f. $H(z)$ as $k \longrightarrow +\infty$? In VI.10 we dealt with some simple cases of the problem of cumulative sums; here we consider the problem in its most general form.

Of course, we must allow for the restriction described in VI.11, by using Definition (6,11,1) of infinitesimal r.v.

DEFINITION (6,12,1). A bounded normalized d.f. H such that $H(+\infty) = 1$, and the corresponding probability law, are said to belong to class C if there exist an infinite sequence X_j ($j = 1, 2, 3,\ldots$) of mutually independent r.v., constants c_k, positive constants a_k ($k = 1, 2, 3,\ldots$), and constants $c_{k,j}$ ($k = 1, 2, 3,\ldots$; $j = 1, 2,\ldots, k$) such that

1) the random variables $X_j/a_k - c_{k,j}$ ($j = 1, 2,\ldots, k$) are infinitesimal when $k \longrightarrow +\infty$;

2) the d.f. H_k of the r.v. Z_k defined by (6,12,1) converge to H as $k \longrightarrow +\infty$. ∎

We denote the set of distribution functions of class C which have finite variance by C_2.

By Theorem (6,11,7), the distribution functions of class C are infinitely divisible ($C \subset D$, $C_2 \subset D_2$). However, we have seen that there are infinitely divisible distribution functions (e.g. Poisson d.f.) which do not belong to class C.

Every d.f. belonging to a closed type belongs to class C.

Let F(x) be an infinitely divisible d.f. (bounded and normalized, such that $F(+\infty) = 1$) with characteristic function $\phi(u)$; $\log \phi(u)$ has the form (6,11,7):

$$\log \varphi(u) = i\mu u - \frac{V}{2}u^2 + \int_{-\infty}^{0} \left(e^{iu\alpha} - 1 - \frac{iu\alpha}{1+\alpha^2}\right) dM(\alpha)$$

$$+ \int_{0}^{+\infty} \left(e^{iu\alpha} - 1 - \frac{iu\alpha}{1+\alpha^2}\right) dN(\alpha). \qquad (6,12,2)$$

F belongs to class C if and only if (1) the functions $M(\alpha)$ $(\alpha < 0)$ and $N(\alpha)$ $(\alpha > 0)$ possess right derivatives $\frac{d^+}{d\alpha}M(\alpha)$, $\frac{d^+}{d\alpha}N(\alpha)$ and left derivatives $\frac{d^-}{d\alpha}M(\alpha)$, $\frac{d^-}{d\alpha}N(\alpha)$ and (2) the functions $\alpha\frac{d^+}{d\alpha}M(\alpha)$, $\alpha\frac{d^-}{d\alpha}M(\alpha)$ $(\alpha < 0)$ and $\alpha\frac{d^+}{d\alpha}N(\alpha)$, $\alpha\frac{d^-}{d\alpha}N(\alpha)$ $(\alpha > 0)$ are nondecreasing.

In particular, if F belongs to D_2, $\log \phi(u)$ may be expressed in the form (6,11,8):

$$\log \varphi(u) = i\nu\alpha + \int_{-\infty}^{+\infty} (e^{iu\alpha} - 1 - iu\alpha)\frac{1}{\alpha^2}\, dK(\alpha). \qquad (6,12,3)$$

Thus $F \in C_2$ if and only if (1) $K(\alpha)$ possesses a right derivative $\frac{d^+}{d\alpha}K(\alpha)$ and a left derivative $\frac{d^-}{d\alpha}K(\alpha)$ for all $\alpha \neq 0$, and (2) the functions $\frac{d^+}{d\alpha}K(\alpha)/\alpha$ and $\frac{d^-}{d\alpha}K(\alpha)/\alpha$ are nondecreasing on $(-\infty, 0[$ and $]0, +\infty)$, respectively. Reference to Example (6,11,2) shows that Poisson laws, which belong to D_2, do not belong to C_2.

There are necessary and sufficient conditions on probability laws of the X_j for the existence of constants c_k and a_k such that the r.v. X_j/a_k are infinitesimal as $k \longrightarrow +\infty$, and for H_k to have a nondegenerate limit H. Under the additional assumption that the X_j are second-order r.v. there are necessary and sufficient conditions for the existence of constants c_k and a_k such that the r.v. $[X_j - E(X_j)]/a_k$ are infinitesimal, H_k converges to a limit H and that the variance of H_k converges to that of H. Naturally, these conditions are rather complicated.

We shall cite only the following, less general theorems, in which F_j denotes the d.f. of X_j:

THEOREM (6,12,1). There exist constants c_k and a_k such that, as $k \longrightarrow +\infty$, the r.v. X_j/a_k $(j = 1, 2, \ldots, k)$ are infinitesimal and H_k converges to the normal distribution function L, if and only if there exist numbers $\rho_k > 0$ such that

1) $\lim\limits_{k\to+\infty} \rho_k = +\infty$;

2) $\lim\limits_{k\to+\infty} \sum\limits_{j=1}^{k} \Pr(|X_j| > \rho_k) = 0$;

3) $\lim\limits_{k\to+\infty} \dfrac{1}{\rho_k^2} \sum\limits_{j=1}^{k} \left\{ \int_{|x|\leq\rho_k} x^2 dF_j(x) - \left[\int_{|x|\leq\rho_k} x\, dF_j(x) \right]^2 \right\} = +\infty .$

For arbitrary $\varepsilon > 0$, we can then take

$$a_k^2 = \sum_{j=1}^{k} \left\{ \int_{|x|\leq\rho_k} x^2 dF_j(x) - \left[\int_{|x|\leq\rho_k} x\, dF_j(x) \right]^2 \right\},$$

$$c_k = \frac{1}{a_k} \sum_{j=1}^{k} \int_{|x|<\varepsilon a_k} x\, dF_j(x).$$

THEOREM (6,12,2). Let the X_j be second-order r.v. and set $a_k^2 = \sum\limits_{j=1}^{k} \mathcal{V}(X_j)$, $c_k = \dfrac{1}{a_k} \sum\limits_{j=1}^{k} E(X_j)$. Then, as $k \longrightarrow +\infty$, the random variables $\dfrac{1}{a_k}[X_j - E(X_j)]$ are infinitesimal and the distribution function H_k of Z_k converges to the normal distribution function L if and only if, for all $\varepsilon > 0$,

$$\lim_{k\to+\infty} \frac{1}{a_k^2} \sum_{j=1}^{k} \int_{|x|>\varepsilon a_k} |x - E(X_j)|^2 \, dF_j(x) = 0. \blacksquare$$

It is easy to verify that this theorem includes Theorems (6,10,1) and (6,10,2) as special cases. Changing the notation $Y_j = X_j/a_k$, and denoting the d.f. of Y_j by G_j, we can derive the following useful propositions from Theorems (6,12,1) and (6,12,2):

THEOREM (6,12,3). Let Y_j be mutually independent random variables with respective distribution functions G_j $(j = 1, 2, 3, \ldots)$; if there exist arbitrarily small positive numbers ε and η, and a positive constant σ such that

1) $\sum_j \Pr(|X_j| > \varepsilon) < \eta$;

2) $\left|\sigma^2 - \sum_j \left\{ \int_{|y|\leq\varepsilon} y^2\, dG_j(y) - \left[\int_{|y|\leq\varepsilon} y\, dG_j(y) \right]^2 \right\}\right| < \eta$;

then the difference between the distribution function of the random variable

$$Z = \frac{1}{\sigma} \sum_j \left[Y_j - \int_{|y|\leq\varepsilon} y\, dG_j(y) \right]$$

and the normal distribution function L is arbitrarily small.

THEOREM (6,12,4). Let Y_j be mutually independent second-order random variables with respective distribution functions G_j ($j = 1, 2, 3, \ldots$); if there exist arbitrarily small positive numbers ε and η and a positive constant σ such that

1) $\left| \sigma^2 - \sum_j \mathcal{V}(Y_j) \right| < \eta$;

2) $\sum_j \int_{|y|>\varepsilon} y^2\, dG_j(y) < \eta$,

then the difference between the distribution function of the random variable

$$Z = \frac{1}{\sigma} \sum_j [Y_j - E(Y_j)]$$

and the normal distribution function L is arbitrarily small.

EXERCISE (6,12,1). Prove Theorems (6,12,3) and (6,12,4) directly, using characteristic functions and their logarithms, by imitating the proofs of Theorems (6,10,1) and (6,11,8).

Special case: identically distributed X_j

It may be proved that, if the X_j all have the same probability law, then H_k may converge to a limit H only if the random variables $X_j/a_k - \frac{c_k}{k}$ ($j = 1, 2, \ldots, k$) are infinitesimal. H must therefore be infinitely divisible. However, one can prove a more precise result:

THEOREM (6,12,5). If the r.v. X_j all have the same probability law and there exist constants c_k and a_k such that H_k converges to a limit H, then H necessarily belongs to a closed type. Conversely, every distribution function H belonging to a closed type is the limit of the distribution functions H_k of the r.v. Z_k defined by (6,12,1), where the mutually independent r.v. X_j $(j = 1, 2, 3, \ldots)$ all have the same law, for a suitable choice of c_k, a_k, and the probability law of the X_j. $\blacksquare$

We shall not prove this theorem; the beginning of VI.10 clarifies the principle involved. The following theorem is also stated without proof:

THEOREM (6,12,6). Let F(x) be the common distribution function of the r.v. X_j; then there exist constants c_k and a_k such that H_k converges to a nondegenerate normal-type distribution function if and only if

$$\lim_{\rho \to +\infty} \frac{\rho^2 \int_{|x|>\rho} dF(x)}{\int_{|x|<\rho} x^2\, dF(x)} = 0 . \qquad (6,12,4)$$

13. *The Gauss law of measurement errors*

Let us examine the results presented in VI.10, VI.11 and VI.12, in particular Theorem (6,12,1), from the viewpoint of their application to random variables of concrete significance (such as physical and economic quantities). In general, such random variables have second and even higher-order moments, their dispersions are bounded and the r.v. are often almost certainly bounded. This means that the conditions for convergence to a normal-type law should often be satisfied; the following should serve as a good illustration.

Let G be a measurable quantity whose exact measure is g.

If we proceed to measure G experimentally, the resulting number s will obviously differ slightly from the true measure g; the difference

$$x - g = e \tag{6,13,1}$$

is the *error*.

This error is due to a variety of "causes of error" $A_1, A_2, \ldots, A_j \ldots$, which, if suitably defined, may always be considered mutually independent. For instance, in measuring a length one cause of error might be an unsuspected variation of temperature producing a certain expansion, while incorrect reading of the ruler would constitute another and obviously independent cause.

We shall adopt a fundamental assumption, almost always realized in practice: *the number of independent causes of error* A_j *is very large.*

Each cause of error A_j, were it to act alone, would produce an error Y_j. We shall call the Y_j *partial* or *elementary errors*, and divide them into two categories.

On the one hand, a cause of error A_j may be *systematic*, i.e. it acts with fixed intensity (for the specific measurement method being used). An example is the use of an overly short standard meter in measuring length; the results are "systematically" too large. The partial errors Y_j stemming from such systematic causes are also called systematic; they are reproduced *identically* upon repetition of the measurement by the same method.

In the sequel we shall assume that there are no systematic partial errors (either because there are in fact none or because we shall be able to eliminate their influence by introducing suitable corrections).

Unsystematic causes of error, on the other hand, are called *accidental*: no element of stability is involved, i.e. there is no reason why they should be identically reproduced in repeated measurements. Thus, an observer taking a reading always makes a certain error. Under normal conditions, this error will always lie within a certain range (which varies with the observer), but within this range it may assume any value, which cannot be predicted in advance. This means that no correction is capable of eliminating these errors, and so they can be regarded as random variables, moreover, as random variables whose probability law is almost unknown.

At the same time, since the causes A_j of the Y_j are by definition independent, the Y_j are *mutually independent* random variables. And since the total error produced by the causes A_j acts as a single entity, it is therefore a definite function λ of the partial errors Y_j:

$$e = \lambda(Y_1, Y_2, \ldots, Y_j, \ldots). \tag{6,13,2}$$

Moreover, if it turns out that A_h is the sole cause of error, then $Y_j = 0$ for $j \neq h$, so that for all h, identically in Y_h

$$Y_h \equiv \lambda(0, 0, \ldots, 0, Y_h, 0, \ldots), \tag{6,13,3}$$

and

$$0 = \lambda(0, 0, \ldots, 0, \ldots). \tag{6,13,4}$$

Nevertheless, the function λ is generally unknown, or at least not explicitly known.

We can obviously assume that the Y_j are second-order r.v., without this assumption being overly restrictive for some specific situation. We must then assume that $E(Y_j) = 0$ ($j = 1, 2, 3, \ldots$), since were $E(Y_j)$, say, positive, this could be interpreted as a systematic tendency of the error Y_j to be positive - a systematic property incompatible with the concept of accidental error.

In many cases the number of Y_j is very large, and is it very probable that they are very small. One may then assume that they satisy conditions 1) and 2) of Theorem (6,12,4). If we also assume that λ is sufficiently regular, the probable smallness of the Y_j justifies the approximation (6,1,2), which, together with (6,13,3) and (6,13,4) clearly gives

$$e = \sum_j Y_j .$$

It therefore follows from Theorem (6,12,4) that, for any fixed α and β, the approximation

$$\Pr(\alpha < e < \beta) = \frac{1}{\sqrt{2\pi}\sigma} \int_\alpha^\beta e^{-\frac{z^2}{2\sigma^2}} \, dz , \qquad (6,13,5)$$

where σ is the constant in the statement of Theorem (6,12,4), is valid. Thus, by (6,13,1), the result x of a measurement satisfies

$$\Pr(\alpha < x < \beta) = \frac{1}{\sqrt{2\pi}\sigma} \int_\alpha^\beta e^{-\frac{(z-g)^2}{2\sigma^2}} \, dz. \qquad (6,13,6)$$

This is Gauss's celebrated "law of measurement errors."

It should be borne in mind that the parameters σ and g figuring in the probability laws of e and x are unknown. σ, the standard deviation of e and x, is a characteristic of the measuring procedure, and may differ for two different procedures used to measure the same quantity. Now e has a greater chance of being small, the smaller σ, and so the ratio $1/\sigma$ is commonly said to characterize the *precision* of the procedure.

If σ and g are unknown, one must remember that the ultimate goal of the measurement is knowledge of g, and that (approximate, at least) knowledge of σ is necessary for effective utilization of (6,13,6). It therefore follows that we must have, if not a rigorous determination of these two quantities (which is impossible), at least an adequate estimate. On the other hand, in deriving (6,13,6) we employed certain hypotheses and approximations which, though apparently plausible in many situations, are nevertheless unverifiable by any but experimental methods. Testing of hypotheses and estimation are problems of Mathematical Statistics, and we shall see how to attack them in Volume II of this text.

Note that (6,13,5) attributes nonzero probability, albeit very small, to very large values (positive or negative) of e, though such large values are usually physically impossible. However, the probabilities which (6,13,5) attaches to large values of e are so small that this paradoxical fact has no practical significance. The origin of this situation is easily seen: Owing to the properties of the Y_j, the probability law of e is very close to a normal-type law; but in stipulating (6,13,5) i.e., in assuming that the d.f. of e actually is a normal-type d.f., we have overstepped our bounds: we have acted as if the number of Y_j were infinite, rather than merely very large.

This discussion of errors in measurement illustrates why and to what degree convergence to a normal-type law is one of the most important types of convergence to a limit law. Many English-speaking writers emphasize this by giving the name "central limit theorem" to any theorem stating sufficient conditions for convergence to a law of the normal type. This expression produces the misleading impression that convergence to a normal-type law is the rule in all concrete applications; in fact, convergence to a Poisson law is almost as important as convergence to a normal-type law in present-day applications of probability.

In qualitative terms, one might say that sums of a large number of independent random variables will not exhibit convergence to a limit law of normal type if some of their terms tend (relatively) to dominate the others. Examples of this situation are neither difficult to envisage nor rarely encountered. One example is the J-shaped distribution of Figure (4,2,2). Despite the fact that the actual time at which a pupil reaches school is the result of a large number of causes, one of these causes - the pupil's respect (voluntary or imposed) for the rules - has a marked influence, which gives the law its essentially non-normal character. Similarly, while it is quite true that experience often bears out Gauss's law (6,13,6), there is no dearth of exceptions.

Normal approximation. On the other hand, even when the conditions for convergence to a law of normal type are fulfilled, certain precautions are necessary.

The possible practical use of this convergence is obvious. Imagine a r.v. Z which is the sum of a large number of independent random variables. It is often difficult, even impossible, to determine its probability law; but we know that if the proper conditions are satisfied, this law is quite close to a normal law. We then substitute this normal law, which is quite easy to deal with, for the true law. This procedure is known as "normal approximation." It is then necessary, however, to know the order of magnitude of the resulting error, and, preferably, even to have a numerically accurate upper bound for this error, which may be larger than at first sight expected.

For a proper approach to the practical problem, one must remember the following points:

1) The procedure involves approximate evaluation of a probability, i.e. a number always between 0 and 1.

2) The absolute error is in general more important than the relative error.

3) Fortunately, there are cases which do not call for a high degree of precision, in which an absolute error of 0.01 (even 0.1) is tolerable.

Let us state the principal results obtained to date, without proof. L(z) will denote the normal d.f., i.e., the integral

$$L(z) = \frac{1}{\sqrt{2\pi}} \int_{-\infty}^{z} e^{-x^2/2}\, dx .$$

Returning to the notation of VI.10, we shall consider a large number k of mutually independent r.v. X_j ($j = 1, 2, \ldots, k$) possessing moments of order up to 3 *inclusive*, with $E(X_j) = 0$. We set

$$a_k^2 = \sum_{j=1}^{k} \mathcal{V}(X_j)\, ,\ b_k^3 = \sum_{j=1}^{k} E(|X_j|^3) \quad (a_k > 0,\ b_k > 0),$$

and let $H(z)$ be the d.f. of $Z = \frac{1}{a_k} \sum_{j=1}^{k} X_j$. Then:

THEOREM (6,13,1) (A. Berry and C. Esseen)

There exists a constant λ such that, for all X_j satisfying the above conditions, and for all k and z $(-\infty < z < +\infty)$,

$$|H(z) - L(z)| \leqslant \lambda \left(\frac{b_k}{a_k}\right)^3 . \qquad (6,13,7)$$

Without additional conditions on the X_j it is impossible to obtain a better estimate, uniform in z, than (6,13,7). The exact value of λ is not known, though it is known to be less than 8 (see C.G. Esseen [1]). ∎

Note that Theorem (6,13,1) includes Theorem (6,10,2). If the X_j all obey the same probability law and therefore have the same standard deviation σ and the same absolute 3rd moment $\beta^3 = E(|X_j|^3)$, then $a_k = \sigma k^{1/2}$ and $b_k = \beta k^{1/3}$ and inequality (6,13,7) gives

$$|H(z) - L(z)| \leqslant \lambda \left(\frac{\beta}{\sigma}\right)^3 \cdot \frac{1}{\sqrt{k}} \leqslant 8 \left(\frac{\beta}{\sigma}\right)^3 \cdot \frac{1}{\sqrt{k}} . \qquad (6,13,8)$$

For example, in the Bernoulli case of the heads-or-tails scheme, set

$$p = \text{probability of heads}, \; q = 1 - p \qquad (6,13,9)$$

Y_j = indicator of the event "heads on the jth toss," $X_j = Y_j - p$

$$R_k = \sum_{j=1}^{k} Y_j = \text{repetition of heads in k tosses},$$

$$Z = \frac{1}{\sqrt{kpq}} \sum_{j=1}^{k} X_j = \frac{1}{\sqrt{kpq}} (R_k - kp) . \qquad (6,13,10)$$

Then

$$E(X_j) = 0, \; \mathcal{V}(X_j) = pq \, , \; \beta^3 = E(|X_j|^3) = pq(p^2 + q^2) \, ,$$

so that, by (6,13,8),

$$|Pr(Z < z) - L(z)| \leqslant 8 \, \frac{p^2 + q^2}{\sqrt{pq}} \cdot \frac{1}{\sqrt{k}} . \qquad (6,13,11)$$

This estimate is very simple. However, it implies that, if we wish to ensure that the error in the normal approximation is at most 0.1 (by no means a prohibitive demand), we must take $k > 6400$, while in practice one must usually be content with a value of k in the range of a few hundreds, sometimes even less.

It is known that the rate of convergence to a normal limit law is in general the rate of convergence of $1/\sqrt{k}$ to 0, which is quite slow. This fact is often overlooked, and the resulting normal approximations involve prohibitive errors. Admittedly, there is often no alternative line of action. However, experience has shown that the error of the normal approximation is often much smaller than the upper bound of (6,13,7). In fact, if the X_j have certain properties, one can prove estimates, uniform in z, stricter than (6,13,7). Moreover, even without additional assumptions concerning the X_j stricter estimates than (6,13,7), *though not uniform in* z, are available.

We shall now consider these estimates, confining ourselves, however, to the case in which the X_j, besides satisfying the above conditions, all obey the same probability law, with d.f. F(x). Assuming throughout that $E(X_j) = 0$, we set $E(X_j^2) = \sigma^2$, $E(X_j^3) = \mu_3$, and define a number ω_k for all k by

$$\underset{-\infty<z<+\infty}{\text{l.u.b.}} \ \sqrt{k}\left| H(z) - L(z) - \frac{\mu_3}{6\sigma^3}(1 - z^2)\frac{e^{-z^2/2}}{\sqrt{2\pi}}\frac{1}{\sqrt{k}}\right| = \omega_k. \quad (6,13,12)$$

Lattice random variables

A random variable X is said to be a *lattice* r.v., with a lattice probability law and d.f., if it is a discrete r.v. all of whose possible values are terms of some arithmetic progression with a positive difference. There are then infinitely many arithmetic progressions with positive difference whose terms include all possible values of X. The greatest of these differences ρ is called the *span* of X (or of its probability law or d.f.). There exists a number α such that all possible values of X have the form $\alpha + h\rho$, where h is some nonzero integer.

Any r.v. obeying a Poisson law or a Bernoulli law is a lattice random variable.

THEOREM (6,13,2). If the common probability law of the random variables X_j is not a lattice law, then

$$\lim_{k\to+\infty} \omega_k = 0 .$$

If the X_j possess a fourth moment, there exists a constant λ, depending on the probability law of the X_j, such that $\omega_k \leqq \lambda/\sqrt{k}$ for all k. ∎

In our discussion of the Bernoulli case in VI.10, we remarked that since the normal law is symmetric it is intuitively clear that the normal approximation is especially good when the common probability law of the X_j is symmetric (about the origin $0 = E(X_j)$). This implies that H(z) is itself symmetric, corroborating Theorem (6,13,2). When the X_j have a symmetric probability law, then $\mu_3 = 0$, and (6,13,10) shows that the error in replacing H(z) by L(z) is not only at least of the order of $1/\sqrt{k}$, as indicated by (6,13,8), but it is of higher order, under certain conditions even of order 1/k.

However, Theorem (6,13,2) is not applicable when the X_j are lattice r.v. The reason for this is easily seen from the Bernoulli case. Formula (6,10,7) shows that the d.f. H(z) of the r.v. Z defined by (6,13,9) has jumps of order $1/\sqrt{k}$, L(z) is continuous; therefore, ω_k cannot converge to 0. One can prove:

THEOREM (6,13,3). If the X_j have a lattice probability law with span ρ, then

$$\lim_{k\to+\infty} \omega_k = \frac{\rho}{2\sigma\sqrt{2\pi}} . \quad ∎$$

It is easy to show that $\rho \leqq 2\sigma$, and the equality $\rho = 2\sigma$ holds, in particular, if

$$\Pr\left(X_j = -\frac{\rho}{2}\right) = \Pr\left(X_j = \frac{\rho}{2}\right) = \frac{1}{2} .$$

In the Bernoulli case, for X_j defined by (6,13,9) (for which $\mu_3 = 0$) and the corresponding r.v. Z defined by (6,13,10), we have

$$\lim_{k \to +\infty} \underset{-\infty < z < +\infty}{\text{l.u.b.}} \sqrt{k}\, |H(z) - L(z)| = \frac{1}{\sqrt{2\pi}} \cdot \qquad (6,13,13)$$

14. *New approach*

We return to the Bernoulli case of the heads-or-tails scheme, with the notation of VI.10. There we saw that $P_k(z_1, z_2)$ converges to $L(z_2) - L(z_1)$, even when z_1 and z_2 vary with k. However, this result is sometimes trivial; for example, if $z_2 = +\infty$ and z_1 converges to $+\infty$ as $k \longrightarrow +\infty$, it is clear that $P_k(z_1, +\infty)$ and $L(+\infty) - L(z_1) = 1 - L(z_1)$ both converge to 0, and our result adds nothing to this trivial statement. In such a case, to obtain new and nontrivial information, one must consider the behavior of the quotient

$$Q_k(z_1, z_2) = \frac{P_k(z_1, z_2)}{L(z_2) - L(z_1)}$$

as $k \longrightarrow +\infty$, or as z_1 and z_2 both tend to $+\infty$, or when $z_2 = +\infty$ and z_1 tends to $+\infty$. The situation is analogous when $+\infty$ is replaced by $-\infty$, but the case of $-\infty$ clearly reduces to that of $+\infty$.

In this connection, note the following result. If $z_2 = +\infty$, and z_1 tends to $+\infty$ as $k \longrightarrow +\infty$, but is nevertheless of order smaller than $k^{1/6}$, then $Q_k(z_1, +\infty)$ converges to 1.

This theorem may be proved in an elementary way by extending the restricted expansion (6,10,7) of $p_h(k)$.

On the other hand, there are examples in which z_1 tends to $+\infty$ to order higher than $k^{1/6}$, but $Q_k(z_1 + \infty)$ does not converge to 1.

It is of course important to study these problems in cases more general than the Bernoulli case; of the many results, due to H. Cramer [3] and others, we mention the following, which is one of the simplest and includes the preceding case (Bernoulli) as a particular case.

THEOREM (6,14,1). Let $\{X_j\}$ $(j = 1, 2, 3, \ldots)$ be a countable sequence of second-order mutually independent r.v. X_j with the same d.f. $F(x)$, such that $E(X_j) = 0$, $E(X_j^2) = \sigma^2$. Let $H_k(z)$ denote the d.f. of $Z_k = \frac{1}{\sigma\sqrt{k}} \sum_{j=1}^{k} X_j$ and $Q_k(z_1, z_2)$ the quotient

$$Q_k(z_1, z_2) = \frac{H_k(z_2) - H_k(z_1)}{L(z_2) - L(z_1)} \quad (z_1 < z_2) .$$

Assume that there exists a number $\eta < 0$ such that the Laplace transform

$$\int_{-\infty}^{+\infty} e^{-ux} \, dF(x) \quad (u \text{ real})$$

exists for all u such that $|u| < \eta$. Then if $z_1 \longrightarrow +\infty$ as $k \longrightarrow +\infty$ but remains of order smaller than $k^{1/6}$, $Q_k(z_1, +\infty)$ converges to 1.

Converge of probability densities to a normal density

Assume that as $k \longrightarrow +\infty$ the d.f. $H_k(z)$ of a r.v. Z_k converges to an absolutely continuous limit d.f. $H(z)$ with density $h(z)$; assume further that the functions H_k themselves are absolutely continuous, with densities $h_k(z) = \frac{d}{dz} H_k(z)$. The fact that H_k converges to H does not necessarily imply that h_k converges to h (cf. Remark (3,4,2)). For example, if X_j is a sequence of mutually independent discrete r.v. with respective probability densities $f_j(x)$ and c_k and $a_k > 0$ are constants, then the d.f. $H_k(z)$ of the r.v. $Z_k = \frac{1}{a_k} \sum_{j=1}^{k} X_j - c_k$ has a density $h_k(z) = \frac{d}{dz} H_k(z)$. Assume that the constants c_k and a_k are such that H_k converges as $k \longrightarrow +\infty$ to the normal d.f. $L(z)$ with density

$$l(z) = \frac{1}{\sqrt{2\pi}} e^{-z^2/2} .$$

Then it does not necessarily follow that $\lim_{k\to+\infty} h_k(z) = l(z)$; the question of the conditions under which $h_k(z)$ converges or does not converge to $l(z)$ is usually not very important. However, there are applications in which this question arises, and it is worthy of some attention; in this connection we confine ourselves to citing the following result (cf. Gnedenko [1]):

THEOREM (6,14,2). Assume that the X_j all have the same *bounded* probability density $f(x)$, $E(X_j) = 0$, $E(X_j^2) = \sigma^2$; set $Z_k = \frac{1}{\sigma\sqrt{k}} \sum_{j=1}^{k} X_j$ and let $h_k(z)$ denote the probability density of Z_k. Then

$$\lim_{k \to +\infty} h_k(z) = l(z) = \frac{1}{\sqrt{2\pi}} e^{-z^2/2},$$

uniformly in z throughout $(-\infty, +\infty)$. $\blacksquare$

This result may be extended to the case of unbounded $f(x)$ (cf. Gnedenko and Kolmogorov [1]).

Laws of the iterated logarithm

Consider an infinite sequence X_j $(j = 1, 2, 3, \ldots)$ of mutually independent r.v. X_j and the partial sums $S_k = \sum_{j=1}^{k} X_j$; The preceding discussion does not provide precise information about the behavior of S_k as $k \longrightarrow +\infty$. For example, if the X_j are second-order r.v. such that $E(X_j) = 0$ and the constant $a_{o,k}^2$ of (6,7,3) is of the same order as k, we know that

$$\mathcal{V}(S_k) \text{ is of the same order as } k. \qquad (6,14,1)$$

This may be rephrased as follows: S_k is "probably" of the same order as $\sqrt{k}$. If Theorem (6,7,2) is applicable, we may supplement (6,14,1) by a statement of the type

$$S_k/\sqrt{k} \text{ converges a.s. to } 0. \qquad (6,14,2)$$

Now, under certain conditions (which are fairly broad) one can prove considerable improvements of (6,14,1) and (6,14,2), known - we shall soon see why - as *laws of the iterated logarithm.* Note that (6,14,1) and (6,14,2) are forms of the law of large numbers; they may well hold in cases where there is no convergence to a normal law. On the other hand, the laws of the iterated logarithm are related to convergence to a normal law, as will soon become clear.

For a more exhaustive investigation of the behavior of S_k, the following approach is natural. Let $\phi(k)$ be any given positive function of k, defined (at least) for positive integers k. By Theorem (6,7,3), the probability P that the inequalities

$$|S_k| \leqslant \varphi(k)$$

will all hold, at least from a certain (random) finite value of k onwards, must be either 0 or 1. Thus, the functions $\phi(k)$ fall into two catagories: the upper class, for which P = 1, and the lower class, for which P = 0.

A satisfactory solution to the problem that we have posed would be to determine a function $\phi_u(k)$ of the upper class and a function $\phi_l(k)$ of the lower class, such that $\phi_u(k)$ and $\phi_l(k)$ have similar orders of magnitude. One such solution follows from the following theorem.

THEOREM (6,14,3). If the variables X_j satisfy the conditions of Theorem (6,13,2) and have the same probability law, with $E(X_j^2) = \sigma^2$, $\beta^3 = E(X_j^3)$, then the function

$$\varphi(k) = \rho\, \sigma \sqrt{2k \log \log k}\,, \qquad (6,14,3)$$

where ρ is a positive constant, is in the upper class if $\rho > 1$, in the lower class if $\rho \leqq 1$.

We shall only hint at the proof of this theorem. Let E_k be the event

$$|S_k| > \varphi(k), \text{ i.e., } \left|\frac{S_k}{\sigma\sqrt{k}}\right| > \rho\sqrt{2 \log \log k}\ .$$

By Theorem (6,13,1)

$$\Pr(E_k) \leqslant \frac{2}{\sqrt{2\pi}} \int_{\rho\sqrt{2\log\log k}}^{+\infty} e^{-z^2/2}\, dz + 8\left(\frac{\beta}{\sigma}\right)^3 \frac{1}{\sqrt{k}}\,,$$

so that by (4,4,21)

$$\Pr(E_k) \leqslant O\left(\frac{1}{(\log k)^{\rho^2}\sqrt{\log\log k}}\right) + O\left(\frac{1}{\sqrt{k}}\right)\,.$$

Assume that $\rho > 1$; let q be any number > 1, and set $k_h = [q^h]$. It follows that

$$\Pr(E_{k_h}) \leqslant O\left(\frac{1}{h^{\rho^2}\sqrt{\log(h \log q)}}\right) + O\left(\frac{1}{q^{h/2}}\right)$$

so that the series $\sum_h \Pr(E_{k_h})$ converges. It follows (cf. Exercise 2.4) that from some random (but a.c. finite) value of h onwards, none of the events E_{k_h} may occur. In other words, $\phi(k)$ is in the upper class with respect to k-values of the form $[q^h]$. It remains to complete the proof for other values of k and to prove that $\phi(k)$ is in the lower class if $\rho \leqq 1$. However, this part of the proof is more delicate and we shall omit it.

REMARK (6,14,1). The result of Theorem (6,14,3) may be proved under less restrictive assumptions than those adopted above. In particular, one can relax the restriction $E(|X_j|^3) < +\infty$ (cf. P. Hartman and A. Wintner [1]).

REMARK (6,14,2). Other functions $\phi_u(k)$ and $\phi_l(k)$, valid under more or less restrictive conditions, have been constructed. Some of these are "stricter" than those given by Theorem (6,14,3), in the sense that they lead to functions $\phi_u(k)$ and $\phi_l(k)$ whose orders of magnitudes are closer together.

However, in most cases the functions $\phi_u(k)$, $\phi_l(k)$ have relatively simple expressions, resembling those of (6,14,3) in that they involve an iterated logarithm; hence the generic term "laws of the iterated logarithm" for results of this type.

IV. GENERALIZATIONS

15. *Addition of independent* n-*dimensional random vectors*

Our discussion of the addition of independent r.v. and their asymptotic laws when the number of addends becomes very large extends in an obvious manner to more general random elements.

Let $\mathfrak{X}$ be an arbitrary space with elements x, y,..., in which a binary operation is defined. Let $[x,y] \in \mathfrak{X}$ denote the "product" of $x \in \mathfrak{X}$ and $y \in \mathfrak{X}$. Let X and Y be two independent r.e. with values in $\mathfrak{X}$; we wish to study the stochastic properties of the r.e. Z = [X,Y] as a function of the stochastic properties of X and Y. This is only of interest when the binary operation is not arbitrary; the very important case in which $\mathfrak{X}$ is a *group* (e.g. a circle with the rotation group) has been studied extensively.

Strictly stationary sequences of random elements

Note that the concept of a strictly stationary sequence may be introduced for any space $\mathfrak{X}$, just as in VI.9. Let $\mathfrak{W}$ be the set of bilateral sequences w of elements w_j of $\mathfrak{X}$ (j = 0, ±1, ±2,...):

$$w = \{\ldots, w_{-1}, w_o, w_1, \ldots\}.$$

Let $w' = \psi(w)$ be the one-to-one mapping of $\mathfrak{W}$ onto itself which maps the element $w = \{\ldots, w_{-1}, w_o, w_1, \ldots\}$ onto the element

$$w' = \{\ldots, w'_{-1}, w'_o, w'_1, \ldots\}$$

of $\mathfrak{W}$ defined by

$$w'_j = w_{j+1} \quad (j = 0, \pm 1, \pm 2, \ldots).$$

Consider a bilateral sequence $W = \{\ldots, Z_{-1}, Z_o, Z_1, \ldots\} = \{Z_j\}$ (j = 0, ±1, ±2,...) of r.v. Z_j with values in $\mathfrak{X}$. W may be regarded as a r.e. with values in $\mathfrak{W}$. This sequence is said to be *strictly stationary* if for any subset $\Omega \in \mathfrak{W}$ (cf. (6,9,6)),

$$\Pr(W \in \Omega) = \Pr[W \in \psi(\Omega)]. \tag{6,15,1}$$

When j is regarded as a time variable, this definition always expresses the invariance of all the stochastic properties of a sequence W under an arbitrary translation of the time axis. Any sequence W of identically distributed and mutually independent r.e. Z_j is strictly stationary.

However, here we shall confine ourselves to a few immediate generalizations of the preceding sections. Since those sections were concerned with sums, "normalized" sums, arithmetic means, etc. of r.v., our discussion may be confined to the case in which the operations in $\mathfrak{X}$ are addition and multiplication by a scalar, that is to say, $\mathfrak{X}$ is a *vector space*. The question of asymptotic laws is then relevant only if $\mathfrak{X}$ has a topology, and so is a *topological vector space*.

We shall only examine one case, which is at first sight very simple. However, apart from the fact that this is the most useful case in applications, we shall try, as far as possible, to phrase the definitions and choose the methods of proof in such a way that the possible extentions to more general cases will be evident (Hilbert spaces, Banach spaces, etc.).

Let $\mathfrak{X}$ be a vector space $\mathfrak{X}_n$ of arbitrary finite dimension n; the zero element will be denoted by θ. Thus the r.e. in question will be n-dimensional random vectors, equivalent to n-dimensional r.v. (cf. Chapter V). Furthermore, we shall assume that $\mathfrak{X}_n$ is *real*, but this is not an essential restriction, since any complex n-dimensional random vector is equivalent to a real 2n-dimensional random vector. We assume that $\mathfrak{X}_n$ is euclidean, in other words:

1) $\mathfrak{X}_n$ has a scalar product; we use the notation x.y for the scalar product of $x \in \mathfrak{X}_n$ by $y \in \mathfrak{X}_n$.

2) $\mathfrak{X}_n$ is *normed* by its scalar product, i.e., for any $x \in \mathfrak{X}_n$, $M_2(x) = x.x$ is non-negative, and $M_2(x)$ vanishes if and only if $x = \theta$.

3) The topology of $\mathfrak{X}_n$ is that defined by the norm $N_2(x) = [M_2(x)]^{1/2}$.

$\mathfrak{X}_n$ is thus a Hilbert space, and we may refer to II.10. As usual, we use the following notation:

1) $< z^* , z >$ is the (numerical) result of applying a linear functional z^* to an element z of any vector space $\mathfrak{Z}$.

2) $N_2(z)$ and $M_2(z)$ are the norm and squared norm, respectively, of the element z of a vector space $\mathfrak{Z}$, derived from a scalar product in $\mathfrak{Z}$ with respect to which $\mathfrak{Z}$ is normed.

3) $\mathfrak{X}_n^*$ is the dual of $\mathfrak{X}_n$ i.e., the vector space of linear functionals on $\mathfrak{X}_n$.

4) $x^*, y^*, z^*, \ldots$ denote elements of $\mathfrak{X}_n^*$; θ^* denotes the zero element of $\mathfrak{X}_n^*$.

5) $\mathfrak{X}_n^{**}$ is the bidual of $\mathfrak{X}_n$, i.e., the dual of $\mathfrak{X}_n^*$. We know (cf. II.10) that there is a one-to-one correspondence $x^* = G(x)$ between $\mathfrak{X}_n$ and $\mathfrak{X}_n^*$, known as the canonical isomorphism, defined by

$$< G(x), y > \equiv x \cdot y \quad \text{for all} \quad y \in \mathfrak{X}_n. \tag{6,15,2}$$

Since $\mathfrak{X}_n$ is real, G is a linear mapping. Through G, the scalar product in $\mathfrak{X}_n$ induces a scalar product $x^*.y^*$ in $\mathfrak{X}_n$:

$$x^* \cdot y^* = G^{-1}(x) \cdot G^{-1}(y).$$

$\mathfrak{X}_n^*$ is normed by the scalar product $x^*.y^*$, and

$$N_2[G(z)] = N_2(z) \text{ for all } z \in \mathfrak{X}_n .$$

Similarly, we may consider the canonical isomorphism G^* of $\mathfrak{X}_n^*$ onto $\mathfrak{X}_n^{**}$, which in turn defines a scalar product by which $\mathfrak{X}_n^{**}$ is normed. $G^*G = H$ is then an isometry of $\mathfrak{X}_n$ onto $\mathfrak{X}_n^{**}$, which may be defined directly by

$$< y^*, x > = < H(x), y^*> \quad \text{for all} \quad y^* \in \mathfrak{X}_n^* . \tag{6,15,3}$$

Recall that, for any $z \in \mathfrak{X}_n$ and $z^* \in \mathfrak{X}_n^*$,

$$\underset{z \in \mathfrak{X}_n}{\text{l.u.b}} \frac{|< z^*, z >|}{N_2(z)} = N_2(z^*). \tag{6,15,4}$$

Let Z be a r.e. (nondegenerate, cf. Remark (2,12,2)) with values in $\mathfrak{X}_n$. Note that $N_2(Z)$, $M_2(Z)$ are real, nonnegative and nondegenerate r.v. In V.3, we defined the m.e. of E(Z) of Z by

$$< z^*, E(Z)> \equiv E(< z^*, Z >) \text{ for all } z^* \in \mathfrak{X}_n. \qquad (6,15,5)$$

It follows immediately that:

THEOREM (6,15,1). If Z_1 and Z_2 are two random elements with values in $\mathfrak{X}_n$ and mathematical expectations $E(Z_1)$, $E(Z_2)$, then the mathematical expectation of their sum $Z = Z_1 + Z_2$ is $E(Z) = E(Z_1) + E(Z_2)$. ■

It may also be seen (cf. the beginning of V.5, in particular (5,5,1)) that:

THEOREM (6,15,2). If the random element Z with values in $\mathfrak{X}_n$ has mathematical expectation E(Z), and λ is any given linear mapping of $\mathfrak{X}_n$ into a real vector space $\mathfrak{Y}_q$ of finite dimension q, then the random element $T = \lambda(Z)$ has mathematical expectation E(T), and $E(T) = \lambda[E(Z)]$. ■

In particular, if Z has mathematical expectation E(Z), $Z^* = G(Z)$ has mathematical expectation $E(Z^*) = G[E(Z)]$.

THEOREM (6,15,3). If $E[N_2(Z)] < +\infty$, the mathematical expectation E(Z) exists, and $N_2[E(Z)] \leqq E[N_2(Z)]$. ■

Assume $E[N_2(Z)] < +\infty$; by (6,15,4), $E(< z^*, Z >)$ exists for all $z^* \in \mathfrak{X}_n^*$ and, as a function of z^*, is clearly a linear functional. In other words $\mathfrak{X}_n^{**}$ contains an element u (and only one) such that

$$< u, z^*> \equiv E(< z^*, Z >) \text{ for all } z^* \in \mathfrak{X}_n^*, \qquad (6,15,6)$$

or, by (6,15,3),

$$< z^*, H^{-1}(u)> \equiv E(< z^*, Z >) \text{ for all } z^* \in \mathfrak{X}_n^*.$$

This establishes the existence (and uniqueness) of $E(Z) = H^{-1}(u)$. Moreover, by (6,15,6) and (6,15,4), $N_2[E(Z)] = N_2(u) \leqq E[N_2(Z)]$.

CONVERSE (6,15,1). Conversely, if E(Z) exists, $E[N_2(Z)] < +\infty$. ■

Let $X_1, X_2, \ldots, X_n$ be the components of z relative to an orthonormal basis in $\mathfrak{X}_n$. The $X_j = < z_j^*, Z >$ are obtained by applying certain linear functionals z_j^* ($j = 1, 2, \ldots, n$) to $Z \in \mathfrak{X}_n$. Thus, the existence of E(Z) implies that

$$E(|X_j|) < +\infty \quad (j = 1, 2, \ldots, n)\ ;$$

but

$$N_2(Z) = \Big(\sum_{j=1}^{n} |X_j|^2\Big)^{1/2} \leqslant \sum_{j=1}^{n} |X_j|,$$

whence the Converse (6,15,1). This statement is not valid for r.e. with values in an infinite-dimensional vector space (cf. Exercise 6.12).

THEOREM (6,15,4). Let $W = \{Z_j\}$ $(j = 0, \pm 1, \pm 2, \ldots)$ be a strictly stationary bilateral sequence of random elements Z_j with values in $\mathfrak{X}_n$. Assume that $E[N_2(Z_j)] < +\infty$ and (this involves no loss of generality) $E(Z_j) = \theta$. Set $Y_k = \frac{1}{k}\sum_{j=1}^{k} Z_j$. Then there exists a random element Y with values in $\mathfrak{X}_n$ such that

1) $E[N_2(L)] < +\infty,\ E(L) = \theta$;

2) $\lim_{k\to+\infty} N_2(Y_k - L) = 0$ a.s.

If the Z_j are mutually independent, L is almost certainly θ. ■

This theorem is an extension of Theorems (6,9,3) and (6,9,4); it may be proved easily by using Theorems (6,9,3) and (6,9,4), defining in $\mathfrak{X}_n$ an orthonormal basis, and working with the components of Z_j and Y_k relative to this basis. However, this method of proof does not extend to infinite-dimensional spaces. An extension of Theorem (6,15,3) to arbitrary separable Banach spaces may be found in E. Mourier [1].

Second-order random elements

We shall say that the r.e. Z with values in $\mathfrak{X}_n$ is a second-order r.e. if

$$E[M_2(Z)] = E[N_2(Z)^2] < +\infty. \tag{6,15,7}$$

Note that if we interpret the probability law of Z as a mass distribution on $\mathfrak{X}_n$, $E[M_2(Z)]$ is in effect its moment of inertia (about θ). The *variance* $\mathfrak{V}(Z)$ of Z is the quantity

$$\mathfrak{V}(Z) = E\{M_2[Z - E(Z)]\}, \tag{6,15,8}$$

that is to say, the moment of inertia of the same mass distribution, but about E(Z). It is easy to verify that

$$E[M_2(Z)] = M_2[E(Z)] + \mathcal{V}(Z). \qquad (6,15,9)$$

Let $\mathcal{H}_{(n)}$ be the set of all second-order r.e. with values in $\mathfrak{X}_n$. We define the *scalar product* of two elements X, Y of $\mathcal{H}_{(n)}$ by

$$\text{scalar product of X and Y} = E(X.Y). \qquad (6,15,10)$$

The fact that, in $\mathfrak{X}_n$,

$$|X \cdot Y| \leqslant N_2(X) \cdot N_2(Y) ,$$

implies that E(X.Y) exists, and in fact has the properties of a scalar product which *norms* $\mathcal{H}_{(n)}$. The squared norm of an element $Z \in \mathcal{H}_{(n)}$ is defined by

$$E(Z \cdot Z) = E[M_2(Z)].$$

It is easy to see (using II.10) that $\mathcal{H}_{(n)}$ is complete, therefore a Hilbert space, with respect to this norm.

Let $W = \{Z_j\}$ be an ordered sequence, either unilateral ($j = 0, 1, 2, 3, \ldots$) or bilateral ($j = 0, \pm 1, \pm 2, \ldots$), of second-order r.e. Z_j with values in $\mathfrak{X}_n$. We shall call W a *second-order stationary sequence* if

1) $E(Z_j) = \theta$;

2) $r_{k-j} = E(Z_k \cdot Z_j)$ depends only on the difference $h = k - j$.

Let $Z^* \in \mathfrak{X}_n$. If the sequence W is second-order stationary, this does not necessarily imply that the sequence of r.v. $< z^* , Z_j >$ is also second-order stationary. In particular, if we consider some fixed basis in $\mathfrak{X}_n$, and k is a fixed integer ($k = 1, 2, \ldots, n$), the sequence of kth components of Z_j relative to this basis need not be second-order stationary.

Using the methods of VI.8, application of Theorem (6,8,1) yields:

THEOREM (6,15,5). If the sequence $W = \{Z_j\}$ ($j = 0, 1, 2, \ldots$, or $j = 0, \pm 1, \pm 2, \ldots$) is second-order stationary, and we set $Y_k = \frac{1}{k} \sum_{j=1}^{k} X_j$, there exists a second-order random element Y with values in $\mathfrak{X}_n$ such that

1) $E(Y) = \theta$;

2) $\lim\limits_{k \to +\infty} E[M_2(Y_k - Y)] = 0.$ (6,15,11)

If the Z_j are mutually independent then $Y = \theta$ a.c. $\blacksquare$

Note that (6,15,11) is equivalent to Hilbert convergence (or convergence in q.m.) in $\mathcal{H}_{(n)}$.

REMARK (6,15,1). Let Z be a r.e. with values in $\mathfrak{X}_n$. Assume that E(Z) exists, and, to fix ideas, that $E(Z) = \theta$. Let Z^* be a r.e. with values in $\mathfrak{X}_n^*$, and, denoting subsets of $\mathfrak{X}_n^*$ by e, let p(e) be its probability law, i.e., p(e) is a measure on $\mathfrak{X}_n^*$. Using (4,6,18), the following formula holds for the r.e. $< Z^*, Z >$:

$$E(< Z^*, Z >) = \int_{\mathfrak{X}_n^*} E(< Z^*, Z > / Z^* = z^*)\, p(dz^*).$$

If Z and Z^* are *independent*, it follows that

$$E(< Z^*, Z >) = \int_{\mathfrak{X}_n^*} E(< z^*, Z >)\, p(dz^*) = 0. \blacksquare \qquad (6,15,12)$$

Consider a sequence $W = \{Z_j\}$ ($j = 1, 2, 3, \ldots$) of second-order r.e. Z_j with values in $\mathfrak{X}_n$, which are also *mutually independent*. Assume, without loss of generality, that $E(Z_j) = \theta$, and set

$$S_k = \sum_{j=1}^{k} Z_j .$$

Then

$$E(S_k) = \theta$$

and

$$M_2(S_k) = S_k \cdot S_k = < G(S_k), S_k > = \sum_{j,h=1}^{k} < G(Z_j), Z_h > .$$

It then follows from Remark (6,15,1) that

$$\mathcal{V}(S_k) = E[M_2(S_k)] = \sum_{j=1}^{k} E(< G(Z_j), Z_j >) = \sum_{j=1}^{k} E[M_2(Z_j)] = \sum_{j=1}^{k} \mathcal{V}(Z_j), \quad (6,15,$$

which is a generalization of Theorem (6,1,1). It then suffices to reason as in VI.7, to prove:

THEOREM (6,15,6). Let $\{Z_j\}$ ($j = 1, 2, 3, \ldots$) be a sequence of mutually independent second-order random elements with values in $\mathfrak{X}_n$, $E(Z_j) = \theta$. Set $Y_k = \frac{1}{k} \sum_{j=1}^{k} Z_j$; then,

1) if $\sum_{j=1}^{k} \mathcal{V}(Z_j) = o(k^2)$,

$$\lim_{k \to +\infty} E[M_2(Y_k)] = 0 ;$$

2) if there exist two real numbers λ and α such that

$$\lambda > 0, \quad 0 \leqslant \alpha < \frac{1}{2} (1 + \sqrt{5}),$$

and

$$\sum_{j=h+1}^{k+h} \mathcal{V}(Z_j) \leqslant \lambda k^{\alpha},$$

for all $h \geqq 0$ and all $k \geqq 1$, then

$$\lim_{k \to +\infty} N_2(Y_k) = 0 \quad \text{a.s.}$$

Convergence to an n-dimensional normal law

The characteristic functional $\xi(z^*)$ of a r.e. Z with values in $\mathfrak{X}_n$ was defined by (5,3,3) and (5,3,4). Under the above conditions, consider the characteristic functional $\xi_k(z^*)$ of the r.e. $S_k/\sqrt{\mathcal{V}(S_k)}$, and set

$$X_j = \langle z^*, Z_j \rangle \quad (j = 1, 2, \ldots, k), \quad u_k = 1/\sqrt{\mathcal{V}(S_k)}.$$

Let $\psi_j(u)$ denote the second characteristic function of X_j:

$$\psi_j(u) = \log E(e^{iux_j}) ;$$

in view of the independence of the Z_j, and therefore of the r.v. X_j, it is immediate that

$$\log \xi_k(z^*) = \sum_{j=1}^{k} \psi_j(u_k).$$

Note that the X_j are second-order r.v., since

$$|X_j| \leq N_2(z^*)\cdot N_2(Z_j), \quad E(|X_j|^2) \leq M_2(z^*)\ \mathcal{V}(Z_j),$$

and $E(X_j) = 0$ since $E(Z_j) = \theta$. We can therefore use (6,10,11), according to which

$$\log \xi_k(z^*) = -\frac{1}{2}\left[\sum_{j=1}^{k} E(|X_j|^2)\right] u_k^2 + \left[\sum_{j=1}^{k} \rho_j(u_k)\right] u_k^2 . \qquad (6,15,14)$$

1) Assume first that the Z_j, and therefore the X_j, have identical probability laws, and set

$$\mathcal{V}(Z_j) = v^2, \qquad E(X_j^2) = \sigma^2 .$$

Let $\phi(u)$ denote the common characteristic function of the X_j, $\psi(u)$ the logarithm of $\phi(u)$, $F(t)$ the common d.f. of the r.v. $M_2(Z_j)$. The functions $\rho_j(u) = \rho(u)$ are independent of j, and defined according to (4,3,10) by

$$\psi(u) = -\frac{\sigma^2}{2} u^2 + \rho(u)\, u^2 . \qquad (6,15,15)$$

Formula (6,15,15) follows from (3,9,11) for the case n = 2, or

$$\varphi(u) = 1 - \frac{\sigma^2}{2} u^2 + \omega(u)\, u^2, \qquad (6,15,16)$$

where

$$\omega(u) = \frac{1}{2}\,[\alpha(u) + i\ \beta(u)],$$

$\alpha(u)$ being defined by (3,9,13) (with n = 2) and $\beta(u)$ by an analogous formula; formula (3,9,13) becomes

$$\alpha(u) = E[(1 - \cos uX_j)\ |X_j|^2] ,$$

so that

$$|\alpha(u)| \leq E(|1 - \cos uX_j|\ |X_j|^2)$$

$$\leq [N_2(z^*)]^2\ E[|1 - \cos uX_j|\ M_2(Z_j)] .$$

Let A^2 be any positive number: then

$$E[|1 - \cos uX_j|\ M_2(Z_j)] = Pr[M_2(Z_j) \geq A^2]\ E[|1 - \cos uX_j|\ M_2(Z_j)/M_2(Z_j) \geq A^2]$$

$$+ Pr[M_2(Z_j) < A^2]\ E[|1 - \cos uX_j|\ M_2(Z_j)/M_2(Z_j) < A^2] .$$

Now,

$$|1 - \cos u X_j| \leqslant 2 \ ; \ \Pr[M_2(Z_j) > A^2] = 1 - F(A^2),$$

$$E[M_2(Z_j)/M_2(Z_j) \geqq A^2] = \frac{1}{1 - F(A^2)} \int_{A^2}^{+\infty} t \, dF(t);$$

thus

$$|\alpha(u)| \leqslant 2 \int_{A^2}^{+\infty} t dF(t) + E[|1 - \cos u X_j| \ M_2(Z_j)/M_2(Z_j) < A^2 \ F(A^2)] .$$

When $M_2(Z_j) < A^2$, $|X_j| \leqslant N_2(z^*) A$. Let B be any positive number and assume that

$$N_2(z^*) \leqslant B . \qquad (6,15,17)$$

Then $|X_j| \leqq B A$. Set

$$\nu(u) = \underset{|x| \leqslant BA}{\text{l.u.b.}} |1 - \cos u x| .$$

Obviously,

$$\lim_{u \to 0} \nu(u) = 0 .$$

It follows that

$$|\alpha(u)| \leqslant 2 \int_{A^2}^{+\infty} t \, dF(t) + \nu(u) \int_0^{A^2} t \, dF(t) .$$

Since

$$\mathfrak{v}(Z_j) = E[M(Z_j)] = \int_0^{+\infty} t \, dF(t) < +\infty ,$$

we see that $\lim_{u \to 0} \alpha(u) = 0$ uniformly in z^* satisfying (6,15,17).

The same may now be proved for $\beta(u)$, therefore for $\omega(u)$, and the final result is:

$\lim_{u \to 0} \rho(u) = 0$, uniformly in z^* satisfying (6,15,17).

It now follows from (6,15,14) that

$$\lim_{k \to +\infty} \xi_k(z^*) = \exp.\left(- \frac{\sigma^2}{2v^2}\right) , \qquad (6,15,18)$$

uniformly in z^* for all z^* satisfying (6,15,17).

Consequently (cf. III.12, in particular Theorem (3,12,1)), we conclude that, as $k \to +\infty$, the probability law of $S_k/\sqrt{\mathcal{V}(S_k)}$ converges to a limit law $\mathcal{L}$ whose characteristic functional $\xi(z^*)$ is

$$\xi(z^*) = \exp.\left\{-\frac{\sigma^2}{2v^2}\right\}; \qquad (6,15,19)$$

σ^2 depends on z^*:

$$\sigma^2 = E(|< z^*, Z_j >|^2). \qquad (6,15,20)$$

Let T be a r.e. with values in $\mathfrak{X}_n$ and probability law $\mathcal{L}$. Let us find the characteristic function $\lambda(u)$ of the r.e. $Y = < z^*, T >$:

$$\lambda(u) = E[e^{iuY}] = E[e^{iu<z^*,T>}]$$

$$= E[e^{i<uz^*,T>}] = \xi(uz^*).$$

It is clear from (6,15,20) that

$$\lambda(u) = \xi(uz^*) = \exp.\left\{-\frac{\sigma^2}{2v^2}u^2\right\},$$

which is the characteristic function of a normal law. Thus (cf. V.5), the law $\mathcal{L}$ of T is an n-dimensional normal law in $\mathfrak{X}_n$.

2) If the Z_j do not obey the same probability law, the problem is more complicated. Assume, for example, that

$$E[N_2(Z_j)^3] < +\infty \text{ for all } j, \text{ and}$$

$$\lim_{k\to+\infty} \frac{\sum_{j=1}^{k} E[N_2(Z_j)^3]}{\mathcal{V}(S_k)^{3/2}} = 0.$$

Since $E(|X_j|^3) \leqslant N_2(z^*)^3 \cdot E[N_2(Z_j)^3]$, we can use the method of Theorem (6,10,2) to prove that, as $k \to +\infty$, $\log\ \xi_k(z^*)$ behaves like

$$C_k = -\frac{1}{2}\sum_{j=1}^{k} E(X_j^2)/\mathcal{V}(S_k),$$

and, moreover,

$$0 \geqslant C_k \geqslant -\frac{1}{2}M_2(z^*).$$

However, there is clearly no reason for the C_k to converge to a limit.

The source of this complication is easily understood; it is evident from the preceding sections that, when one considers the sum of a large number k of independent r.e., one can only expect a (nondegenerate) limit law to exist for a quantity derived from this sum by suitable centering and normalization. For the sums S_k encountered in practice, one assumes centering in advance by imposing the condition $E(Z_j) = \theta$, so that $E(S_k) = \theta$. As for the normalization, any linear transformation $\lambda(S_k)$ of S_k, where λ is a one-to-one linear mapping of $\mathfrak{X}_n$ onto itself, may be regarded as a normalization: the most suitable normalization (which is a priori not necessarily unique) is therefore to be sought in the extensive family of isomorphisms of $\mathfrak{X}_n$ onto itself, or, what is the same, in the family of nonsingular square matrices of order n. We shall not dwell on this problem here; fortunately, in many applications, the required normalization is self-evident.

Fluctuations of concentration in the atmosphere; the blue color of the sky

In the sequel, the same letter Ω will denote both a certain portion of atmospheric space and its volume. Consider a bounded portion Ω of the atmosphere containing k gaseous molecules, which we shall denote by the numbers 1, 2,..., j,...,k. k is of course very large if Ω has sizable dimensions, and this we indeed assume. The positions of the molecules are not fixed; owing to their very complex motion, the entire set of molecules is, so to speak, "reshuffled," so that at any given instant one may adopt the following postulates:

a) For any j (j = 1, 2,...,k) the position M_j of j in Ω is a random point.

b) For any j (j = 1, 2,...,k), the probability law of M_j is the uniform law on Ω (cf. II.11).

c) The different random points M_j (j = 1, 2,...,k) are mutually independent.

In assuming c), we are neglecting the interactions of the molecules. These are only important for closely spaced molecules, and air is not a dense gas (in other words, k, always a very large integer, is not *too* large). At any given instant, pairs of very close molecules are rare. In principle, therefore, we may justifiably disregard them; nevertheless, they become important if we are interested in the number $R(\omega)$ of M_j contained in a very small volume ω of Ω, for, if two molecules are simultaneously present in ω, this implies that they are very close to each other. Postulates a), b), c) facilitate the study of the numbers of molecules situated in given portions of Ω.

Let R be the number of M_j belonging to $\omega \subset \Omega$. The probability that $M_j \in \omega$ is $p = \omega/\Omega$, and consequently R is governed by the Bernoulli law (6,1,17). In other words, if, as usual, we set $q = 1 - p$, then

$$\Pr(R = h) = C_k^h \, p^h \, q^{k-h} \; ; \tag{6,15,21}$$

$E(R) = kp$, $\mathcal{V}(R) = kpq$. It is interesting to consider the number

$$D = \frac{R - kp}{kp} \, , \tag{6,15,22}$$

which in fact represents the *relative deviation* of the effective concentration from the "mean" concentration (in ω). Of course, $E(D) = 0$, but it is easy to see that $\mathcal{V}(D) = q/kp$.

Since k is large, it follows from (6,10,7) and VI.10 as a whole that, very approximately,

$$\Pr(R = h) \# \frac{e^{-\frac{(h-kp)^2}{2kpq}}}{\sqrt{2\pi kpq}} \, . \tag{6,15,23}$$

The probability law of $(R - kp)/\sqrt{kqp}$ is very close to the normal law, and for any numbers α and β, very approximately,

$$\Pr(\alpha < D < \beta) = \frac{1}{\sqrt{2\pi}} \int_{\alpha\sqrt{\frac{kp}{q}}}^{\beta\sqrt{\frac{kp}{q}}} e^{-x^2/2} \, dx$$

$$= \sqrt{\frac{kp}{2\pi q}} \int_{\alpha}^{\beta} e^{-\frac{kpy^2}{2q}} \, dy. \tag{6,15,24}$$

Now partition Ω into n pairwise disjoint parts $\omega_1, \omega_2, \ldots, \omega_n$ such that $\bigcup_{l=1}^{n} \omega_l = \Omega$. Let R_l denote the number of molecules in ω_l, $R = \{R_1, \ldots, R_n\}$. Define random variables $X_{j,l}$ ($j = 1, 2, \ldots, k$; $l = 1, 2, \ldots, n$) by

$$X_{j,l} = \begin{cases} 1 \text{ if } M_j \in \omega_l \,, \\ 0 \text{ if } M_j \notin \omega_l \,. \end{cases}$$

For every j, let $X_j = \{X_{j,1}, X_{j,2}, \ldots, X_{j,n}\}$. Let $\mathfrak{X}_n$ denote the space of these n-tuples. $\mathfrak{X}_n$, which is the nth cartesian (or direct) power of the space of real numbers, is an n-dimensional vector space; the X_j and $R = \sum_{j=1}^{k} X_j$ are random vectors in $\mathfrak{X}_n$.

Direct investigation of R is an easy extension of the investigation of the Bernoulli case, carried out in VI.1 and VI.10; note that the R_l certainly satisfy the nonrandom relation

$$\sum_{l=1}^{n} R_l = k \,. \tag{6,15,25}$$

Set

$$p_l = \frac{\omega_l}{\Omega}, \; q_l = 1 - p_l \;\; (l = 1, 2, \ldots, n) \,.$$

If $h_1, h_2, \ldots, h_n$ are any n given nonnegative integers such that

$$\sum_{l=1}^{n} h_l = k \,, \tag{6,15,26}$$

then, in view of (1,3,3) and the independence of the M_j,

$$\Pr\left[\bigcap_{l=1}^{n} (R_l = h_l)\right] = \frac{k\,!}{h_1\,!\; h_2\,!\; \ldots h_n\,!}\; p_1^{h_1}\, p_2^{h_2} \ldots p_n^{h_n} \,. \tag{6,15,27}$$

In imitation of VI.10, set

$$h_l = k\,p_l + y_l \sqrt{k\,p_l} \,.$$

Assume that as k increases the y_l remain bounded. The h_l are large, of the same order as k. Compute the logarithm of (6,15,27), evaluate k! and $h_l!$ by (1,1,13), and introduce truncated Taylor expansions of the quantities $\log\left(1 + \frac{y_l}{\sqrt{kp_l}}\right)$. In view of (6,15,26), we have the very rough approximation

$$\Pr\left[\bigcap_{l=1}^{n} (R_l = h_l)\right] \# \frac{1}{(2\pi)^{n-1/2}}\, e^{-\frac{1}{2}\sum_{l=1}^{n} y_l^2} \,, \tag{6,15,28}$$

which generalizes (6,10,7) and (6,15,23). If we wish to consider the *relative* deviations D_l,

$$D_l = (R_l - kp_l)/kp_l ,$$

it suffices to note that, if $R_l = h_l$, the value d_l of D_l is

$$d_l = y_l / \sqrt{kp_l} .$$

These are the laws of random fluctuation of atmospheric concentration (cf., e.g., P. Langevin [1]). It is obvious that both method and results are applicable to the study of any fluctuations of concentration (e.g., in solutions), provided that the postulates a), b), c) remain valid. We add that, though these postulates seem intuitively plausible, only experimental verification of their consequences (6,15,23) and (6,15,28) by the methods described in Volume II of this book can justify them entirely.

Owing to the constant agitation of its molecules (and the resulting concentration variations), the atmosphere is a turbulent medium (in the terminology of optics), and partially scatters solar light. The proportion of scattered luminous energy at a given wavelength λ is greater, the greater the heterogeneity of the atmosphere at this wavelength. That is to say, on the whole, it is greater, the greater the relative fluctuation of concentration in a cube with edge λ. If we apply (6,15,23), where ω is the volume λ^3 of this cube, we see that the relative fluctuation D has a probable order of magnitude $\sqrt{\frac{q}{kp}}$, which is greater, the smaller λ; this accounts for the fact that the scattering effect of the atmosphere is most marked at low wavelengths, i.e. for blue and near-blue light.

REMARK (6,15,2). Since $R = \{R_1,\ldots,R_n\}$ is a random vector in $\mathfrak{X}_n$, the sum of a large number k of mutually independent random vectors X_j in $\mathfrak{X}_n$, the direct approach adopted above may be replaced by the previous methods of this section. As an illustration, let us briefly consider this approach as applied to R.

Let θ denote the zero element of $\mathfrak{X}_n$, and p the element of $\mathfrak{X}_n$ defined by

$$p = \{p_1 , p_2 , \ldots , p_n\}.$$

Set

$$v^2 = 1 - \sum_{l=1}^{n} p_l^2 , \quad Z_{j,l} = X_{j,l} - p_l ,$$

$$Z_j = X_j - p = \{Z_{j,1}, \ldots, Z_{j,n}\} .$$

It follows immediately that

$$E(X_{j,h}) = p_h , \quad E(X_j) = \{p_1, \ldots, p_n\} = p , \quad E(Z_j) = \theta ;$$

$$E(R) = k\,p .$$

Define the scalar product x.y of elements $x = \{x_1, \ldots, x_n\}$ and $y = \{y_1, \ldots, y_n\}$ in $\mathfrak{X}_n$ by

$$x \cdot y = \sum_{l=1}^{n} x_l\, y_l .$$

Then

$$M_2(X_j) = \sum_{l=1}^{n} X_{j,l}^2 , \qquad E[M_2(X_j)] = p_1 + p_2 + \ldots + p_n = 1 ;$$

$$\mathcal{V}(X_j) = \mathcal{V}(Z_j) = v^2, \quad \mathcal{V}(R) = kv^2.$$

If $\{u_1, u_2, \ldots, u_n\}$ denote the components (relative to the natural basis) of an element z^* of the dual of $\mathfrak{X}_n$, then, for any element $x = \{x_1, \ldots, x_n\}$ of $\mathfrak{X}_n$,

$$< z^*, x > = \sum_{l=1}^{n} u_l\, x_l .$$

Therefore

$$|< z^*, X_j >|^2 = \sum_{\alpha,\beta=1}^{n} u_\alpha\, u_\beta\, X_{j,\alpha}\, X_{j,\beta} ,$$

$$E(X_{j,\alpha}\, X_{j,\beta}) = 0 \text{ if } \alpha \neq \beta \; ; \; E(X_{j,\alpha}^2) = p_\alpha ,$$

so that

$$E(< z^*, X_j >) = \sum_{l=1}^{n} p_l\, u_l ,$$

$$E(< z^*, X_j >^2) = \sum_{l=1}^{n} p_l\, u_l^2 .$$

If we set

$$\sigma^2 = E(< z^*, Z_j >^2) , \quad T_l = (R_l - kp_l)/v\sqrt{k} ,$$

$$T = \{T_1, \ldots, T_n\} = (R - kp)/v\sqrt{k} ,$$

then

$$\sigma^2 = \sum_{l=1}^{n} p_l\, u_l^2 - \Big(\sum_{l=1}^{n} p_l\, u_l\Big)^2 , \tag{6,15,29}$$

and we see from (6,15,19) that the probability law of T is only slightly different from the n-dimensional normal law with characteristic functional

$$\xi(z^*) = \exp\left\{-\frac{\sigma^2}{2v^2}\right\}, \tag{6,15,30}$$

where σ^2 is given by (6,15,29). However, this normal law is degenerate, since (6,15,25) becomes

$$\sum_{l=1}^{n} T_l = 0 .$$

Conversely, assuming (6,15,30), let us try to prove (6,15,28). To this end, set $Y = \{Y_1, Y_2, \ldots, Y_n\}$, where

$$Y_l = v\, T_l / \sqrt{p_l} \qquad (l = 1, 2, \ldots, n) . \tag{6,15,31}$$

Formula (6,15,25) becomes

$$\sum_{l=1}^{n} \sqrt{p_l}\; Y_l = 0 . \tag{6,15,32}$$

Applying the linear mapping (6,15,31) to T, we see from (5,3,8) and Theorem (5,3,2) that the probability law of Y is very close to the n-dimensional (but degenerate) normal law with characteristic functional

$$\zeta(z^*) = \exp.\left\{-\frac{1}{2}\, H(u)\right\}, \tag{6,15,33}$$

where H(u) is the quadratic form

$$H(u) = \sum_{l=1}^{n} u_l^2 - \Big(\sum_{l=1}^{n} \sqrt{p_l}\; u_l\Big)^2$$

We recognize the form (5,5,41). In view of (6,15,32), it suffices to refer to Example (5,5,1), in particular to (5,5,47), to deduce (6,15,28) from (6,15,33).

BIBLIOGRAPHIC NOTES FOR CHAPTER VI

The fundamental elements of the theory of addition of independent r.v. (Section 1 of Chapter VI) are expounded in all general texts on Probability Theory, in particular in those cited in the bibliographies for Chapters II, IV, V.

For a detailed study of stochastic convergence and the laws of large numbers, see M. Fréchet [1] and W. Feller [1]. A very complete and up-to-date exposition of ergodic theory may be found in N. Dunford and J.T. Schwarz [1]; a very synthetic but less up-to-date exposition of this theory is given by S. Kakutani [1].

The most complete exposition of convergence to a limit law may be found in B. Gnedenko and A. Kolmogorov [1], which also devotes some space to the laws of large numbers. One may also consult M. Fréchet [1], P. Lévy [1], [2], H. Cramer [2], [3], [4]. For laws of the iterated logarithm see W. Feller [2], P. Hartman and A. Wintner [1], P. Lévy [2], R. Fortet [1].

For the Gaussian error law and the method of least squares, see R. Deltheil [1], P. Lévy [1], H. Mineur [1].

U. Grenander [1] has assembled practically all that is known in general of r.e. with values in a group. For r.e. with values in Banach spaces (laws of large numbers, characteristic functional, convergence to a normal limit law), the reader may refer to M. Fréchet [3], E. Mourier [1], Yu. Rozanov [1].

For the case of more general topological vector spaces, see Yu. Prokhorov [1], S. Ahmad [1].

Concepts of topological vector spaces and the relevant chapters of functional analysis may be found in F. Riesz and B. Nagy [1], N. Dunford and J.T. Schwarz [1], N. Bourbaki [3].

For particular cases of addition of random vectors in a finite-dimensional vector space, see, among others, E. Lukacs [2] for multidimensional characteristic functions; M. Dwass and H. Teicher [1], G.N. Sakovich [1].

EXERCISES

6.1. Consider the heads-or-tails scheme in the Bernoulli case, denoting by p the probability of heads and by X the number of the first toss which turns up heads. Show that

$$\Pr(X = k) = p\, q^{k-1} \qquad (q = 1 - p\ ;\ k = 1, 2, 3, \ldots)$$

(cf. Exercise 4.1). Let Y be a r.v., independent of X, with the same law as X. What are the possible values s_h (h = 2, 3, 4,...) of S = X + Y? Compute $\Pr(S = s_h)$ and find the characteristic function of S.

6.2. Consider a plane with two rectangular coordinate axes Ox, Oy, and let M_1 and M_2 be two independent random points, governed by the same probability law as the random point M of exercise 5.1. Let ρ and Θ be the polar coordinates (ρ = radius vector, Θ = polar angle) of the random point N defined by $\overrightarrow{ON} = \overrightarrow{OM_1} + \overrightarrow{OM_2}$. Show that ρ and Θ are independent, find their probability laws and the second moment of ρ .

6.3. An urn contains black and white balls in respective proportions p and q (p + q = 1). k balls are withdrawn at random, one by one, from the urn, each ball being replaced immediately. Of the k balls withdrawn from the urn, R are white. Compute the probability $\pi(k, p, h) = \Pr(R \leqq h)$. Find expressions for the probabilities $\pi_1(k, p, h) = \Pr(R \leqq h)$ and $\pi_2(k, p, h) = \Pr(R \geqq h)$. show that

$$\frac{\partial}{\partial p}\pi_1(k, p, h) = -k\,\pi(k - 1, p, h), \quad \frac{\partial}{\partial p}\pi_2(k, p, h) = k\,\pi(k - 1, p, h - 1).$$

Assuming 0 < h < k and h fixed, show that if p increases from 0 to 1, then $\pi_1(k, p, h)$ decreases from 1 to 0 and $\pi_2(k, p, h)$ increases from 0 to 1.

Let ε be any given number between 0 and 1, and consider the r.v. P_1 and P_2 defined by

$$\pi_1(k, P_1, R) = \frac{\varepsilon}{2}, \quad \pi_2(k, P_2, R) = \frac{\varepsilon}{2}.$$

Show that $\Pr(P_2 < p < P_1) < \varepsilon$.

6.4. Let p, q, r be three positive numbers such that $p + q + r = 1$. With each j ($j = 1, 2, 3, \ldots$), associate a 3-dimensional r.v. $T_j = \{X_j, Y_j, Z_j\}$, such that

α) X_j, Y_j, Z_j can only assume the values 0 or 1;

β) $X_j + Y_j + Z_j = 1$;

γ) $\Pr(X_j = 1) = p$, $\Pr(Z_j = 1) = r$; $\Pr(Y_j = 1) = q$.

Assume that the r.v. T_j ($j = 1, 2, 3, \ldots$) are mutually independent, and set

$$A_k = \sum_{j=1}^{k} X_j, \quad B_k = \sum_{j=1}^{k} Y_j, \quad C_k = \sum_{j=1}^{k} Z_j,$$

$$P(a, b, c) = \Pr[(A_k = a) \cap (B_k = b) \cap (C_k = c)],$$

$$\chi^2 = \frac{(a - kp)^2}{kp} + \frac{(b - kq)^2}{kq} + \frac{(c - kr)}{kr}.$$

1) If a, b, c are given, for what values p_0, q_0, r_0 of p, q, r is P(a, b, c) a maximum?

2) Show that if k is large, then P(a, b, c) is approximately equal to $P_0 \, e^{-\frac{\chi^2}{2}}$, where P_0 does not depend on a, b, c (cf. Example (5,5,1)). In the same approximation, what is the probability density $f(\chi)$ of the r.v. $\boldsymbol{\chi}$ defined by

$$\boldsymbol{\chi}^2 = \frac{(A_k - kp)^2}{kp} + \frac{(B_k - kq)^2}{kq} + \frac{(C_k - kr)^2}{kr} \ ?$$

(Answer: $f(\chi) = \chi \, e^{-\frac{\chi^2}{2}}$).

3) Suppose that, in an experiment for the verification of the Mendelian laws (crossing of mice) with $k = 434$, the values $a = 126$, $b = 62$, $c = 246$ were obtained for A_k, B_k, C_k. Compute the corresponding values of p_0, q_0, r_0. Suppose that theory gives p, q, r the values $p = \frac{4}{16}$; $q = \frac{3}{16}$; $r = \frac{9}{16}$; for these values p, q, r and the above values of a, b, c, compute χ^2 and the probability $\Pr(\boldsymbol{\chi} > \chi)$ (use a table of the function $1(x)$ defined by (4,4,9)). What is the implication of your results for the validity of the theory in this experiment?

6.5. Consider an infinite sequence of mutually independent r.v. X_j ($j = 1, 2, 3,\ldots$) all obeying the uniform probability law on $(-\pi, +\pi)$. Set Y_j, $Z_j = \sin X_j$. Compute the m.e., variances and correlation coefficient of Y_j and Z_j.

Set $U_k = \sum_{j=1}^{k} Y_j$, $V_k = \sum_{j=1}^{k} Z_j$, $W_k = \frac{1}{k}(U_k^2 + V_k^2)$. Compute the m.e. and the variances of U_k, V_k, W_k, and the correlation coefficient of U_k and V_k.

6.6. Consider an infinite sequence of nonnegative mutually independent r.v. Δ_j ($j = 1, 2, 3,\ldots$), all governed by the law

$$\Pr(\Delta_j \geq t) = e^{-t} \quad (t \geqslant 0)$$

on the positive time semi-axis, and let $T_k = \sum_{j=1}^{k} \Delta_j (k = 1, 2, 3,\ldots)$.

Let $N(t)$ denote the number of these instants in the interval $[0,t[$ and $F_k(t)$ the d.f. of T_k.

1) Show that $F_k(t)$ has a density $f_k(t)$. Using induction on k, find the functions $f_k(t)$ (use characteristic functions or moment-generating functions; c.f. IV.3 and Corollary (6,2,1)). Compute $p_k(t) = \Pr[N(t) = k]$.

2) Let t' and t'' be two positive instants, $t' < t''$. Compute $\Pr[N(t') = k, N(t'') - N(t') = l]$, where k and l are arbitrary nonnegative integers. Show that the r.v. $N(t')$ and $N(t'') - N(t')$ are independent.

3) Set $X(t) = 1$ if $N(t)$ is even, -1 if $N(t)$ is odd. Compute $E[X(t)]$, $\Gamma(t, h) = E[X(t)\,X(t + h)]$ ($h \geq 0$ is arbitrary); find

$\lim_{t\to+\infty} E[X(t)]$ and $\lim_{t\to+\infty} \Gamma(t, t + h) = \rho(h)$. Set $r(h) = \rho(h)$ for $h \geq 0$, $r(h) = \rho(-h)$ for $h < 0$. Show that $r(h)$ is positive definite.

6.7. ν balls are withdrawn one by one from an urn containing w white and b black balls. Let A be the hypothesis that each ball withdrawn is returned to the urn before the next ball is taken out, and B the hypothesis that no ball, once withdrawn, is ever returned. Let X be the number of white balls among those withdrawn.

1) Express $\Pr(X = x)$ in terms of w, b, ν, under the two hypotheses A and B.

2) k experiments of the above type are performed, in each of which ν balls are withdrawn from the urn. Let R(x) be the number of experiments among these k for which X = x. An observer is ignorant of the values of w, b, ν, but observes the values r(x) assumed by R(x) (x = 0, 1, 2,...). What means does he have to conjecture the values of w, b, ν, under the respective hypotheses A and B? (Hint: use, in particular, VI.7). Consider, in particular, the case of large k.

Numerical application: $r(0) = r(1) = 0$, $r(2) = 1$, $r(3) = 3$, $r(4) = 19$, $r(5) = 31$, $r(6) = 8$, $r(7) = 1$, $\sum_{x>7} r(x) = 0$.

6.8. k mutually independent r.v. $X_1,\ldots X_k$ obey the same Poisson law with parameter m: $Pr(X_j = h) = e^{-m}\frac{m}{h!}$ (h = 0, 1, 2, ...). Let Z be the k-dimensional r.v. $Z = \{X_1,\ldots,X_k\}$ and S the sum of X_j. Set $P(x_1,\ldots,x_k) = Pr\left[\bigcap_{j=1}^{k}(X_j = x_j)\right]$, and M = S/k,

1) Find the probability law of S and the conditional probability law of Z, given S = s (s = 0, 1,...). What is the limit law of U as $k \to +\infty$?

2) Given any values of the x_j, show that the value of m which maximizes $P(x_1,\ldots,x_k)$ is the value $(x_1 + x_2 +\ldots+ x_k)\frac{1}{k} = \frac{s}{k}$ of M when $X_j = x_j$ for every j.

3) A trial assigns the values x_j to the X_j. An observer records these x_j, but does not know the value of m; he tries to conjecture the latter, using the x_j alone. Show that in so doing the only data that he must take into account is the number $(x_1 +\ldots+ x_k)\frac{1}{k}$. He decides to adopt this number as an estimate of m; to what extent is this decision justified?

4) Numerical application to an experiment of electron emission: k = 10, and the x_j have the following values: $x_1 = 496$, $x_2 = 501$, $x_3 = 513$, $x_4 = 505$, $x_5 = 522$, $x_6 = 499$, $x_7 = 561$, $x_8 = 540$, $x_9 = 592$, $x_{10} = 505$. Using the normal approximation, evaluate

$$Pr(|M - m| > 100).$$

6.9. Recall the statement and notation of Exercise 5.1. Set

$$Z_1 = nY_1, \quad Z_n = n(1 - Y_n).$$

1) Show that $\lim_{n \to +\infty} Y_1 = 0$ a.s.

2) Show that as $n \to +\infty$, the correlation coefficient of Z_1 and Z_n converges to 0, and the probability law of the pair $\{Z_1, Z_n\}$ converges to a limit law; find this limit law.

3) Let $F(x)$ be a continuous normalized d.f. such that $F(+\infty) = 1$. Now assume that the r.v. X_j, rather than being uniformly distributed over $(0, 1)$, obey the law defined by the function $F(x)$, and leave all the other conditions of Exercise 5.2 unchanged. Show that the results of Exercise 5.2 and the answers to 1) and 2) of the present exercise concerning the pair $\{Y_1, Y_n\}$ are also valid for the pair $\{F(Y_1), F(Y_n)\}$ (cf. Exercise 4.9).

6.10. 1) Let X, Y be two real r.v., $\varphi(u, v) = E[e^{i(uX+vY)}]$ the characteristic function of the pair $\{X,Y\}$. Assume that $E(X)$, $E(Y)$, $E(X^2)$, $E(XY)$, $E(Y^2)$ exist. Find the relations between these moments and the partial derivatives $\frac{\partial\varphi}{\partial u}$, $\frac{\partial\varphi}{\partial v}$, $\frac{\partial^2\varphi}{\partial u^2}$, $\frac{\partial^2}{\partial u \partial v}$, $\frac{\partial^2\varphi}{\partial v^2}$ at $u = v = 0$. Hence prove that

$$\varphi(u, v) = 1 + i\,[E(X)\,u + E(Y)\,v] - \frac{1}{2}\,[E(X^2)\,u^2 + 2E(XY)\,uv + E(Y^2)\,v^2] + \omega(u, v),$$

where $\omega(u, v)$ tends to 0 at least as rapidly as u^2 and v^2 as u, v tend to 0 (refer to III.12).

2) Let Z be a r.v. possessing moments up to order 4 inclusive; set $E(Z) = \mu$, $E(Z^2) = \mu^2 + \sigma^2$, $E(Z^3) = \mu(\mu^2 + \sigma^2) + \alpha^3$, $E(Z^4) = (\mu^2 + \sigma^2)^2 + \beta^4$, $X = Z$, $Y = Z^2$. Express $E(X)$, $E(Y)$, $E(X^2)$, $E(XY)$, $E(Y^2)$, in terms of μ, σ, α, β, which are assumed given both here and below.

3) Let $Z_1, Z_2, \ldots, Z_j, \ldots$ be an infinite sequence of mutually independent r.v. Z_j, all having the same probability law as the r.v. Z of 2). Set

$$S_k = \sum_{j=1}^{k} Z_j = k\mu + \sqrt{k}\,\sigma\, U_k,$$

$$T_k = \sum_{j=1}^{k} Z_j^2 = k(\mu^2 + \sigma^2) + \sqrt{k}\,\beta^2\, V_k.$$

Show that $E(U_k) = E(V_k) = 0$, $E(U_k^2) = E(V_k^2) = 1$. Evaluate $E(U_k V_k) = r$. Set $\psi(u, v) = E[e^{i(uU_k+vV_k)}]$. Show that, if we set

$$\log \psi(u, v) = -\frac{1}{2}(u^2 + 2ruv + v^2) + \frac{1}{\sqrt{k}}\Omega_k(u, v),$$

then, as $k \to +\infty$, $\Omega_k(u, v)$ remains bounded, uniformly in u, v for bounded u, v.

Hence deduce that the probability law of the two-dimensional r.v. $\{U_k, V_k\}$ converges as $k \to +\infty$ to a two-dimensional normal law; find the parameters of the latter.

4) Set $H_k = \frac{1}{k-1}\sum_{j=1}^{k}\left(Z_j - \frac{1}{k}S_k\right)^2$. Express H_k as a function of S_k and T_k, then as a function of U_k and V_k. Hence deduce that $E(H_k) = \sigma^2$, $\lim_{k\to+\infty} H_k = \sigma^2$ in q.m., and that the probability law of $\sqrt{k}\,(H_k - \sigma^2)$ converges to a normal law as $k \to +\infty$; find the parameters of the latter.

6.11. Let $X_1, \ldots, X_{2k+1}$ be $2k + 1$ mutually independent r.v., all having the same absolutely continuous d.f. $F(x)$ with continuous and positive density $f(x)$. Let $\bar{\bar{X}}$ be the median of the r.v. X_j, i.e., the number $\bar{\bar{X}}$ such that $F(\bar{\bar{X}}) = 1 - F(\bar{\bar{X}}) = \frac{1}{2}$ (cf. IV.2), and Y_k the median of the $2k + 1$ numbers X_j, i.e., the number X_j appearing in the $(k + 1)$-th position when the X_j are arranged in increasing order of magnitude.

1) Show that the r.v. Y_k has the probability density

$$g(y) = \frac{(2k+1)!}{(k!)^2} F(y)^k\,[1 - F(y)]^k\, f(y).$$

2) Set $U = (Y_k - \bar{\bar{X}})\sqrt{2k}$. Show that the probability law of U converges to a normal law as $k \to +\infty$; what are the parameters of the latter?

3) Show that $\lim_{k\to+\infty} Y_k = \bar{\bar{X}}$ a.s.

(Hint: use the change of variable $z = F(y)$, especially in 2) and 3)).

6.12. Let $\mathfrak{X}$ be a separable Hilbert space, $\{z_j\}$ ($j = 1, 2,\ldots$) an orthonormal basis of this space (cf. II.10), and α a given positive real number. Let Z be a r.e. with values in $\mathfrak{X}$ and totally discontinuous probability law defined by $\Pr(Z = j^{\alpha} z_j) = \dfrac{\rho}{j^{1+\alpha}}$ ($j = 1, 2, 3,\ldots$), where $\rho \cdot \sum_{j=1}^{+\infty} \dfrac{1}{j^{1+\alpha}} = 1$.

Let a be the element of $\mathfrak{X}$ defined by $a = \sum_j \dfrac{\rho}{j} z_j$. Show that, for any *continuous* linear functional z^* on $\mathfrak{X}$, we have $E(< z^*, Z >) = < z^*, a >$. Show that $E[N_2(Z)] = +\infty$, so that the Converse Theorem (6,15,1) is not applicable.

6.13. Let $\{X_k\}$ ($k = 1, 2,\ldots$) be an infinite sequence of r.v. such that a.c. $X_k \leqslant X_{k+1}$ for all k. Show that there exists a (possibly degenerate) r.v. X such that $\lim_{k\to+\infty} X_k = X$ a.s. Show that if the expectations $E(X_k)$ exist and are bounded uniformly in k, then $\Pr(X < +\infty) = 1$, and $E(X) = \lim_{k\to+\infty} E(X_k) < +\infty$ (Hint: interpret the X_k as mappings of the class $\mathfrak{U}$ of trials into the space R of real numbers, and apply Theorem (2,8,5).

6.14. This exercise is in part a generalization of Exercise 6.1. Consider the heads-or-tails scheme in the Bernoulli case, and set

p = probability of heads; $q = 1 - p$;

$$X_k = \begin{cases} 1 \text{ if the k-th toss turns up heads;} \\ 0 \text{ if the k-th toss turns up tails;} \end{cases}$$

$S_k = X_1 + X_2 + \ldots + X_k$.

Let s be a fixed positive integer, and let K_s denote the smallest k for which $S_k = s$.

1) Find $P_s(k) = \Pr(K_s = k)$ (k is any given nonnegative integer.

2) Find the characteristic function $\phi_s(u) = E(e^{iuK_s})$ of K_s.

Hence determine the m.e. m_s and variance v_s of K_s (Hint: if we set

$$f(z) = \sum_{n=1}^{\infty} z^{n-1}, \text{ then}$$

$$\frac{d^{s-1}}{dz^{s-1}} f = \sum_{n=s}^{\infty} (n - 1)(n - 2)\ldots(n - s + 1)z^{n-s}).$$

3) Set $H_s = \frac{p}{\sqrt{sq}}\left[K_s - \frac{s}{p}\right]$. Find an expression for the characteristic function $\psi_s(u)$ of H_s. Find the limit of $\psi_s(u)$ as $s \to +\infty$. What conclusion may be drawn from this result?

4) If p is unknown, it may be estimated by either of the following two methods:

Method 1: Choose a fixed value of k, and adopt S_k/k as an estimate of p.

Method 2: Choose a fixed value of s, and adopt s/K_s as an estimate of p.

We are essentially interested in the probable order of magnitude of the relative error in p. Estimate this order of magnitude for Method 1 and Method 2. Suppose that, although p is unknown, we know that it is small; show that Method 2 is preferable, in that the value of s sufficient for a given relative error may be determined in advance.

6.15. 1) Let $X_1, X_2, \ldots, X_n$ be n mutually independent normal r.v. such that $E(X_j) = 0$, $E(X_j^2) = 1$ ($j = 1, 2, \ldots, n$). Find the probability law of the r.v.

$$\Phi = \sum_{j=1}^{n} X_j^2$$

(cf. Exercise 4.11).

2) Let $Y = \{Y_1, Y_2, \ldots, Y_n\}$ be an n-dimensional normal r.v. such that $E(Y_j) = 0$ ($j = 1, 2, \ldots, n$);

Γ the $n \times n$ covariance matrix of Y;

A an $n \times n$ matrix (a_{jk}) ($j,k = 1, 2, \ldots, n$), which is real. symmetric and positive-definite;

Ψ the r.v. $\Psi = \sum_{j,k} a_{jk} Y_j Y_k$.

Using Γ and A, determine the probability law of Ψ. Using the result, interpret the result of 2) in Exercise 6.4.

NUMERICAL TABLES

[1] Duarte, Table of n! and the common logarithm of n! for $1 \leqq n \leqq 3000$, to 33 decimal places. Geneva, Kundig.

[2] A.M. Legendre, Tables of the logarithms of the complete Γ-function, to 12 decimal places. Cambridge Univ. Press, 1921.

[3] E.S. Pearson, Tables of the logarithms of the complete Γ-function, to 10 decimal places. Cambridge Univ. Press, 1922.

[4] K. Pearson, Tables of the incomplete Γ-function. London, His Majesty's Stationery Office, 1922.

[5] K. Pearson, Tables of the incomplete B-function. Biometrika, 1934, pp. 99 - 494.

[6] Jahnke and Emde, Funktionentafeln. Berlin, Teubner, 1933; reprint: New York, Dover Publications, 1945.

[7] H.E. Soper, Tables of $p_k = e^{-m} \frac{m^k}{k!}$ according to values of k, for various values of m. Biometrika, 10, pp. 27 - 35.

[8] E.C. Molina, Poisson's exponential binomial limit. Tables of $p_k = e^{-m} \frac{m^k}{k!}$ and $\sum_{j=0}^{k} e^{-m} \frac{m^j}{j!}$ for k < 100. New York, Van Nostrand, 1947.

[9] Table of $l(x) = \frac{1}{\sqrt{2\pi}} e^{-x^2/2}$ as a function of x. In: Charlier, Vorlesungen über die Grundzüge der Matematischen Statistik. Lund, 1920.

[10] Table of $\theta(x) = \frac{2}{\sqrt{\pi}} \int_0^x e^{-y^2/2}\, dy$ as a function of x (in hundredths). In: Bertrand [1].

[11] Table of $\Phi(x) = L(x) - \frac{1}{2} = \frac{1}{\sqrt{2\pi}} \int_0^x e^{-y^2/2}\, dy$. In: Bowley, Elements of Statistics, London, 1920.

[12] Table of $L(x) = \frac{1}{\sqrt{2\pi}} \int_{-\infty}^{x} e^{-y^2/2}\, dy$. In: Cramer [4].

[13] Tables of the probability functions, Vol. I: Table of

$\Theta(x) = \frac{2}{\sqrt{\pi}} \int_{0}^{x} e^{-y^2}\, dy$ and $\frac{d}{dx}\Theta(x) = \frac{2}{\sqrt{\pi}} e^{-x^2}$; [0(0.0001) 1(0.001)5.6(various) 5.946]; 15 decimal places. Tables of $\frac{d}{dx}\Theta(x)$ and $1 - \Theta(x)$ [4(0.01)10; 85]. National Bureau of Standards, Washington, U.S. Gov. Printing Office, 1941.

[14] Tables of the probability functions, Vol. II: Tables of

$l(x) = \frac{1}{\sqrt{2\pi}} e^{-x^2/2}$ and $2\Phi(x) = \frac{1}{\sqrt{2\pi}} \int_{-x}^{x} e^{-y^2/2}\, dy$ [0(0.001) 1(0.001)7.800 (various) 8.285; 152]; 15 decimal places.

Tables of $l(x) = \frac{1}{\sqrt{2\pi}} e^{-x^2/2}$ and $P = \frac{2}{\sqrt{2\pi}} \int_{x}^{+\infty} e^{-y^2/2}\, dy$ [6(0.01)10. 75]. National Bureau of Standards, Washington, U.S. Gov. Printing Office, 1948.

[15] Table of x as a function of $P = \frac{2}{\sqrt{2\pi}} \int_{x}^{+\infty} e^{-y^2/2}\, dy$ (P in hundredths). Table 1 in Fischer [1]; similar table in Cramer [4].

[16] Tables of the binomial probability distribution: table of $C_k^h p^h q^{k-h}$, $p = \cdot 01(\cdot 01)\cdot 50$; $q = 1 - p$; $k = 2(1)49$; $h = 0(1)(k-1)$; to 7 decimal places. National Bureau of Standards, Washington, U.S. Gov. Printing Office, 1950.

[17] Tables of the binomial probability distribution: table of $\sum_{j=h}^{k} C_k^j p^j q^{k-j}$, $p = \cdot 01(\cdot 01)50$, $q = 1 - p$, $h = 1(1)k$, $k = 2(1)49$; to 7 decimal places. National Bureau of Standards, Washington, U.S. Gov. Printing Office, 1950.

[18] H.G. Romig, 50-100 Binomial Tables: table of $C_k^h p^h q^{k-h}$ and $\sum_{j=h}^{k} C_k^j p^j q^{k-j}$ for $50 \leq k \leq 100$. New York, Wiley, 1952.

[19] Tables of the cumulative binomial probability distribution: table of $\sum_{j=h}^{k} C_k^j p^j q^{k-j}$ for $k \leqq 1000$. Cambridge, Harvard Univ. Press, 1955.

BIBLIOGRAPHY

N.I. Achieser and I.M. Glassmann

[1] *Theorie der linearen Operatoren im Hilbert Raum* - Berlin, 1954, Akademie Verlag (translated from the Russian).

Actualités Scientifiques

[1] Actualités Scientifiques, No. 735; Paris, Hermann.

[2] Actualités Scientifiques, No. 739; Paris, Hermann.

[3] Actualités Scientifiques, No. 766; Paris, Hermann.

[4] Actualités Scientifiques, No. 1146; Paris, Hermann.

S. Ahmad

[1] *Les éléments aléatoires dans les espaces vectoriels topologiques* - Thèse, Paris 1963; Ann. Inst. Henri Poincaré, in preparation.

L. Bachelier

[1] *Calcul des Probabilités* - Paris, 1912, Gauthier-Villars.

G. Bachman

[1] *Elements of Abstract Harmonic Analysis* - New York, 1964, Acad. Press.

J. Bass

[1] *Eléments de Calcul des Probabilités* - Paris, 1962, Masson et Cie.

Baumgardt

[1] *The quantic and statistical basis of visual excitation* - J. Gen. Physiol. 31, 1948, p. 269.

J. Bertrand

[1] *Calcul des Probabilités* - Paris, 1888, Gauthier-Villars.

A. Blanc-Lapierre, P. Casal and A. Tortrat

[1] *Méthodes Mathématiques de la Mécanique Statistique* - Paris, 1959, Masson et Cie.

A. Blanc-Lapierre and R. Fortet

[1] *Théorie des fonctions aléatoires* - Paris, 1953, Masson et Cie.

G. Bodiou

[1] *Théorie dialectique des Probabilités englobant leurs calculs classique et quantique* - Paris, 1964, Gauthiers-Villars.

S. Bochner

[1] *Vorlesungen über Fouriersche Integrale* - Leipzig, 1932, Akad. Verlagsgesellschaft.

[2] *Harmonic Analysis and the Theory of Probability* - Berkeley, 1955, Univ. of Calif. Press.

S. Bochner and K. Chandrasekharan

[1] *Fourier Transforms* - Princeton, 1949, Princeton Univ. Press.

E. Borel

[1] *Eléments de la Théorie des Probabilités* - Paris, 1924, Hermann.

[2] *Principes et formules classiques du calcul des Probabilités* - Paris, Gauthier-Villars.

[3] *Valeur pratique et Philosophie des Probabilités* - Paris, 1939, Gauthier-Villars.

E. Borel and R. Deltheil

[1] *Probabilités, erreurs* - Paris, A. Colin.

N. Bourbaki

[1] *Théorie des ensembles* - Actualités Scientifiques No. 1212, Paris 1960, Hermann.

[2] *Intégration* - Actualités Scientifiques No. 1175, 1244, 1281, Paris, Hermann.

[3] *Espaces vectoriels topologiques* - Actualités Scientifiques No. 1189 et 1229, Paris, Hermann.

L. Brillouin

[1] *Les statistiques quantiques et leurs applications* - 2 vols. 2nd edition, Paris, 1930, Presses Universitaires.

L. de Broglie

[1] *La statistique des cas purs en Mécanique Ondulatoire* - Rev. Scient., 86ème année, 1949, p. 259.

G. Cagnac, E. Ramis and J. Commeau

[1] *Nouveau cours de Mathématiques Spéciales* - 2 vols., Paris, 1961, Masson et Cie.

R. Carnap

[1] *Logical Foundations of Probability* - London, 1950, Routledge and Kegan Paul.

M. Caullery

[1] *Les conceptions modernes de l'hérédité* - Paris, Flammarion.

A. Cournot

[1] *Essai sur les fondements de nos connaissances* - Paris 1851.

H. Cramer

[1] *Über eine Eigenschaft der Normalenverteilungsfunktion* - Math. Zeit. 41, 1936, p. 405.

[2] *Random Variables and Probability Distributions* - Cambridge tracts, No. 36, Cambridge, 1937.

[3] *Sur un nouveau théorème-limite* - Actual. Scient. No. 736, Paris, 1938, Hermann.

[4] *Mathematical Methods of Statistics* - Princeton, 1945, Princeton Univ. Press.

R. Deltheil

[1] *Erreurs et moindres carrés* - Paris, 1930, Gauthier-Villars.

G. Doetsch

[1] *Handbuch der Laplace-Transformation* - 3 vols., Basle and Stuttgart, 1950, Birkhauser Verlag.

J. Dubourdieu

[1] *Les principes fondamentaux du Calcul des Probabilités* - Paris 1939, Gauthiers-Villars.

N. Dunford and J.T. Schwartz

[1] *Linear Operators* - Part I (General Theory) - New York, 1958, Interscience Publishers.

M. Dwass and H. Teicher

[1] *On infinitely divisible random vectors* - Ann. of Math. Stat., 28, 1957, p. 461.

C.G. Esseen

[1] *Fourier analysis of distribution functions. A mathematical study of the Laplace-Gaussian law* - Acta Math. 77, 1945, p.1.

J. Favard

[1] *Cours d'Analyse de l'Ecole Polytechnique* - 3 vols., Paris 1960, Gauthiers-Villars.

W. Feller

[1] *The fundamental limit theorems in probability* - Bull. Amer. Math. Soc., 51, 1945, pp. 800-832.

[2] *The law of the iterated logarithms for identically distributed random variables* - Ann. Math. 47, 1946, pp. 631 - 638.

[3] *An Introduction to Probability Theory and its Applications* - Second ed., New York 1961, J. Wiley.

B. De Finetti

[1] *La prévision, ses lois logiques, ses sources subjectives* - Ann. de L'Inst. H. Poincaré, 7, 1937, p.1.

R.A. Fischer

[1] *Statistical Methods for Research Workers* - London, Oliver and Boyd.

R. Fortet

[1] *Les fonctions aléatoires du type de Markov* - Journal de Math., 22, 1943, p. 177.

[2] *Calcul des moments d'une fonction de répartition à partir de sa caractéristique* - Bull. des Sc. Math. 68, 1944, p. 117.

[3] *Opinions modernes sur les fondements du Calcul des Probabilités* - In: *Les Grands courants de la Pensée Mathématique*, p. 207; Paris, 1948, Cahiers du Sud.

[4] *Algèbre des tenseurs et des matrices* - Fasc. II; Paris 1953, Centre de Documentation Universitaire.

[5] *Hypothesis testing and estimation for Laplacian functions* - Proc. of IVth Berkeley Symposium on Mathematical Statistics and Probability, Vol. I, p. 289, Berkeley, 1961, Univ. of Calif. Press.

M. Fréchet

[1] *Généralités sur les probabilités, variables aléatoires* - Paris, 1951 (2nd edition), Gauthiers-Villars.

[2] *Les probabilités associées à un système d'évènements compatibles et dépendants* - 2 fasc., Actual. Scient. No. 859 and 942, Paris, Hermann.

[3] *Les événements aléatoires de nature quelconque dans un espace distancié* - Ann. Inst. H. Poincaré, 10, 1948, p. 215.

B.V. Gnedenko

[1] *A local limit theorem for densities* - Dokl. Akad. Nauk. SSSR, 95, 1954, p.5.

B.V. Gnedenko and A. Khintchine

[1] *Introduction à la théorie des Probabilités* - translated from the Russian, Paris 1960, Dunod.

B.V. Gnedenko and A. Kolmogorov

[1] *Limit distributions for sums of independent random variables* - New-York 1954, Addison Wesley (translated from the Russian).

R. Goldberg

[1] *Fourier Transforms* - New York 1961, Cambridge Univ. Press.

E. Goursat

[1] *Cours d'Analyse Mathématique* - 3 vols., Paris 1927 (4th ed.), Gauthiers-Villars.

U. Grenander

[1] *Probability on Algebraic Structures* - Uppsala 1963, Almqvist and J. Wiley.

U. Grenander and M. Rosenblatt

[1] *Statistical Analysis of Stationary Time Series* - Stockholm 1956, Almqvist & Wikseel.

P. Halmos

[1] *Measure Theory* - New York 1950, Van Nostrand.

[2] *Introduction to Hilbert Space* - New York 1957, Chelsea.

P. Hartman and A. Wintner

[1] *On the law of the iterated logarithm* - Amer. J. of Math., 63, 1941, p. 169.

M. Jirina

[1] *Probabilités conditionelles sur des σ-algèbres à base dénombrable* - Czechoslovak Math. J., 4. 1954, p. 372.

S. Kakutani

[1] *Ergodic theory* - Proc. Int. Congress of Math., Cambridge, 1950, Vol. II, p. 128.

O. Kempthorne

[1] *An Introduction to Genetic Statistics* - New York 1957, J. Wiley.

M.G. Kendall and A. Stuart

[1] *The Advanced Theory of Statistics* - 2 vols., London, 1961, (2nd ed.), C. Griffin & Co.

J.M. Keynes

[1] *A Treatise on Probability* - London, Macmillan.

A. Kolmogorov

[1] *Grundbegriffe der Wahrscheinlichkeitsrechnung* - Ergebnisse der Math., Berlin, Springer. English translation: foundations of Probability, New York, 1956, Chelsea (translated from the Russian).

[2] *Ein vereinfachter Beweis des Birkoffkhintchineschen Ergodensätzes* - Rec. Math. de Moscou, 44, 1937, p. 367.

[3] *Algèbres de Boole métriques complètes* - VI Zjazd Mathem. Polsk., 21 Sept. 1948.

B.O. Koopman

[1] *The bases of probability* - Bull. Amer. Math. Soc., 46, 1940, p. 763.

[2] *The axioms and algebra of intuitive probability* - Ann. Math., 41, 1940, p. 269.

[3] *Intuitive probabilities and sequences* - Ann. Math. 42, 1941, p. 169.

A.G. Kurosch

[1] *Gruppentheorie* - German translation, Berlin, 1955, Akademie Verlag.

P. Langevin

[1] *Les progrès de la Physique Moléculaire* - Conf. de la Soc. Fr. de Physiq., Paris 27/11/1913.

E.L. Lehmann

[1] *Testing Statistical Hypotheses* - New York, 1959, Wiley.

A. Lentin and J. Rivaud

[1] *Lecons d'Algèbre Moderne* - Paris, 1961, Vuibert.

P. Lévy

[1] *Calcul des Probabilités* - Paris, Gauthiers-Villars.

[2] *Théorie de l'addition des variables aléatoires* - 2nd ed., Paris 1954, Gauthiers-Villars.

[3] *Sur la division d'un segment par des points choisis au hasard* - C.R. 208, 1939, p. 147.

A. Lichnerowicz

[1] *Algèbre et Analyse linéaires* - Paris, 1956, Masson et Cie.

M. Loève

[1] *Etude asymptotique des sommes de variables aléatoires liées* - Thèse de Doctorat, Paris, 1946.

[2] *Probability Theory* - (3rd ed.), New York, 1963, Van Nostrand.

L.H. Loomis

[1] *Abstract Harmonic Analysis* - New York, 1953, Van Nostrand.

E. Lukacs

[1] *Characteristic functions* - London 1960, Griffin.

[2] *Recent developments in the theory of characteristic functions* - Proc. of IVth Berkeley Symposium on Mathematical Statistics and Probability, II, Berkeley 1961, Univ. of Calif. Press.

G. Malecot

[1] *Les mathématiques de l'Hérédité* - Paris 1948, Masson et Cie.

Marshall Hall

[1] *A survey of combinatorial analysis* - In: *Some Aspects of Analysis and Probability* - New York, 1958, Wiley.

H. Mineur

[1] *Technique de la méthode des moindres carrés* - Paris, 1938, Gauthier-Villars.

R. De Mises

[1] *Wahrscheinlichkeit, Statistik und Wahrheit* - Berlin, 1928, Springer.

P. Montel

[1] *Sur les combinaisons avec répétitions limitées* - Bull. Scienc. Math., 66, 1942, p. 86.

E. Mourier

[1] *Eléments aléatoires dans un espace de Banach* - Ann. de l'Inst. H. Poincaré, 13, 1952-53, p. 161.

J. Neveu

[1] *Bases mathématiques du Calcul des Probabilités* - Paris 1964, Masson et Cie.

E. Parzen

[1] *Modern Probability and its Applications* - New York, 1960, Wiley.

F. Perrin

[1] *Etude mathématique du mouvement brownien* - Ann. de l'Ecole Norm. Sup., 45, 1928, p.1.

Yu. Prokhorov

[1] *Convergence of stochastic processes and limit theorems* - Teoriya veroyatnostei i ee primeneniya, 1, 1956, p. 177.

D.A. Raikov

[1] *On the decomposition of the Gaussian and Poisson laws* - Izv. Akad. Nauk SSSR, seriya mat., 1938, p. 91.

C.R. Rao

[1] *Advanced Statistical Methods in Biometric Research* - New York, 1952, Wiley.

H. Reichenbach

[1] *Les fondements logiques du Calcul des Probabilités* - Ann. de l'Inst. H. Poincaré, 7, 1937, p. 267.

A. Rényi

[1] *Nouvelle construction axiomatique de la probabilité* - Mag. Ind. Akad. Mat. Fiz. Oszt. Közl., 54, 1954, p.369.

[2] *Quelques remarques sur les probabilités des événements dépendants* - Journ. de Math., 37, 1958, p. 393.

[3] *Théorie des éléments saillants d'une suite d'observations* - Ann. Fac. Sci. de Clermont-Ferrand, 8, 1962, p.7.

[4] *Wahrscheinlichkeitsrechnung mit einem Anhang über Informationstheorie* - Berlin, 1962, VEB Deutscher Verlag der Wissenschaften.

F. Riesz and Bela Sz.-Nagy

[1] *Lecons sur l'Analyse Fonctionelle* - Paris, 1955, Gauthier-Villars.

J. Riordan

[1] *An Introduction to Combinatorial Analysis* - New York, 1958, Wiley.

Yu. Rozanov

[1] *Spectral analysis of abstract random functions* - Teoriya veroyatnostei i ee primeneniya, 4, 1959, p. 291.

Saks

[1] *Théorie de l'intégrale* - Collec. Monographies de Mathématiques, Vol. II, Warsaw, 1933.

L.J. Savage

[1] *The Foundations of Statistics* - New York, 1954, Wiley.

L. Schwartz

[1] *Théorie des distributions* - 2 vols., Actualités Scientif. No. 1091 et 1122; Paris, 1950 and 1951, Hermann.

[2] *Méthodes mathématiques pour les Sciences Physiques* - Paris, 1961, Hermann.

G.N. Sakovich

[1] *A simple form for the necessary and sufficient conditions for attraction towards stable laws* - Teoriya veroyatnostei i ee primeneniya, 1, 1956, p. 361.

M.H. Stone

[1] *Linear Transformations in Hilbert Space* - New York, 1932, Am. Math. Soc.

E.C. Titchmarsh

[1] *Introduction to the Theory of Fourier Integrals* - (2nd ed.), Oxford, 1948, Clarendon Press.

Tornier

[1] *Wahrscheinlichkeitsrechnung* - Leipzig 1936, Teubner.

A. Tortrat

[1] *Calcul des Probabilités* - Paris, 1963, Masson et Cie.

C. de la Vallée-Poussin

[1] *Integrales de Lebesgue* - Paris, 1954, Gauthier-Villars.

J. Ville

[1] *Etude critique de la notion de collectif* - Paris, 1939, Gauthier-Villars.

A. Weil

[1] *L'intégration dans les groupes topologiques* - Actualités Scientifiques Nos. 869 - 1145; Paris, 1953, Hermann.

H. Wold

[1] *A Study in the Analysis of Stationary Time Series* - Doctoral Thesis, Uppsala, 1938.

INDEX

Terms belonging to the general mathematical terminology are omitted, for the most part, in this index.

absolutely monotone function 211
addition of random variables 394
" " n-dimensional random variables 480
almost certain number 128
" " random element 128
almost everywhere 67
arrangement 5
" with repetitions 9
axiom of conditional probability 285
" " total probability 62

Bayes' formula, theorem 313, 331
Bernoulli case 398, 443, 475
" law 399
beta function 3
" " , incomplete 4
Bienaymé-Chebyshev theorem, inequality 75, 262
binomial theorem (Newton) 7
Birkhoff's argodic theorem 432
Bochner's theorem 194
Borel-Cantelli theorem 133

Cauchy law, type 406
central value 254, 266
certain number 128
" random element 128
characteristic function 270, 323
chromosome theory of heredity 21
class of trials 48
combination 6
" with repetitions 10
complementary event 33
convergence, almost everywhere 79
" almost sure 416
" in α-th mean 79, 412
" in measure 79
" in probability 413
" in quadratic mean 79, 410
" in "the ordinary sense" 79
" of distribution functions 156, 228
" of measures 122

convergence of point functions 79
" , stochastic 408
convolution 184
" of distribution functions 199, 400
correlation 337
" coefficient 236, 326
covariance 350
" matrix 352
cumulants 274

density 118, 320
" of a distribution function 145, 150
" , Radon-Nikodym 122
deviation, root mean square 262
" standard 263
difference of two events 33
dispersion 254, 261
distribution function (d.f.) 137
" " , absolutely continuous 145, 150
" " associated with a measure 137
" " , Dirac 150
" " , n-dimensional 137, 219
" " , normalized 149
" " of a random variable 246, 320
" " , singular 151
" " , totally discontinuous 145, 150

equiprobable events 36
equivalent distribution functions 144, 148
" mappings 72
" point functions 71
" random elements 128
ergodic theory, theorems 426, 432
estimate 339
" , best linear 342
event 31
" , almost certain (almost sure) 126
" , " impossible 126
" associated with a random element 110
" , certain (sure) 31
" defined by a system 34
" , impossible 31
" , probabilized 61
" , random 31
events, independent 292
" mutually exclusive 32
exponential law 288

Fatou's theorem 84
Fischer-Riesz theorem 106
Fourier transform 176
" " , n-dimensional 229
" " of a bounded distribution function 189
" " of a convolution 185
" " of a function 176
" " of a measure 240
frequency 44

gamma function 2
" " , incomplete 4
Gauss law of measurement errors 466
generating function 218, 275
Glivenko-Cantelli, theorem of 425

heads-or-tails scheme 397, 419, 423
Henry's line 283
Hilbert space 92, 348
" subspace 95
Hölder inequalities 75

inclusion of events 32
independence, mutual 294
independent events 292
" random elements 298
" random variables 334, 348
" σ-algebras 295
indicator of an event 248
" " a set 68
infinitely divisible law, distribution function 452
integral, definite 69
" , indefinite 74
" , Lebesgue 118
" , Riemann-Stieltjes 152, 220
" , with respect to a measure 69
" " " " " distribution function 145
intersection of events 34
isometry 98
isomorphism, canonical 101

Karamata's theorem 217, 243
Kolmogorov's theorem 424

$\mathcal{L}_2$ spaces 105
Laplace transform 208

law of large numbers 408, 420
lattice random variable 473
laws of the iterated logarithm 477
Lebesgue integral 118
" measure 117
" sets 117
" theorem 84
limit law, convergence to a 451, 462
limit of events 56

mass distribution 64
mathematical expectation (m.e.) 255
" " , conditional 309, 331
mean vector of mass distribution 239
measure 59
" , absolutely continuous 118, 122
" , bounded 61
" , induced 112
" , Lebesgue 117
" , totally discontinuous 85
" , uniform 121
" , zero (set of) 67
median 267
Minkowski inequalities 78
moment (absolute, algebraic) of a bounded d.f. 201
" , central 233
" of a random variable 264
" " " 2-dimensional, n-dimensional r.v. 324

normal approximation 471
" density, distribution function, random variable 278, 360, 369, 376, 381, 388
" law, 278
" " , convergence to a 443, 451, 462, 487
" type 282

permutation 6
" with repetitions 11
point function, integrable 70
" " , simple 68
Poisson law 275
" " , generalized 278
positive-definite function 186
probability 35
" , conditional 285
probability law (of a random element) 111
" " , marginal 329
" " , uniform 250

Radon-Nikodym density, theorem 122
random 31
random element (r.e.) 109
random variable (r.v.) 114
" " , complex 268, 346
" " , continuous 250
" " , degenerate 129
" " , discrete 247
" " , n-dimensional 115
" " , normal 278, 360, 381
" " , normal complex 388
" " , reduced 266
" " , second-order 348
" " , uniformly distributed 250
random vector 115
" " , degenerate 129
" " , n-dimensional 115
" " , normal 360
reduced law, random variable 266
regression 332
rank of an n-dimensional random variable 355
" " a normal random variable 369

Schwarz's inequality 76, 88
second characteristic function 271
semi-invariants 274
set function 59
" " , completely additive 59
stationary sequence, second-order 430
" " , strictly 433
statistics of perfect gases, classical 14
" , Bose-Einstein 17
" , Brillouin 18
" , Fermi-Dirac 18
Stirling's formula 3
symmetric distribution function, random variable 252
system of events 34
" " " , exhaustive 34
σ-algebra of subsets 60

trial 31
type of distribution functions 252
" " probability densities 252
" " " laws 251
" " " " , closed 405
" " random variables 251

union of events 32

variance 263
" , conditional 333